WELL CONTROL PROBLEMS AND SOLUTIONS

WELL CONTROL PROBLEMS AND SOLUTIONS

A. VIELMA.
02/84./
HOUSTON.

NEAL ADAMS

PRENTICE AND RECORDS
ENTERPRISES, INC.

Petroleum Publishing Company
Tulsa, Oklahoma

Copyright © 1980 by
The Petroleum Publishing Company
1421 South Sheridan Road/P. O. Box 1260
Tulsa, Oklahoma 74101

Library of Congress Cataloging in Publication Data

Adams, Neal.
  Well control problems and solutions.

  Includes index.
  1. Oil wells—Blowouts. I. Title.
TN871.A312   622'.3382   80–12443
ISBN 0–87814–124–3

Printed in the United States of America

1 2 3 4 5   84 83 82 81 80

TO

Louis R. Records, Sr.—Inspiration and Friend,
Charles M. Prentice—Engineer and Friend, and
Crystal Adams—Wife and Friend.

TABLE OF CONTENTS

APPENDICES

ACKNOWLEDGEMENTS

Personal acknowledgements are given to Crystal E. Adams, the author's wife, for her assistance, support, and kind words during the preparation of this work. Her unselfish donation of the dining room table as a writing table is appreciated and will never be forgotten. In addition, special thanks are given for her patience during the frustrating periods associated with all works of this type.

Professional acknowledgements are expressed to Louis R. Records, Sr., and Charles M. Prentice for the inspiration and guidance provided during the author's association with them. A few selected words are inadequate to describe the experiences and education gained from these professional well control engineers. It goes without saying, however, that this work would not have been possible without them.

A word of thanks goes to Cindy Dupont, typist, who never complained during the many long hours typing this work.

Acknowledgements are given to the Society of Petroleum Engineers for their permission to use copyrighted illustrations as well as the permission to reprint SPE 2560.

PREFACE

Well Control Problems and Solutions is designed to provide answers to anyone who may find himself in a precarious situation resulting from one or more well control problems. The book is aimed primarily at the drilling foreman, engineer, or toolpusher in charge of well control operations but should prove useful to anyone associated with these activities.

The book is based on the experiences of the author and his senior associates in killing or supervising 3,800 well control operations while serving as professional well control consultants, and on the author's experience in teaching well control subjects worldwide. Although the material presented in this book will not provide answers to all well control problems, it will give the reader insight to develop viable solutions for his particular situation.

The text allows a self-study program for the reader. Many example problems and illustrations are included within each chapter. In addition, a section of problems at the end of each chapter is similar to the problems within the text. Solutions for these problems are available from the author at Box 53557 OCS, Lafayette, LA 70505.

This work, *Well Control Problems and Solutions*, is the first in a series dealing with well control problems. Topics to be covered in greater detail in other works are (1) blowout-control, (2) workover well control, and (3) deepwater well control.

NEAL ADAMS, 1980

WELL CONTROL
PROBLEMS AND
SOLUTIONS

1

Well control equipment

MANY FACETS OF WELL CONTROL must be mastered in order to control kicks and prevent blowouts. One of the more important parts is the proper selection and utilization of the equipment that will be used to control the well. This equipment encompasses not only the surface blowout preventers but other items such as the mud, mud monitoring equipment, degassers, and the mud mixing systems. When all of these systems are functioning, proper procedures can be executed simply in efforts to maintain control of the well and prevent blowouts.

Fluid Density Control

The primary well control tool available is hydrostatic pressure of the drilling fluid in the well. The drilling fluid, or mud, is used to prevent a well from kicking and, should a kick occur, the mud is used to kill the kick and regain control of the well. Control of the drilling fluid density is therefore a prime concern of the drilling supervisor.

Hydrostatic pressure is defined as the pressure exerted by a column of fluid. In the drilling industry, the fluid is generally considered to be mud but could also be gas, air, foam, or water. The formula to calculate hydrostatic pressure is:

Hydrostatic pressure = 0.052 × mud density × depth　　　Eq. 1.1

Where   Hydrostatic pressure is in pounds per square inch (psi)
0.052 psi/ft/ppg is a constant,
Mud density is in pounds per gallon, and
Depth is true vertical depth (TVD) in ft.

Example 1.1   Calculate the hydrostatic pressure for each of the following systems.
(a) 10,000 ft of 12.0 ppg mud
(b) 12,000 ft of 10.5 ppg mud
(c) 15,000 ft of 15.0 ppg mud

Solution
Hydrostatic pressure = 0.052 × mud density × depth
(a) Hydrostatic pressure =
0.052 × 12.0 ppg × 10,000 ft = 6,240 psi
(b) Hydrostatic pressure =
0.052 × 10.5 ppg × 12,000 ft = 6,552 psi
(c) Hydrostatic pressure =
0.052 × 15.0 ppg × 15,000 ft = 11,700 psi

A mud gradient can also be used to calculate hydrostatic pressures.
Mud gradient is defined as the hydrostatic pressure for each foot of mud,
and is calculated using the following formula:

Mud gradient = 0.052 × mud density   Eq. 1.2

Where   Mud gradient is in psi/ft.
0.052 psi/ft/ppg is a constant, and
Mud density is in pounds per gallon.

The hydrostatic pressure would be written as,

Hydrostatic pressure = mud gradient × depth   Eq. 1.3

It should be obvious that Eq. 1.1 is a combination of Eqs. 1.2 and 1.3.
Example 1.2 illustrates the usage of the mud gradient.

Example 1.2   Use mud gradients to calculate the hydrostatic pressure exerted by 15,000 ft of 15.0 ppg mud.

Solution
(1) Using Eq. 1.2,
Mud gradient = 0.052 × mud density
= 0.052 × 15.0 ppg
= 0.780 psi/ft

$0.052 \dfrac{PSi}{ft \cdot PPG} \times 15.0\,PPG$

$\Rightarrow 0.780\ ^{PSi}/_{ft}$

(2) Using Eq. 1.3,
Hydrostatic pressure = mud gradient × depth
= 0.780 psi/ft × 15,000 ft
= 11,700 psi

The drilling fluid density is controlled by varying the concentration of high specific gravity solids within the fluid. The fluid density is increased by adding these solids. Density is decreased by either removing the solids or adding a low density fluid to dilute the concentration of the solids. Table 1.1 lists some of the more common materials used to increase fluid density.

Barite is the most commonly used density control material. Its relatively high specific gravity and inert properties make it ideal for use in the mud system. Caution must be taken when using barite from uncertain sources. Poor quality control in some mines may yield a product that is mixed with hydratable clays, which when introduced into the mud system will cause increased mud viscosity properties.

Galena, lead sulfide, will occasionally be used for density control in special applications. Its specific gravity of 6.8 will generate mud weights that will attain high hydrostatic pressures over relatively short columns of fluid. Use of galena muds has been confined to special well control applications, due to the problem of maintaining suspension of the high specific gravity solids in the mud system.

Proper well planning requires that a sufficient quantity of barite be maintained on the drilling location to kill a kick. To calculate this volume of barite properly, many operators have established a one pound per gallon safety measure which means that barite volumes will be maintained at a level sufficient to increase the present mud density by one pound per gallon. (This safety margin is based on statistics from a professional well killing company which show that the average kick would require one-half (0.5) lb/gal increase in mud weight or less. The one lb/gal margin thus incorporates a safety factor of 2 relative to the average kick).[1] The following equation can be used to calculate required barite volumes:

Pounds/barrel = 1490 $(W_2 - W_1)/35.4 - W_2$ Eq. 1.4

Table 1.1 Fluid Density Control Additives

Additive	Specific gravity	Maximum fluid density (ppg)	Remarks
1. Clay	2.3–2.5	11.5	Viscosity effects control the upper density limit
2. Barite (regular)	4.2–4.3	22.0	Barium sulfate
3. Barite (coarse grind)	4.2–4.3	22.0	Removed with 80 mesh screen
4. Galena	6.8	32.0	Lead sulfide: Special applications only.
5. Calcium carbonate	2.7	12.0 (water muds) 11.5 (oil muds)	Ground limestone
6. Sodium chloride	—	10.0	Density is temperature dependent
7. Calcium chloride	—	11.7	Density is temperature dependent
8. Zinc chloride	—	17.0	Extremely corrosive
9. Calcium bromide	—	15.0	—
10. Zinc chloride/ calcium chloride		14.0	—
11. Zinc bromide	—	19.2	—

Where   Pounds/barrel (lb/bbl) is the number of pounds of barite required to increase the density of one barrel of mud,
1490 is the weight of one barrel of barite,
W_2 is the final mud density in pounds per gallon,
W_1 is the original mud density in pounds per gallon, and
35.4 is the density of one gallon of barite.

Example 1.3 illustrates the usage of Eq. 1.4.

Example 1.3   A well is being drilled with 15.0 pounds per gallon mud. The hole volume is 850 barrels and the surface pit volume is 350 barrels. How many sacks of barite should be maintained on the drilling location? (Assume that one sack contains 100 pounds of barite).

Solution
(1) Using Eq. 1.4,

$$\text{Pounds/barrel} = \frac{1490 \ (W_2 - W_1)}{35.4 - W_2}$$

THE 1.0 lb/GAL MARGIN INCORPORATES A SAFETY FACTOR OF 2 RELATIVE TO THE AVERAGE KICK.

$$W_1 = 15.0 \text{ ppg}$$
$$W_2 = W_1 + 1.0 \text{ ppg} = 16.0 \text{ ppg}$$
$$\text{Pounds/barrel} = \frac{1490 \ (16.0 - 15.0)}{35.4 - 16.0}$$
$$= \frac{1490}{19.4} = 76.8 \text{ pounds/barrel}$$

(2) $(850 + 350)$ barrels $\times$ 76.8 pounds/barrel
        = 92,160 pounds of barite
(3) $\dfrac{92,160 \text{ pounds}}{100 \text{ pounds/sack}}$ = 921.6 sacks

Blowout Preventers

When primary control of the well has been lost due to insufficient mud hydrostatic pressure, it becomes necessary to seal the well by some means in order to prevent an uncontrolled flow, or blowout, of formation fluids. The equipment which seals the well is called a blowout preventer. It consists of drill pipe blowout preventers designed to stop the flow through the drill pipe, and annular preventers designed to stop flow in the annulus. Since there are numerous types of elements in a stack of blowout preventers, each element will be described with some of its special design features presented.

Annular Blowout Preventers

The blowout preventer stack is designed to control the flow of fluids in the annulus and may be a composite of several types of annular blowout preventer elements. Some, but not all of these elements, may include spherical preventers, blind and pipe rams, and drilling spools. Each type of element will be discussed with actual blowout preventer stack design criteria presented in later sections. (A listing of various blowout preventer element specifications is presented in the Appendix).

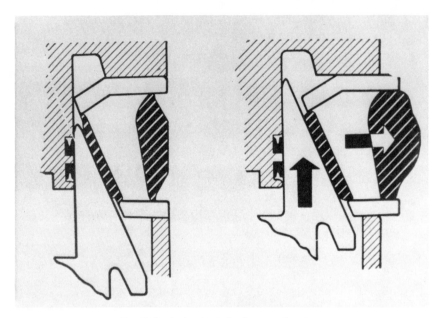

Fig. 1-1   Spherical closing mechanism

Spherical preventers.   The first preventer normally closed when shut-in procedures are initiated is the spherical preventer. The four basic segments of the spherical preventer are the head, body, piston, and steel-ribbed packing element (Fig. 1.1). When the preventer's closing mechanism is actuated, hydraulic pressure is applied to the piston, causing it to slide in an upward direction and force the packing element to extend into the wellbore around the drill string. The preventer element is opened by applying hydraulic pressure in a manner that slides the piston downward and allows the packing to return to its original position.

Caution should be taken when closing the spherical preventer to insure that the preventer is closed only when the drill string is in the hole. When pipe is not present, the rubber packing element is overstressed in attempting to close on itself and will shorten the overall life of the element. When emergencies exist, however, most spherical preventers will seal in this case if necessary.

The initial recommended hydraulic pressure for closing most spherical preventers is 1,500 psi. (A listing of recommended closing pressures for most types of preventers and valves used in well control is presented in the Appendix.) After the preventer is closed, the hydraulic pressure should be reduced to minimize damage to the rubber portion of the element. Several manufacturers have recommended closing pressures depending on the kick casing pressure after the well is shut-in. (Charts and tables for these pressures are presented in following sections in this chapter.) If the recommended closing pressure is not known, reduce the hydraulic pressure until the preventer begins to allow a very small leak around the pipe.

While well killing procedures are in progress, it is not always necessary to exert hydraulic pressure on the preventer in excess of the kick pressure. Most spherical preventers, as well as many ram-type preventers, are designed to utilize wellbore pressures to aid in maintaining closure. In some cases, it may be observed that the preventer will remain closed even when virtually no hydraulic pressure is applied.

The spherical preventer element can be changed without removing the pipe should the element become damaged during well killing operations. When the preventer becomes damaged, the pipe rams below the spherical should be closed and locked. The top plate or cover of the preventer must be unbolted and the rubber element lifted out with the rig hoist line.

With the element out of the body, use a knife and split the rubber from around the pipe. Cut the new rubber between the ribs (Fig. 1.2) and install the element in reverse order from the removal sequence. After the top plate is bolted on the preventers, the spherical can again be used in the killing process. This process is obviously not applicable in subsea applications. If the accumulator is positioned significantly above the preventers, it may be necessary to drain the hydraulic fluid to the spherical before the cover can be unbolted.

A special design feature of the spherical preventer is that it will allow stripping operations to be carried out while maintaining a seal during tool joint passage. The packing element is generally considered superior to the rams for stripping purposes due to its greater abrasion resistance. Although the accumulator pressure regulator will maintain a constant hydraulic pressure on the packing element, caution must be exercised because the slow response of regulators presently in use requires that tool joints be moved slowly through the preventer in order to avoid damage to the packing element.

Hydril Corp. manufactures several models of spherical preventers with different packing elements available for specific types of service. Table 1.2 lists the available packing elements, and Tables 1.3, 1.4, and 1.5 give the recommended closing pressures for the MSP, GK, and GL models, respectively. When the GL model is used in subsea applications, special tables of closing pressures, dependent upon the water depth, are available from the manufacturer.

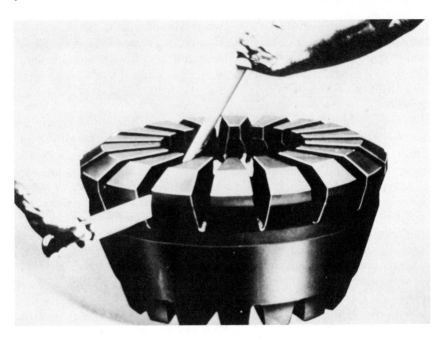

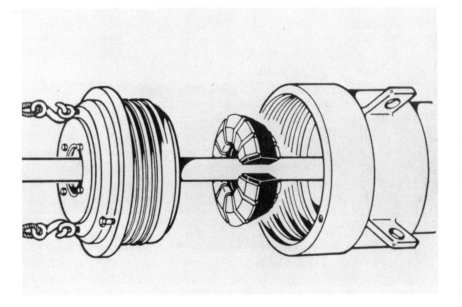

Fig. 1-2 Spherical element replacement

Table 1.2 Hydril packing elements

Packing type	Color code	Letter code	Manufacturer's recommended usage
Natural rubber	Black	R	Water base muds with less than 5% oil; operating temperatures greater than −30°F; applicable for H_2S service.
Synthetic rubber	Red	S	Oil base muds with aniline points between 165°F and 245°F; applicable for H_2S service; operating temperatures greater than 20°F.
Neoprene	Green	N	Oil base muds with operating temperatures between 20°F and −30°F; applicable for H_2S service.

Hydril "MSP" operating features.
(1) Low pressure service.
(2) Will close on open hole (but not recommended).
(3) Primary usage is in diverter systems.
(4) Automatically returns to the open position when the closing pressure is released.
(5) Sealing assistance is gained from the well pressure.

Fig. 1-3 Hydril "MSP" preventer. Courtesy of Hydril Corp.

Table 1.3 "MSP" closing pressures

Pipe OD, in.	Preventer size (in.-psi)			
	6–2,000	8–2,000	10–2,000	20–2,000
5½	—	—	350	500
4½	350	400	450	550
3½	400	450	550	600
2⅞	450	550	650	650
2⅜	500	650	750	700
1.90	600	750	850	800
1.66	700	850	850	900
CSO	1,000	1,050	1,150	1,100

Manufacturers' note: A pressure of approximately 1,500 psi is recommended for closing the "MSP-2000" preventer. A tight pack-off can normally be maintained with closing pressures as shown in the table above. When drilling from a floating vessel, the closing pressure should be preset to the values shown.

Hydril "GK" operating features.
(1) Full range of bore sizes and pressure ratings.
(2) Available with flanges of a higher pressure rating than the preventers to connect with high pressure ram type preventers.
(3) Will close on open hole (but not recommended).
(4) Sealing assistance is gained from the well pressure.
(5) Requires high accumulator pressures when used in subsea installations.
(6) Has provisions to measure piston travel to gauge packing element wear.

Hydril "GL" operating features.
(1) Will close on open hole (but not recommended).
(2) Some sealing assistance is gained from well pressure.
(3) Bolted cover for easier element change.
(4) Available in large bore models only.
(5) Originally designed for subsea applications.
(6) Has a secondary chamber designed to balance riser pipe hydrostatic pressure in subsea systems.

Shaffer, a division of the NL Petroleum Services group, manufactures a spherical preventer that operates on the same principle as the Hydril spherical (Fig. 1.6). Although the preventer has undergone many en-

Table 1.4 "GK" closing pressures

Pipe OD, in.	6-3M	6-5M	7¹/₁₆-10M	8-3M	8-5M	9-10M	10-3M	10-5M	11-10M	12-3M	13⅝-5M	13⅝-10M	16-2M	16-3M	16¾-5M	18-2M
6⅝										450	550		350	450		500
5						350	450	450		500	600		400	500		550
4½	350	350	350	400	450	380	450	450	420	550	650	525	500	550	600	600
3½	400	400	550	450	550	570	550	550	600	600	700	640	600	600	650	650
2⅞	450	450	750	550	650	760	650	650	780	700	750	815	700	700	750	700
2⅜	500	500	850	650	750	860	750	750	870	800	800	885	800	800	850	750
1.90	600	600	900	750	850	950	920	850	960	900	900	990	900	900	950	850
1.66	700	700	1,000	850	950	1,000	950	950	1,000	1,000	1,000	1,050	1,000	1,000	1,050	950
CSO	1,000	1,000	1,150	1,050	1,150	1,150	1,150	1,150	1,150	1,150	1,150	1,150	1,150	1,150	1,150	1,150

Manufacturers' note: The pressures above are a guideline. Maximum packing unit life will be realized by use of the lowest closing pressure that will maintain a seal. For subsea applications, see the appropriate operator's manual for computation of best closing pressure.

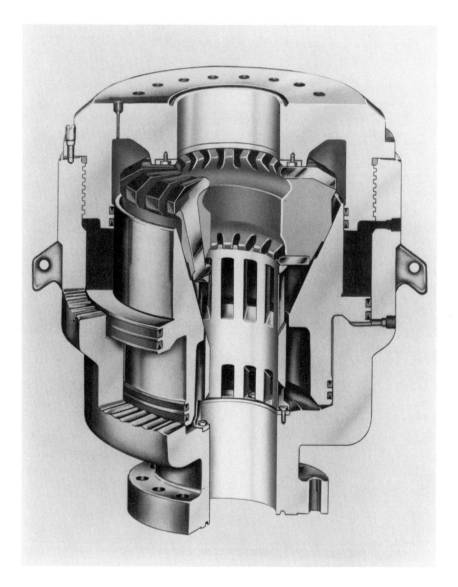

Fig. 1-4   Hydril "GK" preventer

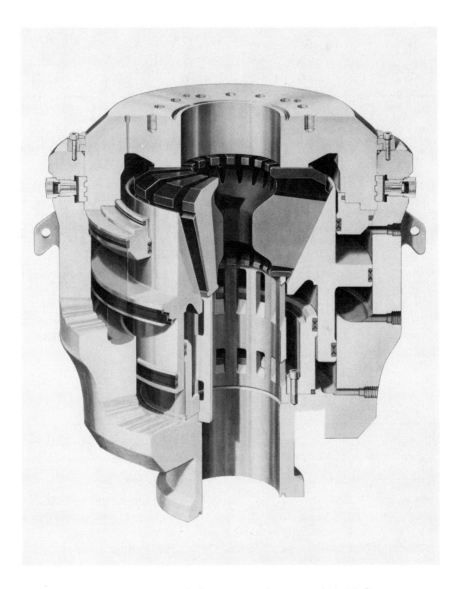

Fig. 1-5 Hydril "GL" preventer. Courtesy of Hydril Corp.

Table 1.5 "GL" closing pressure

	Preventer size, (in. - psi)								
	13⅝-5,000			16¾-5,000			18¾-5,000		
	Well pressure (psi)			Well pressure (psi)			Well pressure (psi)		
Pipe OD, in.	2,000	3,500	5,000	2,000	3,500	5,000	2,000	3,500	5,000
7	900	950	1,100	700	825	950	700	825	950
5	900	1,000	1,100	725	850	1,000	800	900	1,000
3½	1,200	1,200	1,200	800	925	1,050	1,000	1,050	1,100
Full closure	1,400	1,500	1,500	1,400	1,500	1,500	1,500	1,500	1,500

Note: These are closing pressures from the secondary chamber to the opening chamber for surface installations. Consult the manufacturer for subsea applications.

gineering changes within the past few years, rigorous laboratory testing has shown that the new version is durable even when tested under severe pressure and wear conditions. Table 1.6 describes the packing elements available for the Shaffer unit. Fig. 1.7 shows the recommended closing pressures for the Shaffer sphericals.

Table 1.6 Shaffer packing elements

Packing type	Color code	First digit of serial no.	Manufacturer's recommended usage
Natural rubber	Red	1 or 2	Low temperature operations in water base muds; stripping abrasion resistance.
Neoprene	Black	3 or 4	H_2S in water base muds; applicable in oil muds but shortened life compared to nitrile.
Buna (nitrile)	Blue	5 or 6	Oil and water base muds; H_2S in oil base muds.

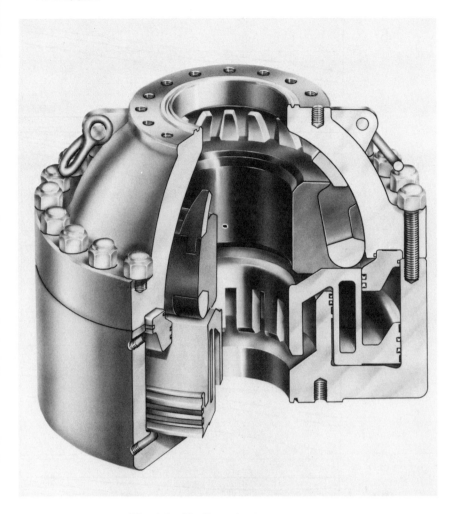

Fig. 1-6   Shaffer spherical preventer

Shaffer spherical operating features.
(1) Will close on open hole. (The manufacturer does not recommend
    against closing on the open hole).
(2) Applicable in subsea service.
(3) Slight sealing assistance is gained from the well pressure.
(4) Element returns to full bore when opened.

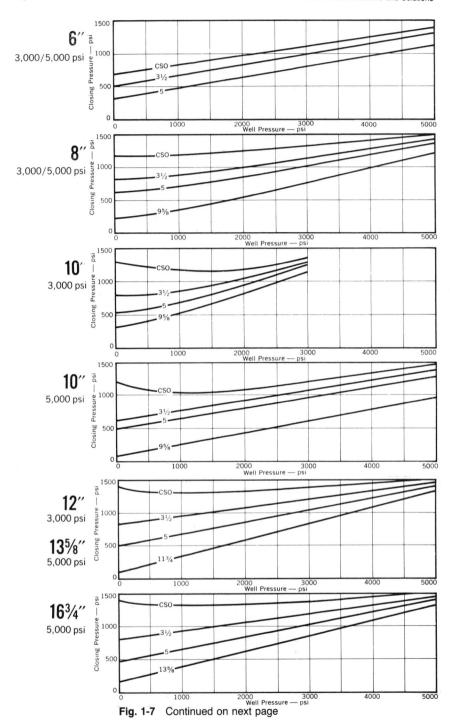

Fig. 1-7 Continued on next page

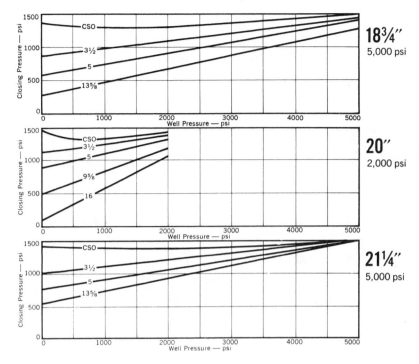

Fig. 1-7 Shaffer recommended closing pressures

Cameron Iron Works manufactures the spherical preventer shown in Fig. 1.8. The Model D, which is a re-engineered version of the original A model, has received extensive field and laboratory testing since 1971 and has exhibited highly desirable well control operating characteristics. The functional mechanism is that of the hydraulic operated, sliding piston type.

Cameron "D" operating features.
(1) Will close on open hole (but not recommended).
(2) Sealing assistance is gained from well pressure.
(3) Quick release top latch allows fast element change.
(4) Low vertical height.
(5) Low element weight.
(6) Requires less fluid to open and close the element than comparable models.
(7) Standard trim suitable for both normal and H_2S service.

Fig. 1-8   Cameron "D" preventer. Courtesy of Cameron Iron Works

Regan Forge and Engineering Co. offers a spherical preventer developed on a different operating mechanism than the other available preventers. The Regan KFL has 3 rubber, non-ribbed packers that are actuated by applying hydraulic pressure on the exterior of the elements, thereby forcing them inward to seal on the pipe in the well. This arrangement is also unique in that it is designed to allow the insert or inner rubber packer to be removed and exchanged without removing the top cover. This is an extremely functional characteristic since the insert packer is the only part of the preventer that receives wear.

Fig. 1.9 shows the Regan KFL spherical preventer. A model KFLR is available with a set of lower hydraulic dogs which provides a retractable seat for use with marine risers that have an inner diameter smaller than the diameter of the standard insert packer and when it is necessary to have a full opening through the KFLR to the preventer stack.

Regan KFL operating features.

(1) Designed for subsea applications but serviceable in surface preventers.

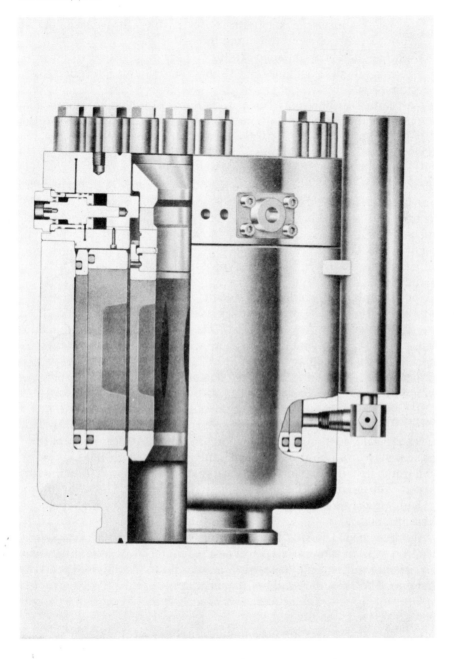

Fig. 1-9 Regan "KFL" preventer

(2) Retrievable insert packer (with retrieving tool) for replacement of worn element without removing preventer stack.

(3) A surge chamber is attached to the preventer to allow greater pipe stripping flexibility than conventional accumulator controlled stripping.

(4) Well pressure does not assist in maintaining the element closed. The manufacturer recommends maintaining hydraulic pressure on the element greater by 500 psi than the casing pressure.

(5) Available in large bore models only.

Ram-Type Preventers. Unlike the operational manner of the spherical preventer, the ram preventers seal the annulus by forcing two elements to make contact with each other in the annular area. These elements have rubber packing seals that effect the complete closure. Other than the sealing mechanism, ram-type blowout preventers differ greatly from spherical preventers in that each type and size of ram has one function and cannot be used in a variety of applications.

For example, ram bodies with 4½-in. pipe rams will seal on 4½-in. pipe and will not seal with any other size of pipe, nor will it seal without pipe in the well. (The exception to this is Cameron's new Variable Bore Ram). Ram-type preventers, however, are generally considered to be more reliable in high pressure service, as well as being more easily serviceable and shorter. The types of rams that warrant discussion are pipe, blind, and shear rams.

Several design features of all ram preventers should be understood by the drilling supervisor. One of these features is the direction of the pressure seal. Most ram preventers are designed to hold pressure from the lower side which means that (1) the preventer will not secure the well if it is installed upside down, and (2) the ram will not pressure test from the top down. The last consideration is important when designing a blowout preventer stack arrangement and the manner in which it will be pressure tested.

Another special design feature is the secondary rod sealing action which is available for most types of ram preventers. Due to routine wear, the primary rod sealing mechanism may begin to leak under excessive pressure. The secondary rod sealing mechanism is thus used to provide an additional measure of protection in sealing the area around the rod that is used to close the preventer.

Most ram preventers are manufactured with a self-feeding action for the rubber sealing section (Fig. 1.10). As the rubber wears, the small

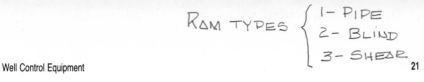

RAM TYPES
{
1- PIPE
2- BLIND
3- SHEAR
}

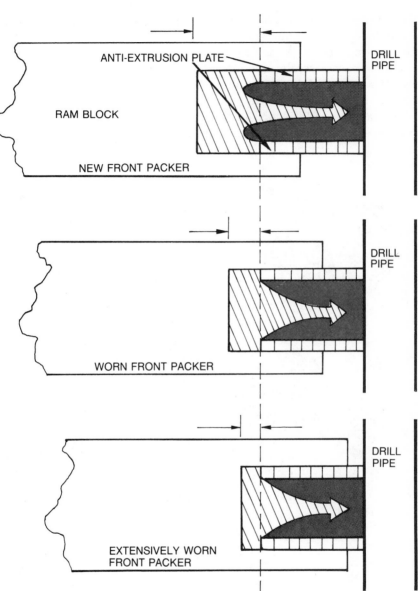

ANTI-EXTRUSION PLATE

DRILL PIPE

RAM BLOCK

NEW FRONT PACKER

DRILL PIPE

WORN FRONT PACKER

DRILL PIPE

EXTENSIVELY WORN FRONT PACKER

Fig. 1-10 Rams self-feeding action

plates shown are forced into the recessed area which allows additional rubber to extend past the ram face and aid in securing a seal. If the rams are used improperly, the self-feeding action will cause the rubber seal to extend an excessive distance into the wellbore which will cause over-stressing and rapid deterioration of the element. Due to this, pipe rams should not be closed routinely if pipe is not in the hole.

Ram bodies are universal in that they will accept either blind ram elements or pipe ram elements. Also, units are available that are com-

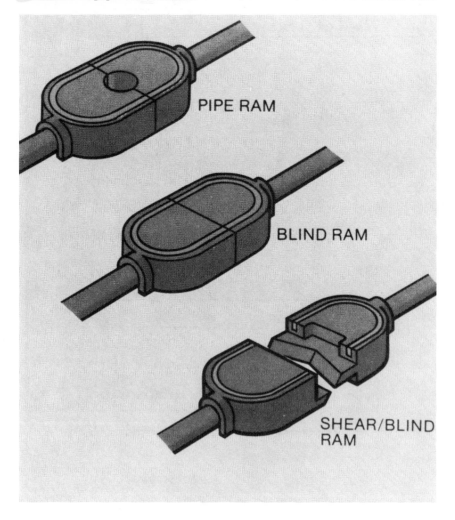

PIPE RAM

BLIND RAM

SHEAR/BLIND RAM

Fig. 1-11   Ram elements

prised of single, double, or even triple ram bodies. In the multiple unit ram bodies, any combination of pipe and blind ram elements may be used.

Fig. 1.11 displays a pipe ram element showing the element, rubber seal, and the pipe guides that center the pipe during closure. (See the discussion of the Cameron Variable Bore Ram for an exception to the rule.)

Caution must be given to ram size selection when aluminum drill pipe is in use. This type of pipe has a tube in the middle section that is slightly smaller than the tube near the tool joint. Regular 4½-in. pipe rams will seal on the middle tube section of 4½-in. aluminum pipe but not near the tool joint, as could be done with steel pipe. The shut-in procedures must be planned accordingly to account for this irregularity.

One special design feature of the pipe ram element is that when closed and locked, the ram can support the weight of the drill string if necessary by hanging a tool joint on the ram. This feature is useful when storm conditions exist or underwater blowouts are impending. This usage is not recommended under normal conditions however, and if this practice is to be routine, special hardness ram elements can be purchased for this function.

Blind rams are designed to seal the well if pipe is not in the hole. The element is flat-faced and contains a rubber section as shown in Fig. 1.11. The rams are not designed to effect a seal when pipe is in the hole, although occasionally the pipe will be cut if the blind rams are accidentally closed. Precautions should thus be taken with the blowout preventer control panel to ensure that the blind rams cannot be accidentally closed.

Shear rams are a specially designed blind ram. As the word "shear" indicates, this type of ram will seal if pipe is in the hole by shearing, or cutting, the pipe and sealing the open wellbore. Since this type of action allows the drill string to be dropped, a set of pipe rams may be installed below the shear rams, and a tool joint set on the pipe rams before the shear rams are activated. When the shear rams are installed in conventional ram bodies, booster power units and larger bonnets may be necessary for efficient operations (Fig. 1.11).

This discussion on some of the models of ram bodies available for usage does not encompass every model available nor does it completely cover the models presented. The manufacturer should be consulted for complete details. A listing of various specifications of ram type preventers is presented in the Appendix.

Cameron Iron Works manufactures five models of ram preventers for well control. These are types F, SS, QRC, U, and the Variable Bore Ram.

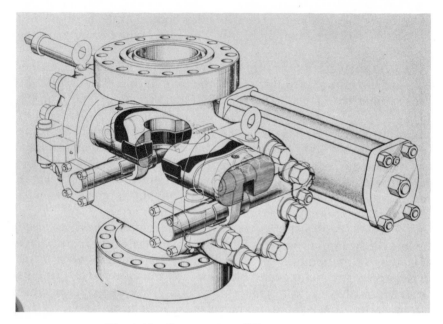

Fig. 1-12 Cameron type "F" ram preventer

Although the types F, SS, and QRC are no longer manufactured, they each still enjoy widespread use.

Cameron type "F" operating features (Fig. 1.12).
(1) Low closing pressures required due to low closing ratios.
(2) Well pressure aids in maintaining rams closed.
(3) Rams can be changed and repaired in the field.
(4) Relatively lightweight.
(5) Only one screw required for manual locking.
(6) The hydraulic operator can be replaced while the preventer is locked in pressure service.
(7) Ram rubbers have self-feeding action.

Cameron type "SS" (Space Saver) operating features (Fig. 1.13).
(1) Low vertical height. (Originally designed for rigs with low substructures).
(2) Ram position can be determined by external observation of the locking screws.
(3) Well pressure assists in maintaining rams closed.
(4) Rams can be changed and repaired in the field.
(5) Ram rubbers have self-feeding action.

Type "SS"

Fig. 1-13 Cameron type "SS" ram preventer

Cameron type "QRC" (Quick Ram Change) operating features (Fig. 1.14).
(1) Rams can be manually locked in the closed position.
(2) Well pressures help to maintain rams closed.
(3) Rams can be changed and repaired in the field.
(4) Ram position can be determined by exterior observation.
(5) Has a secondary operating rod seal.
(6) Ram rubbers have self-feeding action.

Cameron type "U" operating features (Fig. 1.15).
(1) Originally designed to meet subsea specifications, but is applicable for surface operations.
(2) Well pressures help to maintain rams closed.
(3) Rams can be changed and repaired in the field.
(4) Ram position can be determined by exterior observations.

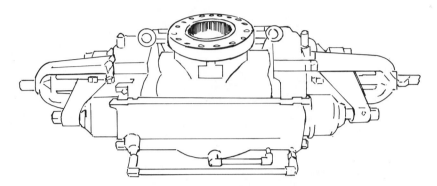

Fig. 1-14 Cameron type "QRC" ram preventer

(5) Has secondary rod seal.
(6) Ram rubbers have self-feeding action.
(7) Rams can be manually locked or hydraulically locked if special options are installed.
(8) Shear rams are available. Bore sizes smaller than 16¾-in. require special bonnets and booster power units.
(9) Available with special trim for H_2S service.

Fig. 1-15 Cameron type "U" ram preventer. Courtesy of Cameron Iron Works

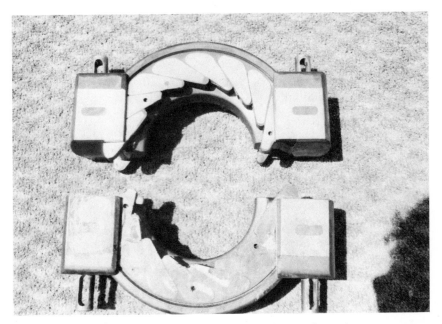

Fig. 1-16 Cameron variable bore ram element

Cameron variable bore ram element operating features (Fig. 1.16).
(1) Will effect a seal around several pipe and kelly sizes. (Table 1.7)
(2) The variable bore ram fits into a standard type "U" ram body.
(3) Steel inserts in the packer rotate inward when the rams close and maintain a continuous ring of steel supporting the rubber which seals against the pipe.
(4) Rubbers have a self-feeding action.

Table 1.7 Variable bore ram closure diameters

Preventer bore, in.	Pipe size range*, in.
11	5–2 ⅞
13 ⅝	7–5
13 ⅝	5–2 ⅞
16 ¾	7–3 ½
18 ¾	7 ⅝–3 ½

* Manufacturers should be consulted for additional pipe size ranges to be marketed.

(5) Well pressures aid in maintaining the ram closed.
(6) Subsea applications.
(7) Ram position can be determined by exterior observation when used in surface operations.
(8) Rams can be manually locked or hydraulically locked if special options are installed.

Shaffer, a division of the NL Petroleum Services group, have manufactured several models of preventers. These models were LWP, LWS, XHP, B, E, and the SL. Types B and E will not be presented in this discussion.

Shaffer "LWS" operating features (Fig. 1.17).
(1) Wellbore pressures help to maintain rams closed.
(2) Rams can be changed and repaired in the field.
(3) Relatively lightweight.
(4) New models have the secondary rod sealing action and old models can have it installed.

Fig. 1-17 Shaffer "LWS" ram preventer

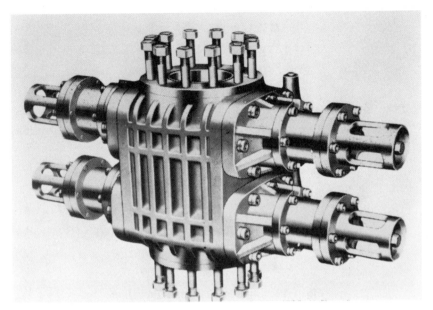

Fig. 1-18   Shaffer "LWP" ram preventer

(5) Ram position can be determined by exterior observation with manual locking system.
(6) External ram locking mechanism or can have an automatic locking system installed.
(7) Ram rubbers have self-feeding action.
(8) Side door opening mechanism requires some additional space.
(9) Shear rams are available.
(10) New model preventers are applicable for H_2S service after changing retracting screws.

Shaffer "LWP" operating features (Fig. 1.18).
(1) Limited pressure ratings.
(2) Relatively lightweight.
(3) Rams can be changed and repaired in the field.
(4) Side door opening mechanism requires some additional space.
(5) Ram position can be determined by exterior observation with manual locking system.
(6) Ram rubbers have a self-feeding action.

Shaffer "XHP" (Extra High Pressure) operating features (Fig. 1.19).
(1) Extreme pressure service.

Fig. 1-19   Shaffer "XHP" ram preventer

(2) Secondary rod sealing action.
(3) Rams can be changed and repaired in the field.
(4) Ram rubbers have a self-feeding action.

Shaffer "SL" operating features (Fig. 1.20).
(1) Wellbore pressures help to maintain rams closed.
(2) Rams can be changed and repaired in the field.
(3) Ram rubbers have self-feeding action.
(4) Rams can support drill string load to 600,000 pounds.
(5) Rams have a maximum Rockwell hardness of 22 for H_2S service.
(6) Shear ram elements are fully interchangeable with pipe rams in any cavity.
(7) Automatic locking system available.
(8) Secondary rod sealing action permits injection of plastic packing.

Hydril Co. manufactures two types of ram preventers. They are the V model for pressures to 5,000 psi and the X model for pressures greater than 5,000 psi.

Hydril "V" operating features (Fig. 1.21).
(1) Available with manual or automatic locking systems.
(2) Cylinder liner and upper seal seat are field replaceable or repairable.
(3) Secondary rod sealing action.

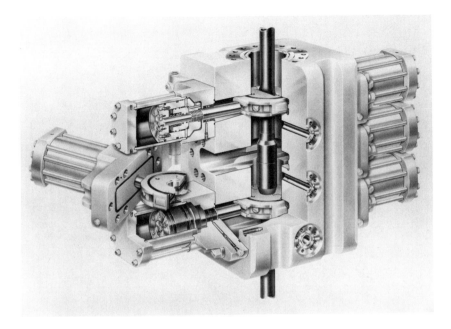

Fig. 1-20 Shaffer "SL" ram preventer

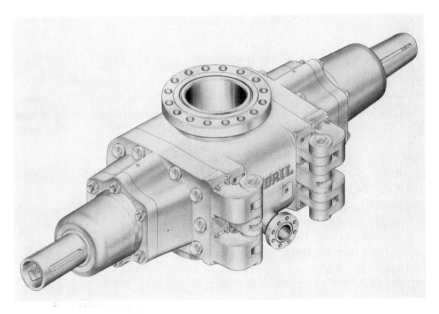

Fig. 1-21 Hydril "V" ram preventer

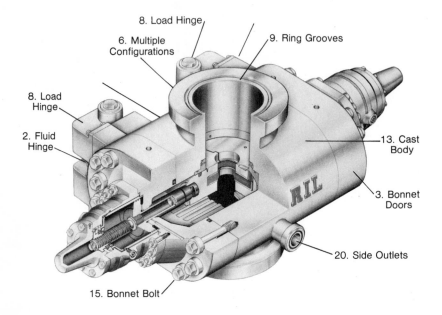

8. Load Hinge
6. Multiple Configurations
9. Ring Grooves
8. Load Hinge
2. Fluid Hinge
13. Cast Body
3. Bonnet Doors
20. Side Outlets
15. Bonnet Bolt

Fig. 1-22   Hydril "X" ram preventer

(4) Rams can be changed and repaired in the field.

(5) Additional room must be allowed for side door openings.

(6) Sloped ram cavity is self draining of mud and sand.

(7) Rams are designed to permit drill pipe hangoff.

(8) Internal manual locking threads.

Hydril "X" operating features (Fig. 1.22).

(1) Available for extreme pressure service.

(2) Available with manual or automatic locking system.

(3) Secondary rod sealing action.

(4) Rams can be changed or repaired in the field.

(5) Additional room must be allowed for side openings.

(6) Sloped ram cavity is self draining of mud and sand.

(7) Rams are designed to permit drill pipe hangoff.

(8) Cylinder liner and upper seal seat are field repairable or replaceable.

(9) Internal manual locking threads.

Pressure Control Equipment

Drilling spools. If blowout preventer elements that have no built-in mud exit lines are used, it becomes necessary to install a drilling spool which is a connector placed within the blowout preventer stack to which mud access lines, termed choke and kill lines, are attached. The spool may be studded, flanged, or clamp-on connected and should meet the following API requirements:[2]

(1) Have a working pressure consistent with that of the remainder of the blowout preventers.
(2) Have one or two side outlets, no smaller than 2 inches in diameter, with a pressure rating consistent with the blowout preventer stack.
(3) Have a vertical bore diameter at least equal to the maximum inner diameter of the innermost casing. If the spool is to pass slips, hangers, or test tools, the bore should be at least equal to the maximum bore of the uppermost casing head, or blowout preventer stack.

Fig. 1.23 illustrates a flanged drilling spool with two side outlets.

Casing head. The basis of all blowout preventer stacks and usually the front component installed is the casing head. The head can be

OUTLET

OUTLET
CAN'T BE SMALLER
THAN 2" IN DIA.

Fig. 1-23 Flanged drilling spool

equipped with flanged, slip-on and weld, or threaded connections for attachment to the casing and the preventer stack and can have threaded or open-face flanged side outlets. The casing head should meet at least the minimum API requirements as follows:[2]

(1) Have a working pressure rating which equals or exceeds the maximum anticipated surface pressure to which it will be exposed.
(2) Equal or exceed the bending strength of the outermost casing to which it is attached.
(3) Have end connections of mechanical strength and pressure capacity comparable to corresponding API flanges or to the pipe to which it is attached.
(4) Have adequate compressive strength to support subsequent casing and tubing weight to be hung therein.

Figure 1.24 is an example of a casing head with threaded lower connections and flanged upper connections.

Diverter bags. In certain cases, proper well control procedures demand that a kick not be shut-in, but rather allow the well to blowout in a controlled manner away from the rig. (Reasons for this procedure will be presented in later sections.) These blowout diversion procedures do not

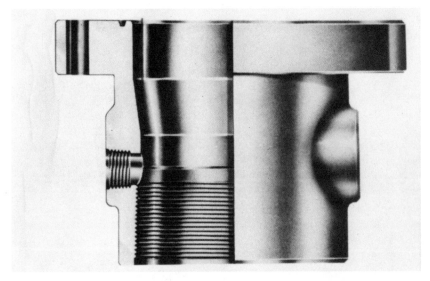

Fig. 1-24 Casing head with threaded lower connections and flanged upper connections

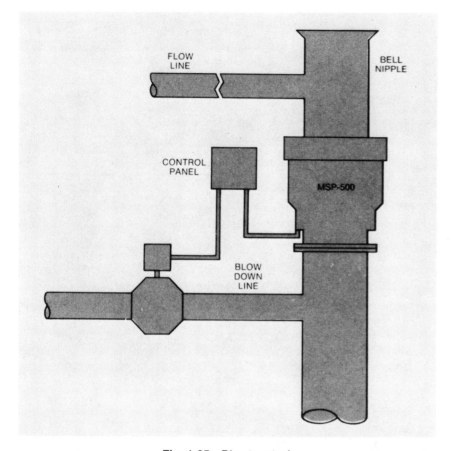

Fig. 1-25   Diverter stack

require a full blowout preventer stack, but instead a diverter bag is used which is a relatively low working pressure tool. Fig. 1.25 illustrates a diverter stack in which a spherical preventer is used as the diverter bag.

Rotating head.   The primary function of a spherical preventer is to provide pressure control while allowing a small amount of pipe movement. Occasionally a tool is needed that will provide greater amounts of pipe movement flexibility at lower service pressures. The rotating head serves this purpose. Rotating heads (Fig. 1.26) have been used in air and gas drilling, controlled pressure drilling, and in reverse circulation operations with well pressures to 2,000 psi and at rotating speeds to 150 rpm. When used in controlled pressure drilling, the head allows the use of lighter muds with increased penetration rates and reduced swabbing. The head also maintains the gas in a kick under pressure to reduce its volume.

Fig. 1-26 Rotating head

Choke and kill lines. In well killing operations, it will generally be necessary to circulate fluid down the drill pipe, up the annulus, and through an exit at the surface. The lines that are attached to the blowout preventers to provide this exit are termed choke and kill lines. The choke line carries the mud and kick fluid from the blowout preventer stack to the choke device. The primary purpose of the kill line is to serve as a back-up choke line. The kill and choke lines may be used to pump mud directly into the annulus if necessary although the kill line usually performs this function.

WHAT HAPPENS NEXT ?

The kill and choke lines may be attached to several members of the blowout preventer stack. These lines could be attached to the outlets of the drilling spool shown in Fig. 1.23, or they could be attached directly to the blowout preventer as indicated in Fig. 1.22. Only under extreme circumstances, and never preferentially, should the choke and kill lines be attached to the casing head, casing spool, or below the lowermost set of rams. (See the section on preventer stack design for a further explanation.)

IS NOT

The kill and choke lines should meet a number of requirements. Some, but not all, are as follows:

(1) The pressure ratings of these lines should be consistent with that of the blowout preventer stack.
(2) These lines should meet all minimum blowout preventer testing requirements.
(3) The lines should have a consistent inner diameter to minimize erosion at the point of diameter changes.
(4) The number of angular deflections within these lines should be kept to a minimum. If the lines are required to make several angular changes between the stack and the choke manifold, it may be advisable to use tees and crosses to absorb the turbulent erosion effects at these points.

Valves

There should be valves of different types attached to the blowout preventer stack to attain certain end results. The valves presented in this section are ball, gate, check, and fail-safe valves. Proper arrangement of these valves will be presented in the following sections on preventer stack design.

Ball valve. The name of this type of valve is descriptive of its construction type. A steel ball (Fig. 1.27) has an opening, or orifice, along a line passing through its center. The valve is open when the ball is

Fig. 1-27 Ball valve (cut-away view)

the first component installed is the casing head. The head can be oriented so the orifice is in line with the fluid flow path. It is closed when the ball is oriented so the opening is perpendicular to the flow path.

The valve can be operated either remotely by using hydraulic action or can be manually operated. In practical applications, it will often be necessary to pressurize the downstream side of the valve before opening if pressure of a sufficient magnitude is acting on the upstream side of the valve. This pressurization is done because (1) the upstream pressure may create a differential pressure sufficient to cause the opening stem of the valve to shear when a sufficient external force is applied, and (2) to avoid a sudden pressure surge on the downstream equipment.

Gate valve.   A gate valve utilizes a closing mechanism different than a ball valve. In the gate valve shown in Fig. 1.28, a blank plate is positioned across the flow path to halt fluid flow. When the valve is opened, the plate is moved in a manner such that a section of the plate containing an orifice is positioned across the flow path which thus allows fluid movement through the orifice. Similar to the ball valve, the gate valve can be operated either manually or through remote hydraulic controls.

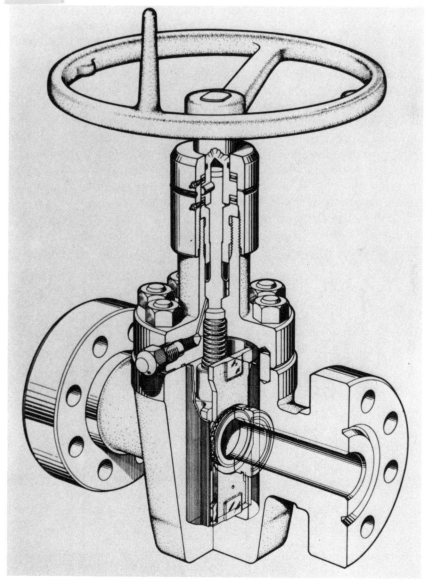

Fig. 1-28 Cameron gate valve

The Cameron Type "F" gate valve shown in Fig. 1.28 has several design features that make it particularly useful in well control operations. First, it has been termed a "WOGM" valve which means that it is serviceable when water, oil, gas, or mud is the circulating fluid. Second, a ratchet device at the orifice positions a new surface to exposure each time the valve is operated. This feature was installed after it was observed that abrasive wear always occurred at the same location on the valve. By exposing a new surface upon each valve actuation, the overall life of the valve is significantly extended.

NICE FEATURE.

Check valve.   When well control conditions warrant pumping fluid directly into the annulus, a check valve on the kill line has applications. The valve, shown in Fig. 1.29, is generally a spring-loaded valve that allows fluid movement in one direction only, i.e., mud can be pumped into the annulus through the valve but cannot reverse flow direction out of the annulus. A special design feature of some models of check valves is that the internal elements can be removed when necessary to allow fluid flow in either direction. A check valve is generally identified in schematic notation by placing an arrow immediately above the valve to indicate the direction of fluid flow.

Fail-safe valve.   A fail-safe valve can be either a gate or ball valve that has been equipped with a pressure sensing device to actuate (open

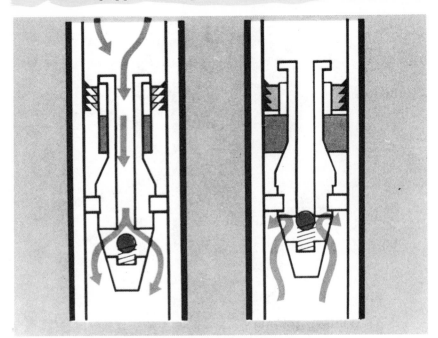

Fig. 1-29   Check valve

FAIL-SAFE VALVE

Fig. 1-30 Cameron fail-safe valve

or close) automatically the valve if pressures change to unfavorable values (Fig. 1.30). These valves can be set according to either internal or external valve pressures, minimum or maximum pressures, or erratic pressure changes. Fail-safe valves are used extensively in subsea applications, and to some extent, on land drilling operations.

Accessory Equipment for Annular Blowout Preventers

Several pieces of accessory equipment are necessary for proper annular blowout preventer operation. This equipment includes flange bolts, clamp connectors, H_2S or sour gas trim, and seal rings. Guidelines for the

Table 1.8 Bolt tension and torque necessary to offset hydrostatic test pressures or 50% minimum yield of bolts, whichever is greater[1]

Bolt size, in.	API STD 6A April 1970 for 6B[2]		API STD 6A April 1970 for 6BX[3]		USAS B16.5-1968 and Aw1 STD 6D March 1970[4]	
	Bolt tension, lb	Nut torque, ft-lb	Bolt tension, lb	Nut torque, ft-lb	Bolt tension, lb	Nut torque, ft-lb
3/4	17,535	112	20,600	131	17,535	112
7/8	24,255	177	25,800	189	24,255	177
1	31,815	265	37,000	308	31,815	265
1 1/8	41,475	380	51,800	475	41,475	380
1 1/4	52,500	507	52,300	505	52,500	507
1 3/8	64,732	702	77,900	845	64,732	702
1 1/2	78,330	915	102,000	1,191	78,330	915
1 5/8	93,450	1,171	126,000	1,579	93,450	1,171
1 3/4	109,200	1,460	145,000	1,939	109,200	1,460
1 7/8	126,525	1,797	188,000	2,670	126,525	1,797
2	145,425	2,187	175,000	2,632	145,425	2,187
2 1/4	185,900	3,127	256,000	4,283	185,900	3,127
2 1/2					233,000	4,290
2 3/4					259,000	5,197
3					305,000	6,630
3 1/4					365,000	8,601

3½	426,000	10,700
3¾	493,000	12,715
4	561,000	15,963
4¼	503,000	15,138
4½	556,000	18,006
4¾	630,000	21,063
5	702,000	24,650
5¼	778,000	28,288
5½	850,000	32,780
5¾	935,000	37,490
6	1,015,000	42,548

1 Coefficient of friction is 0.067; thread Form 10 UNC for ¾" 9UNC for ⅞", 8 UNC for 1", and 8 UN for 1⅛" and larger bolts; and heavy hex nuts,

2 Based upon 50% of minimum yield bolts.

3 Based upon internal hydrostatic test induced stresses. Also are included sizes of bolts used in larger and/or higher pressure flanges than appear in this standard, but which are in use in some areas.

4 Based upon 50% of minimum yield of bolts. Per ASTM 193 Grade B-7 bolts; or ASTM A364 bolts, with heavy hex ASTM A194 nuts up through 2½" bolts minimum yield 105,000 psi, minimum tension strength 125,000 psi; over 2½" through 4" bolts minimum yield 95,000 psi, minimum tension strength 115,000 psi, over 4" bolts minimum yield 75,000 psi, minimum tension strength 100,000 psi.

(Note: Maximum test pressure calculated to bear upon total area to OD of ring groove. This is felt to be the maximum load and will exceed load on raised face and other types.)

proper use of this equipment, though often not understood well by rig personnel, are as important as proper stack design or pressure testing procedures.

Annular preventer connections. When the elements of a blowout preventer stack are connected, they must provide a pressure seal at the connection point consistent with the pressure rating of the preventer elements. One of the most common problems associated with the preventers under pressure service is that of leaking connections. It becomes obvious that flange connections and the make-up and routine monitoring of these connections are of primary importance in well control. The two types of connections available are flange bolts and clamps.

Bolted flanges are probably the most common type of connections used in the industry. The bolts used in these flanges must be properly selected to insure that their tensile ratings are sufficient to withstand the maximum load which may be imposed. Also, the torque applied to the

Table 1.9 Comparison of make-up torque with various lubricants

Lubricant	Coeff. of friction	% of effort to friction	% of effort to tension	Relative torque, ft-lb, 1⅜" bolt & ½ min. yield
Light machine oil, as shipped	0.15	92.5%	7.5%	1,445
Oily machined surfaces, 4140 steel	0.133	91.67%	8.33%	1,290
Graphite with petrolatum	0.100	89.2%	10.8%	998
Tool joint compound	0.08	86.9%	13.1%	820
Select-a-torq. 503 Moly-Graph	0.067	84.7%	15.3%	702
API 5A2 thread lubricant	0.05	77.9%	22.1%	487

nuts and bolts must meet certain values in order to effect the flange pressure seal. Table 1.8 lists the API recommended values for tensile strength and torque requirements for various size bolts under different design conditions.

The lubricant applied to the bolt is of prime consideration when installing the blowout preventers. An analysis of the screw jack formula applied to preventer bolts shows that a major contributing factor to the torque required to tighten the bolt properly is friction incurred. This friction is dependent upon the type of lubricants used on the threads. Table 1.9 compares the effect of 6 lubricants on the coefficient of friction on the threads. This table clearly points out that the choice of lubricants will determine the difficulty incurred in attaining the proper bolt torque.

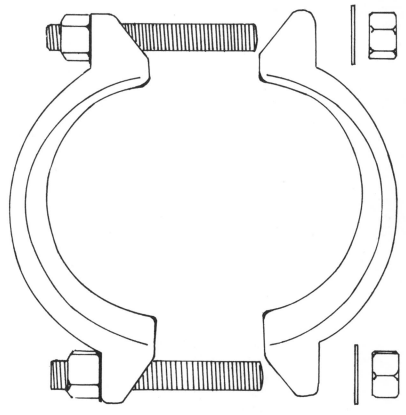

Stud Bolted Clamp.
The Two Halves Are Identical.

Fig. 1-31   Clamp for flange connection

(Note that this discussion is directly applicable to drill pipe tool joint make-up and that the make-up torque is somewhat lubricant dependent.)

Clamp connections are becoming more popular due to the relative ease of operations and time savings when compared to bolted flanges. The preventers have connection ends that mate with a wraparound connector. An example is shown in Fig. 1.31. When the preventers are aligned, the clamp is attached and tightened. This reduces the stabbing time and difficulty. The clamp connections are designed to withstand the same pressure as bolted flanges. Table 1.10 lists dimensional data for the more common sizes of large bore diameter clamps and bolts.

It should be noted that preventers should never be welded together. Welding presents problems such as heat stresses and temporarily irreversible connections. Manufacturers of preventers report that many preventer failures are the result of an operator welding on the body without applying the necessary treating to relieve the heat stresses.

Seal rings and ring grooves. The pressure seal at the connection points or flanges are effected by seal rings and seal ring grooves. The most common ring grooves are the API 6B for working pressures of 2,000 to 5,000 psi and the API 6BX for 5,000 to 20,000 psi working pressures. Fig. 1.32 illustrates each of these ring grooves.

Seal rings for each type of ring groove are also presented in Fig. 1.32. The "R" gaskets service the API 6B flange, the "RX" pressure-energized gaskets service the 6B flange, and the "BX" is designed for the 6BX flange. Types "RX" and "BX" are pressure-energized rings which will use well pressures to a degree to effect a seal and not rely completely on flange bolt torque. Type "R" uses bolt torque exclusively to effect a seal between flanges. The pressure-energizing feature is advantageous since normal preventer vibrations will reduce bolt torque. (Periodic checking and tightening of bolts should be a part of blowout preventer maintenance procedures).

H_2S trim. Drilling in H_2S environments causes not only personnel safety problems but also has adverse effects on the equipment used in well control. Standard metals with a tensile strength greater than 90,000 psi are subject to hydrogen embrittlement, a possible result of H_2S exposure, and may as a result fail. For this reason, preventers must be used that have a tensile strength less than 90,000 psi, or a hardness of 22 or less on the Rockwell "C" measurement scale.

To achieve this prerequisite, it may be necessary to use larger amounts of softer metal in order to contain the required pressures of well control service. (A more complete discussion of H_2S problems will be presented in Chapter 5.)

Table 1.10 Dimensional data for various swing bolt clamps

Size & working pressure	A	Dimensions, in. B	C	D	Torque ft/lb	Bolt size, in.
11" 3,000	$3\frac{1}{8}$	$3\frac{17}{32}$	$1\frac{31}{32}$	$6\frac{11}{32}$	3,750	2
See note 1	$3\frac{7}{8}$	$4\frac{13}{32}$	$2\frac{29}{64}$	$6\frac{53}{64}$	10,500	$2\frac{1}{2}$
13⅝" 3,000	$3\frac{1}{2}$	$3\frac{31}{32}$	$2\frac{13}{64}$	$6\frac{37}{64}$	5,700	$2\frac{1}{4}$
16¾" 2,000	$3\frac{1}{8}$	$3\frac{17}{32}$	$1\frac{31}{32}$	$6\frac{11}{32}$	5,100	2
13⅝" 5,000	$3\frac{1}{2}$	$3\frac{31}{32}$	$2\frac{13}{64}$	$6\frac{37}{64}$	6,900	$2\frac{1}{4}$
16¾" 3,000	$3\frac{7}{8}$	$4\frac{13}{32}$	$2\frac{29}{64}$	$6\frac{53}{64}$	9,375	$2\frac{1}{2}$
13⅝" 10,000	$3\frac{7}{8}$	$4\frac{13}{32}$	$2\frac{29}{64}$	$6\frac{53}{64}$	10,500	$2\frac{1}{2}$
20¼" 2,000	$3\frac{7}{8}$	$4\frac{13}{32}$	$2\frac{29}{64}$	$6\frac{53}{64}$	9,600	$2\frac{1}{2}$
20¼" 3,000	$3\frac{1}{2}$	$3\frac{31}{32}$	$2\frac{13}{64}$	$6\frac{37}{64}$	7,050	$2\frac{1}{4}$
See note 2	$3\frac{1}{2}$	$3\frac{31}{32}$	$2\frac{13}{64}$	$6\frac{37}{64}$	7,050	$2\frac{1}{4}$
16¾" 5,000	$3\frac{7}{8}$	$4\frac{13}{32}$	$2\frac{29}{64}$	$6\frac{53}{64}$	10,500	$2\frac{1}{2}$
30" 1,000	$3\frac{7}{8}$	$4\frac{13}{32}$	$2\frac{29}{64}$	$6\frac{53}{64}$	10,500	$2\frac{1}{2}$
24" 1,000	$3\frac{1}{2}$	$3\frac{31}{32}$	$2\frac{13}{64}$	$6\frac{37}{64}$	6,900	$2\frac{1}{4}$
11" 10,000	$3\frac{1}{2}$	$3\frac{31}{32}$	$2\frac{13}{64}$	$6\frac{37}{64}$	6,900	$2\frac{1}{4}$
36" 1,000	$3\frac{1}{2}$	$3\frac{31}{32}$	$2\frac{13}{64}$	$6\frac{37}{64}$	5,925	$2\frac{1}{4}$
21¼" 10,000	$4\frac{5}{8}$	$5\frac{3}{8}$	$2\frac{61}{64}$	$5\frac{53}{64}$	22,500	3
18¾" 10,000	$4\frac{1}{4}$	$4\frac{51}{64}$	$2\frac{45}{64}$	$6\frac{61}{64}$	18,750	$2\frac{3}{4}$
16¾" 10,000	$4\frac{1}{4}$	$4\frac{51}{64}$	$2\frac{45}{64}$	$6\frac{61}{64}$	18,750	$2\frac{3}{4}$
18¾" 10,000 Forged	$4\frac{5}{8}$	$5\frac{3}{8}$	$2\frac{61}{64}$	$5\frac{53}{64}$	22,500	3

1. For 7¹/₁₆" 15,000, 9" 10,000 and 11" 5,000.
2. For 20¾" 2,000, and 20¾" 3,000.

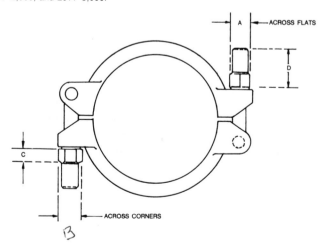

API TYPE R RING-JOINT GASKETS
(For Use In 6B Flanges)

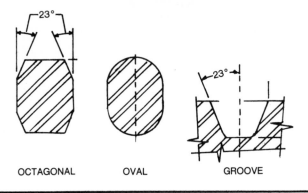

OCTAGONAL   OVAL   GROOVE

API TYPE RX PRESSURE ENERGIZED RING-JOINT GASKETS
(For use in 6B flanges and segmented flanges)

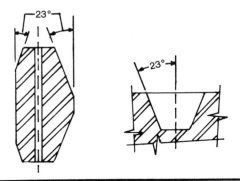

API TYPE BX PRESSURE ENERGIZED RING-JOINT GASKETS
(For use in 6BX flanges)

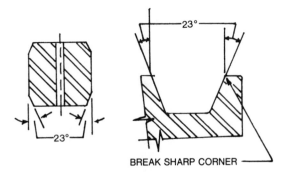

BREAK SHARP CORNER

Fig. 1-32   API seal rings and grooves

Some equipment, as noted, is serviceable in H_2S environments without any alterations. Some pieces require special trim. For complete details on any equipment that is to be used in H_2S service, the manufacturer should be consulted. It may be necessary in some applications to perform remedial work to the equipment such as the addition or replacement of special trim, or heat treating to relieve stresses.

Wear bushing. Routine daily activities with the drill pipe and kelly can cause excessive wear within the blowout preventer stack if certain precautions are not exercised. To avoid this wear, bushings or sleeves are installed in the lowermost section of the preventers as illustrated in Fig. 1.33. These bushings absorb the rotation and trip wear and can be replaced at a fraction of the cost required to rework blowout preventers.

The manner in which the bushing is secured in place should be given some consideration. Some bushings are lowered through the blowout preventer stack and remain unsecured while others can be locked in place with external locking screws. This is an important consideration as occasionally wells have had blowouts because an unsecured wear bushing was blown up inside the preventers stack making it impossible to close several of its members.

Drill Pipe Blowout Preventers

The prevention of blowouts through the drill pipe is an important facet of well control. When a kick occurs, the influx fluid will generally enter the annulus due to the direction of drilling fluid flow during normal drilling circulation. However, if the kick fluid does enter the drill pipe for any of several reasons, the shut in drill pipe pressures will be greater than normal kick conditions due to the vertical column of mud that will be displaced by a relatively small volume of influx fluid. As a result of this, the selection and utilization of the drill pipe blowout prevention equipment is essential for proper kick control.

Several tools contain drill pipe pressures during kicks. The primary tool is the kelly and its associated valves such as the kelly cocks. When the kelly is not in use, drill string valves are necessary to control the pressures. These valves may be automatic or manual-control valves, and may be a permanent part of the drill string or installed when the kick occurs.

Kelly and kelly cock. The kelly, which is used to impart rotary motion to the drill string, is the connection between the drill string and the surface drilling equipment. Valves are generally placed above and below the kelly to provide pressure protection for the kelly and all the surface equipment. These valves, called kelly cocks, should be of a

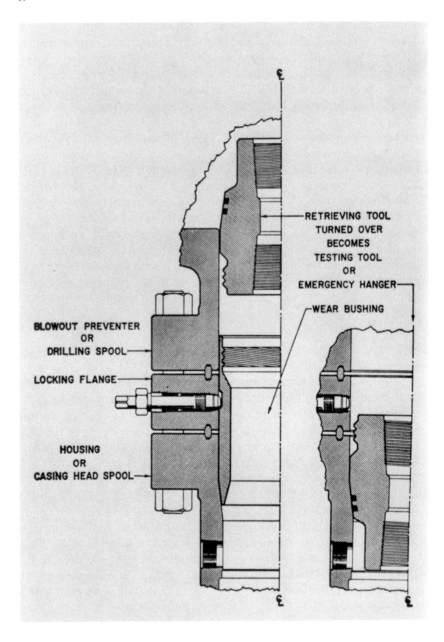

Fig. 1-33   Wear bushing

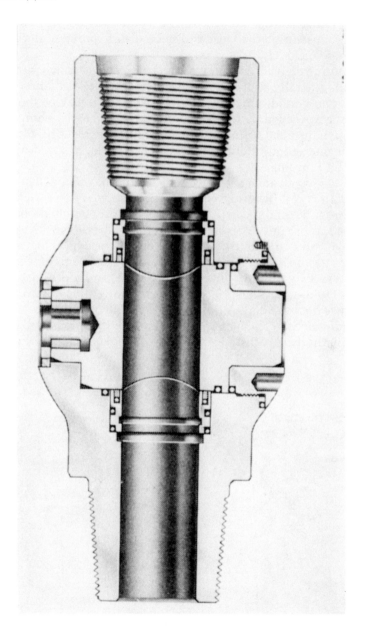

Fig. 1-34 Kelly cock

pressure rating consistent with the remainder of the drill string and capable of sustaining the wear and hook load required of the hoisting equipment (Fig. 1.34).

Automatic valves. An automatic closure, or float valve, in the drill string will generally allow fluid movement down the drill pipe but will not allow flow in an upward direction. The valve may be the flapper type, a spring-loaded ball, or the dart type and may be permanent or pump down installed. Although the valve does prevent drill pipe blow-outs, it is often used to minimize flow back during connections, or to prevent bit plugging.

There is a disadvantage relative to well control when a float valve is installed in the drill string because the basis of proper kick killing procedures is dependent on a drill pipe pressure determination. Since a direct reading of static drill pipe pressures is not possible with a conventional float valve, alternative pressure reading procedures that are more complex must be implemented. This problem can be circumvented if a flapper type valve is used that has small built-in fluid ports to allow pressure build-up at the surface while still preventing a blowout. Fig. 1.35 shows both the conventional and the vented flapper.

Manual valves. The manual valve, commonly called a full opening safety valve, is usually installed on the drill pipe after a kick occurs when

PLAIN
FLAPPER

VENTED
FLAPPER

Fig. 1-35 Regular and vented flapper

Fig. 1-36 Full opening safety valve

the kelly is not in use. The advantage of a manual valve is that it can be in the open position when it is stabbed on the drill pipe and will thus minimize the effect of upward moving mud lifting the valve. The mud will pass through the valve during the stabbing, after which the valve can be closed.

Automatic valves, in some types, can be locked in the open position to achieve this stabbing feature. Closing of the manual valve requires that a wrench be kept on the rig floor accessible to the rig crew (Fig. 1.36).

The manual valve possesses one feature that makes it advantageous over the automatic valve in certain applications. When in the open position, the manual valve has a non-obstructed orifice whereas the automatic valve locked in the open position has the sealing mechanism (flapper, ball or dart) serving as an obstruction. Should it become necessary to do any wireline work, the manual valve can be opened and will allow passage of any tools that have a diameter smaller than that of the inner valve. This cannot be done with the automatic valve.

Regan Forge and Engineering Co. manufactures a quick-coupling attachment for emergency drill pipe closure. The coupling (Fig. 1.37) is designed to drop over the pipe with an open valve attached and automatically latch under and seal off around the collar or tool joint. After the pressure is bled off, the coupling is released by depressing a ring, or individual dogs, depending on the type of pipe. The coupling will accept either manual or automatic valves.

Downhole Blowout Preventers

Blowout preventers in the past have been thought of as surface equipment designed to contain the kick. Downhole blowout preventers have been designed based on a different concept. The downhole preventer, which is generally an inflatable drilling packer, is designed to seal the annulus, contain the kick fluid below the packer, and allow kill mud to be circulated above the packer.

After heavy muds are circulated to the surface, the packer is released and the kick fluid is circulated from the well. The advantage of this procedure is that the wellbore is not exposed to the full stresses of the kill pressures. Fig. 1.38 illustrates the sequence of kill steps using a downhole preventer.

There are disadvantages to the preventers, which are the same basic problems associated with any drilling or open hole packer. Among these are the time required to inflate the packer after the kick has occured, attaining a seat in open hole, and packer malfunction due to prolonged wear. Downhole preventers have never received prominent use.

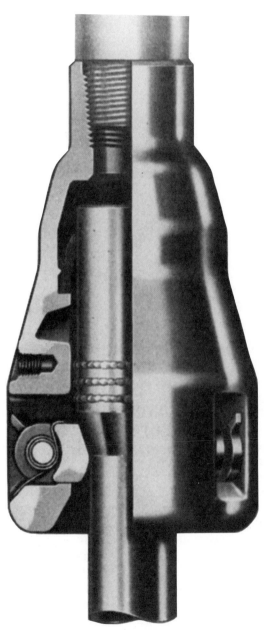

Fig. 1-37   Regan fast shut-off coupling

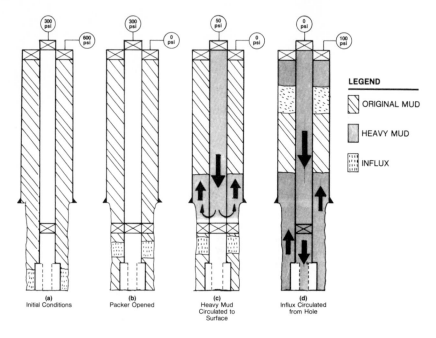

Fig. 1-38 Downhole packer kill procedure

Blowout Preventer Stack Design

There are several considerations in designing an arrangement of annular blowout preventers. Among these are pressure design, component selection and arrangement, subsea related variations, and diverter systems.

Pressure design. There are several well-founded viewpoints related to the pressure requirements that preventer stacks should meet. Some, but not all, of the arguments are that the working pressure needs to be no greater than the burst strength of the exposed casing string, formation fracture pressure of the shallowest exposed zone, or a predetermined maximum allowable surface casing pressure. Upon close inspection, however, it can be seen that all of these guidelines may present serious problems when applied in severe well control situations.

The most common of these guidelines is that the preventers need to be no stronger than the casing string to which they are attached. The inherent fallacy with this guideline is that it assumes that the casing string has been properly designed to withstand kick-imposed stresses. This is quite often not the case. It would follow that if the casing is improperly designed, the preventer pressure rating is also improperly designed.

The safest procedure for designing preventer pressure ratings is to insure that the preventers can withstand the worst pressure conditions that could be imposed. These conditions occur when all drilling fluids have been evacuated from the annulus and only low density formation fluids such as gas remain. This procedure is illustrated in Example 1.4.

Example 1.4   A well is to be drilled to 10,600 ft and has an expected bottom hole pressure (BHP) equivalent to 10.5 ppg mud. What pressure rating should the preventers be? (Assume a gas density of 2.5 ppg.)

Solution
(1) Determine the maximum anticipated formation pressure.
    Pressure = 0.052 × 10.5 ppg × 10,600 ft = 5,787 psi
(2) Determine the gas hydrostatic pressure that will act downward on the zone assuming that the mud is evacuated from the hole.
    Pressure = 0.052 × 2.5 ppg × 10,600 ft = 1,378 psi
(3) The pressure imposed on the preventer would be the difference between the formation pressure and the gas hydrostatic pressure.
    5,787 psi − 1,378 psi = 4,409 psi

The preventers must be able to withstand 4,409 psi. Using API designations (Table 1.11) a 5,000 psi working pressure rating system would be required.

Experience suggests that this method should always be used in shallow well situations where it is possible to achieve a complete mud evacuation. However, as the depth of the well increases, it becomes more unlikely that a full mud evacuation will occur. As a result, a modification based on a percentage of the maximum possible pressure load should be used to determine the preventer pressure rating. This percentage would depend on the operator's experiences in a particular drilling environment. Example 1.5 illustrates the modification of the technique for deep wells.

Example 1.5   A North Sea operator wishes to drill an expected bottom hole pressure of 16.0 ppg at 16,500 ft. The operator's experience dictates that an 80% design factor would account for unexpected eventualities. What pressure rating should the preventers be? (Assume a gas density of 2.0 ppg).

Solution
(1) BHP = 0.052 × 16.0 ppg × 16,500 ft = 13,728 psi
(2) Gas pressure = 0.052 × 2.0 ppg × 16,500 ft = 1,716 psi
(3) Resultant pressure = BHP − gas pressure
= 13,728 psi − 1,716 psi = 12,012 psi
(4) Working pressure = resultant pressure × 80% = 9609 psi
Using the API designations in Table 1.11, a 10,000 psi working pressure stack of preventers would be necessary to control the well properly.

Component design. After the pressure rating for the preventer has been selected, the component arrangement must be considered. The logic will be developed using four components: a spherical, pipe rams, blind rams, and a drilling spool. Logic for the minimum stack can be extended to any stack.

Table 1.11 Blowout preventer pressure rating designations

API class	Working pressure, psi	Service condition	Flange size, in.	Minimum vertical bore, in.	Matching casing sizes, OD, in.
2M	2,000	Light duty	20	20¼*	20 and 18⅝
			16	16*	16
3M	3,000	Low pressure	12	13⅝	13⅜ and 11¾
			10	11	10¾ and 9⅝
			8	9	8⅝ and 7⅝
			6	7 1/16	7⅝ to 4½
5M	5,000	Medium pressure	13⅝	13⅝	13⅜ and 11¾
			10	11	10¾ to 8⅝
			6	7 1/16	7⅝ to 4½
10M	10,000	High pressure	13⅝	13⅝	13⅜ and 11¾
			11	11	10¾ to 8⅝
			7 1/16	7 1/16	7⅝ to 4½
15M	15,000	Extreme pressure	7 1/16	7/16	7⅝ to 4½

*API standard bores; however, other bores are used.

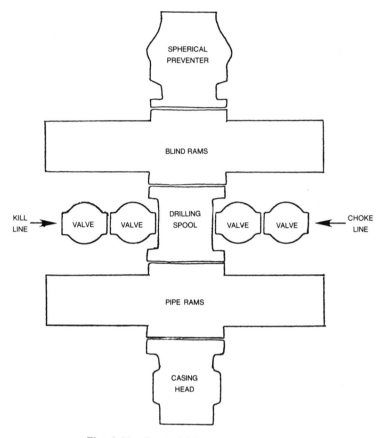

Fig. 1-39 Basic BOP stack arrangement

Fig. 1.39 shows the proper arrangement for this 4-member stack. Should one component fail, there will always be a back-up system. This sequence of operations explains the design.

Step 1. The spherical preventer is closed.

Step 2. If the spherical fails while killing the well, the lower set of pipe rams are closed.

Step 3. One of three emergency procedures are exercised. Either the spherical is changed, the blind rams are changed to pipe rams, or both the spherical and blind rams are changed.

This implies several important points. The lower pipe rams are not for circulation purposes but simply to close in the well while repairs to the upper members are made. Also, a kill or choke line should never be attached below the lowermost set of pipe rams, i.e. ram outlets or casing head valves. Failure of this line will mean a certain blowout since there is no back-up system for proper control.

The valves adjacent to the blowout preventer stack should be arranged based on the back-up system principle. The innermost valve next to the stack should be for emergency use only while the next valve outward is for day-to-day actuation. As a result, the outer valve is generally a hydraulic valve for remote control during kick killing procedures.

In deepwater drilling, the blowout preventers are generally located on the sea floor. This necessitates installing certain built-in safety precautions in the stack. Since component failure cannot readily be repaired, additional preventer elements must be installed to handle any eventualities. The typical subsea stack (Fig. 1.40) illustrates that the same logic developed in the previous section for a minimum stack was utilized by insuring that a built-in back-up system is available under all circumstances. Some of the back-up systems shown in this illustration are the two spherical preventers, two choke lines with the primary line on top and the secondary line on the bottom, fail-safe valves on each choke line, and shear rams at the bottom of the stack to allow for emergency rig departure if necessary.

There are many instances in shallow sections of the hole where it will not be possible to shut-in a well due to an insufficient amount of casing

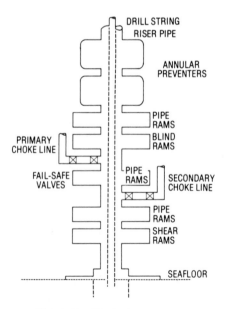

Fig. 1-40 Typical subsea stack

in the well to sustain a kick. When this happens, a blowout must be diverted away from the rig using the typical blowout preventer arrangement shown in Fig. 1.41. As soon as the kick is observed, the diverter line(s) is opened and the annular preventer is closed. Fortunately, most shallow kicks that occur in this situation will deplete the reservoir, or bridge the hole and kill the kick. The important point to remember, though, is that in shallow kicks of this type, a blowout requires special control procedures.

The arrangement shown in Fig. 1.41 has several important features that are recommended for diverter systems. The control panel is designed so that movement of a single control lever in one direction will open the diverter valves and simultaneously close the diverter preventer. Movement of the control lever in the opposite direction will close the valves and open the preventer. Diverter lines should be at 180° angles to each other. When possible, the line used should be that which will take advantage of the wind direction to carry the blowout away from the rig.

The lines should be as large as possible with a suggested minimum ID of 6 in. The United States Geological Survey requires, for outer continental shelf drilling, that a minimum combined cross sectional area of 22 sq. in. be used. This corresponds to two 4-in. lines. Angles and bends should be minimized in the diverter lines to avoid unnecessary restrictions. The preventer may be a low pressure spherical preventer or some type of diverter bag used to direct the flow into the lines.

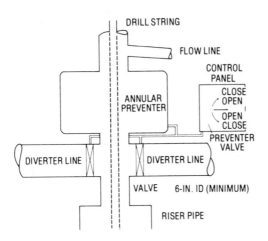

Fig. 1-41 Typical diverter stack

Choke Devices

A choke is any tool used to apply resistance to flow. This resistance creates a back pressure which can control formation pressures when a kick is circulated from the well.

Many types of chokes have been used for well control purposes. Among these are positive or fixed-orifice size chokes, and adjustable chokes manufactured with rubber or steel trim. The primary type of choke used today is the steel adjustable, remote-controlled choke because of its greater durability over the rubber choke, and its ability to change orifice sizes quickly.

Hand Adjustable Steel Chokes

One of the older types of chokes that is still in use is the hand adjustable steel choke. This type of choke creates a back pressure with a stem and beveled seat mechanism. Fluid is allowed to flow through the seat, or orifice. As some alteration in the amount of back pressure is required, the stem is positioned in the seat to create a resistance to flow. Control of the back pressure is attained by the degree to which the stem is forced into the beveled seat or extracted from it.

Certain problems with this choke have been noted from field usage. The stem and seat mechanism has exhibited a tendency towards turbulent erosion thereby reducing its sealing ability. Shale cuttings buildup and resultant choke plugging has been observed due to the inability of the choke to open to a size sufficient to allow passage of the cuttings. Also, the placement of this choke in the manifold requires the operator to be removed from the rig floor during choke operation which increases the difficulty of the well control procedures.

Remote-Control, Adjustable Rubber Chokes

An early hydraulically operated choke consisted of a rubber sleeve in a steel cylinder similar to the schematic drawing shown in Fig. 1.42. Back pressure was generated by applying hydraulic pressure to the pressure plate which in turn compressed the rubber sleeve and decreased the flow orifice. The design of this choke was a significant advancement in well control equipment because it offered variable pressure control. Remote operation allowed the operator to control the choke from the rig floor. And the choke could allow passage of plugging obstructions such as shale cuttings or pieces of rubber pipe protectors.

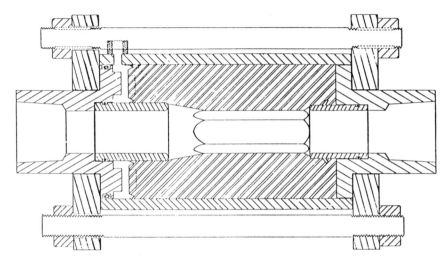

Fig. 1-42 Schematic of rubber choke

The problems associated with the rubber choke eventually led to the development of the steel choke. Some of the problems were low working pressure ratings and susceptibility to hydrocarbon abrasion and cutting while circulating the kick fluids from the well. Cutting of the rubber sleeve was a particular problem with high pressure gas kicks.

Remote-Control, Adjustable Steel Chokes

The advent of the steel adjustable choke gave pressure control in almost all conceivable situations. The steel choke can tolerate all types of kick fluids for long periods of time at high pressures if necessary.

Swaco choke.   The Swaco steel choke, (Fig. 1.43) consists of two tungsten carbide plates with "half-moon" orifices that either allow or prevent fluid flow depending on the relative position of the orifices. The orifices are misaligned in the closed position and, as one plate is rotated on the other, the orifices becomes aligned which allows fluid to flow through the choke.

Swaco choke operating features
(1) The choke design allows for complete closure.
(2) The choke can be operated hydraulically by means of a pump operated by rig air, by a manually operated hand pump, and manually operated with an attached bar.

Fig. 1-43   Swaco "Super Choke"

(3) Variable choke speed control.
(4) No automatic control features.
(5) 10,000 psi minimum working pressure rating.
(6) H$_2$S service model is available.

   Cameron Iron Works.   The Cameron choke utilizes a rod and cylinder system to develop the desired back pressure. Mud is circulated through the cylinder (seat) when the choke is open (Fig. 1.44 and 1.45). As the choke is closed, the rod gate is forced hydraulically into the cylinder to create an obstruction and resistance to flow.

Cameron choke operating features
(1) The choke design does not allow complete closure, resulting in an inability to pressure test with water. The choke will generally seat, however, when mud is used.
(2) The choke can be operated hydraulically by means of a pump and rig air supplies, hydraulically with an attached nitrogen bottle, and manually with a hand operated pump.
(3) Variable choke speed control.
(4) 5,000 psi–20,000 psi working pressure models.

Fig. 1-44   Cameron rod and seat

(5) All models of the choke are rated for H_2S service.
(6) Both gate and seat are made of tungsten carbide.
(7) The gate and seat are reversible for double life.
(8) A "maximum allowable choke manifold pressure" option is available for the 10,000 psi choke that will automatically open the choke when the pressure exceeds a preset value.
(9) The choke control panel can connect to two separate chokes and alternately operate either with a switch on the panel face.

The control panel for choke operation will generally contain the gauges and controls necessary to monitor the well during the kick killing operation. The panel should contain accurate drill pipe and casing pressure gauges, accumulative pump stroke counter, and a choke control lever (knob). Some optional items for inclusion on the panel are rig air supply monitor, variable choke speed control lever (knob), relative choke position indicator, pump stroke rate indicator, and automatic

Fig. 1-45   Cameron drilling choke

control switches. The panel should be installed where the operator can communicate with key personnel such as the driller, in close proximity to a preventer control station, and near adequate lighting to facilitate accurate pressure readings. Fig. 1.46 is an example of a Cameron control panel.

Choke Manifolds

The choke manifold is an arrangement of valves, lines, and chokes designed to control the flow of mud and kick fluids from the annulus during the killing process. Some of the conditions that the manifold may be called upon to work under are a variety of fluids such as mud, oil,

Fig. 1-46 Cameron choke panel

water, or gas, high pressures, upstream flow rates, downstream velocities, and obstructions in the produced fluids such as sand, shale or pipe
protector rubbers. The manifold should control pressures by using one of
several chokes. It should divert flow to one of several areas including a
burning pit, the reserve pit, a mud pit, or overboard a drilling vessel
when applicable. The choke should have pressure ratings at least equal
to the preventer stack, and meet all pressure testing specifications imposed on the preventers. It should be suitably anchored to prevent
movement during the killing operation. The choke should feature easy
access to every manifold component, with all lines constructed as
straight as possible. All lines and valves should have the same consistent
inner diameter to minimize turbulent erosion at diameter changes.

Manifold design. The principle applied to the design of the blowout
preventer stack also will be applied in designing the choke manifold.
The proper procedure is to insure that a back-up system is available
should the primary tool fail. Also, it is a good practice initially to use the
manifold necessary to reach the total depth to avoid installing a different
manifold with each casing setting depth.

Fig. 1.47 illustrates a choke manifold recommended for most drilling
operations. Note that this design meets all of the requirements for choke
manifolds. Buffer chambers are used at the downstream connections to

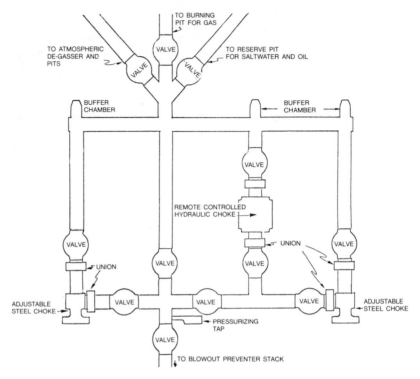

Fig. 1-47 Typical choke manifold design

act as hydraulic cushions and minimize erosion. A tap has been provided to allow for pressurization of the manifold to prevent pressure surges when opening the valves near the stack. Two hand adjustable chokes have been provided due to the high stem and seat erosion rates associated with these chokes and due to the tendency of these chokes to pack off with cuttings. A direct line from the preventer stack to the burning pit for gas has been provided should it become necessary to divert the well temporarily. Note that this design does not constitute a true diverter system.

Blowout Preventer Testing

After the blowout preventer stack has been installed, it must be pressure tested to insure that it can control the designed pressures and periodically retested for a maintenance of pressure integrity. Important considerations in blowout preventer testing are test fluids, pressures, testing equipment, procedures, and frequency of retesting.

Test fluids. Clean water is perhaps the best fluid because of availability and because it will not plug small leaks as will mud. If high pressure gas wells are to be drilled, some operators will test with an inert gas such as nitrogen. Oxygen or hydrocarbon gases should never be used to pressure test a stack.

Test pressures. A high and low pressure testing procedure should be employed. The high pressure test should be to either the working pressure of the preventers or to the maximum anticipated pressure as previously calculated in Examples 1.4 and 1.5. The working pressure of the preventers is the recommended option. This pressure would be used to test all of the blowout prevention equipment (stack, kelly, and manifold) except the spherical preventer. Since the overall life of the spherical

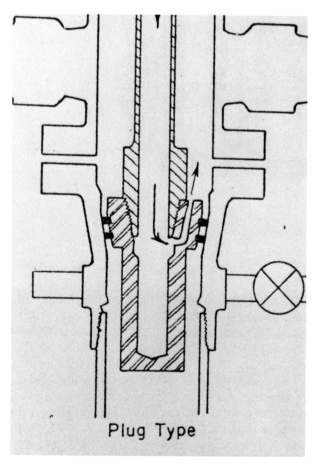

Plug Type

Fig. 1-48 Boll weevil test plug

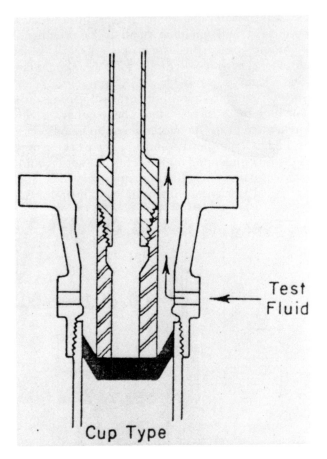

Fig. 1-49 Cup tester

element is dependent on imposed pressure and number of actuations, the test pressures are generally 70% of the pressure used on the remainder of the stack. This action is justified because actual field applications would dictate using rams when kick pressures approach test pressures.

A low pressure test in the range of 100 to 300 psi should also be applied to the stack. Although the stack is generally washed and cleaned with water prior to testing, it is difficult to remove completely existing dried mud particles from a potential leak hole. High pressure tests applied to the mud may pack the mud and effect a seal whereas a low pressure test may allow the leak to occur.

Testing equipment.   The pumps used to generate pressures for preventer testing may be any type that is capable of attaining the desired

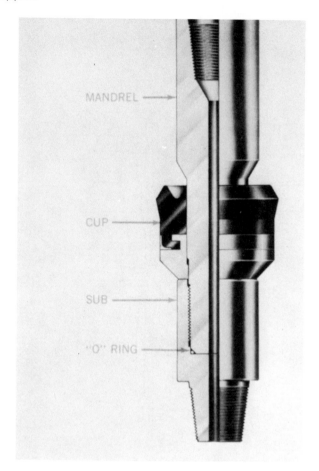

Fig. 1-50 Cameron type "F" cup tester

pressures. However, since most testing pressures will be out of the range of rig pumps, a smaller high pressure pump must be used. In many applications, a cementing-type reciprocating pump is suitable if it is convenient. If a cementing-type pump is not available, several service companies offer preventer testing and provide a small, high pressure reciprocating pump. Often, the accumulator system can be adjusted for rig preventer testing.

While testing the preventers, it is generally not desirable to expose the casing and open hole sections to the test pressures used on the preventers. Some types of test plug must be set in the bottom of the preventers to prevent this occurence. The plugs most commonly used are the well-

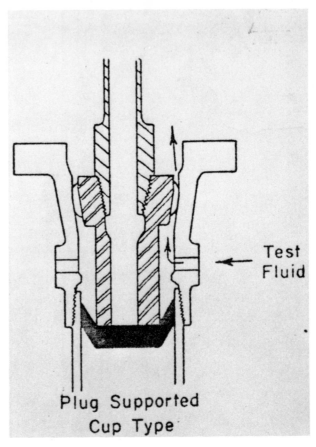

Fig. 1-51   Combination boll weevil-cup tester

head, or boll-weevil plug, cup type plug, or a combination boll-weevil/cup type plug.

The boll-weevil or wellhead plug is designed to seat in the wellhead and each will generally seal in only one type of head (Fig. 1.48). The plug is lowered into the head with a joint of drill pipe or a special test joint to test the pipe rams and spherical preventer. The pipe is removed with the plug resting on the head to test the blind rams.

If a boll-weevil plug is to be used, care must be taken to insure that the plug used is designed for the existing wellhead. Wellheads with the same basic dimensions oftentimes require different plugs. As an example, one manufacturer of 7-in. wellheads has 7 different test plugs for this head due to small head variations.

Cup-type testers (Figs. 1.49 and 1.50) are more universal in that they are designed to effect a seat in the casing and not the head. Although the cup can be placed at any point in the casing, pressure testing specialists recommend positioning it opposite the slips in the casing spool or head.

Since the cup is not supported by the wellhead, the force (cup area × pressure) created by the pressure test must be supported by the drill pipe or test joint. This will often limit the use of drill pipe for testing because its yield strength might be exceeded. The cup cannot be used to test the blind rams.

A combination plug is available that offers the advantages of both the cup and the boll-weevil plug (Fig. 1.51). The plug is supported by the head and allows testing of the blind rams while the cup creates a pressure seal.

Testing procedures.   Although the order of element testing may vary with companies, the basic procedures generally will remain the same. Dawson has developed a schematic representation that systematically provides testing for all elements of the blowout prevention system (Figs. 1.52–1.64). These schematics are useful because of their thorough yet simple presentation.

Specialists have observed that problems during pressure testing are usually the result of a few basic problems.[4] Some reasons for test failure might be that the test plug is too large for the wellhead, or that it will not seat properly because wellhead hanger holddown studs leak. Packing glands around holddown studs can leak. Rams or ram bodies could be inverted. Flow line valves might not hold pressure if they are installed backwards. Finally, ring gaskets and bonnet-seal gaskets might be bad. These problems fall into one of three general categories: (1) mistakes in preparing for the test (wrong plug sizes), (2) errors in installing the equipment, and (3) the lack of maintenance during drilling operations.

If a service company is used to test the preventer, the checklist in Table 1.12 for ordering information will minimize the problems associated in preparing for the test.

Table 1.12 Checklist for preventer pressure testing

---

1. Make, size, and type of wellhead equipment.
2. If a reworked head is used, insure that no changes in type were made without appropriate changes in marking.
3. Working pressure rating of the wellhead and preventers.
4. Size of drill pipe or work string.
5. Type of tool joint.
6. If a mixed string is used, list both sizes.
7. Size and weight of casing.

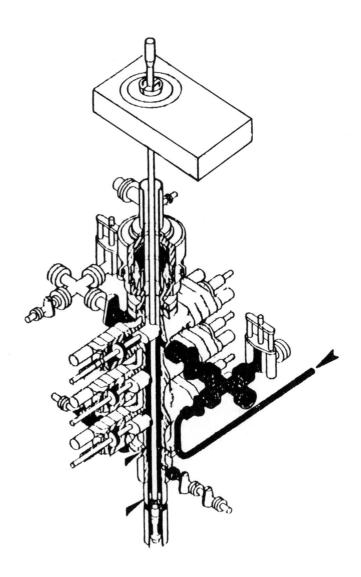

Fig. 1-52   Dawson testing procedures step "A"

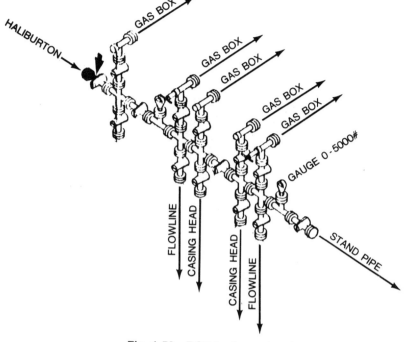

Fig. 1-53 BOP testing—step 1

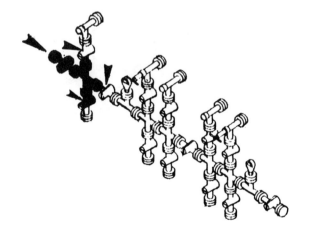

Fig. 1-54 BOP testing—step 2

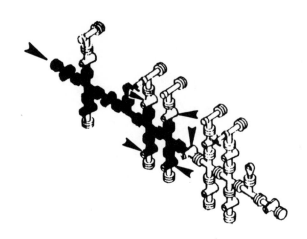

Fig. 1-55   BOP testing—step 3

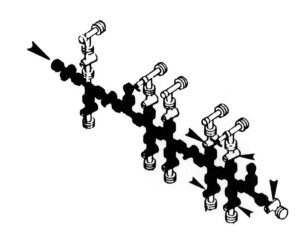

Fig. 1-56   BOP testing—step 4

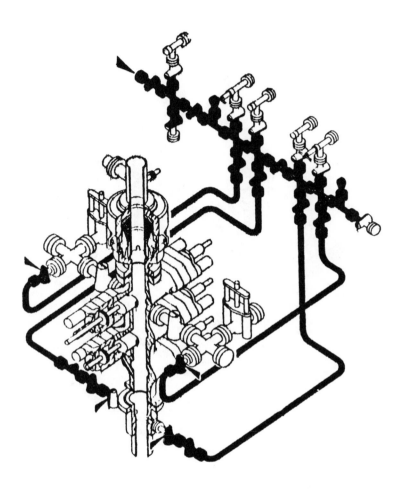

Fig. 1-57 BOP testing—step 5

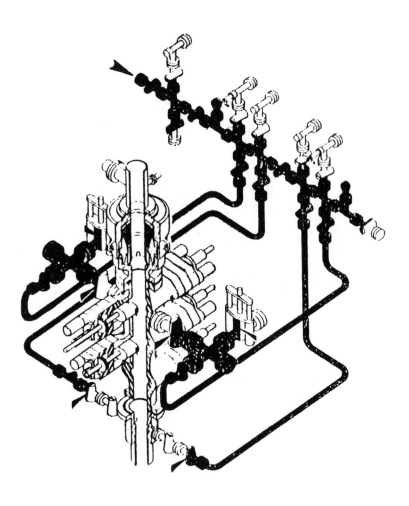

Fig. 1-58 BOP testing—step 6

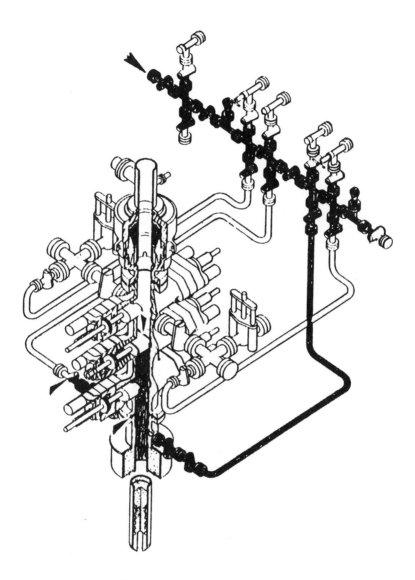

Fig. 1-59 BOP testing—step 7

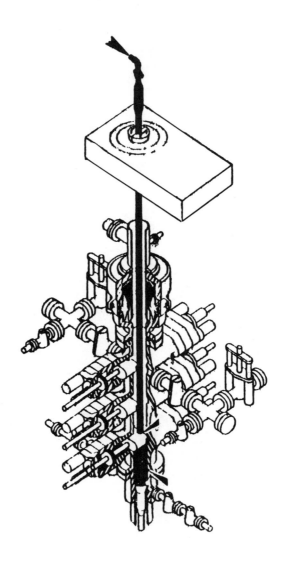

Fig. 1-60 BOP testing—step 8

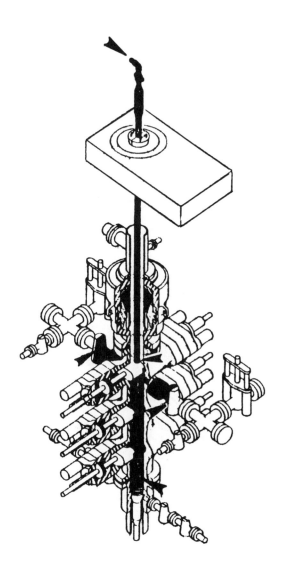

Fig. 1-61 BOP testing—step 9

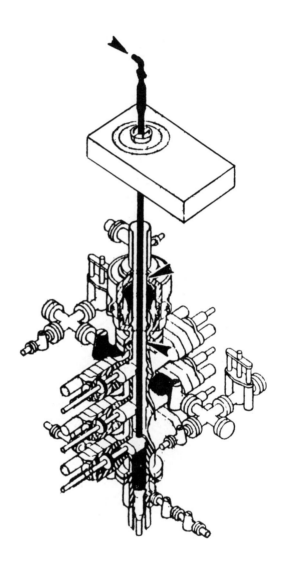

Fig. 1-62   BOP testing—step 10

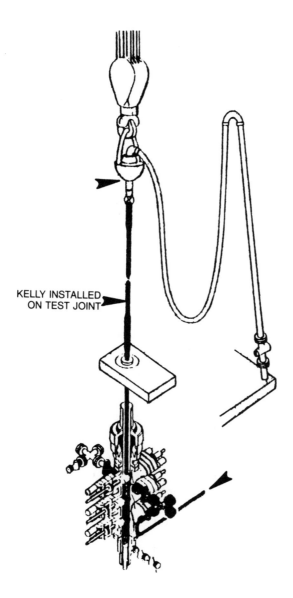

KELLY INSTALLED ON TEST JOINT

Fig. 1-63 BOP testing—step 11

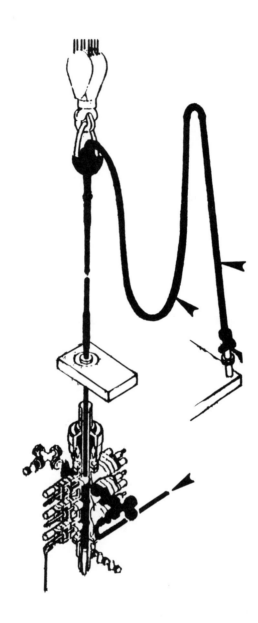

Fig. 1-64   BOP testing—step 12

Well:_____Contractor:_____
BOP test date:_____Well depth:_____
Last test:_____Last casting set at:_____
Test fluid:_____

Equipment		Test pressure, psi	Test period, minutes	Remarks
Spherical preventer:		_____	_____	_____
Pipe rams:	upper	_____	_____	_____
	lower	_____	_____	_____
Blind ram:		_____	_____	_____
Kill line:	valve 1	_____	_____	_____
	valve 2	_____	_____	_____
	valve 3	_____	_____	_____
Choke line	valve 1	_____	_____	_____
	valve 2	_____	_____	_____
	valve 3	_____	_____	_____
Manifold:	valve 1	_____	_____	_____
	valve 2	_____	_____	_____
	valve 3	_____	_____	_____
	valve 4	_____	_____	_____
Choke:	hydraulic	_____	_____	_____
	manual	_____	_____	_____
Kelly cock & kelly		_____	_____	_____
Floor safety valve		_____	_____	_____

Type Test: Initial_____ Weekly_____ Ram Change_____

Fig. 1-65   Sample BOP test report form

Frequency of testing. The preventers should be tested on initial installation. They should be retested each week and after any repair that requires breaking a pressure connection. Many operators require a retest prior to entry into a transition zone. After each test has been completed, it should be entered on the morning report form and a test report should be completed and entered in the well history file. Fig. 1.65 is a sample report form.

Auxiliary Equipment

Accumulator Systems

The purpose of the accumulator system is to provide closing energy to all members of the blowout preventer stack. This is usually done with a hydraulic system designed and built to provide closing power to the equipment in 5 seconds or less and to maintain the required pressures as desired.

The working of the accumulator is a function of hydraulic oil stored under a compressed inert gas, usually nitrogen. As hydraulic oil is forced into a vessel (bottle) by a small volume output, high pressure pump, the nitrogen is compressed and stores potential energy. When the preventers are actuated, the pressured oil is released and opens or closes the preventers. Hydraulic pumps replenish the accumulator with the same amount of fluid as was used to work the preventers. Fig. 1.66 shows an accumulator which includes the bottle, pumps, controls, and a hydraulic oil tank.

A precharge pressure is generally applied to the nitrogen to insure that all the oil can be forced from the bottle when necessary. The precharges may range from 500 to 3,000 psi with the desired precharged pressure being dependent on the service conditions during fluid drawdown. Fig. 1.67 is a drawdown curve for 3 different precharge pressures and is used to size preventers with respect to accumulator pressure.

The accumulator must be equipped with several pressure-regulating devices so that different stages of pressure can be maintained with the

Fig. 1-66 Accumulator

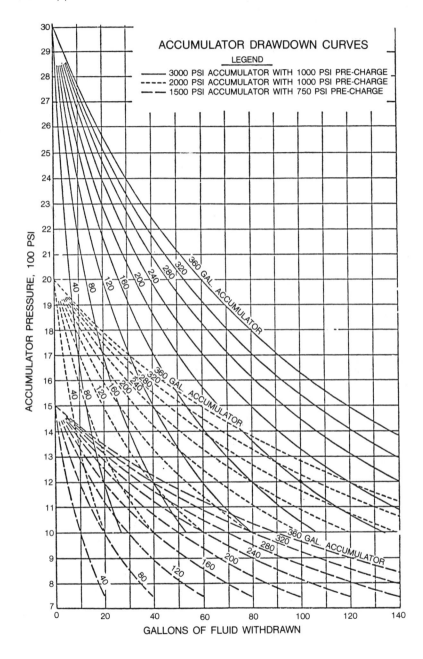

Fig. 1-67 Accumulator drawn down curves

unit. As an example, an accumulator pressure of 3,000 psi is recommended in most cases but the pressure must be regulated to provide 1,500 psi to the spherical preventer since this is the maximum recommended closing pressure for most sphericals. Accordingly, other stack members may require different operating pressures. (Recommended operating pressures are listed in the Appendix.) A bypass valve is built into the accumulator for use should it become necessary to use the full pressure to close the preventers in emergency conditions.

Another purpose of this hydraulic system is to maintain constant pressures when stripping pipe through the spherical preventer. As tool joints are stripped through the packing element, the accumulator must allow the excess fluid pressure to move from the annular closing chamber. Where the tool joint passes through the packing element, the accumulator must force additional fluid back into the spherical preventer to maintain a constant pressure.

Design procedures. The accumulator should have the ability to close a minimum of three members of the stack, one of which must be the spherical, without having to recharge the accumulator. Many operators require that the accumulator close all members of the stack without recharging. A total of 50% of the original fluid should remain as a reserve after accumulator activation. A minimum final pressure of 1,200 psi is required to insure that the preventers remain closed.

Example 1.6   What would be the minimum requirements for an accumulator if the following elements are in use? Use the preventer information in the Appendix.

Element	No.	Type	Size, in.	Pressure rating
Spherical	1	Hydril "GK"	11	10,000 psi
Rams	2	Cameron "U"	11	10,000 psi
Valves	4	Cameron "F"	3	10,000 psi
			(3⅛ bore)	

Solution

Part I. Volumetric requirement

Element	Gallons to Open	Gallons to Close
Spherical (1)	18.87	25.10
Rams (2)	6.40	6.72
Valves (4)	1.12	1.12
	26.39	32.94

Total   26.39 + 32.94 = 59.33 gal

59.33 × 2 = 118.66 gal

The accumulator system should have a minimum volumetric capacity of 118.66 (or 120) gallons.

Part II. Pressure requirement
Using the drawdown curves shown in Fig. 1.67, several options are available. Two of these options are a 120 gal or greater accumulator with a 3,000 psi charge, or a 200 gal or greater accumulator with a 2,000 psi charge.

Mud Mixing Equipment

During the process of killing a kick, it will become necessary to increase the mud density by adding weight materials such as barite to the mud system. It is important to design this mud mixing system so that the mud density can be increased as quickly as possible in order to initiate the kick-killing procedures. The three main components of the mud system involved in this process are the mixing pumps, the hopper, and the barite.

Mixing pumps. The mud mixing pump can be any type that is used to add weight material and chemicals to the mud stream. Although centrifugal pumps are generally used for this purpose, small reciprocating pumps can and have performed efficiently. The centrifugal pump consists of an impeller, suction and discharge lines, and a power source. Mud enters the pump from the suction line, accelerates due to the

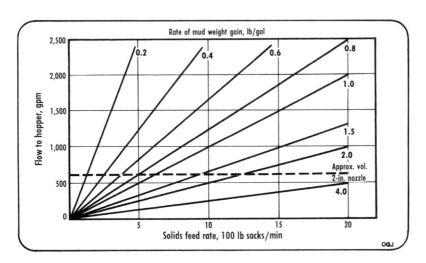

Fig. 1-68 Required mud flow rates for barite addition

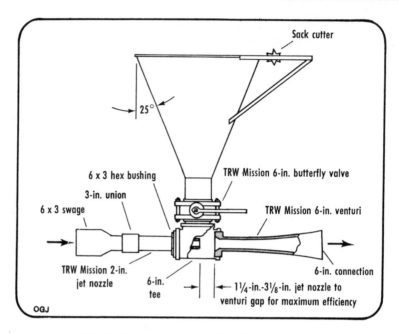

Fig. 1-69   Recommended mud hopper design

centrifuging action of the impeller, and leaves the discharge line at high velocities.

The size of the centrifugal pump will depend on the function that it must serve. Fig. 1.68 shows the flow rates required of the pump to mix various amounts of barite.[5] As an example, if the bulk barite system can feed barite at a rate of 15 sacks (100 pounds per sack) per minute, the centrifugal system should be designed to produce a 1,500 gpm flow rate to the hopper in order to increase the density by a standard design factor of 1.0 ppg. If the centrifugal pumps are called on to do more than mix mud during a kick, either the pump sizes must be increased or more pumps must be used.

Mud hoppers.   The mud hopper is used to add weight material to the mud stream during kick situations. The hopper consists of entrance and mud exit lines, a jet, and one or more valves. Fig. 1.69 illustrates a recommended design for a hopper. The dimensions shown are for optimum efficiency. Significant variations from this design will result in poor hopper performance.

Barite systems.   Barite is marketed commercially in several forms. The particle size is usually a fine grade for drilling purposes but may be obtained in a coarse grade which is removed by the shale shaker during the first circulation. The coarse grade is used when it is not desirable to have solids remain in the mud system. Barite may be obtained in bulk tonnage quantities or in 50 or 100 pound sacks.

The sack weight material is not convenient in well control operations. Since it is usually stacked in 30 sack lots, time and space are consumed in individual sack handling and pallet movement for accessibility.

Bulk weight material is preferred for well control because fewer crew members are necessary for mixing and weight-up time is generally much shorter. Bulk tanks with a capacity of 500 to 1,500 sacks can be used with mud hopper equipment already attached. It is necessary only to attach mud lines to the hoppers and fill the tanks. With this system, large volumes of barite can be added quickly without the necessity of cutting sacks, or moving pallets.

Mud Pumping Equipment

There are several types of rig pumps that can be used during normal circulation and kick killing procedures. The two primary types are the double-acting duplex pump and the single-acting triplex pump. Other pumps are available which are generally based on the same principles as either the duplex or triplex pumps.

The double-acting duplex pump has two liners with four sets of suction and discharge valves. A set of valves is located on each end of the liner so that fluid is pumped when the stroke is in the forward direction as well as the reverse direction. The pump has the characteristic of pumping large volumes of fluid at relatively low stroke rates.

The single-acting triplex pump has three liners with one set of valves located on the forward end of each liner. The pump generally strokes at faster rates than the duplex but the output volume is comparable. Experience has shown that the triplex has better wear resistance and easier maintenance.

Pump crippling.   In many critical well control situations, it may become necessary to pump at low rates to avoid fracturing the formation and inducing an underground blowout. Unfortunately the trend in rig pump design has been to increase the pump size which increases the minimum flow rate of the unit. As a result, it may become necessary to reduce the output by pulling certain valves or rods which cripple the designed efficiency of the pump.

Fig. 1.70 points out the valves to pull in order to gain approximate percentage reductions and still maintain maximum efficiency. The chart is only for the duplex due to its number and arrangement of valves. Crippling of the triplex would be on the same lines as the duplex.

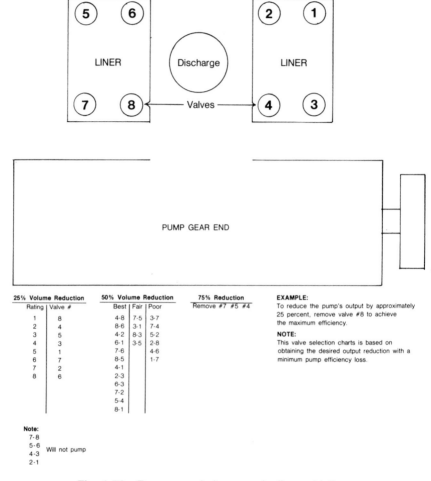

Fig. 1-70 Recommended pump crippling guidelines

Trip Monitoring Equipment

It is necessary to monitor the amount of mud that exits or enters the hole as the drill string is run in or out. The monitoring, or measurement, can be done either by using the rig pumps and calculating the number of strokes required to fill the hole, or by using a trip tank.

A trip tank is any pit or tank in which the mud volume can be measured accurately to within ± 1.0 barrel. As the pipe is pulled from the hole, the mud from the tank is allowed to fill the hole as needed, which at the same time denotes the amount of mud being used. The mud can fill the hole by gravity feed, or by a pump with a return line from the bell nipple to the tank. Fig. 1.71 is an example of a trip tank. The advantage of a trip tank over pump stroke counting is that it is a continuous fill up device and does not require as much of the driller's attention.

Degassers. The purpose of degassers is to remove air or gas entrained in the mud system in order to insure that the proper density mud is recirculated down the drill pipe. If the gas or air is not removed, the mud weight measured in the pits may be misleading. This will result in the addition of unnecessary amounts of weight material thereby giving true mud densities down the hole that are more than desired. The most common type of degassers are the vacuum and atmospheric types.

The atmospheric separator, or poor-boy degasser as it is often called, is probably the first line of defense on gas removal in most well control operations. A typical unit schematic is shown in Fig. 1.72. The mud and gas enters the top and is allowed to separate through gravity segregation. The unit is useful because of its ease of operation, maintenance, and construction as well as its ability to remove large volumes of gas. It should be noted that the vent line should be of a sufficient length to insure that gas is not vented near the rig floor, i.e. to the top of the derrick.

Fig. 1-71 Trip tank

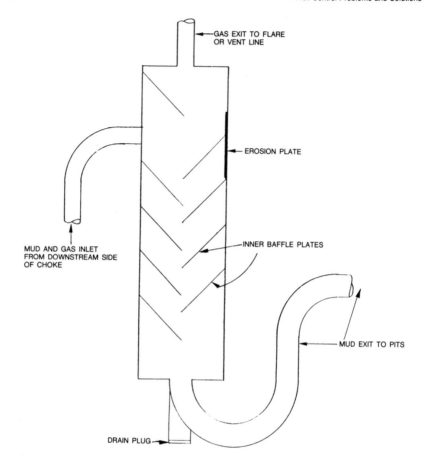

GAS EXIT TO FLARE
OR VENT LINE

EROSION PLATE

MUD AND GAS INLET
FROM DOWNSTREAM SIDE
OF CHOKE

INNER BAFFLE PLATES

MUD EXIT TO PITS

DRAIN PLUG

Fig. 1-72 Schematic of a typical atmospheric degasser

Problems associated with this unit are degasser body construction that is not sufficiently large, small diameter vent lines, or gas flow rates through the degasser that perhaps should be flared at gas-to-surface conditions.

The vacuum degasser (Fig. 1.73) consists of a vacuum generating tank which, in effect, pulls the gas out of the mud due to gravity segregation. Some degassers have a small pump to create a vacuum while others similar to the one shown use the centrifugal mixing pumps to create a vacuum. It is important to note that most degassers regardless of type have a minimum required mud throughput for efficient operation.

There are several other types of degassers available such as the centrifugal spray type or the pressurized separator. The centrifugal spray type is relatively new, and has the desirable characteristic of ease of installa-

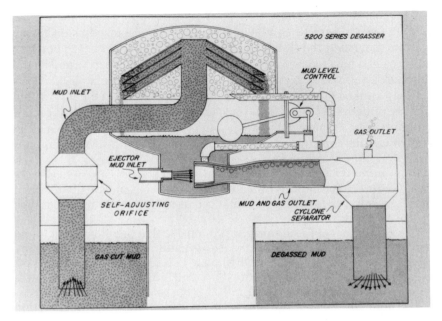

Fig. 1-73 Welco degasser

tion and operation. The pressurized separator is perhaps the best degassing tool for severe gas kick control and has a good service record under these conditions (Fig. 1.74). The unit is somewhat complex in operation and maintenance.

Mud Monitoring Equipment

Monitoring of the mud system is an important task that must be fulfilled in order to maintain safe control of the well. The mud will give warning signs and indications of kicks that can be used to reduce the severity of the kick by early detection and resultant shut in before a large influx is taken. If this system is properly monitored, other drilling problems such as lost circulation can be minimized.

Flow detectors.   When a kick occurs, one of the primary warning signs will be an increased flow rate leaving the well. A flow monitor is designed to gauge the rate of mud flow and, should any abnormal changes occur, the monitor must record the changes and sound an alarm notifying the crew. The flow detector not only warns of kicks, but also of lost circulation should the flow rate decrease.

The most common type of flow detector is a flapper placed in the flow line. A tension spring is attached to the flapper and adjusted to the warning device. If the flow rate increases, the flapper changes position and creates a new tension on the spring which would be recorded by the monitor. The reverse is true when lost circulation occurs.

Pump stroke counting is a viable procedure for filling the hole as the pipe is pulled. The flow monitor can be synchronized with the mud pumps to signal that mud is flowing out of the bell nipple, automatically shut down the pumps, and record the number of pump strokes required to fill the hole.

Pit monitors.   Another key warning sign of a kick is an increased pit volume. As the formation fluid enters the borehole, an equal volume of mud is displaced into the pits which can be recorded by the proper type of detection equipment.

The basis of most pit monitoring systems is that of a float level in the mud pit which has been attached to a calibrated recorder. In many operations, especially floating drilling, the recorder should have a pit volume totalizing (PVT) feature that will compensate for pit level changes due to ship heave and roll.

Gas detectors.   There are several gas detectors available that function on different principles. However, they all generally report the gas content as units of gas in the mud stream. (It is interesting to note that the exact value of one unit of gas is not otherwise defined.) When a certain

Fig. 1-74 Pressurized separator. Courtesy of Pollution Control Rentals

amount of gas has been sensed, an alarm will sound or a light will signal the crew. The disadvantages of gas detectors are maintenance problems, the general inability to function in large concentrations of gas, and a misleading nature in kick detection.

Problems

Author's Note: At the end of each chapter, a set of problems is given similar to those contained within the chapter. Several problems are similar to each example problem. The first of these has a solution provided while the others will be left as exercises for the student.

Problems denoted with an asterisk are advanced in content and are designed to develop a more complete understanding of certain principles. These problems are for more advanced students and should not be routinely assigned to the beginner.

Complete sets of solutions to all of the problems in this book are available from the author.

1.1　Calculate hydrostatic pressures for each of the following systems:
　　　(a) 13,500 ft of 14.0 ppg mud
　　　(b) 8,600 ft of 9.0 ppg salt water
　　　(c) 17,000 ft of 18.5 ppg mud

Solution: (a)　9,828 psi

(b)　4,025 psi

(c)16,354 psi

1.2　A well is 15,000 ft deep. It contains 7,500 ft of 15.0 ppg mud and 7,500 ft of 16.0 ppg mud. What are the hydrostatic pressures for each section and what is the total hydrostatic pressure at 15,000 feet?

1.3　A typical kick situation has developed the following arrangement of fluids in the annulus:
　　　(a) 2,500 ft of 12.0 ppg mud,
　　　(b) 2,500 ft of 8.6 ppg saltwater,
　　　(c) 3,500 ft of 12.0 ppg mud,
　　　(d) 4,000 ft of 13.1 ppg mud.
　　　What is the hydrostatic pressure of each interval and what is the total pressure exerted at the bottom of the hole?

1.4*　An equivalent mud weight is often used to convert a combination of pressures into mud weight units of pounds per gallon in order to compare equivalent systems. The equivalent mud weight (E.M.W.) formula is derived from the hydrostatic pressure equation:

$$E.M.W. = \frac{\text{Total pressures}}{0.052 \times \text{depth}} \quad or \quad \frac{\text{Total pressures} \times 19.23}{\text{depth}}$$

Using the solution from Problem 1.2, what is the equivalent mud weight at 15,000 ft?

Solution: 15.5 ppg

1.5*   What is the equivalent mud weight of the system developed in Problem 1.3 at 2,500 ft? at 5,000 ft? at 8,500 ft? at 12,500 ft?

1.6*   A kick situation has developed that has yielded 500 psi on the annulus pressure gauge. The annulus contains 8,000 ft of 10.0 ppg mud above 1,000 ft of 9.0 ppg saltwater. What is the equivalent mud weight at 8,000 ft?, at 9000 ft?

1.7   If the active mud system of the Louisianan Producer #14 contains 1,260 barrels of 12.5 ppg mud, what would be the number of 100 pound sacks of barite necessary to increase the mud weight to 13.5 ppg?; number of tons?

Solution: 857 sacks; 42.9 tons

1.8   In problem 1.7, if the mud weight was to be increased from 17.5 ppg to 18.5 ppg, would the requirements be the same? If not, how much would be required in 100 pound sacks?

1.9   The AMSCO Oil Co. is drilling at 12,675 feet with 13.2 ppg mud. It becomes necessary to increase the bottom hole hydrostatic pressure by 450 psi. What mud weight is required to achieve this increase? If the active mud volume is 975 barrels, how much barite will be required?

1.10* In a special blowout situation, the Dry Hole Oil and Gas Co. Wildcat #182 must develop a hydrostatic pressure of 7,400 psi over a 5,000-ft interval. What mud weight will be required to achieve this pressure? Assuming that a galena mud must be used, what volume in tons of galena is required if the mud system presently contains 460 barrels of 14.0 ppg mud. (Hint: See Equation 1.4.)

Solution: 28.5 ppg, 279 tons

1.11* The mud density required to kill a particular underground blowout was 26.0 ppg. A total of 500 barrels of 18.0 ppg mud was used as the base fluid. How many tons of galena would be required to weight the mud?

1.12* If the hydrostatic pressure must be increased by 700 psi in a well that contains 12,600 feet of 11.5 ppg mud, how much barite in 100 pound sacks is required? If galena were used instead of barite, how many tons of galena would be necessary? The system volume is 1,200 barrels.

1.13   A shallow well is to be drilled to 5,500 ft in a normally pressured formation (9.0 ppg). In order to safely control the bottom hole pressures, what should be the preventer pressure rating? (Assume a 2.5 ppg gas density in all problems in this chapter.)

Solution: 2,000 psi (1,859 psi)

1.14   An inoermediate depth well is expected to encounter 10.1 ppg formation pressures at 11,050 ft. Will 3,000 psi working pressure preventers be sufficient if a 100% design factor is used? If not sufficient, what pressure rating should be used?

1.15   The Ocean Tide Exploration Co. will drill a development well to 14,600 ft with an expected formation pressure of 16.9 ppg at that depth. The operator's experience in the area dictates that an 80% design factor should be used in preventer selection. What should be the pressure rating of the preventers?

1.16   An operator elects to design an accumulator system to actuate all stack members with a 50% fluid reserve capacity and a final minimum pressure of 1,500 psi. What are the minimum volumes and initial charge pressures acceptable if the following stack is used? (Use the drawdown curves in the text).

Element	No.	Type	Size, in.	Pressure rating
Spherical	1	Shaffer	6	5,000 psi
Rams	3	Shaffer "LWS"	7$\frac{1}{16}$	10,000 psi
Valves	3	Cameron "F"	3 (3 ⅛ Bore)	10,000 psi

Solution: 100 gallons (92.36)
3,000 psi (200 gallon accumulator)

1.17   Using the same requirements as in Problem 1.16, what would be the minimum acceptable accumulator capacities and pressures for the following subsea stack.

Element	No.	Type	Size, in.	Pressure rating
Spherical	2	Hydril "GL"	11	10,000 psi
Rams	5	Cameron "U"	11	10,000 psi

1.18   A particular governing body rules that the accumulator system must be of sufficient size to close the three stack members with the largest fluid requirements and have a 50% reserve fluid capacity. Also, after activation the accumulator must have a minimum final pressure of 1,200 psi. If a well is drilled in the governed domain and utilizes the following stack, what would be the size and initial pressure requirements for the accumulator? What must the minimum operating pressure be on the spherical preventer if the casing pressure is 1,550 psi?

Element	No.	Type	Size, in.	Pressure rating
Spherical	1	Regan "KFL"	13⅝	5,000 psi
Rams	3	Hydril "X"	11	10,000 psi
Rams	1	Cameron "U"	11	10,000 psi

References

1.   Adams, Neal J. "Well Control Manual," February, 1978.

2.   Copied with permission from API 6A, Tenth Edition–March 1974 (Obsolete—superseded by the Eleventh Edition, October 1977)

3.   Copied with permission from API Bulletin D13 (Obsolete—superseded by RP 53)

4.   Personal Communication from Rowe Greene, Greene's Pressure Testing, Inc., Lafayette, Louisiana.

5.   Lee, Henry A., "Good Design Can Improve Mud-Hopper Performance," *Oil and Gas Journal*, Vol. 75, No. 51, December 12, 1977.

2

Well control procedures and principles

WELL CONTROL AND BLOWOUT PREVENTION has become a particularly important topic in the oil industry for a number of reasons. Among these are higher drilling costs, possible loss of life, and waste of natural resources when blowouts occur. One additional reason for concern is the increasing number of governmental regulations and restrictions being placed on the oil industry partially as a result of recent, much-publicized well-control incidents.

For these and other reasons, it is important that drilling people understand the principles of well control and the procedures that must be followed in order to control potential blowouts properly.

Introduction to Kicks

Different drilling problems confront the operator on a day to day basis. Among these are lost circulation, stuck pipe, deviation control, and well control. The drilling problem considered in this discussion is well control. Other drilling problems are presented when related to some aspect of well control.

A kick can be defined as a well-control problem in which the pressure found within the drilled rock is greater than the mud hydrostatic pressure acting on the borehole or face of the rock. When this occurs, the greater formation pressure has a tendency to force formation fluids into the wellbore. This fluid flow is called a kick. If the flow is successfully controlled, the kick has been killed. A blowout is the result of an uncontrolled kick.

The severity of a kick depends upon several factors. One of the most important is the ability of the rock to allow fluid flow to occur. Permeability of rock describes its ability to allow fluid movement. Porosity measures the amount of space in the rock that contains fluids. A rock with high permeability and high porosity has a greater potential for a severe kick than a rock with low permeability and porosity. As an example, sandstone is considered to have a greater kick potential than shale because, in general, sand has a greater permeability and porosity than shale.

Another controlling variable for kick severity is the amount of pressure differential involved. Pressure differential is the difference between the formation fluid pressure and the mud hydrostatic pressure. If the formation pressure is much greater than the hydrostatic pressure, a large negative differential pressure exists. If this negative differential pressure is coupled with high permeability and porosity in a rock, a severe kick can occur.

A kick is labeled in several manners. One label depends upon the type of formation fluid that entered the borehole. Known kick fluids include gas, oil, saltwater, magnesium chloride water, hydrogen sulfide (sour) gas, and carbon dioxide. If gas entered the borehole, the kick would be called a gas kick. Furthermore, if a volume of 20 barrels of gas entered the borehole, the kick could be termed a 20-barrel gas kick.

Another method of labeling kicks is that of the required mud weight increase necessary to control the well and kill a potential blowout. As an example, if a kick required a 0.7 ppg mud weight increase to control the well, the kick could be termed a 0.7 ppg kick. (It is interesting to note at this point that an average kick will require a 0.5 ppg mud weight increase or less.[1])

An additional important consideration in well control is that of the amount of pressure that the formation rock can withstand without sustaining an induced fracture. A measure of this type of rock strength is often called the fracture gradient and is usually expressed in lb/gal equivalent mud weight.

The equivalent mud weight term is the summation of all pressures that are exerted on the borehole wall and can include mud hydrostatic pressure, pressure surges due to pipe movement, friction pressures applied against the formation as a result of pumping the drilling fluid, or any casing pressure caused by a kick.

As an example, if the fracture gradient of a formation is determined to be 16.0 ppg, the well can withstand any combination of the above mentioned pressures that would yield the same pressure as a column of 16.0 ppg mud to the desired depth. This combination could be (1) 16.0 ppg mud, (2) 15.0 ppg mud and some amount of casing pressure, (3) 15.5 ppg mud and a smaller amount of casing pressure, or (4) many other possible combinations. Several methods of fracture gradient determinations will be presented and applied in Chapter 5.

Causes of Kicks

Kicks occur as a result of a formation pressure being greater than the mud hydrostatic pressure which causes fluids to flow from the formation into the wellbore. In almost all drilling operations, the operator attempts to maintain a hydrostatic pressure greater than the formation pressure and thus prevent kicks. On occasion however, and for various reasons, the formation pressure will exceed the mud pressure and a kick will occur. A study of the reasons for this pressure imbalance will explain the causes of kicks.

Insufficient mud weight. Insufficient mud weight is one of the predominant causes of kicks. In this case, a permeable zone is drilled while using a mud weight that exerts less pressure than the formation pressure within the zone. As a result of this pressure imbalance, fluids begin to flow into the wellbore and the kick occurs.

Abnormal formation pressures are often associated with this cause for kicks. Abnormal formation pressures are those pressures that are more than the pressures observed under normal conditions. In well control situations, formation pressures greater than normal are of the most concern. Since a normal formation pressure is that pressure equal to a full column of native water, abnormally pressured formations would be those exerting more pressure than a full column of water. If one of these abnormally pressured formations is encountered while drilling with mud weights insufficient to control the zone, a potential kick situation has developed. Whether or not the kick occurs depends upon the permeability and porosity of the rock.

There are a number of methods that can be used to estimate formation pressures in an effort to prevent this type of kick. Some of these methods are listed in Table 2.1.

Kicks caused by insufficient mud weights would seem to have the obvious solution of drilling with high mud weights to avoid this problem. However, this is not a viable solution for several reasons. First, high mud weights may exceed the fracture gradient of the formation and

Table 2.1 Abnormal pressure indicators

Qualitative methods
   Paleontology
   Offset well log analysis
   Temperature anomaly
   Gas counting
   Mud or cuttings resistivity
   Cutting character
   Hole condition

Quantitative methods
   Shale density
   "d" exponent
   Normalized penetration rate
   Other drilling equations

---

induce an underground blowout. Second, mud weights in excess of the formation pressure may significantly reduce the penetration rates. Also, pipe sticking becomes a serious consideration when excessive mud weights are used. Therefore, the best solution would be to maintain the mud weight slightly greater than the formation pressure until that time that the mud weight begins to approach the fracture gradient requiring an additional string of casing.

Improper hole fill-up during trips. Improperly filling the hole during trips is another predominant cause of kicks. As the drill pipe is pulled out of the hole, the mud level falls because the drill pipe steel had displaced some amount of mud. With the pipe no longer in the hole the overall mud level will decrease, and as a consequence, the hydrostatic pressure of the mud will also decrease. The following example illustrates the hydrostatic reduction when pulling drill pipe and drill collars.

Example 2.1 Calculate the hydrostatic pressure reduction when pulling 10, 93 ft stands of drill pipe from the hole without filling the hole. (Use the tables from Appendix.)

Hole size = 8½ in. (casing ID)
Drill pipe = 4½ in. 16.6 lb/ft
Collars = 7 in. OD with 2.5 in. ID
Drill pipe displacement = 0.00648 bbl/ft

Collar displacement = 0.0415 bbl/ft
Annular capacity (4½ × 8½-in.) = 0.05 bbl/ft
Annular capacity (7 × 8½-in.) = 0.0226 bbl/ft
Drill pipe capacity = 0.01422 bbl/ft
Collar capacity = 0.0061 bbl/ft
Mud weight = 15.0 ppg

Solution
(1) What is the total fluid displaced by 10 stands of pipe?
    10 stands × 93 ft/stand × 0.00648 bbl/ft = 6.0264 bbl
(2) How many feet does 6.0264 barrels fill?
    Annular capacity plus drill pipe capacity = 0.05 + 0.01422 = 0.06422 bbl/ft
    6.0264 bbl/0.06422 bbl/ft = 93.84 ft.
(3) What pressure reduction would be effected?
    93.84 ft × 0.052 × 15.0 ppg = 73.2 psi

Example 2.2   Using the information given in Example 2.1, calculate the hydrostatic pressure reduction when pulling only one stand of collars without filling the hole.

Solution
(1) 1 stand × 93 ft/stand × 0.0415 bbl/ft = 3.8595 bbl
(2) Annular capacity plus collar capacity =
    0.0226 + 0.0061 = 0.0287 bbl/ft
    3.8595 bbl/0.0287 bbl/ft = 134.47 ft
(3) 134.47 ft × 0.052 × 15.0 ppg = 104.8 psi

In this example, note that pulling collars without filling the hole is 10 times more critical with respect to displacement than pulling drill pipe without filling the hole.

It should be obvious from the above examples that it is necessary to fill the hole with mud periodically to avoid reducing the hydrostatic pressure and allowing a kick to occur. Several methods can be used to fill the hole, but all must be able to measure accurately the amount of mud required. It is not satisfactory under any conditions to allow a centrifugal pump continuously to fill the hole from the suction pit since accurate mud volume measurement is not possible. The two methods most commonly used to monitor hole fillup are a trip tank and pump stroke measurement.

A trip tank is any small tank with a calibration device used to monitor the precise volume of mud entering the hole. The tank can be placed level with the preventer to allow a gravity feed into the annulus, or a centrifugal pump may pump mud into the annulus with the overflow returning to the trip tank. The main advantages of a trip tank are that the hole remains full at all times, and an accurate measurement of the mud entering the hole is possible.

Another method of keeping the hole full of mud is to fill the hole periodically with a positive displacement pump such as the rig mud pump. A flow line device can be installed to measure pump strokes required to fill the hole and will automatically shut off the pump when the hole is full. The following example illustrates the usage of the rig pump to fill the hole during a trip.

> *Example 2.3*   Calculate the number of pump strokes required to fill the hole if 10 stands of pipe are pulled from the hole. (Use the data from Example 2.1 and the Appendix.)

Pump = double-acting duplex pump, 6-in. liner × 18-in. stroke
Output = 0.1916 bbl/stroke, or 5.2 strokes/bbl

Solution
(1) From Example 2.1, 10 stands of pipe displacement is 6.0264 bbl.
(2) How many strokes will be required?
      Barrels × strokes/bbl
      = 6.0264 bbl × 5.2 strokes/bbl
      = 31.3 or 32 strokes (per 10 stands)

Swabbing.   Swab pressures are pressures created by pulling the drill string from the borehole. Swab pressure is negative and reduces the effective hydrostatic pressure throughout the hole below the bit. If this pressure reduction is large enough to lower the effective hydrostatic pressure to a value below the formation pressure, a potential kick has developed. Among the variables controlling swab pressures are pipe pulling speed, mud properties, hole configuration, and the effect of "balled" equipment. Some of these effects can be seen in Table 2.2.

Pulling speed is the only variable that can be controlled during the drilling process when a trip is made. In order to reduce the swab pressure, the pulling speed must be reduced.

It is important to remember that the swab pressure aggravates the pressure reduction resulting from not keeping the hole full as pipe is pulled. Also, the swab pressure is exerted at every point throughout the open hole below the bit, even though the drill string may be inside the casing string.

Cut mud.   Gas contaminated mud will occasionally cause a kick although this occurence is rare. The mud density reduction is usually caused by the fluids obtained from the core volume cut by the bit and

Table 2.2 Swab pressures (in psi) in various hole sizes with several pulling speeds for a 14.0 ppg mud, 4½-in. pipe

Hole size, in.	Pulling speeds, seconds/stand					
	15	22	30	45	68	75
8 ½	276	167	124	98	84	75
6 ½	589	344	256	192	159	140
5 ¾	921	524	394	289	231	200

released into the mud system. As the gas is circulated to the surface, it may expand and reduce the overall hydrostatic pressure to a point sufficient to allow a kick to occur. An example illustrates the concept of core volume cutting.

Example 2.4 Using the data given below, what will be the reduction in the mud weight near the surface due to core volume cutting?

Hole size = 12 ¼-in.   Pump rate = 10 bbl/min
Depth = 9,000 ft   Mud weight = 9.3 ppg
Drilling rate = 100 ft/hr   Formation pressure = Normal
Gas zone = 50 ft., 20% porosity, 25% water saturation

Solution
(1) What is the gas volume in the sand?
    (50 ft) $(\pi)/4$ $(12.25)^2/(12)^2$ (0.20) $(1 - 0.25) = 6.1$ cu ft
(2) What will be the expanded volume of this gas at the surface? (The appropriate Z factors are used.)
    $P_1V_1/T_1Z_1 = P_2V_2/T_2Z_2$
    $[(9,000 \times 0.465) + 15]$ $(6.1)/(637)(.94) = (15)(V_2)/(1)(530)$
    $V_2 = 1,516$ cu ft
(3) What volume of mud is mixed with the 1,516 cu ft of gas?
    Pumping time while drilling the sand is 30 minutes
    Pump rate = 10 bbl/min
    Volume = 30 min × 10 bbl/min = 300 bbl
    300 bbl × 5.615 cu ft/bbl = 1,684.5 cu ft of mud
(4) Therefore, 1,516 cu ft of gas will be mixed with 1,684.5 cu ft of mud, and the mud weight will be reduced to approximately 4.4 ppg, or half of its original weight.

Although the mud weight is cut severly at the surface, the total hydrostatic pressure is not reduced significantly since most of the gas expansion occurs near the surface and not at the bottom of the hole.

Lost circulation   Occasionally, kicks are caused by lost circulation
when a decreased hydrostatic pressure occurs due to a shorter column of
mud. When a kick occurs as a result of lost circulation, the problem may
become extremely severe since a large amount of kick fluid may enter the
hole before the rising mud level is observed at the surface. Due to this, it
is a recommended practice to attempt to keep the hole filled with some
type of fluid in order to monitor the fluid level.

Warning Signs of Kicks

There are a number of warning signs and possible warning signs of
kicks that can be observed at the surface. It is the responsibility of each
crew member to recognize and interpret these signs and to take the
proper actions with respect to his well control duties. Although all of the
signs do not positively identify a kick, they do warn of a potential kick
situation. Here, each warning sign is identified as either primary or
secondary relative to its importance in kick detection.

Flow rate increase.   An increase in the flow rate leaving the well
while pumping at a constant rate is one of the primary kick indicators.
The increased flow rate is interpreted to mean that the formation is
aiding the rig pumps in moving the fluid up the annulus by forcing
formation fluids into the wellbore. (Primary indicator)

Pit volume increase.   If the volume of fluid in the pits is not changed
as a result of surface controlled actions, an increase in pit volume
indicates that a kick is occurring. The fluids entering the wellbore as a
result of the kick displace an equal volume of mud at the flow line and
result in a pit gain. (Primary indicator)

Flowing well with pumps off.   When the rig pumps are not moving
the mud, a continued flow from the well indicates that a kick is in
progress. An exception to this is when the mud in the drill pipe is
considerably heavier than that in the annulus as in the case of a slug.
(Primary indicator)

Pump pressure decrease and pump stroke increase.   A pump
pressure change may indicate a kick. The initial entry of the kick fluids
into the borehole may cause the mud to flocculate and temporarily
increase the pump pressure. As the flow continues, the low density
influx will displace the heavier drilling fluids and the pump pressure
may begin to decrease. As the fluid in the annulus becomes less dense,

the mud in the drill pipe will tend to fall and the pump speed may increase. (Secondary indicator)

There are other drilling problems that may exhibit these same signs. A hole in the pipe, called a washout, will cause the pump pressure to decrease, and a twist-off of some portion of the drill string will give the same signs. It is the proper procedure, however, to check for a kick if these signs are observed.

Improper hole fillup on trips.   When the drill string is pulled out of the hole, the mud level should decrease by a volume equivalent to the amount of steel removed. If the hole does not require the calculated volume of mud to bring the mud level back to the surface, it is assumed that a kick fluid has entered the hole and filled the displacement volume of the drill string. Even though gas or saltwater entered the hole, the well may not flow until enough fluid has entered to reduce the hydrostatic pressure to an amount less than the formation pressure. (Primary indicator)

String weight change.   The drilling fluid in the hole provides a buoyant effect to the drill string and effectively reduces the actual pipe weight that must be supported by the derrick. Heavier muds have a greater buoyant force than less dense muds. When a kick occurs and low density formation fluids begin to enter the borehole, the total buoyant force of the mud system is reduced. As a result, the string weight observed at the surface begins to increase. (Secondary indicator)

Drilling break.   An abrupt increase in the bit penetration rate, called a drilling break, is a warning sign of a possible kick. A gradual increase in penetration rate, which is an abnormal-pressure-detection indicator, should not be misconstrued as an abrupt rate increase.

When the rate suddenly increases, it is assumed that the type of rock being drilled has changed. It is also generally assumed that the new rock type has the potential to kick as in the case of a sand, whereas the previously drilled rock did not have this potential as in the case of shale. Although a drilling break may have been observed, it is not certain that a kick will occur, but only that a new formation has been drilled that has the kick potential. (Secondary indicator)

It is a recommended practice that when a drilling break is recorded, the driller should drill 3 to 5 ft into the sand and stop to check for flowing formation fluids.

Cut mud weight.   Reduced mud weight observed at the flow line has occasionally caused a kick to occur. Some of the causes for the reduced mud weight are core volume cutting, connection air, or aerated mud that was circulated from the pits and down the drill pipe. Fortunately, the

Table 2.3 Effect of gas cut mud on the
bottom hole hydrostatic pressure

	10 ppg mud cut to 5 ppg	18 ppg mud cut to 16.2 ppg	18.0 ppg mud cut to 9 ppg
Depth, ft	Pressure reduction, psi	Pressure reduction, psi	Pressure reduction, psi
1,000	51	31	60
5,000	72	41	82
10,000	86	48	95
20,000	97	51	105

lower mud weights due to the cuttings effect are found very near the surface, generally due to gas expansion, and do not appreciably reduce the density of the mud throughout the hole. Table 2.3 shows that gas cutting can have a very small effect on the bottom hole hydrostatic pressure.

An important point to remember about gas cutting is that if the well did not kick in the time required to drill the gas zone and circulate the gas to the surface, there is only a small possibility that it will kick. Generally, gas cutting only indicates that a formation has been drilled that does contain gas. It does not mean that the mud weight must necessarily be increased. (Secondary indicator)

Shut-in procedures

When one or more of the warning signs of kicks are observed, steps should be taken to shut-in the well. If there is any doubt as to whether the well is flowing, shut it in and check the pressures. Also, there is no difference between "just a small flow" and a "full flowing" well because both can very quickly turn into a big blowout.

Under no circumstances should the pipe be run back to the bottom with the annular preventers open and the well flowing.

There has been some hesitation in the past in closing-in a flowing well due to the possibility of sticking the pipe. It can be shown that for all types of pipe sticking (differential pressure, heaving shale, and sloughing shale) it is better to close in the well quickly, reduce the kick influx, and as a result, reduce the chances of pipe sticking. Under any circumstances, it must be remembered that the primary concern at this point is to kill the kick safely, and when feasible, the secondary concern is to avoid pipe sticking.

Some concern has been expressed about fracturing the well and possibly creating an underground blowout as a result of shutting in the well when a kick occurs. If the well is allowed to flow, it will eventually become necessary to shut-in the well, at which time the possibility of fracturing the well will be greater than had the well been shut-in immediately after the initial kick detection. Table 2.4 shows an example of higher casing pressures as a result of a continuous flow.

Initial Shut-in

There has been considerable discussion as to the merits of the "hard" shut-in procedures versus the "soft" shut-in procedures. The hard shut-in procedure is one in which the annular preventer(s) is closed immediately after the pumps are shut down. In soft shut-in procedures, the choke is opened prior to closing the preventers, afterwhich the choke is closed. The two arguments in favor of soft shut-in procedures are that it avoids a water hammer effect due to stopping fluid flow abruptly, and it provides an alternate means of well control (low choke pressure method) should the casing pressure become "excessive."

The water hammer effect has no proven substance. The low choke pressure method will be shown to be an unreliable procedure.

The primary argument against the soft shut-in procedures is that a continuous influx is permitted while the procedures are executed. For these reasons, only the hard shut-in procedures are presented.

There are several types of hard shut-in procedures for well control

Table 2.4 The effect of continuous influx on the casing pressure
as a result of failure to close in the well.*

Volume of gas gained, bbl	Casing pressure, psi
20	1,468
30	1,654
40	1,796

*In a 16,000 ft well with typical geometry.

depending upon the type of rig in use and the drilling operation occurring when the kick is taken.

1. Drilling - immobile rig
2. Tripping - immobile rig
3. Drilling - floating rig
4. Tripping - floating rig
5. Diverter procedures - all rigs (when surface pipe is not set.)

Drilling - immobile rig.   An immobile rig is one that does not move during normal drilling operations. Some types are land and barge rigs, jack-ups, and platform rigs.

Shut-in procedures:
1. When a primary warning sign of a kick has been observed, immediately raise the kelly until a tool joint is above the rotary table.
2. Stop the mud pumps.
3. Close the annular preventer.
4. Notify the company personnel.
5. Read and record the shut-in drill pipe pressure, the shut-in casing pressure, and the pit gain.

Raising the kelly is an important procedure.With the kelly out of the hole, the valve at the bottom of the kelly can be closed if necessary. Also, the annular preventer members can attain a more secure seal on pipe than a kelly.

Tripping - immobile rig.   A high percentage of well control problems occur when a trip is being made. The kick problems may be compounded when the rig crew is preoccupied with the trip mechanics and fails to observe the initial warning signs of the kick.

Shut-in procedures:
1. When a primary warning sign of a kick has been observed, immediately set the top tool joint on the slips.
2. Install and make up a full opening, fully opened safety valve in the drill pipe.
3. Close the safety valve and the annular preventer.
4. Notify the company personnel.
5. Pick up and make up the kelly.
6. Open the safety valve.
7. Read and record the shut-in drill pipe pressure, the shut-in casing pressure, and the pit gain.

Installing a full opening safety valve in preference to an inside-blowout-preventer (float) valve is a prime consideration because of the advantages offered by the full opening valve. If flow is encountered up the drill pipe as a result of a trip kick, the fully opened, full opening valve is physically easier to stab on the pipe than a float type inside blowout preventer valve which would automatically close when the upward moving fluid contacts the valve. (Assume that a manual lock float valve is not in use.)

Also, if wireline work such as drill pipe perforating or logging becomes necessary, the full opening valve will accept logging tools approximately equal to its ID, whereas the float valve may prohibit wireline work altogether. After the kick is shut-in, an inside blowout preventer float valve may be stabbed on top of the full opening valve to allow stripping operations to return to bottom.

Drilling - floating rig. A floating rig moves during normal drilling operations. The primary types of floating vessels are semisubmersibles and drill ships.

Several differences in the shut-in procedures apply to floaters because drill string movement can occur even with a motion compensator in operation. Also, the blowout preventer stack is on the sea floor. To solve the problem of possible vessel and drill string movement and the resultant wear on the preventers, a tool joint is lowered on the closed pipe rams and the string weight hung on these rams.

The problem when the stack is located a considerable distance from the rig floor is to insure that a tool joint does not interfere with the closing of the preventer elements. A spacing out procedure should be executed prior to taking a kick. Close the rams, slowly lower the drill string until a tool joint contacts the rams, and record the position of the kelly at that point.

Shut-in procedures:
1. When a primary warning sign of a kick has been observed, immediately raise the kelly to the level previously designated during the spacing out procedure.
2. Stop the mud pumps.
3. Close the annular preventer.
4. Notify the company personnel.
5. Close the upper set of pipe rams.
6. Reduce the hydraulic pressure on the annular preventer.
7. Lower the drill pipe until the pipe is supported entirely by the rams.

8. Read and record the shut-in drill pipe pressure, shut-in casing pressure, and the pit gain.

Tripping - floating rig. The procedures for kick closure during a tripping operation on a floater is a combination of floating drilling procedures and the immobile rig tripping procedures.

Shut-in procedures:
1. When a primary warning sign of a kick has been observed, immediately set the top tool joint on the slips.
2. Install and make up a full opening, fully opened safety valve in the drill pipe.
3. Close the safety valve and the annular preventer.
4. Notify the company personnel.
5. Pick up and make up the kelly.
6. Close the upper set of pipe rams.
7. Reduce the hydraulic pressure on the annular preventer.
8. Lower the drill pipe until the pipe is supported by the rams.
9. Read and record the shut-in drill pipe pressure, shut-in casing pressure, and the pit gain.

Diverter Procedures - All Rigs

When a kick occurs in a well that has an insufficient amount of casing to control a kick safely, a blowout will occur. Since a shallow underground blowout is difficult to control and may cause the rig to be lost, an attempt must be made to divert the surface blowout away from the rig. (Diverter equipment is discussed in Chapter 1.) Special attention must be given to this procedure to insure that the well is not shut-in until the diverter lines are opened.

Diverter procedures:
1. When a primary warning sign of a kick has been observed, immediately raise the kelly until a tool joint is above the rotary table.
2. Shut down the pumps.
3. Open the diverter line valve(s).
4. Close the diverter bag (annular preventer).
5. Start the pumps at a fast rate.
6. Notify the company personnel.

Crew Member Responsibilities for Shut-in Procedures

Each member of the crew has different responsibilities during the various shut-in procedures. These are listed according to job classification.

Floorhand (roughneck)
1. Notify the driller if any warning signs of kicks are observed.
2. Assist in installing the full opening safety valve if a trip is being made.
3. Initiate his well control responsibilities after the well is shut-in.

Derrickman
1. Notify the driller if any warning signs of kicks are observed.
2. Initiate his well control responsibilities and begin mud mixing preparations.

Driller
1. Shut-in the well immediately if any of the primary warning signs of kicks are observed.
2. If a kick occurs while making a trip, set the top tool joint on the slips and direct the crews in the installation of the safety valve prior to closing the preventers.
3. Notify all proper company personnel.

Obtaining and Interpreting the Shut-in Pressure

Shut-in pressures are defined as pressures recorded on the drill pipe and casing when the well is closed. Although both pressures are important, the drill pipe pressure will be used almost exclusively in killing the well. The shut-in drill pipe pressure is abbreviated SIDPP. Shut-in casing pressure is SICP. (At this point, assume that the drill pipe does not contain a float valve.)

Reading and interpreting the pressures. During a kick, fluids flow from the formation into the wellbore. When the well is closed in to prevent a blowout, pressure begins to build at the surface due to formation fluid entry into the annulus as a result of the difference between the mud hydrostatic pressure and the formation pressure.

Since this pressure imbalance cannot exist for long, the surface pressures will finally build to a point such that the surface pressure plus the mud and influx formation fluid hydrostatic pressures in the well are equal to the formation pressures. Equations 2.1 and 2.2 express this relationship for the drill pipe and the annular side, respectively.

$$
\begin{array}{ll}
\text{SIDPP} + \begin{array}{l}\text{Drill pipe} \\ \text{hydrostatic} \\ \text{pressure}\end{array} = \begin{array}{l}\text{Bottom hole} \\ \text{formation} \\ \text{pressure}\end{array} & \text{Eq. 2.1}
\end{array}
$$

$$
\begin{array}{ll}
\text{SICP} + \begin{array}{l}\text{Annular mud} \\ \text{hydrostatic} \\ \text{pressure}\end{array} + \begin{array}{l}\text{Annular influx} \\ \text{hydrostatic} \\ \text{pressure}\end{array} = \begin{array}{l}\text{Bottom hole} \\ \text{formation} \\ \text{pressure}\end{array} & \text{Eq. 2.2}
\end{array}
$$

Example 2.5 and Fig. 2.1 show how the shut-in pressures are read and interpreted.

> *Example 2.5* While drilling at 15,000 ft, the driller observed several of the primary warning signs of kicks and proceeded to shut-in the well. After the shut-in was completed he called the company man and started to record the following pressures and pit gains. The well was shut-in at 6:00 A.M.

Shut-in time	SIDPP, psi	SICP, psi	Pit gain, bbl
6:00 A.M.	650	950	20
6:05 A.M.	750	1,000	20
6:10 A.M.	775	1,040	20
6:15 A.M.	780	1,040	20

The final shut-in pressures after 15 minutes were recorded as follows:

SIDPP	—	780 psi
SICP	—	1,040 psi
Pit gain	—	20 barrels

Interpretation of the recorded pressures.   An important basic principle can be seen in Fig. 2.1. In this case, it is observed that the formation pressure (BHP) is more than the drill pipe hydrostatic pressure by an amount equal to the SIDPP. In other words, the drill pipe pressure gauge is a bottom hole pressure gauge, assuming that the mud weight in the drill pipe is known. It should be obvious that the casing pressure cannot be considered as a direct bottom hole pressure gauge due to the generally unknown amount of formation fluid in the annulus.

Constant bottom hole pressure concept. Fig. 2.1 can be used to illustrate another very important basic principle. It was stated that the 780 psi observed on the drill pipe gauge was the amount necessary to balance the mud pressure at the hole bottom with the pressure of the gas in the sand at 15,000 ft. A basic law of physics states that formation fluids

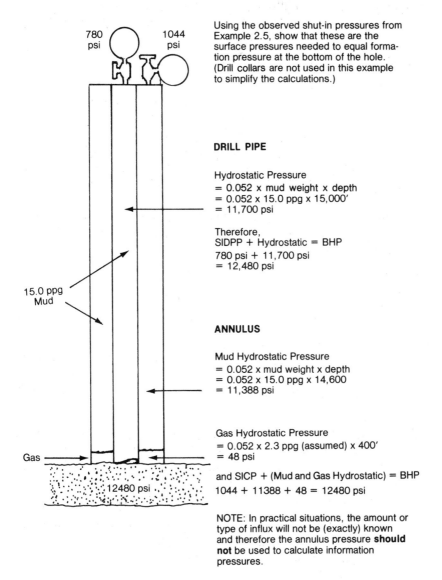

Using the observed shut-in pressures from Example 2.5, show that these are the surface pressures needed to equal formation pressure at the bottom of the hole. (Drill collars are not used in this example to simplify the calculations.)

DRILL PIPE

Hydrostatic Pressure
= 0.052 x mud weight x depth
= 0.052 x 15.0 ppg x 15,000'
= 11,700 psi

Therefore,
SIDPP + Hydrostatic = BHP
780 psi + 11,700 psi
= 12,480 psi

ANNULUS

Mud Hydrostatic Pressure
= 0.052 x mud weight x depth
= 0.052 x 15.0 ppg x 14,600
= 11,388 psi

Gas Hydrostatic Pressure
= 0.052 x 2.3 ppg (assumed) x 400'
= 48 psi

and SICP + (Mud and Gas Hydrostatic) = BHP
1044 + 11388 + 48 = 12480 psi

NOTE: In practical situations, the amount or type of influx will not be (exactly) known and therefore the annulus pressure **should not** be used to calculate information pressures.

Fig. 2-1 Hydrostatic pressure

travel from areas of high pressure to areas of lower pressures only, and do not travel between areas of equal pressures, assuming gravity segregation is neglected.

If the drill pipe pressure is controlled in such a manner that the total mud pressure at the hole bottom is slightly greater than the formation fluid pressure, there will be no additional kick influx entering the well. The concept is the basis of the constant bottom hole pressure method of well control in which the pressure at the bottom of the hole is kept constant, and at least equal to the formation pressure.

Effect of time.   In Example 2.5, a total of 15 minutes was used to obtain the shut-in pressures. The purpose of this time period is to allow the pressures to reach equilibrium, where these pressures will be sufficient to aid in balancing those pressures found in the formations. The amount of time required will depend on such variables as the type of influx, rock permeability and porosity, and the original amount of pressure underbalance. In many places this may take only a few minutes whereas other areas may take several hours. The amount of time required is dependent upon the conditions surrounding the kick.

Several other factors affect the time allowed for pressures to stabilize. Gas migration, which is the movement of low density fluids up the annulus, will tend to build false pressure readings at the surface if excessive time is allowed for the migration. Also, the influx may have a tendency to deteriorate the hole stability and cause either stuck pipe or hole bridging. These problems must also be considered when reading the shut-in pressures.

Trapped pressure.   Trapped pressure is any pressure recorded on the drill pipe or annulus that is more than the amount needed to balance bottom hole pressure. Pressure can be trapped in the system in several ways but the most common are gas migrating up the annulus and tending to expand, or closing the well in before the mud pumps have quit running. It should be obvious that using a pressure reading that contains some amount of trapped pressure, the calculations necessary to kill the well will be in error.

Guidelines help to check when releasing trapped pressure. If they are not properly executed, the well will be much more difficult to kill. These guidelines are listed and explained in Table 2.5.

The basis behind the procedure listed in Table 2.5 is that since trapped pressure is more than the amount needed to balance the bottom hole pressure, trapped pressure can be bled off without allowing any additional influx into the well. However, after all of the trapped pressure is bled off and if the bleeding procedure is continued, more influx will be allowed into the well and the surface pressures will begin to increase.

Table 2.5 Guidelines to check for trapped pressure

---

1. When checking for trapped pressure, bleed from the casing side only. The reasons for this are (1) the choke is located on the casing side, (2) this avoids contamination of the mud in the drill pipe, and (3) it avoids the possibility of plugging the bit jets.
2. Use the drill pipe pressure as a guide since it is a direct bottom hole pressure indicator.
3. Bleed small amounts (¼ to ½ bbl) of mud at a time. Close the choke after bleeding and observe the pressure on the drill pipe.
4. Continue to alternate the bleeding and subsequent pressure observation procedures as long as the drill pipe pressure continues to decrease. When the drill pipe pressure ceases to fall, stop bleeding and record the true shut-in drill pipe pressure and casing pressure.
5. If the drill pipe pressure should decrease to zero during this procedure, continue to bleed and check pressures on the casing side as long as the casing pressure decreases. (Note: This step normally will not be necessary.)

Although these bleeding procedures can be implemented at any time, it is advisable to check for trapped pressure when the well is shut-in initially, and rechecked when the drill pipe is displaced with kill mud if any pressure remains on the shut-in drill pipe. Example 2.6 illustrates the bleeding procedures.

Example 2.6   A kick was taken and shut-in. The SIDPP was read as 525 psi and the SICP was 760 psi. The company representative checked for trapped pressure by bleeding small amounts from the choke and recording the resultant shut-in pressures.

Increment number	Bleeding volume, bbl (approximate)	SIDPP, psi	SICP, psi
0	—	525	760
1	½	510	745
2	½	500	735
3	½	490	725
4	½	480	715
5	½	475	710
6	½	475	710
7	½	475	715

The true pressures were recorded as:

SIDPP = 475 psi
SICP = 715 psi

Drill pipe floats.   A kick can occur while a drill pipe float valve is used. (See Chapter 1 for a discussion of float valves.) Since a float valve prevents fluid and pressure movement up the drill pipe, there will be no drill pipe pressure readings after the well is shut-in. Several procedures can obtain the drill pipe pressure, and each depends on the amount of information known at the time that the kick occurs.

Table 2.6 describes the procedure to obtain the drill pipe pressure if the slow pumping rate (kill rate) is known. Table 2.7 gives a procedure if the kill rate is not known. Examples 2.7 and 2.8 illustrate various uses of these procedures.

Example 2.7   A kick was taken on a well in which a float was used in the drill string. The kill rate and associated pressure taken immediately prior to the kick was 26 spm at 650 psi. The shut-in pressures were 400 psi on the casing (SICP) and zero on the drill pipe. Establish the true SIDPP.

Solution.
1. Instruct the driller to run his pumps at 26 spm.

Table 2.6 Procedure to establish shut-in drill pipe pressure (SIDPP) if the kill rate is known

---

1. Shut-in the well, record the shut-in casing pressure (SICP) and obtain the kill rate either from the driller or the daily tour report.
2. Instruct the driller to start the pumps and maintain the pumping rate at the kill rate (strokes).
3. As the driller starts the pumps, use the choke to regulate the casing pressure at the same pressure that was originally recorded at shut-in conditions.
4. After the pumps are running at the kill rate with the casing pressure properly regulated at shut-in pressure, record the pressure on the drill pipe while pumping.
5. Shut down the pumps and close the choke.
6. The shut-in drill pipe pressure equals the total pumping pressure minus the kill rate pressure, or

SIDPP = Total pressure − kill rate pressure   Eq. 2.3

Table 2.7 Procedure to establish the shut-in drill pipe pressure (SIDPP) if the kill rate is not known

1. Shut-in the well.
2. Line up a low volume, high pressure reciprocating pump on the stand pipe.
3. Start pumping and fill up all of the lines.
4. Gradually increase the torque on the pumps until the pumps begin to move fluid down the drill pipe.
5. The shut-in drill pipe pressure is the amount of pressure required to initiate the fluid movement. This is assumed to be the amount needed to overcome the pressure acting against the bottom side of the valve.

---

2. Operate the choke to maintain the casing pressure at the initial pressure of 400 psi.
3. After the drill pipe pressure has stabilized, record this value as the total pumping pressure. As an example, assume the total pressure to be 870 psi.
4. From Table 2.6,
   SIDPP = total pressure − kill rate pressure
   SIDPP = 870 psi − 650 psi = 220 psi

Example 2.8   A trip was made for a new bit in which the jet sizes were changed. The driller was on bottom and had drilled 5 ft when a kick was taken. The kill rate with the new bit was not known and the drill string contains a float. Establish the shut-in drill pipe pressure.

Solution.
Since the kill rate is not known, the procedure listed in Table 2.7 must be exercised.

1. Line up a low volume positive displacement pump on the stand pipe and fill up all the lines.
2. Pressures were increased on the stand pipe at a low rate with the following results:

Volume pumped, bbl	SIDPP, psi
0	0
1	0
2	0
3	70
4	150
5	220
6	300
7	300

3. The SIDPP was assumed to be 300 psi. A graph of these pressures (Fig. 2.2) will aid in a more precise determination of the desired values.

Table 2.6 is particularly important in another application. Suppose for example that a kick was taken in which the shut-in drill pipe pressure was known (no float valve) but a kill rate had not been established. Step 6 of this table could be modified to read:

Kill rate pressure = total pressure − SIDPP   Eq. 2.4

The procedures presented in Table 2.6 would remain the same with the exception that Eq. 2.4 would be substituted for Eq. 2.3.

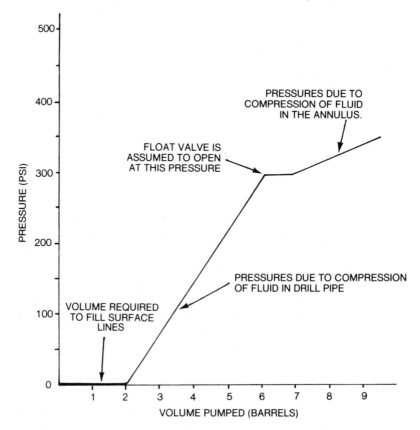

Fig. 2-2   Procedure to establish the shut-in drill pipe pressure on Example 2.8 when the kill rate was not known.

Establishment of the shut-in drill pipe pressure becomes more complex if the kill rate was not previously established and a float valve in the string prohibits pressure readings at the surface on the drill pipe. Table 2.7 must be used initially to determine the SIDPP, after which Eq. 2.4 and Table 2.6 must be implemented to establish the kill rate. Example 2.9 illustrates the application of this technique.

Example 2.9   A kick was taken in which the kill rate was not known and a float valve was used in the drill string. The SICP is 550 psi. Establish both the SIDPP and the kill rate.

Solution

1. After preparing all surface equipment, the company man began pressurizing the drill pipe with the following results:

Volume pumped, bbl	SIDPP, psi
0	0
1	0
2	50
3	150
4	250
5	260
6	265

2. The SIDPP was assumed to be approximately 250 psi.
3. The driller was instructed to bring his pumps to 20 spm (arbitrary value). The company man used the choke to maintain the initial value to 550 psi on the casing.
4. After the pressures had stabilized on the drill pipe, the total pressure was recorded as 950 psi.
5. Using Eq. 2.4, the kill rate was established.

   Kill rate pressure = total pressure − SIDPP   Eq. 2.4
                       = 950 psi − 250 psi
                       = 700 psi (at 20 spm)

Kick identification.   When a kick occurs, it may prove interesting to know the type of influx (gas, oil or saltwater) that entered the wellbore. It must always be remembered however, that the well control procedures developed here are designed to safely kill all types of kicks. The formula required to make this kick influx calculation is as follows:

Influx gradient = mud gradient in drill pipe −   Eq. 2.5
(SICP − SIDPP)/influx height

Where: Influx gradient = the gradient of the formation fluid that entered the wellbore, in psi/ft

Mud gradient = the gradient of the mud in the drill pipe, in psi/ft,

SICP = the shut-in casing pressure, psi,

SIDPP = the shut-in drill pipe pressure, psi,

Influx height = the formation fluid length in the annulus, ft.

The influx gradient in this solution can be evaluated using the guidelines in Table 2.8.

Although SICP and SIDPP can be determined accurately for use in Eq. 2.5, it may prove difficult to determine the influx height since this requires knowledge of the pit gain and the exact hole size. Example 2.10 illustrates Eq. 2.5.

> *Example 2.10*   While drilling, a kick was taken with the following known data. Assume no drill collars in the hole in order to simplify the example. What type of fluid entered the well?

Depth = 15,000 ft   SIDPP = 780 psi

Mud weight = 15.0 ppg   SICP = 1,100 psi

Drill pipe = 4½ in.   Pit gain = 50 bbl

Hole size = 8½ in.

Solution

1. Using the tables in the Appendix, 1 bbl of fluid occupies 20 vertical ft in a 4½ × 8½-in. annulus. Therefore, 50 bbl × 20 ft/bbl = 1,000 ft (influx height)
2. Mud gradient = 0.052 psi/ft/ppg × 15.0 ppg = 0.780 psi/ft
3. Influx gradient = mud gradient − (SICP − SIDPP)/influx height
   = 0.780 psi/ft − (1,100 psi − 780 psi)/1,000 ft
   = 0.780 psi/ft − 0.320 psi/ft
   = 0.460 psi/ft
4. From Table 2.8, it can be seen that the influx is probably salt water or oil.

Kill Weight Mud Calculations

In order to kill a kick, it is necessary to calculate the mud weight needed to balance the bottom hole formation pressure. Kill weight mud is defined as the amount necessary to exactly balance the formation pressure. It will be shown in later sections that it is safer to use the exact required mud weight with no variations.

Table 2.8 Influx gradient evaluation guidelines

Influx gradient	Influx type
0.05 – 0.2	Gas
0.2 – 0.4	Probable combination of gas, oil, and/or saltwater
0.4 – 0.5	Probable oil, or saltwater

Since the drill pipe pressure has previously been defined as a bottom hole pressure gauge, the SIDPP can be used to calculate the mud weight necessary to kill the well. The kill mud formula is as follows:

$$K.W.M. = SIDPP \times 19.23/depth + O.W.M. \qquad \text{Eq. 2.6}$$

Where K.W.M. = the kill weight mud, ppg
       SIDPP = the shut-in drill pipe pressure, psi
       19.23 = the reciprocal of 0.052, ppg/psi/ft,
       Depth = the true vertical bit depth, ft
       O.W.M. = the original weight mud in the drill pipe, ppg

It should be noted that since the casing pressure does not appear in Eq. 2.6, a high casing pressure does not necessarily indicate a high kill weight mud. The same is true for the pit gain since it does not appear in Eq. 2.6. Example 2.11 shows the usage of the kill weight mud formula.

Example 2.11   What will be the kill weight mud for the kick data given below?
True vertical depth = 11,550 ft
O.W.M. = 12.1 ppg
SIDPP = 240 psi
SICP = 1,790 psi
Pit gain = 85 bbl

Solution
K.W.M. = SIDPP × 19.23/depth + O.W.M.
       = 240 psi × 19.23/11,550 ft + 12.1 ppg
       = 0.4 ppg + 12.1 ppg
       = 12.5 ppg

A table of kill mud weights is presented in the Appendix. This table should be used only to check the operational calculations since the table does not list every conceivable combination of depths and pressures.

Kick-Killing Procedures

There have been many kick-killing procedures developed over the years. Some of these have utilized systematic conventional approaches while others were based on logical, but perhaps unsound, principles. The systematic approaches will be presented in this section. A discussion in later sections will be given to explain several non-conventional methods of kick-killing and their inherent faults.

In previous sections, the constant bottom hole pressure concept was developed in which the total of all pressures (mud hydrostatic pressure, casing pressure, etc.) at the bottom of the hole would be maintained at a value slightly greater than the formation pressures to prevent further influxes of formation fluids into the wellbore. Also, since the pressure would be only slightly greater than the formation pressure, this would minimize the possibility of inducing a fracture and an underground blowout. This concept can be implemented in three ways.

- One circulation method. After the kick is shut-in, weight the mud to kill density, then pump out the kick fluid in one circulation using the kill mud. (Alternate names often applied to the approach are wait and weight, engineer's method, the graphical method, or the constant drill pipe pressure method.)
- Two circulation method. After the kick is shut-in, the kick fluid is pumped out of the hole before the mud density is increased. (An alternate name is the driller's method.)
- Concurrent method. Pumping begins immediately after the kick is shut-in and the pressures are recorded. The mud density is increased as rapidly as possible while pumping the kick fluid out of the well.

If applied properly, each of these three methods will achieve the constant pressure at the hole bottom and will not allow any additional influx into the well. However, there are procedural and theoretical differences that makes one of these procedures more desirable for implementation than the others.

One Circulation Method

Fig. 2.3 provides a description of the one circulation method. At Point 1, the shut-in drill pipe pressure is used to calculate the kill weight mud, after which the mud weight is increased to kill density in the suction pit. As the kill mud is pumped down the drill pipe, the static drill pipe pressure is controlled to decrease linearly, until at Point 2 the drill pipe pressure would be zero. This results from heavy mud having killed the drill pipe pressure.

Point 3 illustrates that the initial pumping pressure on the drill pipe would be the total of the SIDPP plus the kill rate pressure, or 1,500 psi in Fig. 2.3. While pumping kill mud down the pipe, the circulating pres-

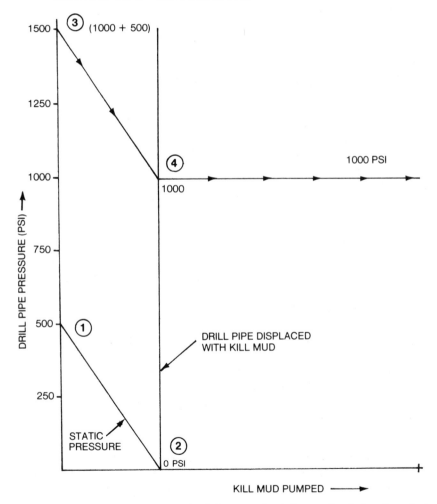

Fig. 2-3 Drill pipe pressure graph of the one circulation method of well control

sure should decrease until at Point 4, only the pumping pressure remains. From the time that the kill mud reaches the bit until the kill mud reaches the flow line, the choke controls the drill pipe pressure at the final circulating pressure while the driller insures that the pump remains at the kill speed.

Two Circulation Method

In the two circulation method, kill mud is not added in the first circulation which implies that the drill pipe pressure will not decrease during this period (Fig. 2.4). The purpose of this circulation is to remove the kick fluid from the annulus.

In the second circulation, the mud weight is increased and causes a decrease from the initial pumping pressure at 1 to the final circulating pressure at 2. This final circulating pressure is held constant thereafter while the annulus is displaced with kill mud.

Concurrent Method

This method is the easiest to comprehend while it is the most difficult to execute properly. As soon as the kick is shut-in, pumping begins immediately after reading the pressures and the mud density is in-

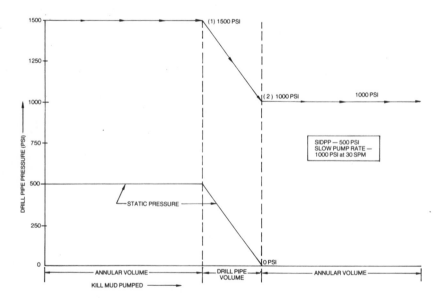

Fig. 2-4 Drill pipe pressure graph of the two circulation method of well control

creased as rapidly as the rig facilities will allow. The difficulty is encountered in determining the mud density being circulated and its relative position in the drill pipe. Since this position determines the drill pipe pressures, the rate of pressure decrease may not be as consistent as seen in the other two methods (Fig. 2.5). As a new density arrives at the bit or some predetermined depth, the drill pipe pressure is decreased by an amount equal to the hydrostatic pressure of the new mud weight increment. When the drill pipe is completely displaced with kill mud, the pumping pressure is maintained constant until kill mud reaches the flow line.

Constant Bottom Hole Pressure Methods

Determining the best well control method suitable for the most frequently encountered situations involves several important considerations. Some of these considerations are (1) the time required to execute the complete kill procedure, (2) surface pressures arising from the kick, (3) the complexity of the procedure itself relative to ease of implementation, and (4) the down hole stresses applied to the formation during the

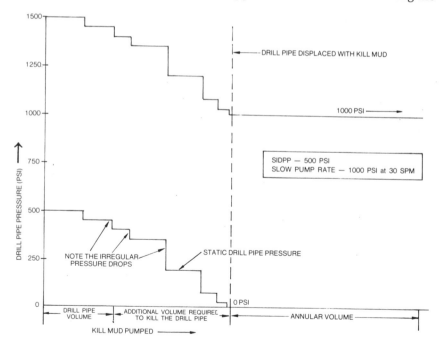

Fig. 2-5 Drill pipe pressure graph of the concurrent method of well control

kick killing process. All of these points must be analyzed before a procedure can be selected.

In an analysis of the kick-killing procedures, major emphasis will be placed on the one and two circulation methods. Inspection of the procedures will show that these are the opposite approaches while the concurrent method falls somewhere in between.

Time. There are two important considerations relative to time required for the kill procedure. The first of these is the time required to increase the mud density from the original weight to the final kill weight mud. Since a few operators are more concerned with the pipe sticking during this time than killing the kick, the well control procedure is often chosen that will minimize the waiting time required to increase the mud density. The procedures with the least amount of initial waiting time are the concurrent method and the two circulation method. In both of these procedures, pumping begins immediately after the shut-in pressures are recorded.

The most important time consideration, however, is not the initial waiting time but the overall time required for the complete procedure to

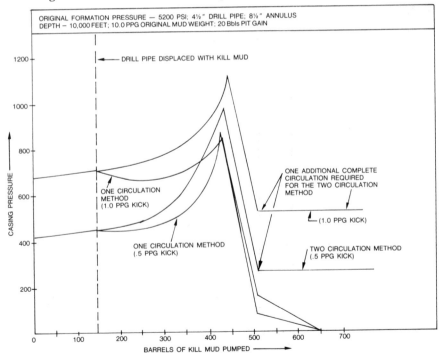

Fig. 2-6   Annular pressures for one circulation method vs. two circulation method in a 10,000-ft well.

be implemented. Fig. 2.3 shows that the one circulation method requires one complete fluid displacement (drill pipe and annulus), while the two circulation method (Fig. 2.4) requires that the annulus be displaced twice in addition to the drill pipe displacement. In certain situations, the extra time increment required for the two circulation method may be a serious matter with respect to hole stability or preventer wear.

Surface pressures. During the course of a well killing process, the surface pressures may approach values that cause alarm. This may be a particular problem in gas kicks due to the volume expansion phenomena of the fluid near the surface. The kill procedure with the least surface pressure required to balance the bottom hole formation pressure is of major importance here.

Figs. 2.6 and 2.7 point out the different surface pressure requirements for several different kick situations utilizing the one circulation and the two circulation methods. The first major difference is noted immediately after the drill pipe is displaced with kill mud. The necessary casing pressure begins to decrease as a result of the increased kill mud hydro-static pressure in the one circulation procedure. This decrease is not

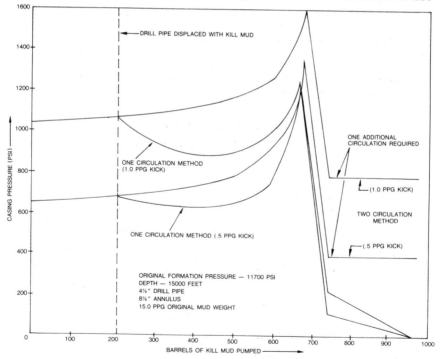

Fig. 2-7   Annular pressures for one circulation vs. two circulation method in a 15,000-ft well.

seen in the two circulation method since this procedure does not circulate kill mud initially. In fact, the casing pressure increases as a result of the gas bubble expansion displacing mud from the hole.

The second major surface pressure difference is observed as the gas approaches the surface. The two circulation procedure again has higher pressures as a result of circulation of the lower density original mud weight. It is interesting to note at this point that these high necessary casing pressures suppress the gas expansion to a small degree resulting in a later arrival of the gas at the surface.

After one complete circulation has been made, it can be observed that the one circulation method has theoretically killed the well resulting in zero surface pressures. The alternate method has pressure remaining on the casing exactly equal to the drill pipe pressure. It will be necessary at this point to introduce kill weight mud and complete another circulation.

Procedure complexity. The degree of suitability of any process is partially dependent upon the ease with which it can be executed. The same principle holds true for well control. If a kick killing procedure is difficult to comprehend and implement, its reliability is diminished accordingly.

The concurrent method of well control falls into the category of reduced reliability due to procedure complexity. To perform this procedure properly, the drill pipe pressure must be reduced according to mud

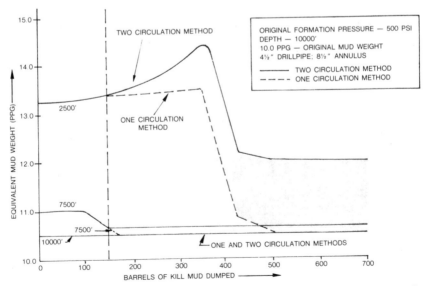

Fig. 2-8 Equivalent mud weight comparison for the one circulation method vs. the two circulation kill procedure (0.5 ppg kick at 10,000 ft).

weight being circulated and its position in the pipe. This implies that (1) the crew will immediately inform the operator when a new mud weight is being pumped, (2) the rig facilities can maintain this increased mud weight increment, and (3) the mud weight position in the pipe can be determined by pump stroke counting. As a result of the complex nature of this method, many operators have discontinued its use.

One circulation and two circulation methods are receiving prominent use because of their ease of application. In both of these procedures, the drill pipe pressure remains constant for long intervals of time. Also, when displacing the drill pipe with kill mud, the drill pipe pressure decrease is virtually a straight line relationship and not staggered as in the concurrent method (Fig. 2.5).

Downhole stresses. Although all of the considerations to this point are important, the primary concern should be the stresses that are imposed on the borehole wall. If the kick-imposed stresses are greater than the formation can withstand, an induced fracture will occur, thus allowing the possibility of an underground blowout. The procedure which imposes the least downhole stresses while still maintaining the constant pressures on the kicking zone should be considered the procedure most conducive to safe kick killing.

Equivalent mud weights are a useful tool to measure downhole stresses. The equivalent weights are defined as the total of all pressures to a depth and converted to pounds per gallon mud weight.

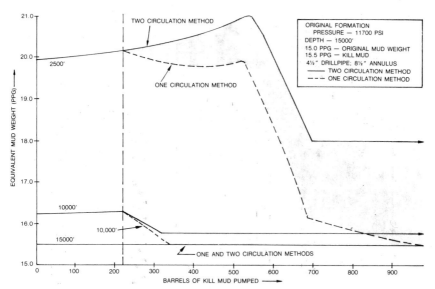

Fig. 2-9 Equivalent mud weight comparison for the one circulation method vs. the two circulation kill procedure (0.5 ppg kick at 15,000 ft).

Equivalent mud weight =
(Total pressures × 19.23)/depth   Eq. 2.7
Where   Equivalent mud weight is in ppg,
         Total pressures in psi to the depth of interest,
         19.23 is a constant with units of ppg/psi/ft,
         Depth is in ft.

The equivalent mud weights for the systems in Figs. 2.6 and 2.7 were
calculated and are presented in Figs. 2.8 and 2.9. The one circulation
method has consistently lower equivalent mud weights throughout the
killing process after the drill pipe has been displaced. However, the
procedures generally exhibit the same maximum equivalent mud weight
which occur from the time that the well is shut-in until the drill pipe is
displaced.

Figs. 2.8 and 2.9 illustrate a very important principle. It can be seen
that the maximum stresses occur very early in the circulation for the
deeper depth and not at the maximum casing pressure intervals. This
can be interpreted to mean that the maximum lost-circulation possibili-
ties will not occur at the gas-to-surface conditions as might seem logical
to the casual observer. For all practical considerations, it can be stated
that if a fracture is not created at shut-in, it will probably not occur
throughout the remainder of the process. A full understanding of this
behavior may negate any unfounded concerns of the operator about
formation fracture as the gas approaches the surface.

Variables Affecting Kill Procedures

Although variables that affect kick-killing do not necessitate a change
in the basic structure of the procedure, they may cause an irregular
behavior that perhaps can mislead an operator into making erroneous
judgements. A study of some of these variables will give insight into well
control problems. Since the one circulation method has previously been
shown to be the safer of the kill procedures, it will be used to demon-
strate the effect of many of these variables.

Influx type.   The type of influx that enters the wellbore as a result of a
kick plays a key role in casing pressure behavior. Although the influx can
range from a heavy oil to fresh water, the most common is gas or saltwater. Each has a pronounced casing pressure curve and different downhole effects.

Gas kicks are generally more dramatic than any other type of influx.
Some of the reasons for this are (1) the rate at which gas will enter the
wellbore, (2) the high casing pressures resulting partially from the low
density fluid, (3) the ability of the gas to expand as it approaches the
surface, (4) fluid migration up the wellbore, and (5) the flammability of

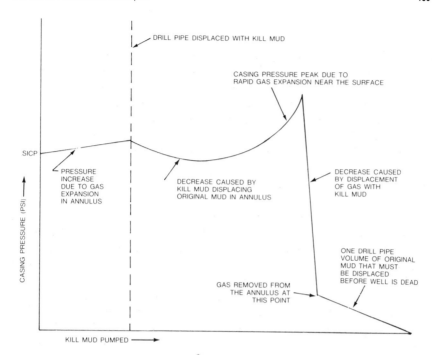

Fig. 2-10 Typical gas kick casing pressure curve for the one circulation method.

the fluid. A typical gas kick casing pressure curve is shown in Fig. 2.10.

Gas expansion as a result of decreased confining pressures as the fluid is pumped up the wellbore affects the kick killing process. Fig. 2.10 illustrates one of these effects. As the gas begins to expand near the surface, the previously decreasing casing pressure begins to increase at an increasing rate. This higher casing pressure may give the false impression that another volume of kick influx is entering the well. Also, immediately after the gas-to-surface conditions, the casing pressure decreases rapidly which may give the impression that lost circulation may have occurred because of the high casing pressures.

Both of these casing pressure changes are expected behaviors and neither indicate an additional influx nor lost circulation. It should be remembered that the possibility of lost circulation is less at gas-to-surface conditions than at the initial shut-in conditions (Figs. 2.8 and 2.9).

When gas expands, the increased gas volume in the hole will displace fluid from the well into the pits resulting in a gain. Fig. 2.11 shows the pit gain for the problem illustrated in Fig. 2.6. This pit gain is in addition to the volume increase from the addition of weight materials. Also, since

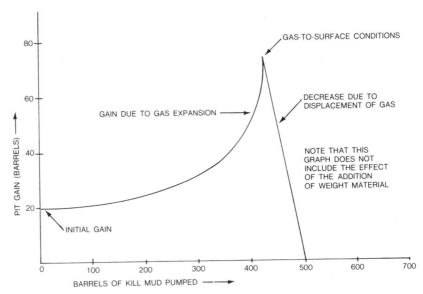

Fig. 2-11 Pit gain curve for the 1.0 ppg kick in Fig. 2-6.

the pit gains in volume, it would be logical to assume that the flow rate exiting the well would increase as is shown in Fig. 2.12.

Gas migration in the annulus may cause special problems. There have been numerous recent studies of this gravity segregation phenomenon in an effort to quantify a migration rate. Although these studies have not yielded a usable solution, field data from a professional well killing corporation suggests a rate of 7 to 15 ft/min in mud systems.[2] Regardless of the rate, the migration effect must be considered because of the gas expansion potential. If the fluid is not allowed to expand properly during the migration period, trapped pressure will be generated at the surface. If excess expansion occurs, additional gas will enter the well from the formation. Example 2.12 illustrates the gas migration phenomenon by using an actual field case.

Example 2.12 While drilling a development well from an offshore platform, a kick was taken. The SIDPP was 850 psi and the SICP was 1,100 psi. Storm conditions forced the tender (barge) to be towed away from the platform to avoid damage to the tender or platform legs. The removal of the tender caused all support services to the platform to be severed, including the mud and mud pumps.

The engineer on the platform knew that the kick would become a severe problem due to gas migration up the annulus. To satisfy the

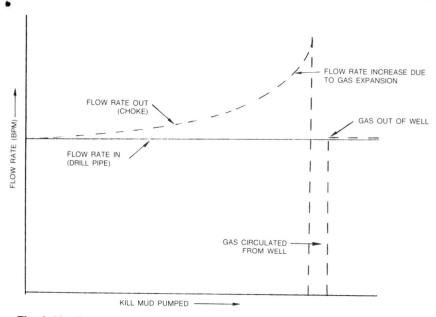

Fig. 2-12 Typical representation of flow rates in and out during a kick killing.

situation, he allowed the migration to build the pressure on the drill pipe to 900 psi which he utilized as a 50 psi safety margin. Thereafter, the migration was allowed to build the SIDPP to 950 before he would carefully bleed a small volume of mud from the annulus to reduce the drill pipe pressure to 900 psi. Since bottomhole pressure was still 50 psi more than the formation pressure, no additional influx occurred. This procedure was continued until all of the gas reached the surface, at which time the pressures ceased to increase and remained at 900 psi. When the support services were restored to the rig, the gas was pumped from the well and kill procedures were initiated.

This example points out the manner in which gas migration can be safely controlled by utilizing the concept of the drill pipe pressure as a bottom hole pressure indicator.

Saltwater kicks do not pose the severe problem of gas kicks because volume expansion does not occur. Also, since saltwater is more dense than gas, the casing pressures necessary to balance the formation pressures are less than for a comparable volume of gas. Fig. 2.13 is a typical saltwater casing pressure curve. It should be noted that shut-in pressures for the 50 bbl saltwater kick are approximately the same as that seen in Figure 2.6 for a 20 bbl gas kick under the same conditions.

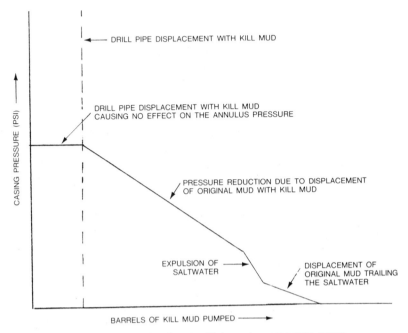

Fig. 2-13 Typical saltwater kick casing pressure curve

Hole stability and pipe sticking considerations are generally more severe with a saltwater kick than a gas kick. The saltwater fluid has a tendency to cause a freshwater mud filter cake to flocculate and create high pipe sticking tendencies and unstable hole conditions. The severity of these problems increases with large kick volumes and with extended waiting periods before the fluid is pumped from the hole.

Volume of influx.   The volume of fluid that enters the well is a controlling variable on the magnitude of the casing pressure throughout the kill process. Increased influx volumes give rise to higher initial SIDPP as well as even greater pressure differences at the gas-to-surface conditions. Fig. 2.14 illustrates a typical influx volume to casing pressure relationship. This representation points out the importance of quick closure rather than hesitation caused by any uncertainties.

Kill weight increment variations.   The original mud density must be increased in most kick situations to kill the well. The incremental density increase has some effect on the casing pressure behavior as seen in Fig. 2.15. In this illustration the usual gas-to-surface pressure conditions which are higher than the original shut-in pressures are observed for 0.5 ppg and 1.0 ppg kicks. However, the 2.0 ppg and 3.0 ppg mud weight increases do not show this tendency with the 3.0 ppg kick having a lower

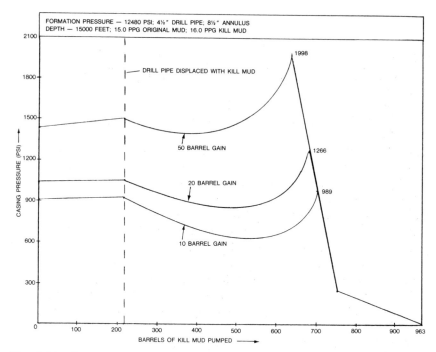

Fig. 2-14 Comparison of casing pressure curves for 10, 20, and 50 bbl kick volumes.

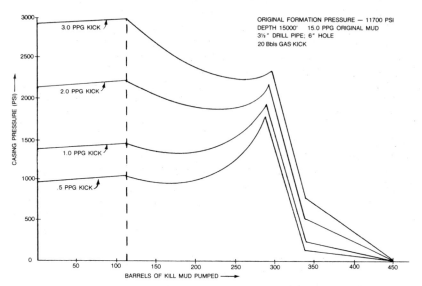

Fig. 2-15 Comparison of several kill mud weight increments

U-Tube Principle — The pressure at the bottom of the casing side must always balance the pressure at the bottom of the drill pipe.

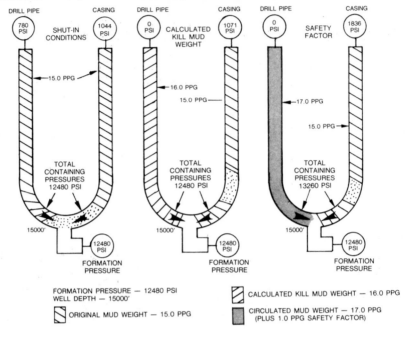

Fig. 2-16 The effect of safety factors (1.0 ppg in this example) causes higher casing pressures than the proper calculated kill mud density

gas-to-surface pressure than at initial closure. This lower pressure is due to the high pressures required which suppress gas expansion, thereby minimizing the associated pressures. This trait is generally observed in kicks requiring greater than a 2.0 ppg incremental increase. (It is interesting to note that the largest known mud weight increment recorded was an 8.5 ppg kick which occurred in South Alabama.)

Another important mud weight variation is the difference between the calculated kill mud weight necessary to balance bottom hole pressure and the mud weight that is actually circulated. If the circulated mud is less than kill mud weight, the casing pressure will be higher than if kill mud had been used because of the necessity to maintain a balanced pressure at the hole bottom. This relationship was observed in Fig. 2.6 and 2.7. In addition to the higher casing pressures, the equivalent mud weights will also be greater which increase the possibility of formation fracture.

On the other hand, circulated mud weights greater than the calculated

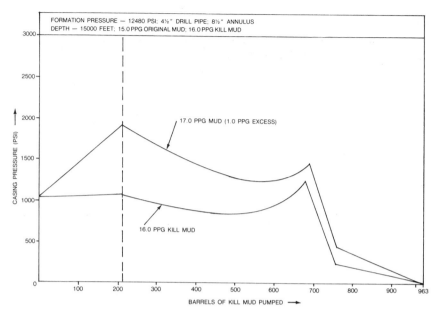

Fig. 2-17 Effect of excess mud weight on annulus pressure

kill mud weight do not decrease the casing pressure. The situation is synonymous with the addition of mud weight safety factors and is often termed "overkill." As the extra heavy mud is pumped down the drill pipe, the casing pressure will increase due to a U-tube effect, as illustrated in Fig. 2.16. This basic principle of the U-tube states that the pressures on each side of the tube must always be equal. This relationship in a kick situation is shown in Fig. 2.17. It must be remembered that these higher casing pressures have associated downhole stresses that increase formation fracture potential.

There have been several attempts throughout past years to achieve the benefits of "safety factors" while avoiding the ill effects of high casing pressures caused by the U-tube effect. The most common attempt at this effort is to subtract the hydrostatic pressure supplied by the extra mud weight increment from the final circulating pressure thereby creating a net zero effect from the added mud weight.

In a static situation, the casing pressure is reduced by an amount equal to the safety factor hydrostatic pressure which would again result in a zero net effect. From the theoretical standpoint, the approach is based on sound principles. However, field experience has shown that this procedure is not practical due to the complexity involved, and the necessary reductions in casing pressure which will often times allow additional influxes of formation fluids. This procedure is not necessary for proper well control and should be restricted to usage only by the most experienced well control engineers.

Hole geometry variations. In practical kick killing situations, there will be hole and drill string size changes that causes the kick fluid geometry to be altered accordingly. This is a particular problem in deep tapered holes where several pipe and hole sizes are used. The influx may occupy a large vertical space at the hole bottom which would create a high casing pressure. As the fluid is pumped into the larger annular spaces, the vertical height is decreased, thus increasing the overall height of the mud column and resulting in lower necessary casing pressures.

Special attention must be given to the kill procedures in this case. Fig. 2.18 (a), (b), and (c) show a typical tapered hole and the associated casing pressure and drill pipe pressure curves.

Implementation of the
One Circulation Method

To implement the one circulation method, certain guidelines must be followed to insure a safe kick-killing exercise. Although the procedure is relatively simple, its mastery demands a basic knowledge of the practical steps taken during the process. Fortunately, there are check points throughout which will indicate any potential problems.

A kill sheet is normally employed during conventional operations. A kill sheet should contain certain prerecorded data, formulas for the various calculations, and a graph or other means of determining the required pressures on the drill pipe as kill mud is pumped. Although many operators employ complex kill sheets, it is necessary for the kill sheet to contain only the basic required kick-killing data. A kill sheet is shown in an example problem in following sections.

Here is a summary of the steps involved in proper kick killing. The sections not directly applicable to deep water situations are noted.

1. When a kick occurs, shut-in the well immediately using the appropriate shut-in procedures.
2. After pressures have stabilized, read and record the shut-in drill pipe pressure, the shut-in casing pressure and the pit gain. (If a float valve is in the drill pipe, use the established procedures to obtain the shut-in drill pipe pressure.)

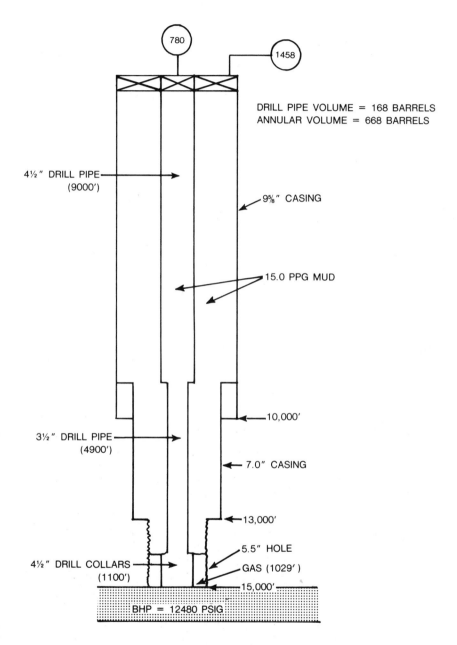

Fig. 2-18 (a) Tapered hole diagram

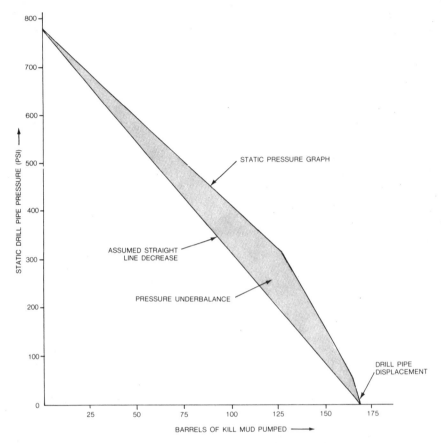

Fig. 2-18 (b) Static drill pipe pressure graph for a typical tapered string. Note the pressure underbalance if a linear depth vs. pressure relationship is assumed.

3. Check the drill pipe for trapped pressure.
4. Calculate the exact mud weight necessary to kill the well and prepare a kill sheet.
5. Mix the kill mud in the suction pit. (It is not necessary to weight up the complete surface volume of mud initially.)
6. After the kill mud has been mixed, initiate circulation by adjusting the choke to hold the casing pressure at the shut-in value while the driller starts the mud pumps. (Not applicable in deep water.)
7. As the driller is displacing the drill pipe with the exact kill mud weight at a constant pump rate (kill rate), use the choke to adjust the pumping pressure according to the required pressures from the kill sheet.

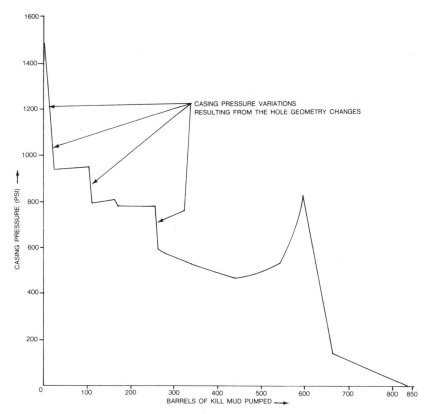

Fig. 2-18 (c) Effect of hole size changes on casing pressure

8. When the drill pipe has been displaced with the kill mud, shut down the pumps, close in the choke and record pressures. The drill pipe pressure should be zero and the casing should have pressure remaining. If the pressure on the drill pipe is not zero, execute the following steps:

(a) Check for trapped pressure using the established procedures.

(b) If the drill pipe pressure is still not zero, pump 10 − 20 additional bbl to ensure that kill mud has reached the bit. The pump efficiency may be reduced at the low circulation rate.

(c) If pressure remains on the drill pipe, recalculate the kill mud weight, prepare a new kill sheet, and return to the initial steps of this procedure.

9. In order to displace the annulus with the kill mud, maintain the drill pipe pumping pressure and pumping rate constant, by using the choke to adjust the pressures as necessary.

10. After the kill mud has reached the flow line, shut down the pumps and close in the choke. The well should be dead at this point. If pressure still remains on the casing, continue circulation until the annulus is dead.

11. When the pressures on the drill pipe and casing are zero, open the annular preventers, circulate and condition the mud, and add a trip margin. In subsea applications, the riser pipe must be killed by reverse circulation down the choke line and up the riser before the annular preventers can be opened.

The author's experience suggests that well control learning experiences are often best accomplished by observation of an actual kick problem. Example 2.13 has been provided for this purpose.

Example 2.13

1. *Pre-kick considerations.* While drilling the R. B. Texas #1 in the Offshore Louisiana Gulf Coast area, ths company representative, Mike Smith, carried out his normal drilling responsibilities related to well control in the event that a kick should be taken. Some of the items that Mike did are listed below:

   a. He read U.S.G.S. Outer Continental Shelf Order #2 and complied with all of the provisions therein.

   b. The barite supplies were checked to insure that a sufficient amount of barite was on board to kill a 1.0 ppg kick if necessary.

   c. The driller recorded on the driller's book that the kill rate was 21 spm and 800 psi pump pressure.

   d. Mike calculated the drill string volume as follows:
   4½-in. drill pipe to 14,000 ft, and 6½ × 2-in. drill collars to 15,000 ft.
   4½-in., 16.6 lb/ft pipe capacity =
   0.01422 bbl/ft × 14,000 ft = 199 bbl
   6½ × 2-in. collar capacity =
   0.0039 bbl/ft × 1,000 ft. = 3.9 bbl
   Total = 199 + 3.9 = 202.9 bbl
   He recorded this information on the kill sheet.

2. *Shut-in and weight up procedures.* The drillers on the rig had just changed tours when a drilling break was observed. The well was checked for flow at which time a flow was recorded with the pumps off. The following steps were immediately taken:

a. The kelly was raised until a tool joint cleared the floor. (A jack-up rig was in use.)
b. The pumps were shut down.
c. The annular preventer was immediately closed.
d. Mike was notified that the well was shut-in.
e. The driller notified his crew in the mud room to stand by in case the mud weight had to be increased.

Mike went to the floor and read his pressures as follows:

SIDPP = 240 psi
SICP = 375 psi
Pit gain = 31 bbl

After checking for trapped pressures, he recorded this information on his kill sheet. From the kill sheet he calculated that he needed to raise his mud weight from the 13.1 ppg original weight to 13.4 ppg.

Mike walked to the mud room to tell the derrickman that he needed 13.4 ppg kill mud when he noticed that the pits were almost full. He knew that the needed barite would raise the mud level, so he instructed the derrickman to pump off a foot of mud, section off the suction pit and increase the weight to 13.4 ppg. Mike did not particularly want to pump off mud but he felt that if would be better to do so at this time than after the killing operation was started.

3. *Pump rates.* The pump output was read from the mud engineer's report as 5.2 strokes per barrel for the 6 × 18-in. duplex mud pump. The volumetric output at 21 spm was 0.1916 bbl/stroke × 21 spm = 4.0 bbl/min. Mike knew that he could cripple his pumps according to the chart previously provided to him but he felt that 4.0 bbl/min was not much more than the recommended 1 – 3 bbl/min as a kill rate.

4. *Kill sheet preparation.* Mike prepared his kill sheet as shown in Figure 2.19.

5. *Working the pipe.* While the mud weight was increased and the kill sheet was being prepared, the driller was instructed to work the pipe every 10 minutes by moving it up and down and then to slowly rotate it. He was also instructed not to move a tool joint through the annular preventer.

6. *Displacing the drill pipe.* After the mud was weighted to 13.4 ppg, Mike was ready to displace the drill pipe. He instructed the driller to start his pumps and run them at 21 spm. Mike cracked open the choke slightly and held his casing pressure at 375 psi until the driller had the pumps at the kill rate. The choke was used to control the drill pipe to decrease gradually according to the values that were on his kill sheet. The pressures were maintained as follows:

PRE-RECORDED DATA

ORIGINAL MUD WEIGHT = 13.1 P.P.G.
SLOW PUMP RATE = 21 SPM AT 800 PSI
DRILL PIPE VOLUME = 203 BARRELS
ANNULUS VOLUME = 700 BARRELS
PUMP OUTPUT = .1916 BARRELS/STROKES

$$\text{DRILL PIPE STROKES} = \frac{\text{DRILL PIPE VOLUME (Bbls)}}{\text{PUMP OUTPUT (Bbl/STROKE)}}$$

$$= \frac{203\ \text{Bbls}}{.1916\ \text{Bbl/STROKE}}$$

$$= 1060\ \text{STROKES}$$

KILL MUD DATA

$$\text{MUD WEIGHT INCREASE} = \frac{\text{SIDPP} \times 19.23}{\text{DEPTH}} = \frac{240 \times 19.23}{15000} = .3\ \text{PPG}$$

KILL MUD WEIGHT = ORIGINAL WEIGHT + INCREASE = 13.1 PPG + .3 PPG = 13.4 PPG

KICK DATA

SIDPP = 240 PSI
SICP = 375 PSI
PIT GAIN = 31 BARRELS
TRUE VERTICAL DEPTH = 15000 FEET

PUMP PRESSURE

INITIAL DRILL PIPE PRESSURE = SIDPP + SLOW PUMP PRESSURE
= 240 PSI + 800 = 1040 PSI

$$\text{FINAL DRILL PIPE PRESSURE} = \frac{\text{KILL MUD WEIGHT} \times \text{SLOW PUMP PRESSURE}}{\text{ORIGINAL MUD WEIGHT}}$$

$$= \frac{13.4\ \text{PPG} \times 800\ \text{PSI}}{13.1\ \text{PPG}}$$

$$= 820\ \text{PSI}$$

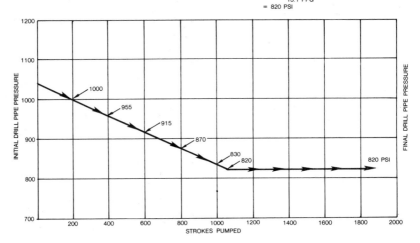

Fig. 2-19   Kill sheet

Strokes	Pressures, psi
200	1,000
400	960
600	915
800	870
1,000	830
1,055	820

When the drill pipe had been displaced, the pump was shut down and the choke was closed. The pressures were then:

SIDPP = 0 psi
SICP = 350 psi

The pressure on the drillpipe told Mike that the heavier kill mud weight was sufficient to kill the well. If it had not been at sufficient density, some amount of pressure would still have remained on the drill pipe.

7. *Displacing the annulus.* Mike was now ready to displace the annulus with kill mud. He initiated pumping by adjusting his choke in order to maintain 350 psi on the casing while the driller started the pumps. After the pumps were running at 21 spm, Mike used the choke to maintain the drill pipe pressure constant at the final circulating pressure of 820 psi. He held this pressure constant until 13.4 ppg mud was observed at the shaker, at which time he closed in the well. Upon reading that both the drill pipe and casing had zero pressure, the choke and then the annular preventer were opened. The well was dead.

8. *Post-kick considerations.* There are several items that Mike considered after the well was dead. He circulated and conditioned the mud in the hole and added a trip margin to the mud weight so that he could make a short trip. Additional barite was ordered from the mud company to resupply the bulk tank. Mike also took the time to inspect all of his equipment to insure that no damage had been sustained from the kick.

Non-Conventional Well Control Procedures

There have been many attempts to develop well control procedures based on principles other than the constant bottom hole pressure concept. These procedures may be based on specific problems peculiar to a geological area. An example of this is low permeability, high-pressured formations contiguous to structurally weak rocks that cannot withstand hydrostatic kill pressures. Oftentimes, however, non-conventional procedures are developed to overcome problem situations encountered as a result of poor well design. These non-conventional procedures, developed for whatever reasons, are not applicable in most situations. Their implications should be thoroughly understood and their use restricted.

Low Choke Pressure Method

The low choke pressure method of well control is based on a simple technique. During a kick killing operation, if the casing pressure should tend to rise above a pre-determined fixed value, the choke will be adjusted, as necessary, to control the pressure at or below the fixed value.

Also, during the initial closure, if the shut-in pressure should rise above a fixed value, immediate pumping will begin and the choke will be adjusted to control the pressure at or below the fixed value. It is intended that the below-necessary low choke pressure will be sufficient to slow the continuing influx into the wellbore until the hydrostatic pressure needed to control the well can be reached through the circulation of the heavier mud.

There is a danger inherent in the low choke pressure method. At any time during a kick-killing operation, if the surface pressure needed to maintain a constant bottom hole pressure equal to formation pressure is reduced in order to avoid exceeding a pre-determined maximum value, an underbalanced situation may occur which will allow a further influx into the annulus. If this underbalanced situation is allowed to continue, the entire annulus will eventually become filled with influx-contaminated mud, which will necessitate high surface pressures if the well is to be killed. These pressures will be higher than any pressures that would have been seen had the constant bottom hole pressure method been followed (Fig. 2.20).

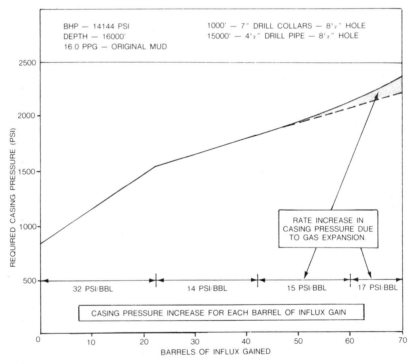

Fig. 2-20 Casing pressures required to close-in various influx volumes

Many variables affect the low choke pressure method and the possible pressure underbalances resulting from its use. Therefore, since a thorough comprehension of each is needed in order to make qualified judgements on the low choke pressure method, a discussion is presented on each variable and its relationship to the method.

Casing pressure to be allowed. The reason for selecting some maximum allowable casing pressure is the basis for the inception of this method of well control. The two criteria usually used in determining this maximum casing pressure are the casing burst pressures and formation fracture pressures. These two pressures are calculated and the lower of the two pressures is designated as the maximum allowable casing pressure.

The calculation of formation fracture pressures take several variables into account. Among these variables are such items as casing depth, formation compressive strength, and mud weight in use. From several authors' works, the formation strength can be approximated for a given set of conditions.[3,4] From this determination, the amount of surface pressure that can be safely maintained without exceeding the compressive strength and fracturing the formation can be approximated according to the following formula:

Calculated maximum allowable pressure =
 (formation fracture pressure) − (hydrostatic pressure)   Eq. 2.8

Example 2.14   If the following data is known, calculate the maximum allowable casing pressure.
Depth = 3,000 ft
Fracture pressure = 13.4 ppg (equivalent)
Mud weight = 10.0 ppg

Solution   (using Eq. 2.8)
Calculated maximum allowable pressure =
 (13.4 ppg) (0.052 psi/ft/ppg) (3,000 ft) −
 (10.0 ppg) (0.052 psi/ft/ppg) (3,000 ft)
 = 2090 psi − 1560 psi
 = 530 psi

If this pressure is lower than the casing burst pressure, it will be labeled as the maximum allowable casing pressure during a kick killing operation.

The calculations of the maximum allowable surface pressures for burst considerations are made similar to those calculations for formation fracture considerations. The maximum allowable surface pressure is the difference between the casing burst rating obtained from a casing catalog minus the resultant hydrostatic pressure. (The resultant hydrostatic pressure at some depth of interest is equal to the hydrostatic pressure of the mud in the casing minus the back-up fluid outside the casing.) The pressure can be calculated according to the following formula:

Calculated maximum casing pressure =
(casing burst rating) − (resultant hydrostatic pressure)   Eq. 2.9

Example 2.15   Using the following data, calculate the maximum allowable casing pressure with respect to casing burst.

Depth = 3,000 ft
Casing burst pressure = 2,730 psi
   (13 ⅜-in., K-55, 54.5 lb/ft)
Mud weight = 10 ppg
Back-up fluid = 9.0 ppg

Solution
(1) Resultant hydrostatic pressure =
    (10.0 ppg) (0.052 psi/ft/ppg) (3,000 ft) −
    (9.0 ppg) (0.052 psi/ft/ppg) (3,000 ft) = 156 psi
(2) Maximum allowable casing pressure
        = 2730 psi − 156 psi
        = 2574 psi

If the burst pressure is lower than the formation fracture pressure, the burst pressure is labeled as the maximum allowable casing pressure during a kill operation.

There may be occasions when the working pressure rating of the surface preventer equipment is the criteria for the pressure limitations. When this is the case, the maximum allowable pressure is equal to the preventer working pressure rating. It should be remembered, however, that if the preventer design criteria presented in Chapter 1 is exercised, it is unlikely that the preventer pressure rating will ever be the limiting criteria.

Well depth.   The depth of the well will effect the low choke pressure method by causing changes of the maximum allowable surface pressure. These pressure changes result from the use of deeper and stronger strings of casing. The depths that these casing strings are set usually have higher fracture pressures which result in the allowance of more surface pressure before formation fracturing will occur. Also, the burst pressures of the deeper strings of casing are normally higher resulting in the allowance of more surface pressure before the burst pressure is exceeded. The variations in the maximum allowable surface pressures due to changes in the well depth are most apparent when considering surface casing in shallow depths versus longer casing strings at deeper depths.

Surface casing in shallow depths have inherent problems that increase the danger of placing allowable choke pressures limitations. The detection of a shallow depth kick is often difficult due to the fast penetration rates and fluctuating pit volumes associated with shallow depth drilling. This difficulty in kick detection may result in a large gain which will require a high casing pressure in order to close in the kick.

On the other hand, the formation fracture pressure usually associated with surface casing is low; consequently the maximum allowable surface pressure is low. If the casing pressure resulting from the large gain exceeds the low allowable surface pressure, the well would not be completely closed in, and the influx would continue. This continuous influx creates a dangerous situation.

The problems associated with deeper wells and longer casing strings are not as severe as the problems associated with shallow depth kicks. Rig personnel are usually more kick conscious when drilling deeper wells, thus resulting in kicks being closed in with less gain in the pits. Also, the maximum allowable surface pressure limitations are higher due to the increased burst pressure ratings of the casing and the increased formation fracture pressures.

However, there are problems peculiar to deeper wells. As the depth increases, the hole size generally decreases with each deeper casing string, resulting in a larger annular ft/bbl ratio. An influx in a small diameter hole would necessitate higher surface pressure than the same volume influx in a large diameter hole due to the longer influx length.

Regardless of the hole geometry, the maximum allowable surface pressure will remain the same since the formation fracture pressure and the casing burst pressure will remain relatively the same. Therefore, a small diameter hole is potentially more dangerous than a large diameter hole since a smaller volume of influx is required to exceed some maximum allowable pressure (Fig. 2.21).

Influx type and effects. The type of influx that enters the well bore will greatly affect the low choke pressure method. Saltwater or oil may tend to promote the effectiveness of this method during a kick, whereas influxes of gas may increase the danger of the potential blowout. Influx type affects the low choke method because of the physical characteristics of the influx itself and the mud contamination it causes. The physical characteristics include volume expansion capabilities, the rate of entry

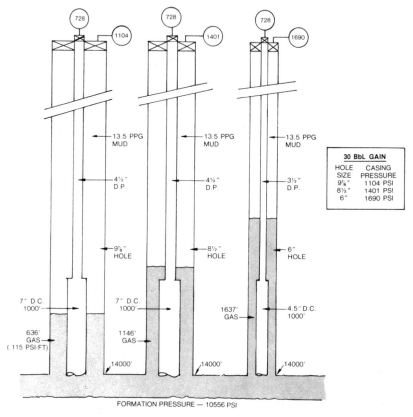

FORMATION PRESSURE — 10556 PSI

Fig. 2-21 Effect of hole size on casing pressures

into the wellbore, and influx density. The mud contamination effects involved are the friction losses that occur as a result of drilling fluid contamination.

Annular friction pressure is the pressure required to move the fluid in the annulus up the borehole. This pressure not only acts upward against the fluid in the annulus, but also against the formation wall. The annular friction pressure increases the total of the pressures on the formation, thus reducing the surface pressure necessary to balance formation pressure. Since annular friction losses are difficult to calculate during a kill operation due to several variables, this pressure safety factor is seldom taken into account.

The annular friction pressure may aid in the kill operation if the low choke pressure method is applied. When the casing pressure is reduced below the amounts needed to balance the formation pressure, as is the case when a surface pressure restriction is imposed, an additional influx

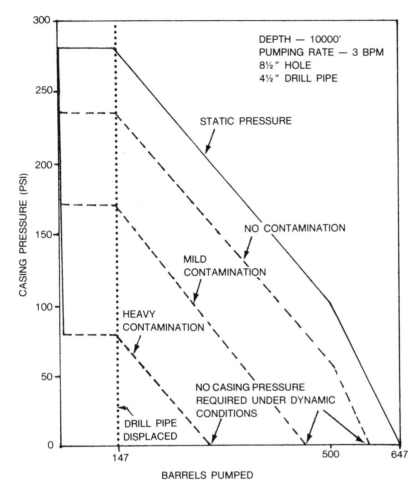

Fig. 2-22   Static and dynamic back pressure curves for saltwater kicks with various degrees of friction loss. The friction loss reduces the necessary casing pressure, in a dynamic system, to zero at some point before the circulation is completed.

will enter the borehole and contaminate the drilling fluid. Contaminated drilling fluids generally have higher friction pressure losses than uncontaminated fluids. (Fig. 2.22). Therefore, if the influx should contaminate the entire mud system, the friction losses may increase to large amounts that may overcome the deficit from the low surface pressure being maintained. This factor may aid in the kill operation, although it should not be considered due to the inability to calculate its magnitude.

The rate at which an influx enters the wellbore and the rate at which any succeeding influxes enter the wellbore directly affect the applicability of the low choke pressure method. Under the same pressure and permeability conditions, salt water and oil will enter the wellbore at a lower rate than gas. Therefore, detection of a salt water or oil kick based on flow rate detection should allow a closure with less total mud pit gain than with a gas kick, due to the lower entry rates. Also, secondary or continuing influxes of salt water or oil will be less during the underbalanced pressure situations than will the continuing influx of gas.

In summation, salt water and oil kicks are less of a danger than a gas kick when the low choke pressure method is applied. This is due to the increased friction pressures, higher fluid pressure gradient, nonexpandable volume, and low entry rates. However, any influx is extremely dangerous, regardless of the type.

Influx volume. The concept of minimizing influx volumes is the single most important factor in determining the applicability, or practicality, of the low choke pressure method of well control. During a kick-killing operation, if the required casing pressure is not allowed to increase as necessary due to surface pressure limitations, a new influx will occur. The size of these successive gains, or influxes, depends upon the influx type, the formation permeability, the amount of pressure underbalance at the formation face, and the duration of the pressure underbalance. Since these factors cannot be determined accurately during the killing operation, it would be extremely dangerous to make the assumption that successive gains would be small, or that these gains can be controlled.

Original kill method applied. The original kill method applied will determine, to a degree, the amount of casing pressure necessary to kill the well. The two workable variations of the constant bottom hole pressure method are the one circulation method and the two circulation method. It has been shown that the one circulation method using the kill mud with no safety factors is the safer method due to lower necessary casing pressures and downhole stresses. Thus, at any time that the low choke pressure method is applied during the use of the two circulation method, the degree of underbalance will be greater than if the one circulation method had been used, due to the higher casing pressures inherent with the two circulation method.

Conclusion. The variables associated with the low choke pressure method are wide ranging and generally not related. Actions of the variables when taken independently, are difficult to predict. Combined, the variables become extremely complicated. Therefore, it is erroneous to assume that a variable such as friction pressure loss would work in a positive manner and nullify any negative actions of variables such as volume expansion, hole geometry, or influx sizes.

Also, the basis of the constant bottom hole pressure method is that pressure will be maintained on the kicking formation equal to or greater than the formation pressure. The low choke pressure method disregards this approach and offers no acceptable substitute. Therefore, this method cannot be considered as a feasible method of well control suitable for general use.

By definition, the low choke pressure method will be applied at any time that the casing pressure should exceed a maximum pressure limitation. This implies that the method will be implemented upon closure if the shut-in casing pressure tends to rise above the fixed value. If this action is taken upon closure, several additional complications must be taken into account. First, the two circulation method must be applied originally since time is not allowed to weight the mud before circulation is begun. Second, no shut-in pressures can be recorded and, consequently, the kill weight mud cannot be accurately calculated. Third, the influx is continuous into the wellbore (Fig. 2.23). As a result of these complications, the low choke pressure method is completely unpredictable if it is used before complete closure.

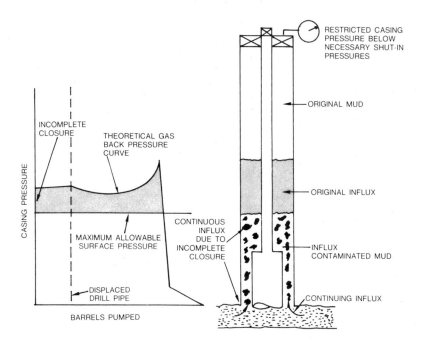

Fig. 2-23 Continuous influx due to implementation of the low choke pressure method before complete closure.

Alternate Solutions

It is not good engineering practice to contradict an attempt at a safety procedure, such as the low choke pressure method, without offering positive alternate solutions. Some solutions are for application prior to drilling. Some are for use before and during an actual kick-killing operation.

Well planning is a vital part of well control. The casing string should be designed with the maximum loading concept, which encompasses all conceivable loads during kick-killing operations as well as during the drilling phase.[5] Also, casing strings should be set in zones that can withstand the stresses imposed by kick conditions. The most important phase of well planning is to insure that drilling will not occur under conditions where a kick cannot be killed safely.

During the kill operation itself, the safest approach is to follow the procedure as dictated by the constant bottom hole pressure method. This minimizes necessary casing pressures and follows a systematic process. Any diversion from this procedure creates more problems and does not permanently solve any existing problems.

Outrunning a kick. Many kicks occur when the drill string is not at the bottom of the well. In these cases, the kill options available to the operator are to strip back to bottom and circulate kill mud, or circulate a very heavy mud at the shut-in position to kill the well and then strip back to bottom. An important point to remember is that the original mud weight was enough to control the well prior to the kick. Therefore, the necessary kill mud weight will be equal to only the original mud weight.

Realizing that the drill string must finally be returned to bottom and that stripping is a tedious operation, occasionally an operator will try to "run to bottom" before shut-in procedures are used. This allows a continuous influx which will necessitate high casing pressures as well as force the handling of large volumes of kick fluids when they are circulated to the surface. Even the running of a few stands of pipe is dangerous and should not be attempted.

The success of this method is dependent on formations with low permeabilities that will allow only small volumes of influx into the well. This is generally an unknown variable and should not be used as the basis of a well control procedure.

Constant pit level method. The constant pit level method of well control stems from a logical principle. During a well killing procedure, the choke was adjusted in a manner to avoid any changes in the mud pits at the surface. It was believed that if the pits were not allowed to gain, no additional influx would occur. Also, since the pit volume would not be allowed to decrease, lost circulation problems would be avoided.

This procedure worked successfully when the influx was saltwater or oil. Since these fluids have minimum expansion capabilities, the pit volumes would not change appreciably. However, gas kicks should have associated gains in the pits. If these gains are not allowed, the casing pressure must be increased to force gas compression, which may result in lost circulation caused by the high casing pressures.

Pumping a kick back into the formation. Occasionally an attempt is made to "pump a kick back into the formation" to avoid the necessity of implementing other kick killing procedures. This should not be misconstrued as practical in all situations since it implies that the formation must be fractured before the pumping can occur. It is not probable that the kick fluid will reenter the original zone unless clear water is the circulating fluid since pore channel plugging with barite and bentonite will occur when mud is in use.

However, in limited situations such as certain hydrogen sulfide kicks, it may be advisable to pump the fluid into a formation rather than circulate it to the surface in the event that proper well planning has not been performed.

Problems

2.1 Calculate the pressure reduction when 450 ft of 4½-in. 16.6 lb/ft pipe is pulled without filling the hole. The hole diameter is 7⅞-in. and 15.6 ppg mud is in use.

Solution: 43.1 psi

2.2 Using the data in Problem 2.1, what would be the pressure reduction if 450 ft of 6.0 × 2.0-in. collars are pulled without filling the hole?

2.3 A well is drilled to 13,000 ft with 13.2 ppg mud. The bottom hole pressure in a gas sand at that depth is 8,710 psi. The intermediate casing is 43.5 lb/ft, 9⅝-in. pipe set to 11,000 ft. The drill pipe is 4½-in., 16.6 lb/ft, and the collars are 7 × 2-in. diameter. The operator requires that the hole be filled after 5 stands of drill pipe or drill collars are pulled. Will the well kick when pulling the drill pipe? Drill collars? (Assume no swabbing effects.)

2.4 Calculate the pump strokes required to fill the hole in Problems 2.1, 2.2 and 2.3 for each of the following pumps. Assume a 90% efficiency in all cases.
   (a) 6½ × 18-in. duplex
   (b) 7 × 14-in. duplex
   (c) 4 × 10-in. duplex
   (d) 3 × 6-in. triplex
   (e) 3 × 10-in. triplex
   (f) 4 × 8-in. triplex

Solution: (a) 2.1 — 14 strokes

2.5 In an effort to measure the output of a 3 × 10-in. triplex pump under actual conditions, a 20 bbl trip tank was filled with mud. A total of 1,274 strokes was required to empty the tank. What is the pump output? What is the efficiency percentage of the pump?

2.6 A drill string contains 12,000 ft of 4½-in., 16.6 lb/ft drill pipe and 1,000 ft of 7 × 2.5-in. collars. How many strokes would be required to displace the pipe if a 4½ × 10-in. triplex pump is used at 90% efficiency? At 80% efficiency?

2.7 A kick was taken on a well in which these pressures were recorded:

Shut-in time, minutes	SIDPP, psi	SICP, psi	Pit gain, bbl
0	250	375	18
5	290	415	18
10	290	415	18
15	295	415	18

(a) What is the true shut-in drill pipe pressure?

(b) The drill string contains 10,600 ft (TVD) of 12.1 ppg mud. What is the bottom hole pressure?

Solution: (a)   290 psi

(b) 6,960 psi

2.8 Using the data given below, what are the true shut-in pressures?

Increment no.	Bleeding volume, bbl	SIDPP, psi	SICP, psi
0	–	760	1,160
1	1	700	1,100
2	1	640	1,040
3	½	630	1,020
4	½	630	1,020
5	½	630	1,030
6	½	630	1,045

Solution: SIDPP =   630 psi

SICP = 1,045 psi

2.9 Given the solution from Problem 2.8, what is the bottom hole pressure for the following situations?

Well no.	True vertical depth, ft	Mud weight, ppg
1	10,750	10.4
2	13,500	14.6
3	8,300	9.5
4	15,000	11.7
5	5,500	9.9

2.10* After a kick had been taken, bleeding procedures were implemented to check for trapped pressure. Using the results given below, what are the true shut-in pressures?

Increment no.	Bleeding volume, bbl	SIDPP, psi	SICP, psi
0	–	250	690
1	1	175	615
2	1	100	540
3	1	50	490
4	1	20	470
5	1	0	450
6	1	0	450
7	1	0	460

2.11* Using the answer from Problem 2.10, what is the bottom hole pressure if the mud weight was 12.1 ppg and the well depth was 13,130 ft?

2.12 A kick was taken on a well in which a float valve was used in the drill string. The SIDPP was read as 0 psi and the SICP was 675 psi. The kill rate and associated pressure was 32 spm at 700 psi. After implementing the procedures to establish the true SIDPP (Table 2.5), the total pumping pressure was 1,150 psi. What was the SIDPP?

Solution: 450 psi

2.13 Using the data given below, calculate the bottom hole pressure.
SIDPP = 0 psi (float valve in the drill string)
SICP = 400 psi
Kill rate = 45 spm at 750 psi
Total pumping pressure (initial) = 900 psi
Depth (T.V.D.) = 11,000 ft
Mud weight = 12.9 ppg

2.14 While drilling the Texas Rover #1, a kick was shut-in with a SIDPP of 400 psi and a SICP of 550 psi. A kill rate had not been established previously. The pumps were started and run at 21 spm while the casing pressure was maintained at 550 psi. The total drill pipe pressure was observed to be 1,250 psi. What is the pumping pressure at 21 spm?

2.15 Using the same conditions given in Problem 2.14, the pumps were run at 35 spm which attained a total pumping pressure of 1,500 psi. What is the kill pump pressure at 35 spm?

2.16 While killing the kick in Problem 2.14, pump #1 washed out a valve while displacing the drill pipe with kill mud. The well was shut-in and the pressures recorded as 175 psi SIDPP and 525 psi SICP. Rather than repair pump #1, pump #2 was started at 45 spm while the casing pressure was held at 525 psi. The total drill pipe pressure at 45 spm was observed to be 1,475 psi. What was the kill rate for pump #2?

2.17 A kick occurred on a well in which the kill rate was not known and a float valve was used in the drill string. The SIDPP was 0 psi and the SICP was 500 psi. A low volume, high pressure pump was connected to the stand pipe and pressure applied. The following results were obtained. What was the SIDPP?

Volume pumped, bbl	SIDPP, psi
0	0
1	0
2	40
3	90
4	140
5	190
6	210
7	215
8	215
9	220

2.18 Using the results from Problem 2.17, the rig pumps were subsequently run at 25 spm with a total pressure of 950 psi. The casing pressure was held constant throughout this procedure. What was the kill rate pressure?

2.19 What is the probable kick influx fluid using the following data: (Assume the kick fluid to be around the drill collars.)
SIDPP = 400 psi
SICP = 600 psi

Pit gain = 25 bbl
Mud weight = 12.0 ppg
Drill collars = 6-in. OD
Hole size = 9⅞-in.

Solution: Gas (0.14 psi/ft)

2.20 Using the data from Problem 2.19, what is the probable influx if the original mud weight was 17.0 ppg?

2.21 If the following data is known, what fluid type entered the well? (Assume no drill collars.)
SIDPP = 800 psi
SICP = 1,400 psi
Pit gain = 20 bbl
Mud weight = 15.0 ppg
Drill string = 3½-in.
Hole size = 6 in.

2.22* Using the same data from Problem 2.21, what fluid entered the well if the SICP was 1,100 psi?

2.23 A kick was taken on a well in which the following data was known. What was the kill mud weight?
SIDPP = 250 psi
SICP = 475 psi
Measured depth = 12,750 ft
True vertical depth = 12,000 ft
Mud weight = 13.4 ppg

Solution: 13.8 ppg

2.24 Calculate the kill mud weight for Problem 2.7.

2.25 If the information from Problem 2.8 is known, what must be the mud weight increase if the depth (TVD) is 12,300 ft?

2.26 Calculate the kill mud weights for the situation developed in Problems 2.8 and 2.9.

2.27* What mud weight increase is necessary for Problem 2.10?

2.28* If the following data is known, what is the mud weight increase necessary to control each situation?

	SIDPP, psi	SICP, psi	Pit gain, bbl	Depth, TVD ft
(a)	300	500	30	14,200
(b)	300	4,750	108	14,200
(c)	450	800	18	21,630
(d)	300	500	unknown	14,200
(e)	300	unknown	35	14,200
(f)	unknown	500	35	14,200
(g)	0	500	35	14,200

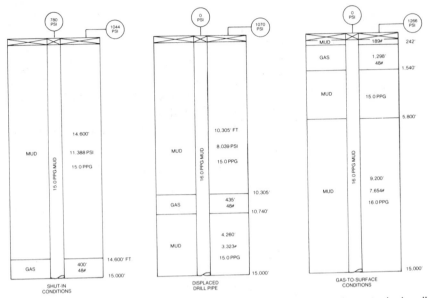

Fig. 2-24 Fluid arrangements for various pumping intervals during a typical well control operation

2.29* Using Fig. 2.24, calculate the equivalent mud weights at 1,000, 2,500, 5,000, 7,500, 10,000, 12,500 and 15,000 ft for each pumping interval.

2.30 Prepare a kill sheet using this kick data:
Original mud weight = 13.6 ppg
Slow pump rate = 25 spm at 750 psi
Drill pipe = 11,000 ft, 4½-in., 16.6 lb/ft
Drill collars = 1,000 ft, 6.0-in. OD × 2.0-in. ID
Depth = 12,000 ft (measured depth)
          11,500 ft (TVD)
Pump = 6½-in. liner, 16-in. stroke, 2.5-in. rod, 90% efficiency
                                                            (duplex)

SIDPP = 300 psi
SICP = 600 psi
Pit gain = unknown

2.31 Calculate and plot a drill pipe pressure schedule (kill sheet) for the following kick:
Original mud weight = 10.5 ppg
Slow pump rate = 30 spm at 900 psi
Drill pipe = 8,000 ft, 3½-in., 13.3 lb/ft
Drill collars = 500 ft, 5.5-in. OD × 2.0-in. ID

Depth = 8,500 ft (TVD)
Pump = 3-in. liner, 8-in. stroke, 80% efficiency (triplex)
SIDPP = 250 psi
SICP = 400 psi
Pit gain = 15 bbl

2.32* Prepare a static drill pipe pressure graph for the following circumstances.
SIDPP = 800 psi
Depth = 14,000 ft (TVD)
Drill pipe = (1) 12,000 ft, 4½-in., 16.6 lb/ft
(2) 1,500 ft, 3½-in., 13.3 lb/ft
Drill collars = 500 ft, 5.0-in. OD × 1.5-in. ID
Pump = 6-in. liner, 18-in. stroke, 90% efficiency (duplex)

References

1. Personal Communication from Louis Records, Louis Records and Associates, Inc., Lafayette, Louisiana.

2. Personal Communication from Charles M. Prentice, Prentice and Records Enterprises, Inc., Lafayette, Louisiana.

3. Eaton, Ben A.: "Fracture Gradient Prediction and its Application in Oilfield Operation," *Journal of Petroleum Technology,* (October, 1969) 1353–1360.

4. Matthews, W. R. and Kelly, J., "How to Predict Formation Pressures and Fracture Gradient," *Oil and Gas Journal*, February, 1967.

5. Prentice, Charles M., " 'Maximum Load' Casing Design," SPE. 2560 presented at the Fall Technical Conference, Denver, Colorado, 1969.

3

Special problems

Deepwater Well Control

WELL CONTROL PROBLEMS CAN OCCUR during deepwater drilling. Although the mechanics of well control are not altered, the methods of implementation may be significantly different due to the equipment involved when drilling in deepwater.

Most of the deepwater problems can be divided into the categories of (1) kicks occurring without a sufficient amount of casing to allow implementation of conventional killing procedures, and (2) those kicks occurring with a sufficient amount of casing to allow closure of the well. The two problems that fall into both of these categories are kick detection and reduced fracture gradients.

Kick Detection

Two of the early warning signs of kicks are an increase in flow rate and an increase in pit volume. These signs are difficult to detect when drilling in deep water due to the nature of the drilling vessel and natural wave motion. Generally, a floating vessel does the drilling in deep water. Wave action moves the drilling vessel and this creates pit level fluctuations even though the total pit volume may remain constant. The same difficulty is observed in detecting flow rate changes leaving the well.

Equipment that compensates for heave-caused fluctuations will minimize these problems in early kick detection. A pit volume totalizing (PVT) tool will detect and report overall pit gains by using multiple pit monitors and resolving the individual losses and gains reported by each monitor into a single value. Fig. 3.1 compares a single monitor system to a multiple monitor system and the response given by each under several conditions.

Fracture Gradients

In well control operations it is essential that the open hole formation exhibits a competency to allow the well to be killed without losing circulation. This implies that the formation fracture gradients must be greater than the equivalent mud weights gradients during the kill operations to prevent formation fracture which induces an underground blowout.

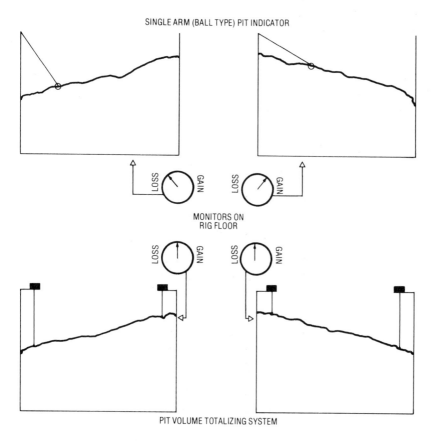

Fig. 3-1 Pit level fluctuations due to rig heave and associated pit monitor readings (mud volume is the same in all pits). The PVT system accounts for the level changes and indicated true pit volumes.

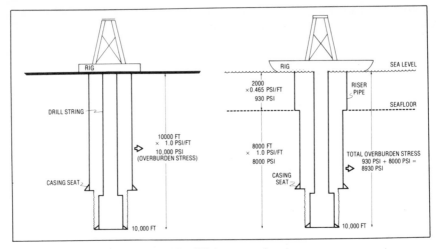

Fig. 3-2 Offshore overburden

Fracture gradients in deep water are effectively less than those observed on land or in shallow water at equivalent drilling depths. This reduction is due in part to the lower overburden stresses in the deepwater case because of the water depth and the depth gradation from clay-silt and sand to shale and sandstone.

Fig. 3.2 illustrates this concept. The overburden stress at 10,000 ft is 10,000 psi. In the deepwater case at the same depth it is 8,930 psi. Although this example does not encompass all parameters, it does illustrate the fracture gradient reduction problem.[1]

Several authors have examined the procedures for the calculation of fracture gradients.[2,3] Until recently, however, their work has been confined to land operations. Charts and equations are presently available to estimate fracture gradients in various water depths. Fig. 3.3 shows a chart prepared by Kendall to estimate fracture gradients in these situations.[4] Calculation procedures for fracture gradients in deepwater situations will be presented in Chapter 5.

Surface Hole (Shallow) Kicks

The most severe well control problem of drilling in deepwater is that of a kick occurring when no protective casing is set, i.e., when drilling the surface hole. At these shallow depths, the formation fracture gradient is low and generally will not withstand shut-in pressures without incurring an underground blowout. Kicks of this type present several unique problems not generally associated with kicks that occur when a protective casing string is set.

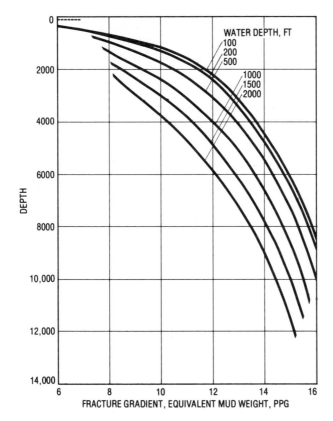

Fig. 3-3 Effect of water depth on fracture gradient

The most dangerous of these problems is possible loss of the vessel due to (1) loss of buoyancy, (2) aerated water entering open sections of the vessel resulting in flooding or (3) gas fires. If an underwater blowout does occur, gas may be expelled directly beneath the vessel which will reduce its mobility.

When kicks have occurred in this situation, it has been observed that the mud in the riser pipe conduit from the sea-floor to the ship may be evacuated and replaced with low density gas. The collapse resistance of the riser pipe, lessened by pipe tension and bending, may fall below the seawater hydrostatic pressure that acts on the OD of the riser and may result in the collapse of the pipe (Fig. 3.4).

To offset this problem, many operators have established policies to insure that the riser pipe is not completely evacuated either by allowing the mud pumps to continue running or by introducing seawater into the riser through a valve at the bottom of the riser.

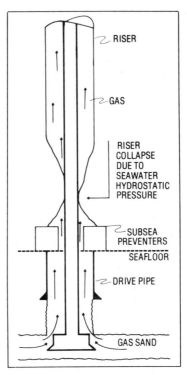

RISER

GAS

RISER
COLLAPSE
DUE TO
SEAWATER
HYDROSTATIC
PRESSURE

SUBSEA
PREVENTERS

SEAFLOOR

DRIVE PIPE

GAS SAND

Fig. 3-4   Riser pipe collapse during shallow gas blowout

Reservoir description.   Naturally occurring abnormal pressures require a pressure entrapment strata above the reservoir. Since this type of strata is generally not found at these shallow depths, the pressure in the reservoir is assumed to be normal for its respective depth. This does not hold true for zones that are pressure-charged as a result of underground blowouts in the immediate area.

The volume of the reservoir has been observed to be small in comparison to commercial gas reservoirs. Its volume, however, is generally sufficient to blowout from several hours to long periods of time in some cases. Since the permeability of the rock in these zones is usually high, a small volume of gas can cause a problem due to the rates at which it can feed into the wellbore.

Emphasis has been placed upon prior detection of these zones, which until recently was done by trial and error. Early seismic work had difficulty in locating the potential trouble zones largely due to the small areal extent of the reservoir and poor seismic resolution at shallow depths. In a history matching effort, seismic data in known blowout cases has often shown small anomalies that are assumed to represent the zone of interest. Extrapolation of the information into future work should tend to minimize the problems. Since seismic data resolution in shallow areas has improved, the reservoirs can be more accurately defined.

Well control measures. Kicks in shallow zones cannot be shut-in and killed using conventional procedures due to the low fracture gradient of the formation. The well control procedure often used in these cases is that of diversion of the kick to allow the reservoir to deplete or bridge. This method has been successful due to the relatively small reservoir volumes that will deplete in a short amount of time.

The equipment used for diverting the kick has been presented in Chapter 1. The basic components of the system are a flow diverting mechanism which is usually an annular preventer or diverter bag, and one or more large diameter diverter lines. When two lines are used, they are generally placed opposite each other to take advantage of the wind direction and divert gas away from the rig. The diverter lines should be as large as possible. Six inch diameter lines are considered a minimum. Valves controlling these lines should be opened before the diverter bag or annular preventer is closed.

Another technique proposed to kill a surface kick is to spot a high density slug of mud on the hole bottom immediately upon detection of the kick. This procedure is designed to kill the kick without exceeding the fracture gradient. Although this technique has been used numerous times on land operations, the parameters involved in deepwater drilling may complicate the procedure.

Inflatable drilling packers have been used to enclose the kick at the flowing zone. The packer is placed on lower sections of the drill string and is designed to be expanded and seal the borehole in the event of a kick. This procedure prevents an uncontrolled fluid flow into the wellbore without fracturing shallow formations. The problems associated with packers are the difficulty in attaining a packer seal in washed out zones, and the time delay between the detection of the kick and the inflation of the packer.

A technique to detect potential problem zones and to avoid kicks of this type has been to drill small diameter pilot holes. The procedure developed was based on the causes for most shallow kicks. Since it has been theorized that most formations at those depths are normally pressured, the blowouts must occur as a result of large quantities of core volume, gas cut mud at the surface. This effectively reduces the hydrostatic pressure a sufficient amount to allow formation fluids into the wellbore. To avoid creating large quantities of core volume gas in the mud stream, a small diameter pilot hole is drilled at low rates. After the pilot hole is completed, the hole diameter is increased with a reaming tool while the operator exercises the necessary precautions.

Kicks Below Protective Casing

When protective casing is set at a depth sufficient to achieve fracture gradients in excess of equivalent mud weights caused by the kick, conventional procedures can be implemented to kill the potential blowout. There are procedure differences however that cause deepwater well control to be more complicated and demand precise control techniques.

Shut-in procedures in deepwater drilling are different than the procedures used with stationary rigs such as jack-ups or land rigs. In deep waters, a floating drilling vessel moves vertically according to wave motion. This ship motion may cause drill string movement which will eventually cause wear failure of stationary subsea preventers. (Motion compensation tools are generally not considered in well control situations even though they are essential to safe drilling.)

To offset this problem, a procedure has been developed in which a set of pipe rams is closed and a drill string tool joint is allowed to rest on the rams. These rams will support the full weight of the drill string while holding full test pressures. This allows the drill string to remain motionless resulting in no damage to the preventers while the killing operation is in progress.

Basic blowout preventer design logic, as presented in Chapter 1, has been extended to subsea preventers. However, since repair and replacement of worn elements is impractical during a killing operation, several components are added to the stack to provide a back-up system should the primary element fail.

The basic stack shown in Fig. 1.40 includes two spherical preventers rather than a solitary unit. Primary and secondary choke lines are used. Shear rams are placed near the bottom of the stack should it become necessary to shear the drill string before moving the drilling vessel from the location in the event of an impending blowout. The lower pipe rams are used should the upper choke lines or flanges fail which would result in the need to close in the well without necessarily shearing the drill string.

Choke line length.   The length of the choke line has become a critical consideration in deepwater drilling for several reasons such as associated friction pressures, and small diameters and volumes which can allow complete fluid displacement to occur in a short time. Preventers used on land and in shallow water often have the same problems but to a much smaller degree. These problems in deepwater systems will neces-

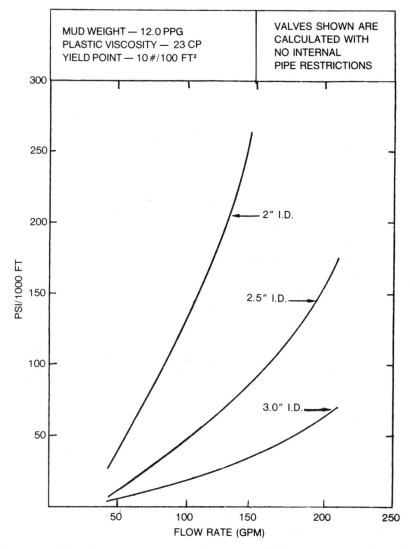

MUD WEIGHT — 12.0 PPG
PLASTIC VISCOSITY — 23 CP
YIELD POINT — 10 #/100 FT²

VALVES SHOWN ARE
CALCULATED WITH
NO INTERNAL
PIPE RESTRICTIONS

2" I.D.

2.5" I.D.

3.0" I.D.

PSI/1000 FT

FLOW RATE (GPM)

Fig. 3-5 Effect of choke line size and flow rate on friction pressure gradients. No allowances are made for internal restrictions.

sitate slight alterations in the kill procedure and force the operator to exercise caution during well killing operations.

The friction pressures in long choke lines has been called the "hidden choke effect". As can be seen from Fig. 3.5, it would take 233 psi to pump at a rate of 200 gpm through a 2.5-in. ID choke line used in 1,500 ft of water under the given conditions. This pressure is required at the base of

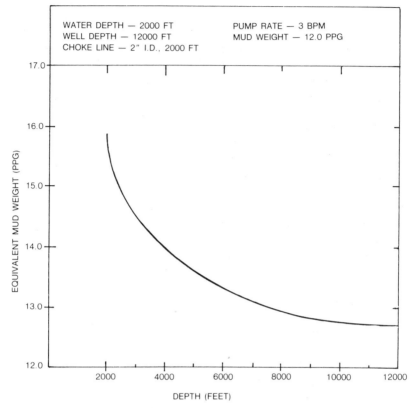

Fig. 3-6   Equivalent mud weights due to choke line friction

the choke line but is expended within the line and is never recorded at the surface choke gauge.

Friction pressure is applied at every point below the preventers, as can be seen in Fig. 3.6. This increases the possibility of formation fracture. The friction pressure consideration becomes more severe when one accounts for the restrictions in the choke line caused by non-internal flush connections.[5]

A procedure should be developed to ensure that the "hidden choke" friction pressures are not transmitted to the formation during kick killing operations. This is usually done by establishing the kill rate through the open riser and not through the choke line. This procedure effectively subtracts the choke line friction pressures from the total circulating pressures while still maintaining the balanced bottomhole pressure concept.

Although the kill rate will be taken through the riser, it is advantageous to know the choke line friction pressures. Should a kick occur, these pressures must be known to determine a pumping pressure properly if one had not been previously established. If the friction pressures are not known, use of the procedures presented in Chapter 2 to determine a pumping pressure will yield a circulating pressure that includes these pressures.

It is recommended that several kill rates be established daily. The rig pumps should be used to determine pressures at low rates in the range of 1–3 bbl/min and also at the rate of approximately ½ bbl/min. Since some rig pumps will not run properly at such low rates, it is advisable to determine these rates with another pump (such as a cementing unit). In portions of the kill procedure, it may become necessary to pump at these low rates. A recommended practice concerning the cementing pumps when used in well killing operations is to ensure that a remote pump throttle and stroke counter is placed on the rig floor to allow better communications between the pump and choke operators.

Due to its length, choke line displacement is perhaps the most severe problem encountered in deepwater well control. As fluids of one density

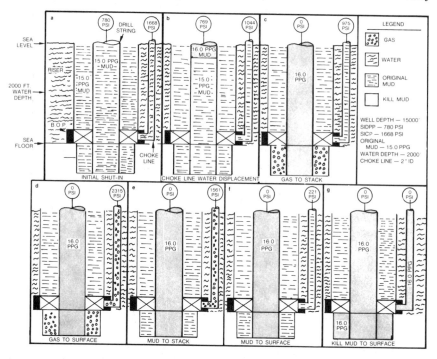

Fig. 3-7 Choke line displacement with mud and gas. Static pressures are shown for illustration.

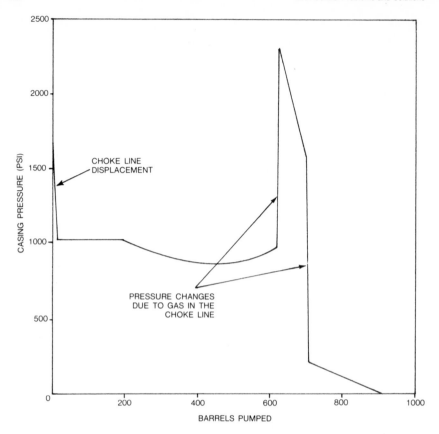

Fig. 3-8 Complete casing pressure curve for the situation shown in Fig. 3-7.

displace fluids of another density in the choke line, the effective hydrostatic pressures in the annulus will change rapidly. This will necessitate sudden and drastic choke adjustments. This is essentially a two part problem based on (1) the frequently large casing pressure adjustment that must be made, and (2) the short time allowed to make the choke adjustments due to the rate at which the line is displaced.

Some of the practical solutions offered to minimize the severity of this problem are

- to displace the line at extremely low rates using a low volume output pump,
- to use both choke lines to increase the displacement time,
- to employ large diameter lines that will increase the volume to be displaced, and

- use a subsea choke and allow the effluent to enter the riser and divert at the surface.[6]

Several of these solutions present associated problems that also must be solved.

Fig. 3.7 illustrates the displacement problem. The initial displacement occurs when pumping commences and mud enters the choke line. This can be remedied, however, by reverse circulation of the choke line with

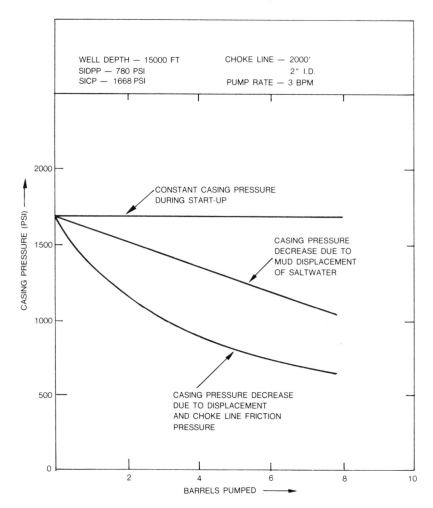

Fig. 3-9   Effect of holding the casing pressure constant during the initial stages of the circulation shown in Fig. 3.8

the lower pipe rams closed before the initiation of the kill operation. The major problems though are noted near the end of the kill procedure when gas enters and then is displaced from the line. The severity of the short time allowed for the pressure changes can be reduced by the procedures previously presented, but in most cases the magnitude of the pressure changes cannot be lessened.

A casing pressure curve of the example from Fig. 3.7 is shown in Fig. 3.8. In this graph, the effect of the mud initially displacing the saltwater in the choke line is shown for illustration. Although the theoretical curve shows large pressure changes at various points, the problem is more severe in field cases because the mud and gas do not enter the choke line as fluids with distinct leading and trailing boundaries as shown in Fig. 3.8. Since the invading fluid will probably be mixed with the mud, numerous displacements may occur and result in several severe casing pressure changes.

Casing pressure changes that occur from the displacement may necessitate a variation in the initial circulating procedures. In normal operations in shallow water or land operations, the casing pressure is often held constant for a short time upon starting the rig pumps to allow the drill pipe pressure to stabilize before beginning the drill pipe pressure control procedures. If this technique is employed in deepwater, large overbalances would quickly occur which would tend to cause formation fracture. As an example, in Fig. 3.9 the amount of overbalance would be 943 psi if the casing pressure were held constant for 2 min.

Fig. 3.10 shows another problem inherent to deepwater well control. After the well is dead, the annular preventer cannot be opened immediately because the riser contains original density mud. It is necessary to close the bottom set of pipe rams and reverse circulate kill mud density through the choke line and riser in order to completely kill the system. A secondary gas bubble may be circulated out during this process if the preventer stack trapped gas above the primary choke line during the initial kill procedure.

Kicks Following Cementing

Kicks and blowouts have been experienced on wells immediately following apparently successful cementing operations. Some of the many problems associated with this type of kick are the seemingly random nature with which the kick occurs, the identification of the flowing zone, and the inability to circulate a heavy fluid and kill the well. Although surface blowouts in this situation are dramatic, underground blowouts resulting from gas-through-cement kicks can also

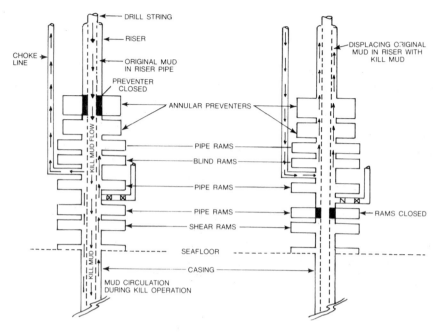

Fig. 3-10 Reversing procedures to completely kill the well

cause problems such as pressure charging other zones that will affect offset drilling and the loss of hydrocarbon reserves.

One operator used a noise log to estimate that as much as 300 Mcfd transferred between zones as a result of the problem in one field case.[7] These kicks, however, are not random in nature and, by exercising certain precautions, their occurrences can be minimized.

Gas-through-cement kicks. Assuming that a cement operation was completed without such problems as previously existing gas cut mud, lost circulation, or cement channeling, then the circumstances to allow gas migration and resulting kicks to occur have been established by several authors.[8,9] These circumstances can be illustrated by Fig. 3.11.

One or more permeable zones should exist above a gas-bearing interval with the upper zone(s) having pressures lower than those in the gas zone. The cement must set on the upper zones and support the drilling fluid hydrostatic pressure. After the cement has set, the pressure within the lower interval must be reduced below formation pressure at which time gas flow may occur. If the seal from the upper set cement interval cannot prevent permeation by the gas, the flow will continue at an increasing rate due to the reduction in overall hydrostatic pressure in the annulus.

The causes for the cement setting on the upper zones prior to setting on

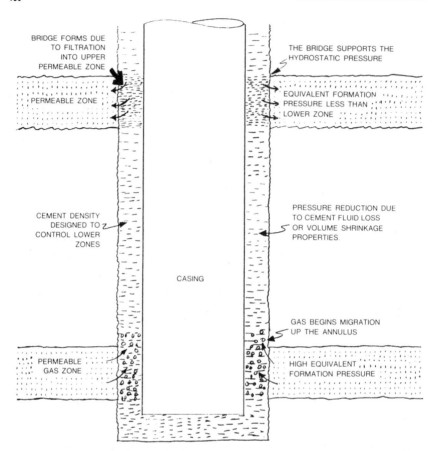

Fig. 3-11 Illustration of the circumstances involved in gas-through-cement kicks

the lower zones must be understood in order to establish guidelines for reducing the gas flow. Cement setting is a function of many variables, one of which is temperature. Studies have shown that the maximum circulating temperatures are not at the bottom of the well as might be suspected but rather at about ⅓ the distance from the bottom.[10] This would tend to cause the cement to set initially above the hole bottom. If formation pressures increase as shown in Fig. 3.12, the larger differential pressures on the upper zone would promote setting of the cement.

Once the cement has set on the upper zone and begins to support the hydrostatic pressure partially, pressures within the cement must be reduced. A common cause for this reduction is the water lost form the slurry. If the water loss is sufficient to allow the fluid pressure to fall below formation pressure, a kick can occur.

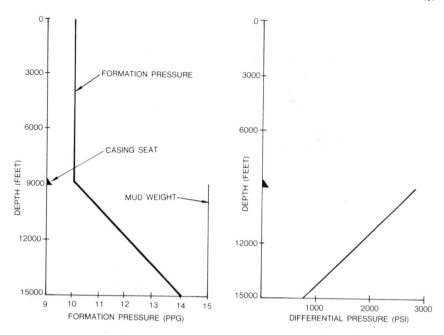

Fig 3-12 High mud weights to control formation pressure may cause large amounts of differential pressure at shallow intervals

Another cause for a pressure reduction in the cement is due to the expansive-shrinkage characteristics of the cement. Tests have shown that cement may have a tendency for initial shrinkage when downhole conditions of temperature and pressure are applied [11] (Fig. 3.13). If shrinkage occurs, the pressure will be reduced because the cement volume will be less than the cement-sealed fixed borehole volume. To offset the shrinkage, commercial cement additives are available that cause the cement to expand from initial conditions without any shrinkage. (Fig. 3.14).

Reducing gas-through-cement kicks. Proper cementing procedures should be exercised during the operation to ensure that the kicks are not a function of mechanical problems. These procedures include the use of cementing aids such as centralizers, proper hole conditioning prior to cementing, and the release of surface pressure after cement placement to avoid the formation of a microannulus. The cementing operation should be monitored to ensure that lost circulation does not occur.

The cement slurry should be tailored with chemical additives to avoid the causes of kicks that were presented in the previous section. Retarding agents should be blended with the cement batches to ensure that the

182

Well Control Problems and Solutions

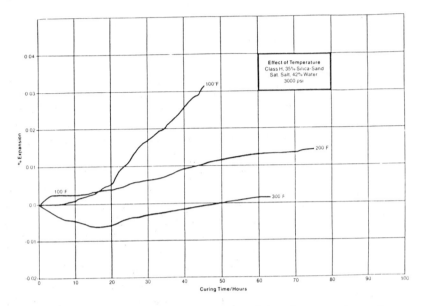

Fig. 3-13 Expansive-shrinkage characteristics of a cement sample.

slurry will set from the bottom to the top. Additives should be used to prevent cement volume shrinkage. The fluid loss from the cement should be reduced as low as possible.

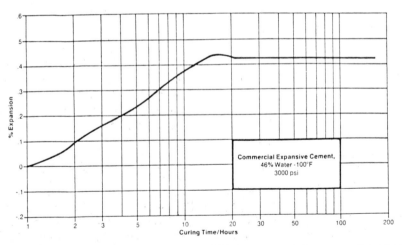

Fig. 3-14 Effect of certain commercial additives on expansive properties of cement.

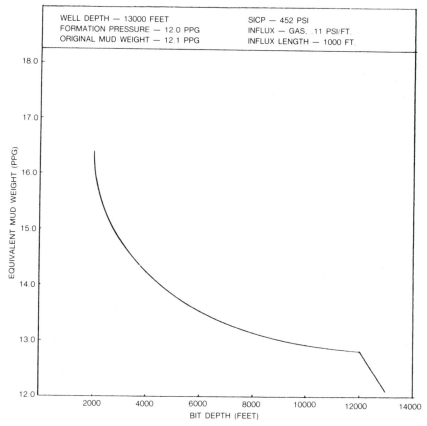

Fig. 3-15 Required kill mud weights at various bit depths for a kick occurring while tripping.

Stripping and Snubbing Operations

When the drill string is pulled from the hole, kicks can occur as a result of swabbing or improper hole fillup. Since the bit is not on the bottom in these cases, kill mud densities may be high and in some cases will be in excess of practical ranges (Fig. 3.15). When this occurs, it is necessary to lower the drill string into the well while maintaining the proper surface pressures to avoid an additional influx of formation fluids. The proper well control procedure under these circumstances is to "strip" or "snub" into the well with the blowout preventer closed. Extreme caution must be exercised if an attempt is made to run the pipe to the bottom of the hole without closing the preventers.

Stripping and snubbing is the process in which the drill string is moved within the well to achieve some specific purpose. The general case occurs when the pipe is lowered into the well to kill an induced kick, while some instances will demand that the pipe be pulled from the hole to perform some operation. Whatever the case may be, special equipment and techniques must then be used to control the well.

The difference between stripping and snubbing is based on the manner in which the kick pressures act on the drill string and the amount of drill string in the well before the kick was closed in. If the resultant upward force exerted by the kick pressures acting on the horizontal surfaces of the drill string exceeds the weight of the drill string, the pipe must be snubbed into or out of the well. If the weight is greater than the upward force, the drill pipe must be stripped (Fig. 3.16).

Snubbing Equipment

In snubbing operations into the well, downward forces must be applied to the drill string and will require the use of special equipment. This equipment may be mechanical or hydraulically operated and will often consist of special blowout preventers for pressure control.

Mechanical snubbers.   This type of equipment utilizes the rig system to force the pipe into or out of the hole (Fig. 3.17). The snubbing equipment consists of a set of traveling snubbers to force pipe movement

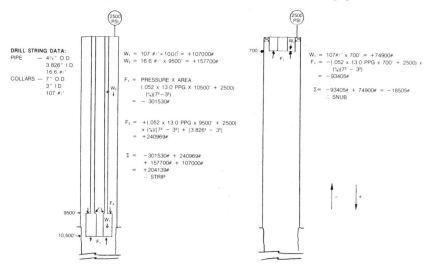

Fig. 3-16   Illustration showing the difference between stripping and snubbing. Note that the spherical preventer friction forces have been neglected to simplify the calculations

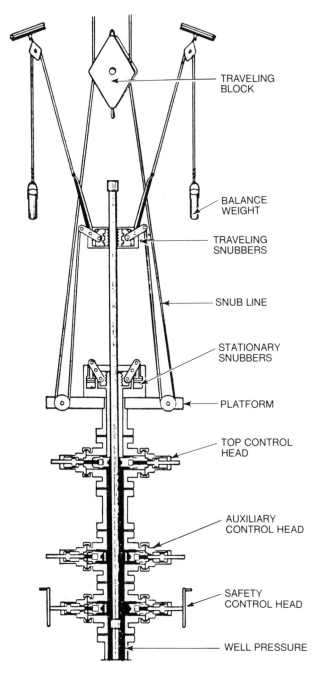

Fig. 3-17 Mechanical snubbers

under well pressure and a set of stationary snubbers to prevent pipe movement when the traveling snubbers are released. A wire line (snub line) passes over the center sheave of the traveling block to clamp to the handle of the traveling snubbers. This line is used to transmit a balanced force downward to the traveling snubbers as the block is raised and eliminates the tendency to bend the pipe by applying equal downward forces. Balancing weights automatically raise the traveling snubbers for a new section of pipe when the traveling block is lowered.

The stationary snubbers are the lowermost section on the snubbing assembly and are bolted to the upper control head. The stationary snubbers are designed, as are most blowout preventers, to take advantage of the well pressure. As the well pressure increases the upward force applied to the pipe, the grip of the snubbers increases accordingly.

The control head assembly which is used to provide a pressure seal between the casing and pipe consists of an upper control head, an auxiliary control head, a safety control head, and a control manifold. The control heads, which are essentially ram type preventers, are hydraulically operated for rapid opening and closing of the rams. The upper and auxiliary control heads employ Garloc Fivrous Hycar packing because it has less friction drag during pipe movement and is easier to replace than the standard rubber rams used in the safety control head.

The manner in which the control head assembly is arranged depends on the configuration of the stack of drilling blowout preventers in use. The control head assembly will usually be added to the preventer stack in place with the safety control head being flanged to the spherical preventer. An auxiliary control head and a spool are flanged above the safety control head with the top control head above the spool. The spool is used to provide space between the upper and auxiliary control heads to allow for the lubrication of drill pipe tool joints through the head and may vary in length to permit lubrication of special tools such as packers.

The control head assembly is designed to prevent the inadvertent opening of all the control heads at the same time. The rams in either the auxiliary control head or the safety contol head must be closed and the pressure between that head and the upper control head released before the upper control head can be opened. The upper control head must be closed and the pressure equalized below it before the lower heads can be opened.

The operation of the mechanical snubbing equipment to allow pipe movement is through the "hand over hand" principle. After the equipment has been installed and pressure tested, the first joint of drill pipe with a back pressure valve is lowered into the upper control head and the rams are closed. The stationary and traveling snubbers are both engaged on the pipe.

The control assembly is pressurized and the drilling preventers are opened. The traveling snubbers are disengaged from the pipe and moved to a point approximately four ft above the stationary snubbers and engaged. The stationary snubbers are released and the pipe is forced into the well by raising the block. When the traveling snubbers reach a point immediately above the stationary snubbers, the lower snubbers are engaged and the traveling snubbers are moved up the pipe to grip a new section. This process is repeated to continue pipe movement into the well.

When the upper end of a tool joint is encountered, the traveling snubbers are used to position the tool joint immediately below the stationary snubbers and above the upper control head. The auxiliary rams are closed, and the upper rams are opened before the tool joint is forced into a position between the two rams. The upper rams are closed and the lower rams are opened to allow for continued stripping until another tool joint is encountered.

The process of snubbing is continued until the weight of the pipe in the hole equals the upward force caused by well pressure acting on the cross sectional area of the pipe. At this point, often called the "snubbing point," the pipe moves easily into the well without having to be snubbed. The snubbing assembly is removed, and the block and elevator are then used to proceed with the stripping process.

Hydraulic snubbers. The hydraulic snubbing unit was developed for application in areas where snubbing was necessary for well control but when a drilling rig was not on the well. The hydraulic unit attains the same end result as the mechanical snubber but is self-contained and therefore does not require any rig assistance (Fig. 3.18).

The pipe movement capabilities of the unit are supplied by a hydraulic jack that generates approximately 350,000 lb lifting capacity and 100,000 lb snubbing capacity. The jack may be a single, large cylinder system or a multi-cylinder system, depending on the manufacturer, and may have a stroke length of 10 – 15 ft for the short stroke jack and approximately 36 ft for the long stroke jack. Units are available that can safely handle pipe diameters from ¾ in. to 7⅝ in.

Two basic types of sealing devices are in common use for sealing the OD of the pipe while working under pressure with the hydraulic snubbing unit. These are ram type blowout preventers and the solid rubber element strippers. Stripper elements are generally considered adequate for pressure control up to 2,500 – 3,000 psi. These elements are con-

Concentric Hydraulic Workover Unit Operating Schematic.

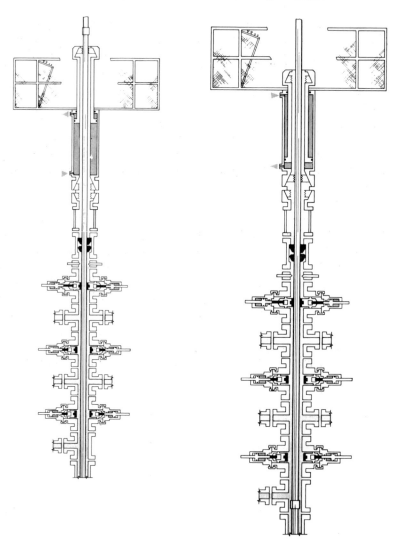

Piston extended and traveling
slips closed prior to forcing pipe
into well.

Piston retracted and traveling slips
open before piston is
again extended.

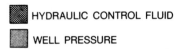 HYDRAULIC CONTROL FLUID

WELL PRESSURE

Fig. 3-18   Hydraulic snubbing unit (Courtesy Otis Corp.)

structed of solid synthetic rubber compounds. They have the ability to stretch as couplings, and some downhole tools are stripped through them. The useful life of the stripper elements will depend on the external condition of the tubing and may range from 10,000 – 20,000 lineal ft at pressures of 3,000 psi or less.

Stripping preventer rams are of solid steel construction, incorporating rubber seals with fiber or polyurethanic replaceable stripper inserts. Since it is necessary to open these rams in order to allow couplings or other diameter changes to pass through the blowout preventers, a second set of rams is employed to seal the well while the other is opened to allow tool joint passage. This process of alternately opening and closing the working blowout preventers requires deliberate stop and start motions by the operator.

In a typical operation, the lower working rams are closed on the pipe and the upper rams are opened. Well pressure is held by the closed rams while the joint is lowered by the hydraulic unit. When the coupling enters the equalizing chamber, the upper ram is closed. The equalizing chamber has two valves hydraulically operated from the operator's console. One of these valves vents the chamber to the casing pressure below the lowermost standby preventer. The other valve vents the chamber to the atmosphere through a bleed-off line. Before opening the lower working ram to allow the tool joint to pass, the proper valve on the chamber is operated and casing pressure is vented into the chamber. The pipe can again be moved downhole.

Snubbing and Stripping Control Procedures

Regardless of the equipment used during the pipe movement process, proper pressure control procedures must be exercised to ensure that additional formation fluids are not allowed into the well and also that excessive pressures that would fracture the formation or burst the casing are not maintained. The two processes most often used are the volumetric and the pressure methods. Although the volumetric method is easier to comprehend initially, the pressure method may be more accurate due to its ease in implementation.

Volumetric method. As the pipe is moved into the well, the pressures observed at the surface will tend to increase due to the compression of fluids in the sealed wellbore. If the compression is allowed to continue, the pressures will eventually build to a value that will fracture the formation. To compensate for this compression, a volume of mud that is equal to the volume of pipe forced into the well is allowed to escape at the surface. The volume of pipe will be the displacement of the string as well as the capacity since a back pressure valve is used that prevents fluid movement up the pipe (Example 3.1). As a section of pipe is lowered into the well, the choke is used to allow a volume of mud from the well equal to the total displacement of the pipe, which is the displacement plus the capacity.

> *Example 3.1* A kick developed on a well while the pipe was being pulled for a new bit. If the pipe is to be stripped into the well using the volume method for well control, what amount of mud should be allowed to escape for each 93-ft stand of pipe run into the well; each 31-ft joint? Use the data given below.

Pipe size $= 3\frac{1}{2}$ in.
Pipe weight $= 13.3$ lb/ft
Pipe displacement $= 0.0054$ bbl/ft
Pipe capacity $= 0.007421$ bbl/ft
Solution
(1) Total displacement $=$ displacement + capacity
$\qquad\qquad\qquad\qquad\quad = (0.0054 + 0.007421)$ bbl/ft
$\qquad\qquad\qquad\qquad\quad = 0.012821$ bbl/ft
(2) For each 93-ft stand
$\qquad\qquad\qquad\qquad\quad = 93$ ft $\times\ 0.012821$
$\qquad\qquad\qquad\qquad\quad = 1.192$ bbl/93-ft stand
(3) For each 31-ft stand
$\qquad\qquad\qquad\qquad\quad = 31$ ft $\times\ 0.012821$
$\qquad\qquad\qquad\qquad\quad = 0.397$ bbl/31-ft stand

Example 3.1 shows the problems associated with the volumetric method. In the example, running the 93-ft stand required that 1.192 barrels of mud leave the well to provide proper control. The difficulty arises in maintaining the mud exit at this value and not some slightly larger amount due to improper choke operation. In some reported field cases, a volume 50% larger than calculated volume was allowed to escape from the well causing additional influxes into the well and resultant larger surface pressures.

Pressure method.   Surface pressures are those needed to balance the bottom hole formation pressure and prevent further influxes of formation fluids. The pressure method for stripping and snubbing employs the same concept with the exception that dynamic pressures are substituted for the static pressures imposed by the blowout preventers. This method provides more accurate fluid control as well as being applicable for pipe

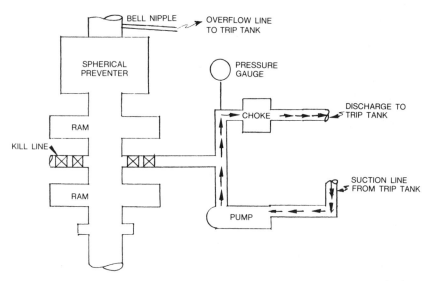

Fig. 3-19   Diagram of equipment used in the pressure method for stripping

movement into and out of the hole, while the volume method is most applicable when going into the hole.

An equipment diagram necessary for the pressure method is shown in Fig. 3.19. The equipment used in this procedure should be available on most drilling rigs and therefore requires only a small amount of special preparation. The pump output is directed through the choke and is designed to supply the pressures necessary for control purposes. The downstream side of the choke returns the mud to a trip tank where it is picked up the pump. A small return line may be connected between the bell nipple and the trip tank to retrieve any mud that escapes through the preventers during tool joint passage.

The procedure is implemented by starting the pump and using the choke to control pressures at a value slightly greater than the well pressure. The control valve is opened and pipe movement commences. Since the confining pressures are greater or equal to those necessary to control the well, no additional influx occurs. The pressure method can be monitored throughout the process to assure its effectiveness by recording the volume increases in the trip tank. These increases should be exactly equal to the calculated total pipe displacement as shown in Example 3.1.

The primary advantage of the pressure method over the volume method is the manner in which the fluid is allowed to escape at the surface. In the volume method, the procedure alternates between a static and dynamic state, whereas the pressure method is in a dynamic state throughout the process.

The pressure method is applicable for pipe movement in either vertical direction. The volume method is essentially limited to pipe movement into the well.

Fluid migration up the hole must be taken into account when the kick fluid is gas. The migration will result in volume expansion causing additional fluid expulsion at the surface and larger necessary confining pressures. To compensate for this with the pressure method, the choke pressure should be increased by small increments (± 50 psi) when it is

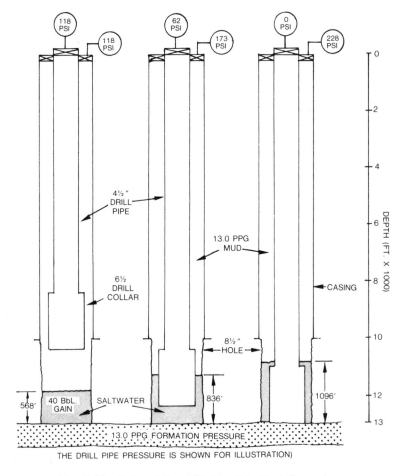

Fig. 3-20   Entry of the drill string into the influx column

observed that the original casing pressure is no longer sufficient to control the well. If the volume method were used, the same pressure increases would be noted.

Entry into the influx column will cause the confining pressures to increase due to elongation of the fluid column (Fig. 3.20). When this entry is noted at the surface, pipe movement into the well should continue until the pressures approach a safe maximum level or until the pipe is moved through the influx, at which time the kick fluid should be circulated from the well.

When stripping to the bottom has been completed, the drilling fluid should be circulated to clean the hole. The density necessary to control the well should be the same as that used prior to the occurrence of the kick. Although a trip margin of mud weight is necessary to control swab pressures, no additional mud weight for pressure control is needed. This point is often misunderstood and has resulted in losing many wells due to fractured formations resulting from the circulation of excessive mud weights after the stripping process was successful.

Handling Drill Pipe Under Pressure

Occasionally pressure will be trapped in the drill string. Some of the more common problems are (1) removal of the kelly under pressure to install a valve or additional drill pipe, (2) below a fish when pulling a joint of pipe with possible pressure trapped below the tools or (3) removal or repair of malfunctioning equipment under pressure. Handling of the drill string under these conditions becomes dangerous and must be approached with caution. The processes often used to solve these problems are the valve drilling and hot tap process and the freeze process.

Valve drilling and hot tap. When equipment such as valves or sections of the drill string has pressure trapped beneath or within, some means must be used to bleed the pressure before safe handling techniques can be used. If conventional equipment cannot be made operable under these conditions, special tools must be employed. The valve drilling and hot tap process is designed to meet this requirement by drilling entry ports into the pressured equipment. The term "hot tap" means entry under pressure.

The equipment often used in this special service is shown in Fig. 3.21. The tool consists of a bit drive shaft adaptable to hand operation or power tools; a ratchet assembly to apply pressure to the bit by transmitting a downward pull on the drill shaft; a rod clamp acting as a pressure point on the rod shaft; a stuffing box to pack off the drive shaft; a bleed-off valve through which pressure can be equalized, bled off, or for circulation, a quick union for ease in make-up and disassembly; a full opening plug valve that can be used to close upon removal of bit and drive shaft; and a saddle clamp to adapt to concentric objects.

An alternate clamp can be used with the tool to drill gate valves, plug valves, or ball plugs that have malfunctioned. Standard tapping equipment as shown is rated to 10,000 psi and has experienced 9,000 psi in field service. The valve drilling equipment has a 15,000 psi rating and has been exposed to 12,500 psi in the field.[12]

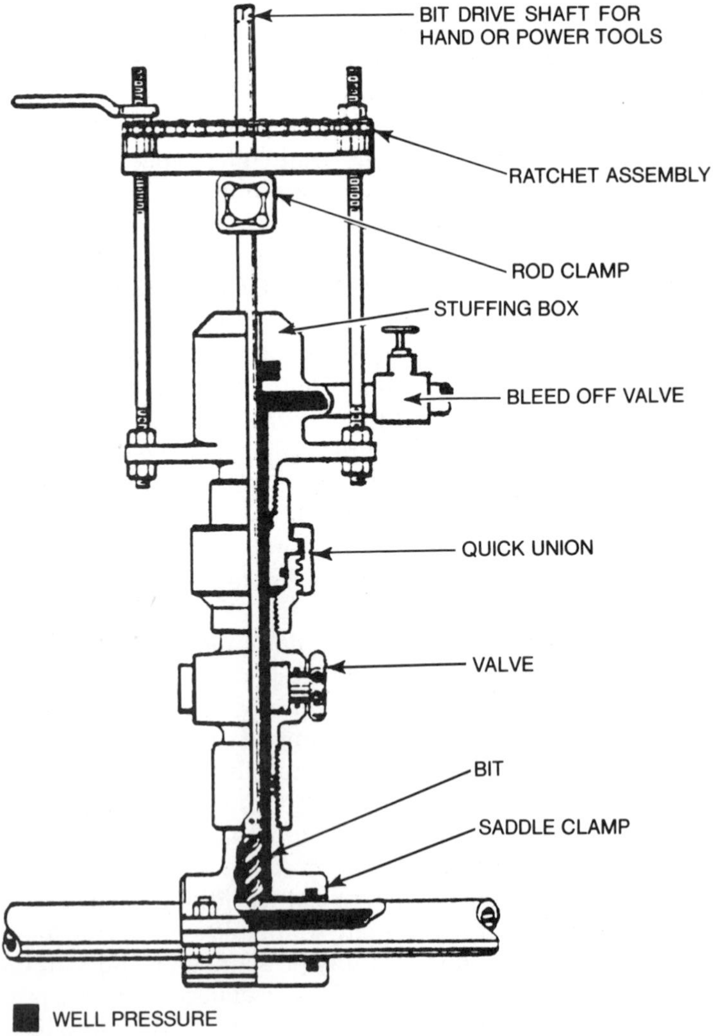
BIT DRIVE SHAFT FOR
HAND OR POWER TOOLS
RATCHET ASSEMBLY
ROD CLAMP
STUFFING BOX
BLEED OFF VALVE
QUICK UNION
VALVE
BIT
SADDLE CLAMP
WELL PRESSURE

Fig. 3-21 Schematic of valve drilling equipment

Freeze process. In some cases, simple entry into the pressured equipment does not offer the complete solution. An example is the case of drill pipe under extreme kick pressures, since it would be impossible to bleed off the pressures through the hot tapping process. The freeze process has been developed to offer additional solutions to some of these problems. The process has been used successfully in such cases as (1) the need to remove the kelly to install a valve, (2) below a pressured fish or (3) below blowout preventers which have failed partially or have been damaged.

The process involves the use of specially prepared dry ice in a sufficient quantity to freeze a solid plug or a bridge of ice to allow for the safe removal of equipment above the plug. The procedure involves wrapping the pipe with a container of dry ice and allowing the fluid within the pipe to freeze. A test sample is often obtained and frozen in a pup joint on the rig floor to determine the proper setting time. The dry ice causes the fluid in the pipe to reach approximately $-142°F$, which will develop a plug that can withstand as much as 15,000 psi differential pressure.

Some of the primary requirements and recommendations for the successful execution of this process are that the pipe must contain a static water-based fluid, the pipe should not be frozen in tension unless necessary, and plastic coated pipe should not be frozen. Specialists in this field should be consulted before attempting the process.

Drill Stem Testing

When drilling a well, it is often desirable to test various formations for potential productivity to determine the future course of the well. This testing, generally known as drill stem testing, can aid in the selection of casing setting depths and also can be used to avoid setting expensive and unnecessary strings of pipe. Drill stem testing may be used to collect formation fluid samples or to perform pressure testing on a zone to determine its extent and flow capabilities. Although the testing is beneficial in decision-making, it does pose special well control problems.

The equipment used in drill stem testing usually consists of the drill pipe, a packer, and surface control systems such as chokes and special manifolds. The packer is designed to isolate the formation of interest from upper open hole sections and should have the capabilities of allowing flow from the formation up the drill pipe or between the drill pipe and annulus with the formation isolated. The manifold is used to change the direction of the flow as necessary to either the drill pipe, flare area, or pumping units.

The principles used in drill stem testing rely on a hydrostatic pressure reduction below formation pressure of the test zone while maintaining a constant pressure on the other open hole sections. This is generally accomplished by placing a packer immediately above the test zone and then reducing the hydrostatic pressure in the drill pipe by one of several methods. The pressure reduction will allow the formation fluids to enter the wellbore and flow through the drill pipe to the surface. After the test is completed, the packer is manipulated in a manner that isolates the formation and allows the formation fluids in the drill pipe to be displaced with drilling fluid (Fig. 3.22).

Flow testing through the drill pipe presents several problems. As low density formation fluids displace the pipe, the static surface pressures must increase to maintain control of the well (Example 3.2). In actual flow testing of the well, the equipment may begin to leak and expose formation fluids at the surface. The problem of safe handling of large volumes of oil and gas at the surface must be taken into account.

Example 3.2 Assume an oil-bearing zone exists at 13,000 ft with a formation pressure equivalent to 14.0 ppg mud. If the oil has a specific gravity of 0.5, what will be the static surface pressure during a drill stem test?

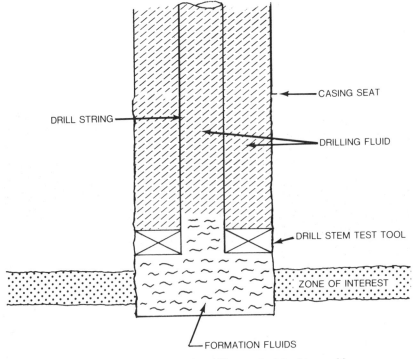

Fig. 3-22 Diagram of a drill stem test tool in position

Solution
1. Formation pressure = 0.052 × 14.0 ppg × 13,000 ft
   = 9,464 psi
2. Oil hydrostatic pressure
   = 0.052 × 0.5 × 8.33 ppg × 13,000 ft
   = 2,815 psi
3. Surface pressure = formation pressure − oil hydrostatic
   pressure
   = 9,464 psi − 2,815 psi = 6,649 psi

After the testing has been completed, the well must be killed before drilling is resumed. The kill procedure involves displacing low density fluids from the drill pipe as well as below the packer. The most difficult part of the procedure is displacing the fluid below the packer because of the inability to circulate drilling fluid in this interval. If this fluid is not completely removed from the well while the packer is in the hole, it may cause a kick while pulling the packer from the hole or when drilling resumes.

Drill String Impairment

During the course of normal kick killing operations, drilling fluid is circulated down the drill string and up the annulus. This flow path uses the high strength drill pipe to withstand the circulating pressures from pumping kill weight mud. Problems may occur that impair the usefulness of the drill string and necessitate immediate remedial procedures. These problems may be a hole, or washout, in the drill pipe or bit and an obstruction or plug that prevents circulation.

Hole in the drill string. A hole in the drill string is commonly called a washout. The hole is generally the result of drilling fluid erosion of the metal and may occur at tool joint connections that were pitted or not properly made up, jet ports in the bit, an isolated metal impurity within the string, or a fracture or crack not previously detected. Proper equipment inspection will locate most of the causes for washouts before they become a serious problem.

The most common indicator of a washout while drilling is not suitable for use in well control. When a washout occurs during normal drilling, a pump pressure decrease will be noted. This indicator is not as applicable in well control because of the fluctuating pressures used during the well control procedures. As an example, an increase in casing pressure necessitated by a washout may be erroneously considered to be caused by gas expansion or by a change in hole geometry.

In well control, continued growth of a washout may eventually sever the drill string. This is a serious problem particularly if the pipe should part near the surface.

Another problem presented by washouts is that of the reduction of the circulation path for the kill mud. If the hole occurs near the surface early in the kill procedure, the heavy mud will exit the drillstring at this point and flow up the annulus. This exit leaves the annular volume below the hole filled with original mud and kick fluids, with no procedure to displace either. If the hole occurs near the bottom of the well, the seriousness of the problem is diminished.

If the washout continues to grow in size, control procedures will be more difficult and dangerous. The increased hole size will decrease the amount of required circulation pressure for a given rate. To control the well properly, a new kill rate and pressure must be established periodically to account for the hole size change. If this is not done, excess drill pipe pressure will be used and will tend to promote formation fracture.

Remedial procedures. An important step in solving this situation is locating the washout. If the point of interest is near or at the bit, the primary concern is using the correct kill pumping pressure as the washout erodes. On the other hand, if the washout is near the surface, the operator must immediately execute procedures to avoid a severed drill string as well as the reduction in the circulation path.

One of the more common methods used to determine the depth of the hole is by observing the number of strokes of kill mud pumped when the mud prematurely reaches the surface. Although there are cases in which this procedure will indicate depths below the actual washout, it is the most accurate of the available methods that can be used without logging tools. Example 3.3 illustrates the procedure.

Example 3.3 During the course of kick killing operations on a well, the kill mud reached the surface after 1,200 strokes had been pumped. Using the following data and the tables in the Appendix, what is the deepest possible depth of the washout?

Pump = 16-in. × 5-in. duplex (90% efficiency)
Drill pipe = 4½-in, 16.6 lb/ft, 12,000 ft
Drill collars = 7 × 2-in., 1,000 ft
Casing = 9⅝-in., 40.5 lb/ft, 10,000 ft
Open hole = 8½-in., 3,000 ft

Solution
1) From the Appendix:
   - the pump output is 0.1020 bbl/stroke.

- the drill pipe capacity is 0.0142 bbl/ft.
- the annular capacity is 0.0562 bbl/ft.
2) The volume pumped when the mud reached the surface was:
1,200 strokes × 0.1020 bbl/stroke = 122.4 bbl
3) The drill pipe capacity plus the annular capacity above the washout will equal the volume pumped.
(0.0142 + 0.0562) bbl/ft × depth of interest = 122.4 bbl
or depth of interest = 122.4 bbl/0.0704 bbl/ft = 1,738 ft

Another field-developed procedure used to locate the zone is to note the number of strokes required to pump a plugging material to the washout. The plug material is usually a rope or the equivalent that can be entangled on the sharp edges of the washed out hole. When the plug reaches the hole, it should cause a pump pressure increase which can be recorded at the surface.

Solving the problem of a washout in the drill string is difficult. The solution depends on the location of the hole, the severity of the problem, the time at which it occurs in the overall kill process, and the type of kick influx. Control procedure decisions must be made immediately upon detection of the washout to minimize the possibility of further complications.

A hole in the drill string near the surface may be remedied by stripping the pipe from the hole and replacing the bad joint(s). If the drill pipe does not contain a back pressure valve, it will be necessary to plug the pipe with a mechanical plug or a pill of viscous, high density mud before stripping operations are initiated. If the hole is located some distance below the surface, the time required to strip out and replace the joint must be weighed against the rate of migration of the influx fluid.

Many operators have attempted to solve the problem by plugging the washout with rope or a similar material. The plugging material is pumped down the drill pipe where it will attempt to exit at the hole. The material may snag on the hole and plug the washout which will then allow fluid movement down the drill pipe. If the plug is successful, it is advisable to use low pump rates through the remainder of the kill procedure to avoid pumping the plugging material out of the hole.

Plugged drill pipe. Occassionally the drill string will become completely plugged during the kill process. The most common cause is due to barite plugging as a result of adding large volumes of the weight material at the surface without adding sufficient suspension agents such as gel (bentonite). When this occurs, it is necessary either to remove the plug or provide a circulation path other than through the bit.

When plugging occurs, it may be possible to reverse flow up the drill pipe and free the plug. If this is not successful, the plug can perhaps be jarred loose by surging the pipe with pressure from the pumps. Although this procedure may form a tighter plug, it is occasionally successful.

If attempts to remove the plug fail, an alternate fluid circulation path must be provided. The drill string may be perforated or severed with a cutter to provide the path. The primary concern is to perforate or cut at the deepest possible interval. This will usually be at the top of the drill collars. Caution must be given to the size of the perforation charge when the perforating interval is opposite a casing string to avoid perforating the drill pipe and the casing.

Many operators have exercised initial preventive measures relative to jet plugging by using a primer cord charge on wireline to blow the jets out of the bit. This procedure is useful when coarse materials are to be pumped and time permits a logging unit to rig up and shoot the drill string.

Problems

3.1 What amount of mud should be allowed to escape from a well during a stripping operation for each 93 ft stand of 3½-in. OD, 15.5 lb/ft drill pipe?

Solution: 1.182 bbl

3.2 If 7-in. OD × 2.5-in. ID drill collars are to be snubbed into a well, what amount of mud should escape for each foot of collars forced into the well?

3.3 A kick was shut in after the complete drill string was removed from the well. Using the following data, what would be the total volume of mud displaced by the drill string when it reaches the bottom of the hole?

Drill collars = 6-in. OD × 2-in. ID, 1,200 ft
Drill pipe = 4½-in. OD, 20.0 lb/ft
Well depth = 13,600 ft

3.4* A kick occurs when the bit is at 4,500 ft. In order to return to bottom, will it be necessary to strip or snub? (Assume no spherical preventer frictional forces in the following problem.)

SICP = 4,500 psi
Drill pipe = 4-in. OD, 14.0 lb/ft, 3,600 ft
Drill collars = 6-in. OD × 2-in. ID
Mud weight = 13.0 ppg

3.5* In Problem 3.4, what would be the "Snub point," or the depth at which the summation of vertical string forces is zero?

3.6* Using the following data, is stripping or snubbing required if the SICP is 1,500 psi?; 3,000 psi?; 6,000 psi?
Drill pipe = 4½-in. OD, 16.6 lb/ft, 6,400 ft
Drill collars = 6.5-in. OD × 2-in. ID, 900 ft
Mud weight = 14.5 ppg

3.7 What would be the maximum surface pressure during a drill stem test under the following conditions?
Formation pressure = 11.5 ppg (equivalent)
Depth = 8,600 ft
Formation fluid = Oil, 0.6 specific gravity

Solution: 2,908 psi

3.8 Using the conditions given in Problem 3.7, what would be the surface pressure if the formation fluid had a gradient of 0.2 psi/ft?

3.9 In many North Sea areas, government agencies require that a gas gradient of 0.11 psi/ft be used in all well planning. Using this design criterion, what API pressure rating equipment must be used in a drill stem test under the following conditions?
Formation pressure = 16.6 ppg (equivalent)
Depth = 15,800 ft

3.10 During a kick killing operation, kill mud reached the surface after 800 strokes had been pumped. Using the following data, what is the deepest possible depth of the washout?
Pump = 5-in. × 18-in. duplex (95% efficiency)
Drill pipe = 4½-in., 16.6 lb/ft, 11,600 ft
Drill collars = 6-in. × 2-in., 1,000 ft
Casing = 10¾-in., 54.0 lb/ft, 9,500 ft
Open hole = 8⅜-in., 3,100 ft

3.11 Using the data from Problem 3.10, what is the deepest possible depth if 5,750 strokes had been pumped?

3.12 Rework Problem 3.11 using a 6-in. × 14-in. duplex pump at 95% efficiency.

3.13* Warning signs of a washout were noticed during a kill operation after the drill pipe had been displaced with mud. A piece of rope was pumped down the drill pipe in an effort to determine the depth of the hole. A pump pressure increase was noted after 260 strokes were pumped. Using the following data, what was the approximate depth of the washout?
Pump = 5.5-in. × 18-in. duplex (95% efficiency)
Drill pipe = 4-in., 14.0 lb/ft, 13,500 ft

References

1. Adams, N.J., "Deepwater Poses Unique Well Kick Problems," *Petroleum Engineer*, May, 1977.

2. Eaton, Ben A., "Fracture Gradient Prediction and Its Application in Oilfield Operations," *SPE Journal*, p. 1353, October, 1969.

3. Matthews, W.R., and Kelly, John, "How to Predict Formation Pressure and Fracture Gradient," *Oil and Gas Journal*, February 20, 1967.

4. Kendall, H.A., "Why You Lost Your Offshore Rig," *Ocean Engineering*, February, 1977.

5. Rehm, B., "Deepwater Drilling Poses Special Pressure-Control Problems," *Oil and Gas Journal*, May 3, 1976.

6. Ilfrey, Alexander, Neath, Tannich, and Eckel, "Circulating Out Gas Kicks in Deepwater Floating Drilling Operations," SPE 6834, October, 1977.

7. Garcia, J.A. and Clark, C.R., "An Investigation of Annular Gas Flow Following Cementing Operations," SPE 5701, January, 1976.

8. Cook, C. and Carter, L.G., "Gas Leakage Associated With Static Cement," *Drilling-DCW*, March, 1976.

9. Carter, G. and Slagle, K., "A Study of Completion Practices to Minimize Gas Communication," *Journal of Petroleum Technology*, September, 1972.

10.   Raymond, L.R., "Temperature Distribution in a Circulating Drilling Fluid," *Journal of Petroleum Technology*, March, 1969.

11.   Beirute, R. and Tragesser, A., "Expansive-Shrinkage Characteristics of Cements Under Actual Well Conditions," SPE 4091, *Journal of Petroleum Technology*, August, 1973.

12.   Wilder, Odis, "Handling Drill Pipe Under Pressure," presented to the Gulf Coast School of Drilling Practices, October, 1977.

4

Blowouts

A BLOWOUT IS AN UNCONTROLLED FLOW of formation fluids. The flow may be to an exposed formation and would be termed an underground blowout, or it may flow uncontrolled at the surface and would be termed a surface blowout. Regardless of the type of blowout, remedial control procedures are expensive, difficult to implement, and not always successful. This chapter discusses the different types of blowouts and their associated control procedures.

Numerous problems in blowout control normally are not considered in conventional well control. One of the primary concerns is that of environmental protection (with consideration given to governmental supervision and intervention). A surface blowout will unleash large volumes of potentially dangerous formation fluids to the atmosphere and adjacent surface surroundings. These formation fluids can include oil, gas, and salt water but may also include such toxic gases as hydrogen sulfide. In the case of toxic gases, the safety of life including human life becomes a serious and potentially paramount consideration.[1]

Loss of hydrocarbon reserves is another problem associated with blowouts. In a recent case in Sumatra, the reservoir is reported to have lost over 450 MMcfd for approximately three months before control could be regained. Although this example is abnormal, blowouts can cause reservoir depletion and/or productivity impairment to a point where the zones are no longer commercial.

The drilling rig equipment is generally destroyed during a blowout.

Fig. 4-1 Drilling rig before and after a blowout

Equipment may be lost by fire and/or cratering around the rig. Fig. 4.1 shows a large drilling rig after a blowout.

The cost factors of a blowout can be large, although seldom given primary consideration during the actual event. Cost variables involved include losses of equipment and hydrocarbons, personnel and equipment costs incurred in regaining control of the well, and the expenses involved in protection of the surrounding areas. A recent blowout in the Persian Gulf is reported to have cost over $100 million before control could be regained.

Surface Blowout Kill Procedures and Equipment

A surface blowout is an uncontrolled flow of formation fluids observed at the surface. The most common case is a blowout that occurs through the casing and surface equipment, although many wells have blown out around the outside of the casing. Also, it is not uncommon for this problem to cause smaller blowouts in adjacent water wells as a result of formation fluids flowing through permeable zones common to the water well and the original blowout.

Initial Preparations For Blowout Control

Planning steps must be taken before an actual killing plan can be implemented. These include organizing personnel, selecting the most feasible kill procedures for the circumstances involved, and gathering the equipment necessary for the kill procedure that was selected. Proper planning in all phases of these operations is essential to a quick, successful kill.

The most important initial step in any kill operation is making the decision on whether to set fire to the blowout if not previously ignited by sparks, electrical wires, or other common rig equipment. It is generally preferable to avoid setting fire to the well either intentionally or by accident. Lack of a fire will aid the kill operation by reducing the temperatures under which the personnel must work.

However, if the blowout fluid contains significant quantities of toxic gases such as hydrogen sulfide, the well must immediately be ignited to burn the gas and minimize the associated dangers of exposure by the workers and area residents. All wells drilled in suspected hydrogen sulfide environments should be equipped with flare guns for this purpose.

After the well is originally ignited, personnel equipped with flare guns and protective breathing apparatus must be stationed as near as possible to the well to reignite the blowout should the fire be extinguished accidentally. Caution must be exercised even after the well is ignited because sulfur dioxide, a byproduct of burning hydrogen sulfide, is also toxic if found in sufficient quantities.

Organization of personnel. A special task force should coordinate all activities during the kill operations. This group will usually consist of personnel from the operating company and may contain specialists such as firefighters, directional drillers, cementing service companies, and special services logging experts. The task force may also include members to research available literature on blowouts to aid in the determination of the proper kill procedures.

Planning sessions should be held at least once each day to coordinate operations. These sessions should include a discussion of the results of recent activities, plans should be presented for the next phase, and long range activity outlines should be developed to allow for the gathering of long lead-time equipment. Also, coordination of personnel is an important objective of the planning sessions since several hundred persons may be indirectly involved in certain phases of the kill process.

It is advisable to establish a public affairs, news reporting liaison between the actual operations and the local and national news services. This individual(s) has charge of all news releases and should hold conferences on a scheduled daily basis to report the activities of the task force. The purpose of the individual is to inform the outside community accurately of the state of the activities in order to minimize the development of unfounded and often dangerous rumors. The operator is better protected when all comment comes from a single source.

Kill technique selection. Experts agree that all blowouts are different to a certain degree.[2] As a result, there is no single technique applicable in all cases. Each blowout must be studied to determine the circumstances involved and then decide on the kill procedure most suitable for that application. There are certain guidelines, however, that can be used to evaluate each particular case.

When a well is blowing out at the surface, the kill techniques available to the operator are all based on pumping a sufficient quantity of fluids such as mud or cement into the well to overbalance the formation pressure of the flowing zone(s). To achieve this, the operator may elect to use specialists to shut-in the blowout at the surface and pump mud into the well at this point, or to attempt to pump into the wellbore below the surface by drilling an intersecting relief well. The requirements for each of these are different and must be understood by the operator before an appropriate decision can be made.

Shut-in procedures at the surface have prerequisites, the most important of which is the accessibility of a casing string. In land based operations where the blowout is through the casing and not through the formations surrounding the casing, access can usually be gained directly at the surface or by digging a pit around and below the casing head. The pit is necessary if the primary control string of casing fell or was cut below the surface.

If a casing string is accessible, the next considerations are the burst strengths of the casing and formation casing seat. If both of these can withstand the imposed pressures of a complete shut-in, surface control procedures can be implemented. If either or both cannot withstand the pressures, an alternative to complete shut-in must be considered.

Diversion of the blowout to safe areas is a method occasionally used to avoid complete shut-in. The procedure involves the installation of special equipment that will force the blowout away from the well area. With the well blowing out a safe distance from the wellhead area, a snubbing unit or other special equipment can be installed that will either force pipe or tubing into the well or tie in with existing tubulars. After this step has been completed, kill mud can be pumped down the pipe and up the wellbore annulus.

A relief well is considered necessary when any of the preceeding requirements cannot be satisfied. The most common cases are blowouts erupting outfide of the casing, fires with such heat intensity that workers cannot or choose not to approach and extinguish the blaze, or offshore wells where the casing string is below the water line and not accessible to control specialists. The relief well will require much time and should be considered only after exhausting the possibility of surface control procedures.

In many blowouts, the formation has bridged the open hole and killed the well. Although this has proven to be a viable kill technique, it should not be considered as a primary control tool because of the inability to predict its occurrence. Hole bridging often does create the lesser but still serious problem of underground blowouts.

Special equipment. If an attempt is made to kill the well from the surface, several types of special equipment may be necessary during the initial phases of the operations. This equipment may be a special design fabricated for one-time use or, more often, conventional designs used in special applications. Blowout specialists will generally attempt to use materials that are readily available when possible.

Explosives may be used in several manners when attempting to cap the well. The most common use is in extinguishing a fire on a well that does not have any significant quantities of toxic gases. This action is taken to allow the surface area near the wellhead to cool so that personnel can more readily perform their well control duties. Explosives may be used to aid in the removal of debris although this application will often set fire to the well. Explosives used are normally those that are readily available in the area and may include dynamite, TNT, or nitroglycerin.

In the cases where it is not desirable or feasible to extinguish the fire, equipment must be designed to cool and protect the personnel from the heat. The most common method used is to spray the workers with a continuous flow of water. Large pumps are designed to pick up water from a holding pit and inject it in the areas near the wellhead. In certain regions, it may be necessary to lay a lengthy source line to reach the quantities of water required for the job.

Sheets of corrugated iron (roofing "tin") are used as individual heat shields to protect against radiated heat. The sheets are equipped with a handle and a small viewpoint. The tin is of a size easily handled by the workers, yet sufficiently large to shield them from the blaze. Tractors, cranes, and bulldozers used in blowout control may also be fitted with the tin to protect the equipment operator.

Asbestos suits may be used when it is not possible to obtain sufficient heat protection from the water or tin sheets. Although the suits are used when necessary, they are considered a last resort because of the increased difficulty in performing the manual operations necessary for the removal or installation of equipment at the wellhead.

Kill techniques often require the removal of debris before work begins. If the drilling rig is intact and located immediately above the well, all connections or obstructions such as the mouse hole, rat hole, or the kelly in the hole must be broken or severed. This is usually done with a hacksaw although explosives are occasionally employed. After the rig is free from movement restrictions, it can be skidded away from the well.

If the blowout should ignite, the rig quickly will be reduced to a mass of debris. When this occurs, special equipment must drag the remaining structure away from the wellsite. This equipment will usually consist of a hook or rake-type device attached to a portable boom that is maneuvered by a crawler tractor. This process will often require many hours of tedious efforts until all the obstructing scrap metal has been removed.

Techniques Employed to Kill an Annular Blowout

The procedures involved in killing an annular blowout are based on installing new control equipment or repairing the existing blowout preventers. After the equipment is made operational, the well is shut-in or capped and mud is pumped into the well. The steps required to prepare the control tools for service will vary with each application.

Existing equipment can often be used to cap the well. If pipe remains in the hole, the ram preventers can be repaired or converted to pipe rams so that the well can be shut-in. Blind rams can be installed if pipe is not in the hole. Before changing rams, a valve below the rams should be opened to determine if a positive or negative pressure exists within the preventers. If a positive pressure is found, the bonnets or doors should not be opened because the well is likely to blow out at that point. If a negative pressure exists, caution should be exercised when opening the bonnets or doors because it is still possible that the blowout will exit through the opening.

It may be determined that the existing equipment is not satisfactory to cap the well due to an insufficient pressure rating to contain the flow. In this case, it may be possible to use the equipment to divert the flow away from the wellhead while preparations are made to pump mud into the well. Diverter lines can be attached to the outlets on the preventers or drilling spool and directed to a designated flare area. This procedure applies some pressure to the preventers although it will be an amount smaller than complete shut-in pressures.

Utilization of existing equipment is the expedient approach if feasible. Rig-up time is decreased and special control equipment is not necessary. Although the well may be on fire, the preventers may still be operable since a lack of oxygen within the preventer body prevents inner combustion.

If the blowout preventers are not useable, new equipment for capping must be installed. The damaged preventers must be removed by unbolting, cutting, or blasting. If the wellhead remains intact, a special-design diverter/capping assembly is stabbed and made up (Fig. 4.2). The diverter section is used to direct the fluid flow until hydraulic lines can be connected to the capping section. This is usually a set of blind rams. After all lines are connected, the valves on the diverter lines and the blind rams are closed.

In cases where the wellhead assembly was damaged or is determined to be unuseable, a new wellhead must be installed before the capping assembly can be used. To remove the existing wellhead, it is generally necessary to cut one or several casing strings. This may be accomplished with special tools such as a pipe line cutter or a long cable assembly used as a saw. Once the control string of pipe is exposed, a slip-and-weld head is installed, and the capping assembly is attached.

Whichever procedure is used to cap or divert the blowout, the next step in the kill process is to pump heavy mud into the annulus. Assuming no pipe is in the hole, pressure limitations on casing and the preventers will generally prohibit bullheading or directly pumping mud into the well. It may be necessary to lubricate mud into the annulus by (1) bleeding small amounts of formation fluids from the well, (2) quickly pumping small amounts of mud into the annulus, (3) allowing time for the mud to fall, and (4) again bleeding formation fluids.

Although lubrication is a tedious process, it will eventually kill the

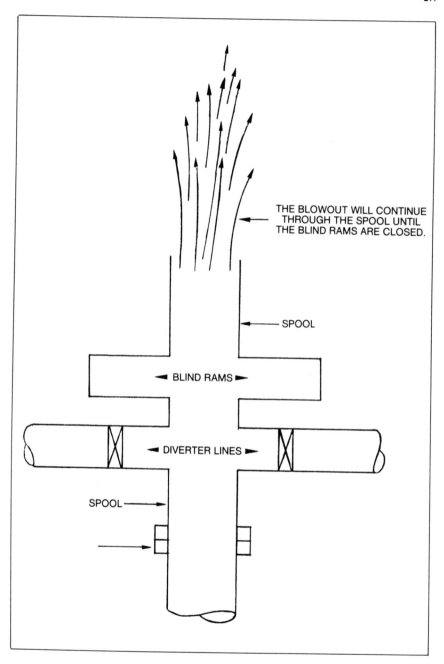

THE BLOWOUT WILL CONTINUE
THROUGH THE SPOOL UNTIL
THE BLIND RAMS ARE CLOSED.

SPOOL

BLIND RAMS

DIVERTER LINES

SPOOL

Fig. 4-2 Typical diverter—capping assembly for annular blowouts

well or at least reduce the surface pressures to the point where other techniques can be employed. If drill pipe is in the well, it should be possible to pump mud directly into the hole and kill the kick.

   In cases where drill pipe is not in the well and mud lubrication is not possible, attempts should be made to snub pipe, tubing or coiled tubing. The equipment necessary for the snubbing process can be rigged directly to the top of the capping assembly which should have a flange for this

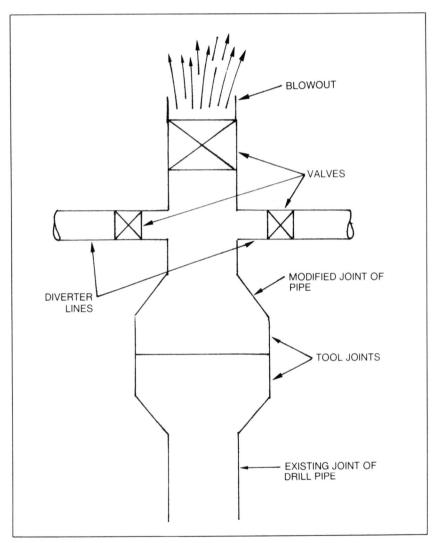

Fig. 4-3 Typical diverter—capping assembly for drill pipe blowouts

purpose. The well can be routinely killed using conventional procedures after pipe is in the hole.

A suitable choice of kill mud weights must be made before the kill process begins. In many field cases, the well was successfully capped and mud circulation begun only to fracture the formation due to using excessive mud weights. The natural tendency at this time is to overcompensate and use high mud weights to prevent further trouble. It should be noted, however, that the worst possible conditions exist before pumping begins and that the pressures should diminish thereafter. In most cases, the kill mud weight will be only slightly greater than the amount used when the well blew out.

Techniques Employed to Kill a Drill Pipe Blowout

The techniques used to kill drill pipe blowouts are based on the same principles used to kill annular blowouts. A diverter/capping assembly shuts in the well until heavy mud can be pumped down the drill pipe. A modified drill pipe connection may be used or a pipe ram locking system can be employed (Fig. 4.3).

Relief Well Kill Procedures

A relief well is designed to kill a blowout below the surface by pumping control fluids directly or indirectly into the annulus of the blowout well. A relief well is necessary when a surface kill procedure is not possible. A relief well may also be drilled to intersect and kill a possible underground blowout as in the case of the Shell blowout in Rankin County, Mississippi.[3] Problems associated with relief well drilling require engineered solutions before the project can be safely undertaken.

Type of directional well. A relief well is directionally controlled and designed to intersect the blowing well. The types of directional plans most commonly employed are the straight kick or "S" curve (Fig. 4.4). The selection of the most appropriate type of curve will depend on formation type, degree of difficulty in directional control, type of kill procedure to be used, and the desired point of intersection between the two wells. Most blowout cases will require the straight kick curve because of the difficulty in returning the hole to a vertical position at deeper intervals with the "S" curve.

The desired point of intersection restrains the types of directional program used. The two cases to be considered are intercepts at either the bottom of the hole or at some upper hole section.

A bottom intersection is preferable to the upper hole plan. The advan-

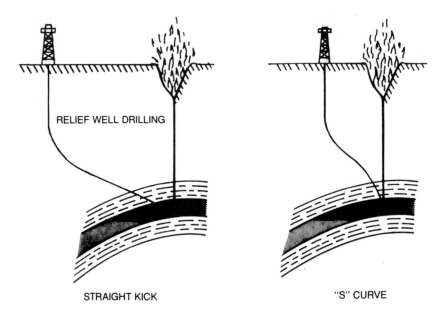

RELIEF WELL DRILLING

STRAIGHT KICK   "S" CURVE

Fig. 4-4   Directional plot for a "straight" kick and an "S" curve

tages include larger hydrostatic pressures due to a longer column of kill fluid, the ability to inject fluids at or near the producing interval, and higher formation fracture gradients that allow greater surface injection pressures and rates. The ability to inject fluids at the bottom is perhaps the primary advantage since the attack will be aimed at the likely producing interval. This will tend to minimize the influx dilution of the

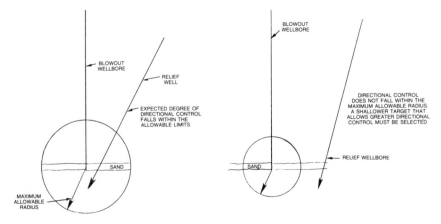

Fig. 4-5   Maximum allowable radius of intersection will aid in determining the most effective directional program

control fluids and therefore will build confining pressures at a greater rate.

Questions that must be answered before the bottom intersection plan can be applied include the degree of difficulty in attaining proper directional control at deep intervals. As normal compaction of the rock increases the strength of the formation, the maneuverability of the bottom hole assembly decreases. Since several precise directional changes are often required to align the wells, difficulty in using the bottom intersection plan is increased. This may lead to adoption of the upper hole intercept plan where greater directional control is attainable.

If some degree of directional control is possible at the bottom of the hole, a maximum allowable distance between wells must be established that will allow effective communication channels to be formed. The distance depends on the fracture strength of the rock and the permeability. If this distance is less than that thought to be within the range of the limited directional control, then again the upper hole plan must be chosen (Fig. 4.5).

Although the upper hole intersection plan is better for directional control, it does have the basic problem of attempting to kill the well at some point above the pressure source. This takes large volumes of control fluids designed to reduce the flow of the blowout gradually until the formation pressure can be overbalanced (Fig. 4.6). The difficulty in killing the well increases as the intercept depth decreases.

Surface location. Establishing the site location for the relief well in a safe area is important, particularly when the formation fluid contains toxic components such as hydrogen sulfide. The blowout has created hazard sectors which are subject to change depending on such factors as wind direction. Meteorologists should be consulted for specific forecasts in a local area since wind direction can change with time of day, temperature, and major frontal movements, or be influenced by local terrain. After all the information has been gathered, a well location should be chosen that is in the probable upwind direction from the blowout.

The location of the well will also be controlled by the type of directional program selected, the depth of intersection and the directional drilling conditions. In most wells, an angle greater than 10–12° is easier to control than a lower inclination. Therefore, the kick off point and the well location must be appropriately chosen to utilize an angle of this magnitude or greater in attaining the proper intersection depth.

Casing Program

Development of a suitable casing program for the relief well can aid in achieving a successful kill procedure. Improper selection of casing sizes,

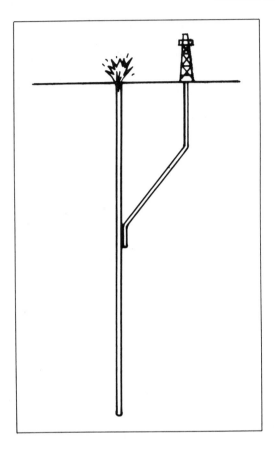

Fig. 4-6   An upper hole intersection between the relief well and the blowout well

strength, or setting depths can lead to an inability to pump control fluids at a rate sufficient to kill the blowout. Although conventional design criteria should be employed in the relief well program, special considerations must be made for the problems created by the blowout in underground formations. Only the special considerations are presented in this discussion.

The diameter of the kill string must be sufficient to allow large volumes of control fluids to be pumped at high rates without excessive friction pressures. To insure that this objective is achieved, the kill string diameter must be established before all other drilling casing strings are selected. Several typical design programs are presented in Fig. 4.7.

Another important consideration is shown by Fig. 4.7. If a drilling problem in the relief well necessitates the placement of an additional

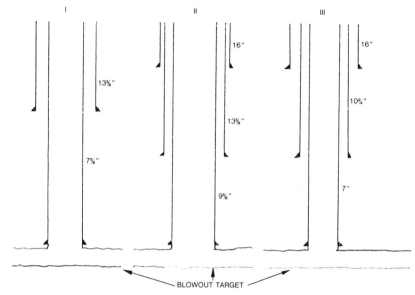

In each of these 3 designs, allowances have been made for at least 1 additional casing string if it becomes necessary.

Fig. 4-7 Typical casing size selections for efficient relief well drilling

casing string before reaching the final depth, the diameter of the kill string may be reduced to a size that will not allow the required volumes of control fluids to be pumped due to excessive friction pressures. To avoid this, the well plan should include allowances for at least one additional casing string should it become necessary. Fig. 4.8 illustrates this concept with a design that made no allowances for an additional string, and the resultant plan that became necessary when a problem was encountered.

The most common drilling problem in relief wells that requires an additional casing string is caused from pressure-charging of shallow zones by the blowout. The pressure within these zones may eventually build to a value that cannot be controlled with reasonable mud weights. When this occurs, a casing string must be run below this interval to allow continued safe drilling throughout the remainder of the well.

In the cases where the pressure charged zones cannot be drilled with conventional mud weights, it may be necessary to use a rotating head and diverter system to drill the interval. While drilling, the well is flared to a pit area until either the zone pressure is depleted or the interval is drilled and a casing string is run.

Flow rates. After the casing sizes and setting depths have been

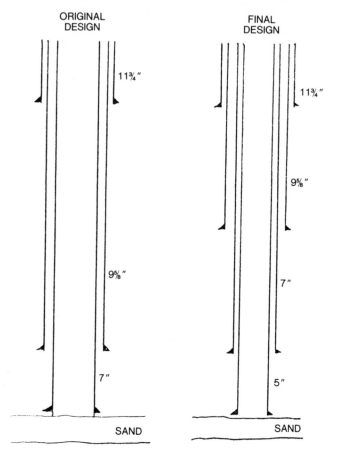

The original design used 7″ casing as the kill string. When a drilling problem was encountered, the 7″ casing had to be set shallower than desired. This necessitated using
the 5″ casing as the kill string.

Fig. 4-8   Effect of unexpected drilling problems on an improperly designed casing
string

selected, the next step is to establish the pressure requirements for the
flow string. It is possible that this string will be exposed to high burst
pressures from such sources as complete displacement of the control
fluids with formation fluids in the annulus, injection of the control
fluids into the formation, or friction pressures.

The pressures tending to collapse the casing may also be greater than
normal due to pressure charging by the blowout of cased intervals, and
partial or complete evacuation of drilling fluids when the producing
zone or borehole is entered. Although it is difficult to account for all

possible occurrences, certain considerations must be made for the pressures that are certain to be encountered.

Pressure due to injection of control fluids into the producing formation is generally the primary consideration when designing the flow string. The maximum allowable injection pressure can be calculated by using Equation 4.1.

$$P_{smax} = P_{frac} - P_h + P_{fric} \qquad\qquad \text{Eq. 4.1}$$

where:

P_{smax} = maximum surface pressure, psi
P_{frac} = formation fracture pressure, psi
P_h = control fluid hydrostatic pressure, psi
P_{fric} = friction pressure, psi

Equation 4.1 implies that the maximum surface pressure will be less than or equal to the injection pressure of the formation and will not create an induced fracture. Example 4.1 illustrates the use of this equation.

Example 4.1 A casing string is to be designed for a relief well. Using Eq. 4.1 and the following data, what is the maximum allowable injection pressure that the casing must withstand?

Fracture mud weight at casing seat = 16.5 ppg
Mud weight = 12.0 ppg
Depth of casing = 10,000 ft
Friction pressure (calculated) = 1,100 psi

Solution
1. Convert the mud weight terms to psi:
   Fracture pressure = 0.052 × 16.5 ppg × 10,000 ft = 8,580 psi
   Hydrostatic pressure = 0.052 × 12.0 ppg × 10,000 ft = 6,240 psi
2. Substitute the appropriate values into Eq. 4.1.

$$
\begin{aligned}
P_{smax} &= P_{frac} - P_h + P_{fric} \\
&= (8,580 - 6,240 + 1,100)\ \text{psi} \\
&= 3,440\ \text{psi}
\end{aligned}
$$

Friction pressure can be an important concern in the design of a flow string. The control fluids used to kill the blowout may have high densities and viscosities that will require significant pumping pressures during the injection phase of the operation. These pressures can be calculated using the equation:

$$\Delta P = \rho^{0.75}\, v^{1.75}\, \mu_p^{0.25}\, L/1800\, d^{1.25} \qquad\qquad \text{Eq. 4.2}$$

Where:

ΔP = friction pressure drop, psi

ρ = mud weight, ppg

v = average flow velocity, ft/sec

L = casing section length, ft

d = pipe ID, in.

μ_p = plastic viscosity, cp

Example 4.2 shows how these pressures are calculated and applied in Equation 4.1.

Example 4.2 What is the maximum allowable injection pressure for a well using the following data?

Fracture mud weight at the casing seat = 16.1 ppg

Kill mud weight = 13.0 ppg

Plastic viscosity = 27 cp.

Casing depth = 12,000 ft

Casing ID = 8.5 in.

Flow rate (maximum) = 50 bbl/min.

Solution

1. The average fluid velocity is calculated as follows:

v = Flow rate/2.448 (Diameter) ^{2}

= (50 bbl/min $\times$ 42 gal/bbl)/2.448 (8.5 in) ^{2}

= 11.87 ft/sec

2. Using Eq. 4.2, the pressure drop is calculated

ΔP = (13.0) $^{0.75}$ (11.87) $^{1.75}$ (27) $^{0.25}$ (12,000) / 1,800 (8.5) $^{1.25}$

= 543.9 psi

3. From Equation 4.1

P_{smax} = (10,046 − 8,112 + 544) psi

= 2,478 psi

A special problem in casing design is presented when the blowout contains significant quantities of corrosive fluids such as hydrogen sulfide. The danger resulting from hydrogen embrittlement failure may require that all casing strings with possible exposure to the blowout be designed according to accepted corrosion-resistant standards. The flow string may be exposed to the fluids if for any reason the control fluids are evacuated from the relief well. Also, other casing strings may require special pipe if pressure charging from the blowout is present in upper zones.

Directional control. The relief well is a directionally drilled hole designed to intercept the blowout wellbore or a formation immediately adjacent to the wellbore. To achieve this objective, the approximate location of the blowout well must be known and the direction of the relief well must be controlled. The final required maximum distance between the two wells will be determined by (1) the methods used to establish inter-well communications, (2) formation fracture pressure, (3) permeability and porosity, and (4) control fluid injection capabilities.

The initial concern in designing the directional plan for the relief well is to establish the location of the projected intercept interval of the blowout. The location can be approximated if complete directional records are available for the well. These records will contain surveys taken at periodic depth intervals which give the degree of inclination and direction. The accuracy in determining the wellbore's location will then depend only on the frequency in which the surveys were taken.

A problem arises if the well was drilled using surveys that record only the degree of inclination with no azimuth determination. In this case, two options are available to the operator. The first plan is to use the surveys as if they were all in the same direction. This procedure will yield the maximum possible deviation from the vertical position.

The second option is to assume that the wellbore approximates an elliptical helix if viewed in a three-dimensional plot.[4] This plan implies that the wellbore will be within a small radius from a point on a vertical line dropped from the surface location. The relief well would be drilled to intercept the line at the desired depth after which a new direction would be determined. This procedure is generally not applicable if the original well was directionally drilled.

Well Logging

Several types of logs are run in the relief well for special applications. Some of the logs are conventional tools while others are especially adapted for blowout control purposes. The reasons for using these tools range from an inference of pressure charged zones to a determination of the location and distance of the blowout wellbore to the relief well. The circumstances surrounding each blowout and the associated control efforts will determine the logs that should be run.

An electrical survey of some type, usually an induction tool, is run in the relief well. Its conventional applications include lithology determination and formation pressure evaluation. An additional use for the electrical survey is for an inference of pressure charged zones. This is accomplished by detecting abnormal resistivities in a zone as a result of underground formation fluid transfer. The most common case is higher than normal resistivities due to gas buildup in the zone. The tool is also useful in determining potentially troublesome zones by detecting gas accumulations which are perhaps not yet sufficient to increase the formation pressure.

Several logging tools are available to indicate the relationship of the relief well to the blowout. These include ULSEL, long normal, Magrange II, sound survey, and temperature log. Most of these tools function on different principles which may restrict their usages in certain situations.

Ultra-long-spaced-electric log (ULSEL). The ULSEL logging system was designed for detecting and mapping the profile of resistive anomalies such as salt domes in the vicinity of the wellbore. In the case of relief well drilling, the casing or drill pipe tubulars in the blowout well serve as the anomalies to current flow. The tool uses ultra-long-spacing normal devices to obtain deep-investigation readings which are influenced by the anomaly.

A standard resistivity log such as the ISF is used for the construction of a layered model of the formation which can be used to compute the ULSEL readings to be expected if no anomaly were present. Significant and consistent departures of the actual ULSEL values from these expected values serve to indicate the presence of resistive or conductive anomalies. Dipmeter data is also used in the interpretation and computation.[5]

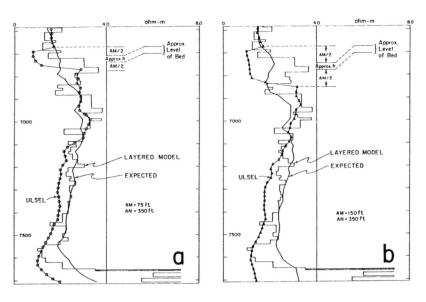

Fig. 4-9   Illustration of the ULSEL Log used to detect a resistivity anomaly

Digitized induction log readings are used in a computer program to arrive at a multilayered model of the formation near the borehole. Layer boundaries are selected on the basis of electrical reflection coefficients (i.e. resistivity contrasts). Each layer of the model is given a constant resistivity equal to the average induction log resistivity of the corresponding interval.

The multilayered model is used in a computer program to determine the ULSEL readings to be expected in the absence of any remote anomaly. Anomalies are detected and evaluated by comparison of the various ULSEL readings with these predicted no-anomaly values (Fig. 4.9).

For the interpretation of distance to a saltdome, the following ratio is computed for each spacing:

Ratio = Corrected ULSEL resistivity/corresponding ULSEL resistivity expected if no anomaly is present   Eq. 4.3

When these ratios deviate from unity by an appreciable amount and in a consistent manner, an anomaly is indicated. The general approach is to interpret the anomalous resistivity ratio in terms of the apparent distance to the subject of interest.

For the purpose of locating a nearby cased well from measurements made in an intercept well, use is made of shorter available ULSEL spacing (e.g., AM = 75 ft, AN = 350 ft, and AM = 150 ft, AN = 350 ft) and a 20-ft normal (AM = 20 ft., AN = 70 ft, 10 in.). The ULSEL devices will detect a 9⅝'' casing at distances up to 70–80 ft. If the distance to casing is definitely known to be less than 20 ft, only the 20-ft normal is required.

A computer produces interpretation charts to be used for the existing conditions such as spacing of the ULSEL or normal device used, casing size and weight, approach angle between intercept well and target casing, average formation resistivity, and anisotropy coefficient of the formation. Interpretations are made from these computer-produced charts by the ULSEL analyst using the relative-resistivity ratios from the computer output.

This technique measures distance only and has no capability by itself to detect the direction of the casing. A procedure to determine direction by detecting magnetism of the casing has been described by several authors.[6] Unfortunately, the method requires that casing magnetism be known before the casing was set in the well.

Magrange II. The Magnetic Range Detector (Magrange II) has the advantage of determining direction as well as distance to the relief well. The equipment used in the Magrange II system consists of a downhole instrument, a winch and seven-conductor cable, a surface electronic unit, a programmable calculator, and a plotter. The downhole instrument contains magnetic field sensors arranged in a noninterfering orthogonal configuration and also in a gradiometric measurement configuration. The sensors, along with their associated electronics and signal

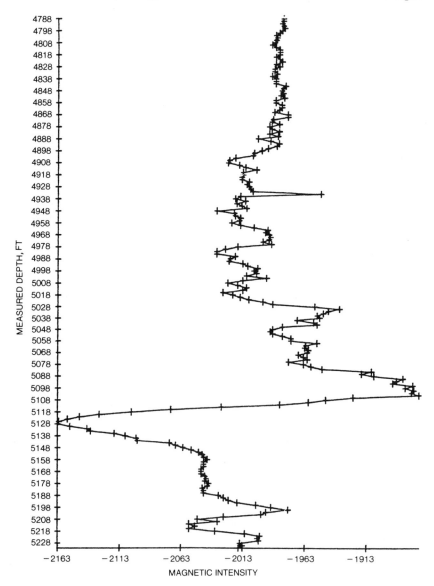

Fig. 4-10 Typical raw data plot for the Magrange II axial gradient sensor

conditioning circuiting, are housed in a nonmagnetic cylindrical container.[7] Experience has shown that under the optimum conditions, Magrange II can detect targets at a range of 100 ft. The direction from the relief well to the target well can be determined to within a few degrees.

Although the tool is a recent development, it has proved successful in a number of field applications. In a blowout in West Cameron, offshore Louisiana, the Magrange II system was used to guide the relief well to an adjacent point near the blowout well that allowed the completion of successful pumping kill operations.[8] A typical plot of raw data gathered from the axial gradient sensor is shown in Fig. 4.10.

Temperature survey. The circumstances required for optimum results with ULSEL and Magrange II systems may restrict their effectiveness in certain applications. Both tools function in uncased boreholes of the relief well and are designed to detect casing in the blowout well. If the blowout well has no casing or if interference is encountered from casing in the relief well, alternative tools such as the temperature or sound survey may be used.

The temperature log is designed to detect either absolute temperature or differential temperature. The tool is used in relief wells to determine if the borehole encounters abnormal temperatures due to fluid flow in the blowout well. Although the temperature change may be a less-than-normal reading due to gas expansion, the most common case is abnormally high temperatures resulting from heat transfer of the flowing formation fluids.

The primary objection to its use it that is gives neither direction nor distance from the relief well to the blowout. Also, the readings from the temperature survey depend on the rate and type of flowing fluids in the well.

Sound survey. The conventional sound survey, or noise log, can provide a qualitative estimate of the proximity of the two wells. The tool, which consists of sonic wave sensors, detects sounds created by fluid flowing in the blowout well. The disadvantages of the tool are its inability to provide direction or distance and the manner in which it can be influenced adversely by surface rig noises. An exception is the Borehole Audio Tracer Survey which is reported to have the capability of determining the distance between the relief well and the blowout well.[9]

Establishing Communications

Communications can be established between wells by several different procedures. The most direct method is to drill the relief well into the actual blowout wellbore. This method is not usually employed, however, because it may necessitate the immediate implementation of the kill procedure without allowing sufficient preparation time. The most common procedure utilized is perforating between the two wells or into the producing formation to allow pumping of kill fluids between the two wells.

A novel method that has been studied involves the detonation of explosives in the wellbore and is designed to either establish communication between the wells or cause the formation surrounding the blowout well to collapse and kill the blowout. The selection of the most appropriate method of establishing communications require planning and a working knowledge of each available option.

Perforation. Considerations when perforating include orientation of the perforations, pressure drop when pumping kill fluids, and formation damage and plugging of the perforations.

Orientation of the perforations is important when the desired communication paths are between the two wells and not from the relief well into the producing formation. In this case, conventional 360-degree perforations would result in a majority of the shots being expended in directions not useful to the kill. Methods most commonly employed to direct the shots properly involve detection of the existing casing in the blowout well or placement of radioactive tracer shots for reference points in the formation opposite the blowout wellbore. Logging service companies can supply the necessary expertise for orientation of the perforations.

The friction pressure required to pump control fluids through perforations can be a controlling variable in the selection of the size and number of shots to be used. This is particularly important when high kill pumping pressures are required due to low permeability formations, excessive distance between the control well and the blowout well, or kill procedures requiring large volumes of control fluids as a result of high blowout rates. Several authors have prepared reference material that can be consulted if necessary.[10]

Formation damage and perforation plugging may be important in cases involving either a limitation of the number of shots that can be used or high pressure losses through the perforations under ideal conditions. When these problems are important to the success of the kill operations, steps must be taken to avoid or minimize the severity of the damage or plugging. Formation damage can be minimized using conventional procedures such as clean perforating fluids and perforating underbalanced. When these procedures are used, caution must be exercised to insure that excessive surface pressures are not imposed on the relief well subsequent to establishing the communication paths. Plugging of the shot tunnels with debris from the perforation gun can be avoided by using material that will shatter upon detonation such as glass-encased shots.[11]

Explosives. The feasibility of using explosives of the conventional and atomic types has been studied to determine if they are acceptable to either create communication channels or kill the blowout.[12] Although explosives have not been used in the United States, they have been successfully used to kill several gas well blowouts in the Soviet Union.[13] This procedure is primarily a kill technique but it will be presented in this section because of its potential to establish communications between the two wells.

McLamore and Suman presented the following criteria to determine if the explosive techniques are applicable:[14]

1. Lack of proper directional data on the blowout well.
2. High reservoir pressures in the blowout well that would require large volumes of high density control fluids to kill the well.
3. Hazardous conditions such as the presence of hydrogen sulfide gas that may restrict entry into the producing interval.
4. Extreme depth or temperature that necessitates excessive time to drill a conventional relief well.
5. Other conditions such as subsurface blowouts that require solutions that can be implemented faster than conventional methods.

They also identified the formation failure mechanisms that would establish the communication channels or cause hole bridging to occur. These mechanisms could serve to set the objectives of the directional drilling programs.

1. Near-field mechanism. In this case, the relief well and explosive charge would be positioned so that the cavity created by the detonation would encompass the casing of the wild well. The casing would be destroyed and the rubble created by the detonation would bridge the well.
2. Mid-field mechanism. The relief well would be positioned relative to the wild well such that the attendant ground motion would collapse the open hole and shatter the casing.
3. Far-field mechanism. The stress wave generated by the detonation would dynamically collapse the casing.

The authors recognized that the first two mechanisms would probably kill the well while the third was uncertain due to the inability to determine the collapse resistance of tubular goods by dynamic stress waves.

Kill Procedures

The most common kill procedure for blowouts involves pumping large volumes of control fluids into the producing formation or wellbore to subdue the well with hydrostatic pressure. The control fluids may be drilling mud, water, cement, or a combination of several of these in distinct pumping stages. The actual kill operation will require careful planning and execution of the pumping to control the blowout successfully.

Each kill operation may take special equipment, selection of the control fluids, and reservoir analysis to determine the expected pressure behavior while pumping. Each blowout will have characteristics that require a tailored set of these variables.

Fracture pressure. Prior to initiating the pumping operations, a maximum allowable pumping pressure must be established to avoid fracturing the formation. This will ensure that the control fluids will enter the desired interval and not exit the relief well at an ill-placed induced fracture. If the exposed interval is in a shale section, procedures presented in Chapter 5 can be used to determine the fracture pressure. Example 4.1 illustrated the method used to calculate the maximum allowable pumping pressure if the fracture pressure is known. In sand sections, estimations of the fracture pressure must be based on a knowledge of such variables as grain cementation.

Equipment. The initial step in selecting equipment for pumping is to determine the required hydraulic horsepower. The design criteria are often based on a flow rate of 50 bbl/min into the relief well and a

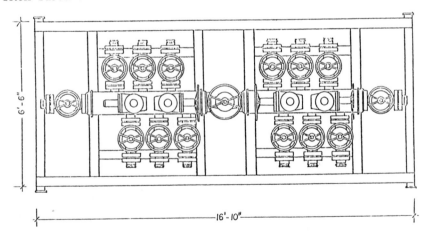

Fig. 4-11 Illustration of a kill manifold

maximum pressure as calculated with the procedures presented in Example 4.1.[15] After the hydraulic horsepower is determined, a safety margin is applied for adequate output if mechanical failures occur. Example 4.3 shows the manner in which these calculations are made.

Example 4.3 Using the data given in Example 4.1, what amount of hydraulic horsepower is required if a design factor of 1.5 is used?

Solution
1. HHP = (Flow volume) (pressure) / 1,714   Eq. 4.4
      = (50 bbl/min × 42 gal/bbl) (3,440 psi) / 1,714
      = 4,214
2. Applying the design factor,
   4,214 HHP × 1.5 = 6,321 horsepower

A suction manifold will be necessary to supply the large volumes of control fluids for the pumping units to aid in attaining maximum output pump efficiency. It should use centrifugal pumps capable of a 50 bbl/min output at 50 psi minimum pressure. Precaution should be taken to arrange the suction manifold in a fashion that will allow for several fluid routes in case of plugging in the primary passage.

A kill manifold connects the injection pumps to the relief well. Fig. 4.11 shows the manifold used by Tenneco Oil Co. to kill a blowout at West Cameron Block 165 offshore Louisiana. This manifold contains ten inlets, six outlets and two gauge points.

Special tanks supply the required volumes of control fluids. These tanks may be of conventional mud pits, mud barges, or vacuum trucks. If the control fluid is cement, special mixing facilities must also be installed.

The pump-in wellhead assembly should be designed to meet several criteria. Among these are (1) multiple pump lines, (2) check valves, (3) sufficient blowout preventers to insure control if equipment failure occurs, and (4) capability to utilize the rig pumps if within their pressure limitations. All lines should be clearly marked for ease of identification. Fig. 4.12 shows the assembly used in the referenced Tenneco blowout.

Control fluids. Determination of the proper control fluids to kill the well will depend on (1) formation pressure, (2) depth of intersection between relief well and blowout well, (3) formation permeability and porosity, (4) formation fluids type, and (5) availability of possible control fluids such as seawater or cement. Selection of the control fluids will be based on an analysis of each of these variables and the actual well conditions. The most common type of control fluids are fresh water,

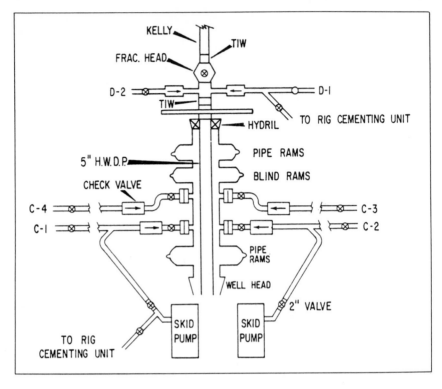

Fig. 4-12   Pump-in wellhead assembly

seawater, or drilling fluid. Cement or acid may be used to achieve certain special results.

Water of some type is usually pumped to establish communications between the two wells and to reduce the flow in the blowout well. This fluid is used because it is available, it can minimize perforation plugging and formation damage, and it costs little. Most wells can be killed with water if the reservoir pressure is normal or sub-normal.

A high density fluid must be used to kill the well when the formation pressure is greater than normal. The fluid will usually be drilling mud due to its overall lower cost as compared to weighted water. Pumping water prior to the mud will reduce the formation fluid flow and allow a faster kill after the mud is introduced. When drilling fluid is used as the control fluid, it may be advisable to have a volume of acid available should the communication paths begin to plug during pumping.

Cement is occasionally used as the primary control fluid. When this procedure is used, proper selection of the density is important because an initial failure in the kill attempt will probably leave the hole partially plugged with cement while still flowing formation fluids. A formation pressure buildup analysis must be made to estimate the maximum bottomhole pressure at a time slightly greater than the set time for the cement. The cement density must be able to control this maximum pressure. Chemical additives such as retarders and accelerators for the cement may be necessary to make the cement set as desired.

Prior to and during pumping, a knowledge of the expected reservoir behavior will aid in determining the effectiveness of the kill operation. Miller and Clements have shown the effectiveness of several calculations in achieving this goal.[16] By assuming a limiting bottomhole injection pressure, Darcy's law yields an equation for the maximum rate of water injection for a given size of water bank.

$$i_{wmax} = \frac{0.00707 \ k_w h \ (P_{iwfmax} - p)}{\mu_w ln \ \left(\dfrac{r_b}{r_{we}}\right)} \qquad \text{Eq. 4.5}$$

where

$$r_{we} = r_w e^{-s} \qquad \text{Eq. 4.6}$$

Also, the bottomhole pressure can be calculated as follows by assuming a constant injection rate,

$$P_{iwf} = p + \left[141.4 \ i_w \mu_w ln \left(\frac{r_b}{r_{we}}\right)\right] / k_w h \qquad \text{Eq. 4.7}$$

The cumulative volume of water injected can be expressed as:

$$W_i = \pi h \ (1 - S_{wc} - S_{or}) \ (r_b^2 - r_w^2) \ / \ 5.615 \qquad \text{Eq. 4.8}$$

The dimensions for each variable are as follows:
h = net thickness, ft
i_w = water injection rate, bbl/day
k_w = effective permeability to water, md
$\overline{p}$ = average static reservoir pressure, psia
p_{iwf} = injection well bottom-hole pressure, flowing, psia
r_b = radius of injected water bank, ft
r_w = wellbore radius of injector, ft
r_{we} = effective wellbore radius of injector, ft
s = skin effect factor

S_{wc} = connate water saturation, fraction
S_{or} = residual oil saturation, fraction
W_i = cumulative water injected, bbl
μ_w = water viscosity, cp
ϕ = porosity, fraction

Underground Blowouts

An underground blowout occurs when formation fluids flow from one zone to another. The receiving zone could be either a permeable/porous interval, a fractured formation or behind ruptured casing which exposes a weaker formation. Occasionally these blowouts will occur without detection at the surface, although this case is rare.

The direction of the fluid flow in the blowout is an important concern when choosing a control procedure. The cause of the underground transfer will often indicate the direction of the flow. Since most under-

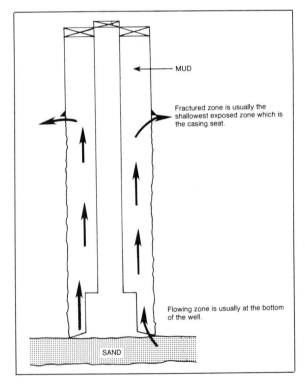

Fig. 4-13 Typical location of the flowing and fractured zones in an underground blowout

ground blowouts are thought to occur after the blowout preventers have been closed on a kick taken while drilling, the flow will normally be from the bottom of the hole to some upper exposed zone. This is based on the assumption that shallow zones will fracture prior to lower zones, and that the initial kicking zone will be the primary source of formation fluid flow (Fig. 4.13).

The flow can be directed from shallow to deeper zones if the lost circulation, or a thief zone, is at or near the bottom of the well. This is a common occurrence when lost circulation is encountered at the bit while drilling. The zone may be naturally fractured, structurally weak, or an unsealed fault plane. When the zone is encountered, the fluid level in the well may fall and decrease the hydrostatic pressure sufficiently to allow an upper zone to flow. The common case is for the fluid to flow downward although the reverse occasionally occurs. (Fig. 4.14).

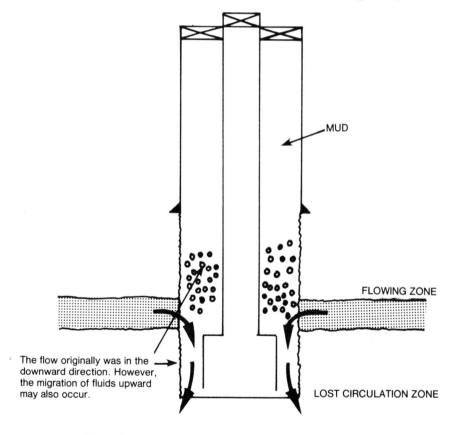

Fig. 4-14   Effect of encountering the lost circulation zone at the bottom of the well

Well Control Problems and Solutions

Underground blowouts can cause a serious problem by pressure charging exposed formations. Many cases have been reported in which water wells have blown out as a result of pressure charged zones occurring from underground blowouts in adjacent oil and gas wells. These high pressures can be a serious problem in drilling offset relief and/or new wells because conventional abnormal pressure detection techniques and principles cannot be applied to this situation. Drilling of these offset wells often requires the use of rotating heads to drill and deplete the intervals.

Underground blowout indicators. There are several indicators used to identify underground blowouts. One or several of these may be observed or none at all may be noticed. Close observation of the indicators may aid in determining the most appropriate kill procedure.

One of the primary indicators is an initial drill pipe and casing pressure buildup and subsequent reduction. When the kick is detected, the preventers are closed and the surface pressures begin to build to the values necessary to balance bottom hole pressure. If the casing pressures create equivalent mud weights greater than formation fracture gradients, a fracture will be formed that will relieve the wellbore pressures and reduce the casing pressure. The reduced pressure may be stable or fluctuating. See Example 4.4 for a typical pressure response to an underground blowout.

Example 4.4 A well was shut-in after warning signs of a kick were observed. These pressure data were observed.

Time	SIDPP, psi	SICP, psi
3:15	350	1100
3:18	475	1300
3:20	510	1360
3:22	525	1380
3:24	475	1340
3:26	475	1110
3:28	425	1090
3:30	350	1090
3:40	0	1100
3:50	125	1250
4:00	140	1200
5:00	130	1120

Monitoring the initial pressures on the drill pipe and casing may be an important factor when killing the blowout. Although the initial pressure on the drill pipe before the reduction does not give a reading that could be used to determine bottomhole pressure, it does indicate the minimum pressures on the formation that must be achieved in order to control the kick. In Example 4.4, the minimum kill mud weight would be calculated with a SIDPP of 525 psi.

Fluctuating or unstable pressure readings indicate an underground blowout. The fluctuating pressures can result from unsteady fluid flow from one or several formations, or from the fractured formation tending to open and close as the exposed pressures change on the interval. Both the drill pipe and casing pressures may fluctuate uniformly or independent of each other. If the annular formation collapses or bridges on the drill string, the casing pressure may stabilize while the drill pipe pressure continues to change.

The drill pipe pressure may be higher than the casing pressure in an underground blowout. This is usually the result of formation fluids, particularly gas, entering the drill pipe after the blowout has been initiated. The drill pipe mud may overbalance the cut annular fluid and U-tube out of the pipe. This leaves a void that can be occupied by gas. Although this same phenomenon can occur in the annulus, the larger feet-per-barrel ratio in the drill pipe gives rise to higher pressures.

The reverse of high pressure may be noticed in some cases. If the mud falls out of the drill pipe, lower than required pressures or even zero pressure may be recorded if no other fluids enter the pipe. It has often been observed that no other fluids will enter the pipe due to either jet plugging of the bit or severe mud flocculation in the pipe.

In most cases of underground blowouts, there will be little or no direct communications between the drill pipe and annulus. In other words, the casing pressure may change without affecting the drill pipe, or the drill pipe pressure may change with no uniform reflection on the casing pressure. The lack of communication between these two points is due to the loss of integrity of the U-tube or borehole.

Kill Procedures for Underground Blowouts

Unlike conventional well control, there are no established kill procedures for underground blowouts that will work in most situations. Although there are kill procedures that work more effectively sometimes, these are based on adequate knowledge of the cause of the underground blowout, the location of the thief zone(s) and flowing zone(s), formation pressures, and the limitations of the proposed procedures.

The most common case in which underground blowouts occur is that of a kick occurring in a deep zone when only surface casing is set in the well.[17] This situation exposes large sections of the hole to high equivalent mud weights which results in formation fracture (Fig. 4.15).

Although it is often advisable to set the intermediate casing string as deep as possible in order to achieve maximum fracture gradients (if this is in accordance with the well design), precautions must be taken in the selection of the surface casing setting depth to account for the deep intermediate casing seat and prevent the situation in Fig. 4.15.

Proper setting depth selection will vary for every well that exhibits different formation pressure characteristics. Note that if a casing depth of

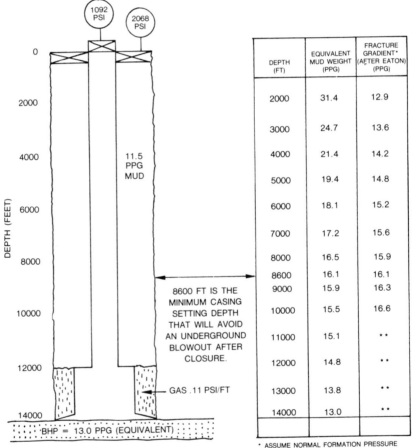

DEPTH (FT)	EQUIVALENT MUD WEIGHT (PPG)	FRACTURE GRADIENT* (AFTER EATON) (PPG)
2000	31.4	12.9
3000	24.7	13.6
4000	21.4	14.2
5000	19.4	14.8
6000	18.1	15.2
7000	17.2	15.6
8000	16.5	15.9
8600	16.1	16.1
9000	15.9	16.3
10000	15.5	16.6
11000	15.1	**
12000	14.8	**
13000	13.8	**
14000	13.0	**

* ASSUME NORMAL FORMATION PRESSURE TO A DEPTH OF 10000 FT.

** DEPENDENT ON FORMATION PRESSURE.

Fig. 4-15   Effect of equivalent mud weight vs. depth during a typical pre-underground blowout situation

8,600 ft had been chosen in Fig. 4.15, the underground blowout would have been avoided at initial closure.

The most successful kill procedure for this type of blowout is the spotting of a heavy slug of mud in the open hole below the point of lost circulation. The objective is to generate high bottomhole pressure from the hydrostatic pressure of the high and low density muds in the well and from any existing casing pressure, such that the combination will overbalance the formation pressure. Although the heavy slug will usually have a density greater than the equivalent fracture gradient at the casing seat, this does not present any lost circulation problems since the mud will not be circulated above this point.

Calculation of the required density to kill the well can be estimated from the initial shut-in pressures before breakdown, and drilling and penetration rate data immediately prior to the kick. The volume of heavy mud required to build the necessary hydrostatic pressure depends on mud density, hole geometry (washout), and rate of formation influx into the wellbore. A rule of thumb used successfully in many applications is to build a volume of mud three times the calculated volume to allow for mud cutting and hole washout. Example 4.5 illustrates this kill procedure.

Example 4.5 An underground blowout is occurring on a particular well. An attempt is to be made to kill the well using the heavy slug method. With the following known data, what amount of mud should be used? Use the pressure data from Example 4.4.

Well depth = 10,000 ft
Casing seat = 3,500 ft
Original mud weight = 11.0 ppg
Hole size = 8½ in.
Drill pipe = 4½ in.
Drill collars = 6½ in.

Solution
1. Bottomhole pressure (minimum)
   $(0.052 \times 11.0 \text{ ppg} \times 10,000 \text{ ft}) + 525 \geq 6,245 \text{ psi}$
2. Assume that the total annular pressure (mud + SICP) is equivalent to 11.0 ppg mud. (This is an arbitrary assumption.) Therefore, the heavy slug must exert at least 525 psi more than an equal height column of 11.0 ppg mud. Any of the following options would accomplish this objective.

Mud weight, ppg	Column height, ft
15	2,524
16	2,019
17	1,682
18	1,440

3. The operator elects to mix and pump a volume of 15.0 ppg mud that will give a minimum height of 2,500 ft in the annulus. Drill collars were disregarded in the calculations.
   (Fluid height) (annular capacity) = volume
   (2,500 ft) (0.05 bbl/ft) = 125 bbl
4. Using a safety margin of 3 to reduce the effects of mud cutting, the operator pumped 375 bbl of 15.0 ppg mud to kill the well.

After the blowout has been killed, steps must be taken to restore the well to a drillable condition. The fractured zone(s) must be cemented to resolve the lost circulation problem. The original mud in the annulus must be conditioned to a density sufficient to control formation pressures but less than the fracture gradient of all exposed zones. The heavy

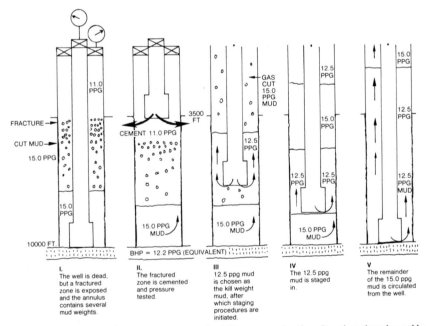

Fig. 4-16 Steps required to complete the kill process for the situation developed in Example 4-5

mud used to kill the well must be circulated from the well in several stages to avoid refracturing the lost circulation zone. These steps are illustrated in Fig. 4.16.

Another procedure often used to kill underground blowouts is that of spotting a barite plug. This technique is designed to form a barite bridge in the hole that will seal the blowout and allow a heavier mud to be circulated above the plug. The technique is not based on well control through a hydrostatic pressure increase as was the heavy slug procedure but rather a bridging effect.

The density and composition of the barite plug is the key to its success. A typical mix composition is given in Table 4.1. Recent field applications show that a density of 18–22 ppg yields optimum settling and bridging characteristics. A small amount of SAPP (sodium acid pyrophosphate) is often added to reduce the viscosity and channeling effects of any flocculated mud in the annulus, increase the setting rate of the barite by thinning any contamination of the plug, and form a tight, low permeability filter cake. The volume of plug normally used is that which will give a 500-ft column.

As seen in Table 4.1, the barite plug contains no suspension agent for the barite. As a result, surface mixing facilities and plug placement must be continuous and rapid. If the mixing or pumping is halted for even a short time, settling in the pits or plugging of the drill pipe may occur. Mixing of the plug will generally require a cementing type pump and agitation system since normal surface rig facilities are not suitable due to their low mixing rates.

Squeezes with diesel oil as the base fluid have proved effective in solving the lost circulation problems in underground blowouts when the thief zone is not caused by low fracture gradients. Applications could include porous, permeable zones, unsealed fault planes, poor cement jobs, or others of this type.

The diesel oil is used as a transport agent for water reactive additives such as bentonite or cement. The mixture is pumped into the formation where it mixes with mud or formation water and hydrates or sets. The

Table 4.1 Typical Barite Plug Composition

Formula for a barrel of 22 ppg slurry.

750 lb of barite
21 gal of fresh water
½ lb SAPP
¼ lb caustic soda

Table 4.2 Typical composition for a diesel oil-bentonite squeeze

300 pounds of bentonite/bbl of diesel
15 pounds of mica (or walnut hulls)/bbl of
    diesel for additional plug strength
Diesel oil as required

Recommended slurry volumes range from 20–150 bbl.

slurry, often called a gunk squeeze, forms a high viscosity pill in the formation that restricts or retards the fluid flow. See Table 4.2 for typical composition of the diesel oil-bentonite squeezes.

Since the oil squeeze contains additives not soluble in diesel oil, the mixture has an extended pumping time unless it becomes contaminated with mud. This feature reduces the possibility of premature setting in the drill string before placement is complete. Also, if it becomes necessary to drill the set plug, sidetrack tendencies are reduced since the final plug does not form a bridge as firm as cement.

Remedial procedures other than the diesel squeeze can be used to reestablish circulation. These include lost circulation materials, cement, polymer pills and numerous commercial specialty products designed to seal the thief zone. When these products are used, it may be advisable to remove or blast the jets of the bit with primer cord on an electric wireline to avoid the possibility of plugging.

Time may be a key factor in killing an underground blowout. The hole may bridge due to chemical sloughing or mechanical heaving if the flow is allowed to continue. Also, the kicking reservoir may deplete or reduce its pressure to a point such that lower mud weights can regain control of the well. It should be noted, however, that the operator cannot rely solely on time since many underground blowouts have shown the capabilities of flowing for extended periods.

Certain techniques can be used in conjunction with various kill procedures to aid in controlling the blowout. One of these is commonly termed bullheading. This involves injecting into the formation by either pumping fluid into the annulus with the drill pipe closed, or by pumping mud into the drill pipe with the blowout preventers closed. Although this technique is normally not recommended for conventional well control since it implies that a lost circulation zone be established, the procedure is applicable in underground blowouts because the zone is already present.

Bullheading may be a quicker and more economical approach in cases such as wells with shallow surface casing. The mud that would be lost by pumping down the casing into the formation would be considerably less than pumping down the drill string and up the annulus. As with all procedures and techniques related to underground blowouts, bullheading must be evaluated at the rig site to determine its applicability.

Another procedure that can be used to aid in implementing blowout control procedures is stripping out of the open hole. This avoids the possibility of the hole bridging on the pipe and also allows the operator to apply lost circulation remedies directly to the thief zone if it is located at the casing seat. If stripping procedures are followed and a float is not included in the drill string, fill the pipe with heavy, viscous mud to prevent a drill pipe blowout after the kelly is removed.

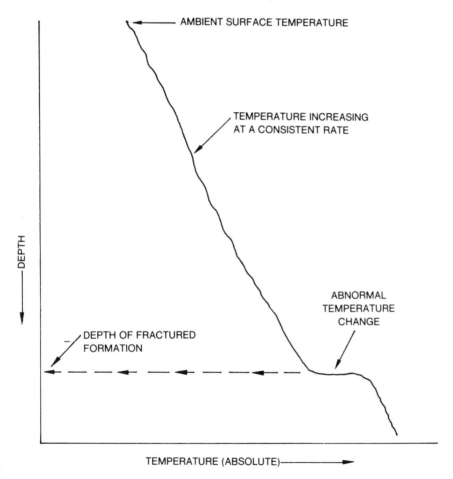

Fig. 4-17 Illustration of the expected results from a temperature log used to locate the loss zone during an underground blowout.

Lost circulation zone detection techniques. It is important to determine the thief zones' location in order to calculate volumes and densities of kill fluids and the position at which they should be spotted. The history of the well or the field may supply the information necessary to locate the zone. A structurally weak zone or an unsealed fault plane may be present. The field may contain one or more depleted sands that are causing the loss. Also, the conditions under which the blowout occured may indicate that the zone is either at the bottom of the well or at the casing seat.

The most common tool used to define the interval is the temperature log. This log is not generally used to record absolute temperature but rather differential temperature. As the logging tool is lowered down the

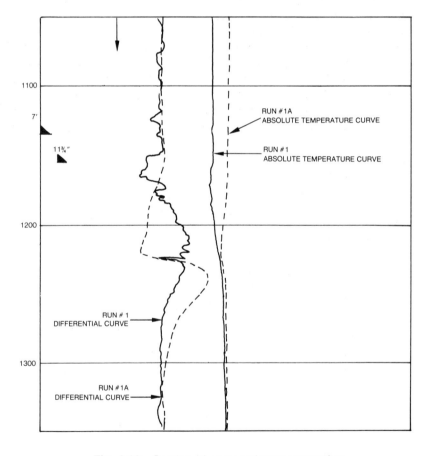

Fig. 4-18   Section of an actual temperature log

drill pipe, it will read an abnormal change at the loss zone if the underground flow is continuous (Fig. 4.17).

The tool senses the heat from the fluid that is greater than it should be for the depth at which it is encountered. In some cases, the temperature change has been reported as a cooling effect supposedly due to gas expansion. Nonetheless, a temperature change is the key. Fig. 4.18 shows a section from an actual temperature log indicating a fluid exit at approximately 1,225 ft.

Fig. 4.17 showed the case in which the formation fluids were moving. If the log gave the results shown in Fig. 4.19, the indications are that the thief zone is at the bottom of the well or that a static situation exists in which the fluid is no longer flowing.

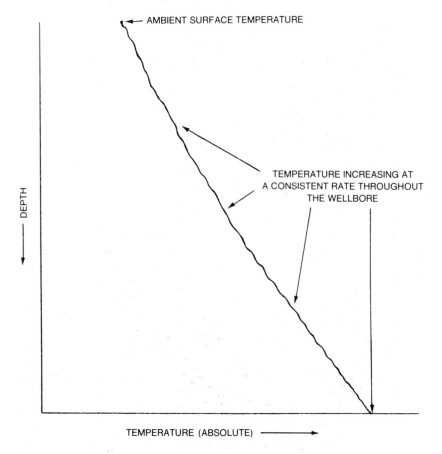

Fig. 4-19 Readings from a temperature log when the well bore fluids are static.

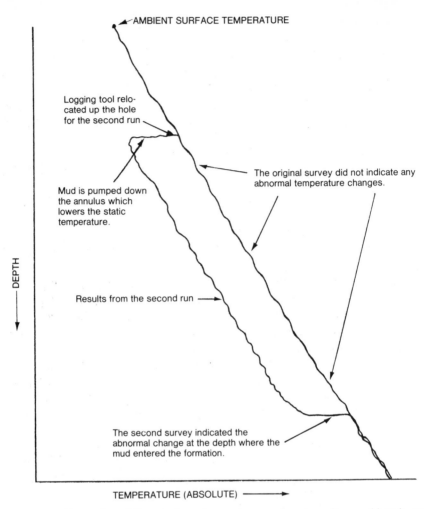

AMBIENT SURFACE TEMPERATURE

Logging tool relocated up the hole for the second run

Mud is pumped down the annulus which lowers the static temperature.

The original survey did not indicate any abnormal temperature changes.

Results from the second run

DEPTH

The second survey indicated the abnormal change at the depth where the mud entered the formation.

TEMPERATURE (ABSOLUTE)

Fig. 4-20 Illustration of the readings from a temperature log after mud has been pumped down the annulus

The temperature log can be used in this case by pulling the tool up the drill pipe, pumping a volume of mud into the annulus, and then running the logging tool again. The tool will read a fluid that has a lower than normal temperature until it reaches the thief zone where it will record normal temperatures (Fig. 4.20). If the results on the second attempt are similar to those shown in Fig. 4.19, it is assumed that the lost circulation zone is at the bottom of the well.

A radioactive tracer tool is often used to isolate the lost circulation problems. A radioactive material is pumped into the well through the mud system after which a logging tool, usually a gamma ray detector, is used to determine areas of high radioactive concentration. The depths at which the high concentrations are noted is assumed to be the point at which the fluid entered the formation. Fig. 4.21 shows an actual case in which a radioactive tracer was used to determine the lost circulation interval.

A recent innovation in logging that can be used to determine the lost circulation zone is the noise log. The tool is a sonic detector that records the sounds created by fluid movement. As the fluid flow from the blowout continues in the annulus, the tool will delineate the static fluid column above the loss zone from the moving fluid below the zone. However, the sensitivity of the tool may be a deterrent to its usage. Field case reports have shown an interference from surface rig vibrations transported through the tubular goods in the well. Regardless, the tool can be used as a qualitative indicator.

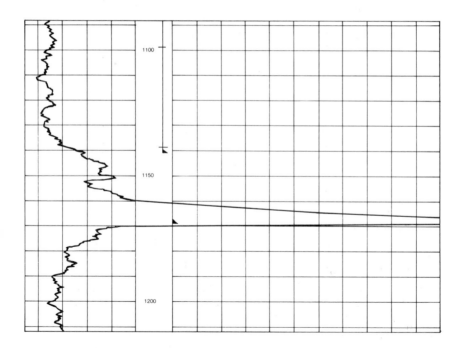

Fig. 4-21   Results of a radioactive tracer log

Problems

4.1   What is the maximum allowable injection pressure for the follow-
       ing set of circumstances?
       Depth of casing = 8,500 ft
       Fracture mud weight (at 8,500 ft) = 15.8 ppg
       Mud weight = 9.0 ppg
       Friction pressure during pump-in (calculated) = 400 psi
                                                 Solution: 3,405 psi

4.2   Calculate and plot the maximum allowable injection pressures for
       Problem 4.1 using mud weights of 9.5 ppg; 10.0 ppg; 10.5ppg; and
       11.0 ppg.

4.3   Calculate and plot the maximum allowable injection pressures for
       Problem 4.1 using friction pressures of 300 psi; 500 psi; 700 psi;
       and 900 psi.

4.4*   Calculate and plot the maximum allowable injection pressures for
       the following well.
       Casing depth = 13,000 ft
       Casing ID = 7.0 in.
       Fracture mud weight at the casing seat = 15.9 ppg
       Kill mud weight = 12.7 ppg
       Plastic viscosity = 25 cp.
       Flow rate = 30 bbl/min

4.5*   Using the data from Problem 4.4, what are the friction pressures
       for pumping rates of 40 bbl/min; 50; 60; and 75 bbl/min? Calculate
       the maximum injection pressures for these rates.

4.6   Use the following data to determine the required hydraulic horse-
       power for a pump-in operation.
       Flow rate = 40 bbl/min
       Maximum pressure = 4,000 psi
       Design factor = 1.3
                                                 Solution: 5,096 HHP

4.7*   Calculate and plot the required hydraulic horsepower for Prob-
       lem 4.5. Use a design factor of 1.5.

4.8   Use the following data to calculate the maximum rate of water
       injection during a pump-in operation.
       Radius of injected water bank = 9 ft
       Effective wellbore radius = 0.75 ft
       Water viscosity = 1.1 cp.

Permeability = 600 md
Net formation thickness = 25 ft
Average reservoir pressure = 2,000 psia
Flowing injection bottomhole pressure = 3,000 psia

4.9* In Problem 4.8, what would be the maximum rate if the radius of the injected water bank were 29 ft?

4.10* What would be the expected cumulative volume of injected water at the front of the water bank under the following circumstances?
Connate water saturation = 0.5
Residual oil saturation = 0.2
Radius of injected water bank = 25 ft
Radius of injector well = 0.5 ft
Net formation thickness = 40 ft
Formation porosity = 0.3

4.11* Calculate and plot the expected cumulative injected volumes for the following water bank radii: 20 ft; 30; 40; 60; and 70 ft. Use the data from Problem 4.10.

4.12 The following pressure and hole data were observed on a kick that preceeded an undbgground blowout. What is the minimum bottom hole pressure? If a heavy slug of mud is to be used that is 3 ppg (equivalent) greater than the bottom hole pressure, what vertical column height is necessary? Assume a 2.5 volume safety margin to reduce the effects of mud cutting.
Well depth = 13,000 ft
Casing seat = 2,000 ft
Original mud weight = 12.9 ppg
Hole size = 12¼ in.
Drill pipe = 5.0 in.
Drill collars = 7½ in., 1,200 ft
Total annular pressure = 12.9 ppg

Time	SIDPP,psi	SICP, psi
1:00	650	930
1:05	670	950
1:10	690	970
1:15	700	910
1:25	700	880
1:40	660	750
2:00	650	750
3:00	650	760

References

1. Personal communications with Boots Hansen of Boots and Coots Inc., Houston, Texas.

2. Personal communications with Coots Matthews of Boots and Coots Inc., Houston, Texas.

3. Davenport, H.H., Bulpard, B.J. and Cashman, J.A.: "How Shell Controlled its Gulf of Mexico Blowouts," *World Oil* (Nov., 1971) 71–73.

4. Barnett, R.D., "A Logical Approach to Killing an Offshore Blowout West Cameron 165 Well No. 3, Offshore Louisiana," SPE 6903 presented at the 1977 Fall Technical Conference, Denver, Colo.

5. *Schlumberger Log Interpretation/Applications*, pp. 110-116.

6. Robinson, J.D., and Vogiatzis, J.P.: "Magnetostatic Methods for Estimating Distance and Direction from a Relief Well to a Cased Wellbore," *Journal of Petroleum Technology*, (June, 1972) 741–749.

7. Morris, F.J., Waters, R.L., Roberts, G.F., and Costa, J.P.: "A New Method of Determining Range and Direction From a Relief Well to a Blowout Well," SPE 6781 presented at the 1977 Fall Technical Conference, Denver, Colorado.

8. Lewis, J.B.: "The Use of the Computer and Other Special Tools for Monitoring a Gas Well Blowout During the Kill Operation — Offshore Louisiana," SPE 6836 presented at the 1977 Fall Technical Conference, Denver, Colo.

9. Britt, E.L.: "Theory and Applications of the Borehole Audio Tracer Survey," presented at the SPWLA Seventeenth Annual Logging Symposium, June, 1976, Denver, Colo.

10. Jones, L.G., Blount, E.M., Glaze, O.H.: "Use of Short-Term Multiple Rate Flow Tests to Predict Performance of Wells Having Turbulence," SPE 6133.

11. Bruist, E.H.: "A New Approach in Relief Well Drilling," Journal of Petroleum Technology, June, 1972, pp. 713–722.

12. Reynolds, M., Jr., Bray, B.G. and Nann, R.L.: "Project Rulison: A Preliminary Report," Conference 700101, Symposium on Engineering with Nuclear Explosives, American Nuclear Society, 1970, pp. 597–626.

13. "Survey of Possible Uses of Nuclear Explosions for Peaceful Purposes Within the National Economy of the Soviet Union," paper presented at General Conference of the International Atomic Energy Agency, Vienna, September, 1970.

14. McLamore, R.T., and Suman, G.O., Jr.: "Explosive Termination of a Wild Well - Evaluation of a Concept," SPE 3591 presented at the 1971 Fall Technical Conference, New Orleans, La.

15. Lewis, J.B., Mabie, G.J., Harris, J.Z., and Barnett, R.D.: "New Innovations for Fighting Blowouts," OTC 2766 presented at the 1977 Offshore Technology Conference held in Houston, Tex.

16. Miller, R.T., and Clements, R.L.: "Reservoir Engineering Techniques Used to Predict Blowout Control During the Bay Marchand Fire," Journal of Petroleum Technology, March, 1972, pp. 234–240.

17. Personal communication with Louis R. Records, Sr. of Louis Records and Associates and Charles M. Prentice of Prentice and Records Enterprises. Inc., of Lafayette, Louisiana.

5

Well planning

SAFE DRILLING PRACTICES TO PREVENT blowouts requires that the planning stages prior to spud of the well must encompass all problem areas that could be encountered. These problems may include weak formations, toxic gas-bearing formations, and high pressured zones. Preventive measures must be installed in the initial planning stages to compensate for the possible problems.

Many subjects should be addressed when planning a well, but only well control topics are presented here. These include:

- fracture gradient determinations,
- casing setting depth selection,
- casing design,
- hydrogen sulfide considerations, and
- contingency planning.

Abnormal pressure detection is presented in Chapter 6.

Fracture Gradient Determination

Well planning demands a knowledge of the pressures required to initiate a fracture into a formation. Fracture gradient calculations, as they are termed, are essential in minimizing or avoiding lost circulation problems and in the proper selection of casing seat depths.

Theoretical Determination

A number of theoretical and field-developed equations have been used to approximate formation fracture gradients.[1] Many of these are suitable for immediate application in a given area while some require a hindsight approach based on density (or other) logging measurements taken after the well has been drilled.

A common base for most fracture gradient determination principles is the assumption that the geological area in question is a tectonically relaxed basin containing plastic-like shales with interbedded sandshale sequences.

As noted by Pilkington,[2] "None of the relationships discussed are valid in either brittle, or naturally fractured formations including limestones, dolomites and shales. Returns may also be lost in vugular formations regardless of the fracture gradient in the overlying formations."

Calculation procedures for these areas rely on either a history of the field or geologic structure, or on field determinations utilizing leak off tests or logging methods.

Hubbert and Willis. In the paper, "Mechanics of Hydraulic Fracturing," [3] Hubbert and Willis explored the variables involved in initiating a fracture in a formation. According to the authors, the fracture gradient is a function of overburden stress, formation pressure, and a relationship between the horizontal and vertical stresses. They believed this stress relationship to be in the range of ⅓ to ½ of the total overburden. Therefore, fracture gradient determination according to Hubbert and Willis would be as follows:

$$\frac{P}{Z} \text{ (min)} = \frac{1}{3} \left(\frac{S_z}{Z} + \frac{2P}{Z} \right) \qquad \text{Eq. 5.1}$$

or

$$\frac{P}{Z} \text{ (max)} = \frac{1}{2} \left(\frac{S_z}{Z} + \frac{2P}{Z} \right) \qquad \text{Eq. 5.2}$$

Where: P = fracture pressure, psi
Z = depth, ft
S_z = overburden at depth Z, psi
p = pore pressure, psi

If an overburden stress gradient (S_z) of 1 psi/ft is assumed, then Eq. 5.1 reduces to:

$$\frac{P}{Z} = \frac{1}{3}\left(1 + \frac{2P}{Z}\right) \qquad\qquad \text{Eq. 5.3}$$

and likewise for Eq. 5.2.

These procedures can be used in a graphical form for a quick solution. In Fig. 5.1, enter the ordinate with the mud weight required to balance the formation. With a horizontal line, intersect the formation pressure gradient line and then construct a vertical line from this point to the minimum and maximum fracture gradients. Read the fracture mud weight from the ordinate. From the example shown in Figure 5.1, the fracture mud weight for a 12.0 ppg formation pressure could range from 14.4 to 15.7 ppg.

Hubbert and Willis made the assumption in these equations that the stress relationships and the overburden gradients were constant for all

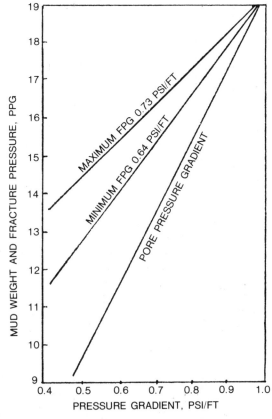

Fig. 5-1 Graphical determination of fracture gradients as proposed by Hubbert and Willis

depths. Since this has been proven to be untrue in most cases, subsequent methods have attempted to account for one or both of these variables more accurately.

Matthews and Kelly.   In realizing that the cohesiveness of the rock matrix will usually be related to the matrix stress and will vary only with the degree of compaction, Matthews and Kelly[4] developed the following equation for calculating fracture gradients in sedimentary formation.

$$F = \frac{P}{D} + \frac{K_i \sigma}{D}$$ Eq. 5.4

Where: P = formation pressure at the point of interest, psi
          D = depth at the point of interest, ft
          σ = matrix stress at the point of interest, psi
          K_i = matrix stress coefficient for the depth at which the value
             of σ would be the normal matrix stress, dimensionless.
          F = fracture gradient at the point of interest, psi

The matrix stress coefficient relates the actual matrix stress conditions of the formation of interest to the conditions of matrix stress if the formation were compacted normally. The authors believed that the conditions necessary for fracturing the formation would then be similar to those for the normally compacted formation.

The stress coefficient versus depth is presented in Fig. 5.2. As can be seen from this illustration, the authors believed that the coefficient would vary with different geological conditions. The values shown were obtained by substituting actual field data of breakdown pressures into Eq. 5.4 and solving for K_i.

The procedure for calculating fracture gradients using the Matthews and Kelly technique is as follows:

1. Obtain formation-fluid pressure, P. This can be done by any satisfactory method.
2. Obtain the matrix stress by using Equation 5.4 and assuming a gradient of 1.0 psi/ft for the overburden.
$$\sigma = S - P$$ Eq. 5.5
3. Determine the depth, D_i, for which the matrix stress, σ, would be the normal value. Make use of the assumption that the overburden pressure is 1.0 psi/ft. From this it follows that
$$0.535 D_i = \sigma$$ Eq. 5.6
   from which the value of D_i can be found.

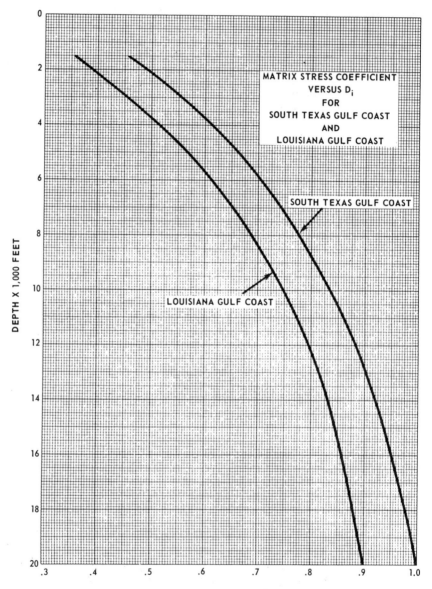

Fig. 5-2 Matrix stress coefficients of Matthews and Kelly

4. Use the value of D_i and apply it to Fig. 5.2 to obtain the corresponding value of K_i.

5. Using the values of D, σ, P, and K_i thus obtained, calculate the value of the fracture gradient F.

Example 5.1 Casing was set on a Texas Gulf Coast well at 7,200 ft. It was estimated that formation pressure was equivalent to 11.0 ppg mud. What is the fracture gradient immediately below the casing seat? Use the Matthews and Kelly procedure.

Solution
1. $P = (11.0 \text{ ppg}) (0.052) (7,200 \text{ ft})$
   $= 4,118 \text{ psi}$
2. $\sigma = S - P$
   $= 7,200 - 4,118$
   $= 3,082 \text{ psi}$
3. Depth of interest, D_i:
   $0.535 \, D_i = \sigma$
   $$D_i = \frac{\sigma}{0.535} = \frac{3,082}{0.535} = 5,760 \text{ ft}$$
4. From Figure 5.2, $K_i = 0.695$
5. $F = \dfrac{P}{D} + \dfrac{K_i \sigma}{D}$
   $$\frac{4,118}{7,200} + \frac{(0.695)(3,082)}{7,200}$$
   $= 0.571 + 0.298$
   $= 0.869 \text{ psi/ft}$
   $= 16.2 \text{ ppg}$

A graphical solution to the Matthews and Kelly technique is presented in Fig. 5.3. Note that the curved lines on the graph represent actual formation pressures and not mud weight in use. Unfortunately, these are often erroneously interchanged. To solve for fracture gradients with Fig. 5.3, enter at the desired depth and read horizontally until the actual formation pressure line is intersected. Plot a vertical line from this point and read the fracture gradient in ppg.

Eaton. The concepts presented by Matthews and Kelly were extended by Eaton[5] to introduce Poisson's ratio into the expression for the fracture pressure gradient.

$$F = \frac{S - P}{D} \left(\frac{v}{1-v} \right) + \frac{P}{D} \qquad \text{Eq. 5.7}$$

Where: Pw = wellbore pressure, psi
   D = depth, ft
   S = overburden stress, psi
   P = formation pressure, psi
   v = Poisson's ratio

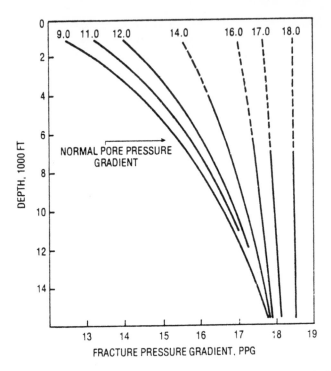

Fig. 5-3 Graphical determination of fracture gradients using the Matthews and Kelly approach

Eaton assumed that both overburden stress and Poisson's ratio were variable with depth. Using actual field fracture data and log derived values, the author prepared graphs illustrating these variations (Fig 5.4 and 5.5). Upon a suitable choice for each variable, the nomograph prepared by Eaton, et. al., (Fig 5.6) can be used to calculate a fracture gradient.

A graphical presentation for the Eaton approach provides a quick solution. The chart (Fig. 5.7) is used in the same manner as the Matthews and Kelly chart (Fig. 5.3). Although the Eaton chart (Fig. 5.7) uses the curves in Fig 5.4 and 5.5, a similar chart can be established for a different area if the overburden stress or Poisson's ratio values differ greatly.

Eaton's method or its modifications is perhaps the most widely used procedure in the industry. It has proved successful both on and offshore and throughout the world. Fertl[6] has shown that Eaton's predictions agree within an acceptable range to actual field derived values from such places as Delaware Basin fields in west Texas, offshore Louisiana, and Green River Basin of Wyoming.

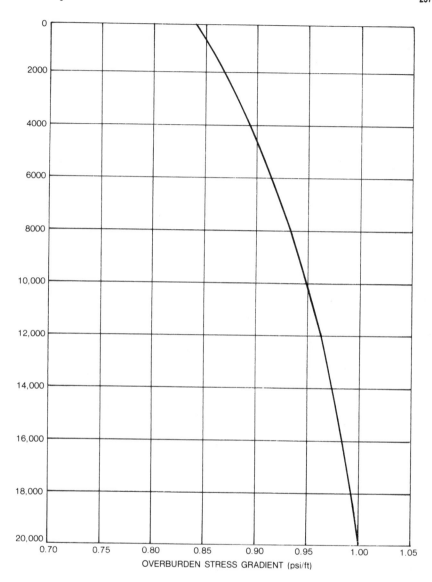

Fig. 5-4 Variable overburden stress by Eaton

Christman. The previously discussed fracture gradient determination procedures assume that overburden stress consists of rock matrix stress and formation fluid stress. In deepwater environments, the distance from the rig to the mudline has no rock matrix and therefore causes the fracture gradients to be reduced when compared to land operations at

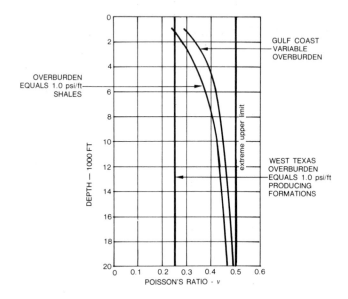

Fig. 5-5 Variable Poisson's ratio with depth as proposed by Eaton

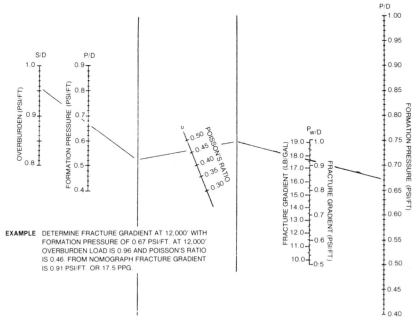

EXAMPLE DETERMINE FRACTURE GRADIENT AT 12,000' WITH
FORMATION PRESSURE OF 0.67 PSI/FT. AT 12,000'
OVERBURDEN LOAD IS 0.96 AND POISSON'S RATIO
IS 0.46. FROM NOMOGRAPH FRACTURE GRADIENT
IS 0.91 PSI/FT. OR 17.5 PPG.

Fig. 5-6 Nomograph determination of fracture gradients as proposed by Eaton

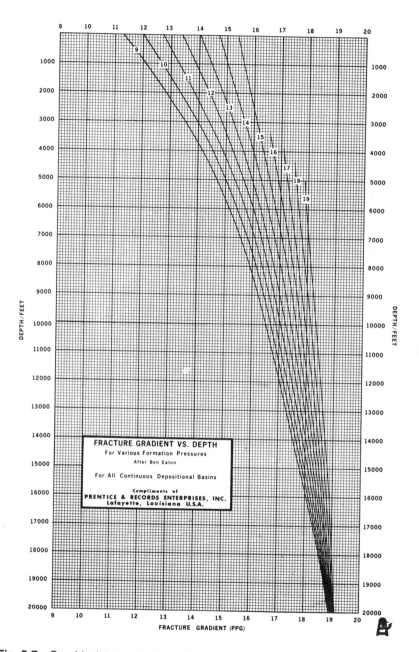

Fig. 5-7 Graphical determination of fracture gradients using the Eaton approach

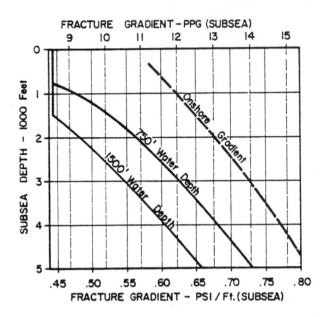

Fig. 5-8   Offshore fracture gradients as a function of depth. Subsea indicates depth below floor.

equivalent depths. Christman[7] has accounted for the effect of water depth in the final form of his total overburden gradient equation.

$$G_{ob} = \frac{1}{D} (0.44 \, D_w + P_b \, D')$$ Eq. 5.8

Where: G_{ob} = total overburden gradient, psi/ft
  D = depth below datum, ft
  D_w = water depth, ft
  P_b = average bulk density, gm/cc
  D' = depth below the mudline, ft

The effect of water depth on fracture gradients can be seen in Fig. 5.8.

Prentice[8] has developed a procedure to calculate deepwater fracture gradients. The procedure utilizes the techniques as established by Christman and data collected by Eaton. Example 5.2 illustrates the procedure for a well drilled in 1,000 ft of water.

Example 5.2   In the illustration in Fig. 5.9, what is the effective fracture gradient at the casing seat?

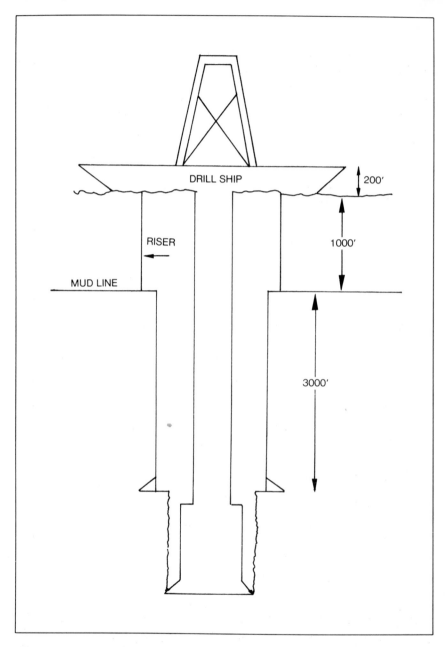

Fig. 5-9 Typical effect of water depth on fracture gradient (see Example 5-2).

Solution
1. Convert the water depth to an equivalent section of formation.
   1,000 ft × 0.465 psi/ft = 465 psi
2. From Eaton's overburden stress chart in Fig. 5.4, the stress gradient at 4,000 ft equals 0.89 psi/ft.

$$\frac{465\text{psi}}{0.89 \text{ psi/ft}} = 522 \text{ ft, equivalent}$$

3. Calculate and convert apparent fracture gradient to actual fracture gradient.
   522 ft + 3,000 ft = 3,522 ft, equivalent
   From Eaton's fracture gradient chart, the gradient at 3,522 ft = 13.92 ppg. or
   Fracture pressure = 0.052 × 13.92 ppg × 3,522 ft
   = 2,549 psi
4. The effective fracture gradient from the mud flow line at the drill ship deck to the casing seat is:
   2,549 psi × 19.23/(200 + 1,000 + 3,000) ft = 11.67 ppg

Field Determination of Fracture Gradients

It is a common practice to pressure test each new casing seat in field applications to determine the exact minimum fracture gradient. The primary reason for this practice is due to the inability of any theoretical procedure to account for all possible formation characteristics. As an example, several authors have noted wells that exhibited lower than expected fracture gradients due to abnormally low bulk densities in the rock.[9]

The most common procedure used for the field determination of fracture gradients is the leak off test (often called the pressure integrity test). The procedure used in the test is to close the blowout preventers and then gradually apply pressure to the shut-in system until the formation initially accepts fluid. These results of the test would be similar to those shown in Fig. 5.10. Example 5.3 illustrates the procedure.

Example 5.3 Casing was set at 10,000 ft in a well. The operator wishes to perform a leak off test to determine the fracture gradient at 10,000 ft. If the mud weight in the well was 11.2 ppg, what is the fracture gradient at the casing seat?

Solution
1. Close the blowout preventers and rig up a low volume output pump.

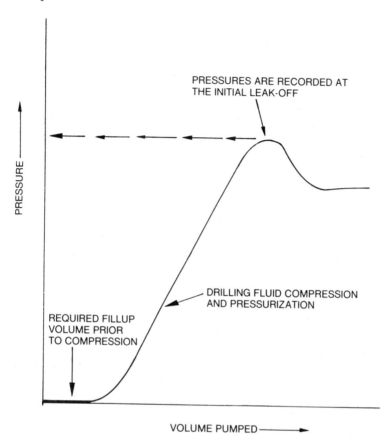

PRESSURES ARE RECORDED AT
THE INITIAL LEAK-OFF

DRILLING FLUID COMPRESSION
AND PRESSURIZATION

REQUIRED FILLUP
VOLUME PRIOR
TO COMPRESSION

VOLUME PUMPED ⟶

Fig. 5-10 Typical results from a leak off test

2. Apply pressure to the well and record the results as follows:

Volume pumped, bbl	Pressure, psi
0	0
1	45
1½	125
2	230
2½	350
3	470
3½	590
4	710
4½	830
5	950
5½	990
6	1010

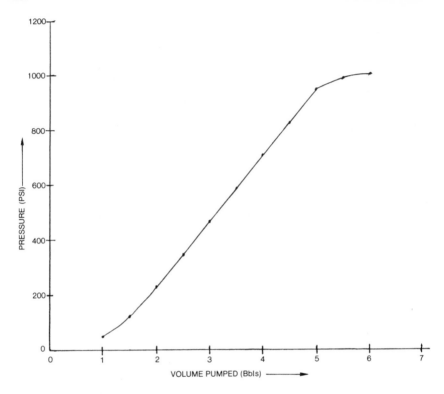

Fig. 5-11 Results of leak off test data from Example 5-3

3. The results are plotted in Fig. 5.11.
   From these results, it appears that the formation will begin to fracture when 950 psi is applied.
4. Fracture gradient

$$= [(11.2 \text{ ppg}) \ (0.052) \ (10,000 \text{ ft}) \ + \ 950 \text{ psi}]/ \\ 10,000\text{ft}$$
$$= 6,774 \text{ psi}/10,000 \text{ ft} = 0.6774 \text{ psi/ft}$$
$$= 13.02 \text{ ppg equivalent}$$

Casing Setting Depth Selection

The importance of selecting proper depths for setting casing cannot be overemphasized. Many wells have been engineering and economic failures because the casing program had specified setting depths too shallow or too deep. The application of a few basic drilling principles combined with a basic knowledge of the geological conditions in an area can aid in determining where casing strings should be set to ensure that drilling can proceed with minimum difficulty.

Casing seat depths are directly affected by the geological conditions for the well that is to be drilled. In some areas, the prime criterion for selecting casing seats is to cover exposed, severe lost circulation zones. In others the seat selection may be based on differential sticking problems, perhaps resulting from pressure depletion in a field. In deep wells, however, the primary consideration is usually based on controlling abnormal formation pressures and preventing their exposure to weaker, shallow zones. The design criteria of controlling formation pressures is generally applicable to most drilling areas.

Selecting casing seats for pressure control purposes starts with knowledge of geological conditions such as formation pressure and fracture gradient. This information is generally available within an acceptable degree of accuracy due to recent advances in conventional logging and seismic interpretation and in fracture gradient determination through theoretical calculations. These pre-spud calculations and the actual drilling conditions will determine the exact locations for each casing seat.

The principle used to determine setting depth selection can be adequately described by the adage that "hindsight is 20-20." The initial step is to determine the formation pressures and fracture gradients that will be penetrated in the well. After these have been established, the operator

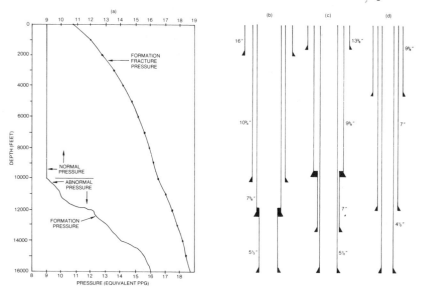

Fig. 5-12 Several casing designs for a typical set of conditions

must design a casing program based on the assumption that he already knows the behavior of the well even prior to its being drilled.

This principle is used extensively for infill drilling where the known conditions dictate the necessary casing program. Using these guidelines, the operator can select the most effective casing program that will meet the necessary pressure requirements and minimize the casing cost.

An example can be seen in Fig. 5.12 (a–d). In (a) the anticipated formation pressures and fracture gradients are shown for a well that is to be drilled. Fig. 5.12 (b–d) illustrate several casing programs that would satisfy all of the basic design requirements. The actual selection of one of these programs would be based on further design criteria of the individual operator. He would consider casing availability, potential for deepening the well, or other specific drilling problems.

Setting depth selection for intermediate or deeper strings. Setting depth selection should be made for the deepest casing strings to be run in the well and then successively designed from the bottom string to the surface. Although this procedure may appear to be reversed upon initial inspection, application in this manner avoids the necessity of several time-consuming iterative procedures. Surface casing design procedures are based on other criteria.

The initial design step is to establish the projected formation pressures and fracture gradients. Using Fig. 5.13 (a) as an example, it can be seen that a 17.2 ppg (equivalent) formation pressure exists at the hole bottom. To reach this depth, wellbore pressures greater than 17.2 ppg will be necessary and must be taken into account.

These pressures may include a trip margin of mud weight to control swab pressures, an equivalent mud weight increase due to surge pressures, and a safety factor. Once these total pressures have been established, the fracture gradient curve in Fig. 5.13 (a) must be used to locate those zones that cannot withstand the imposed pressures and thereby must be covered with casing. Example 5.4 will illustrate this process.

Example 5.4   Using Figure 5.13, select the setting depth for a casing string that will allow the lowermost hole section to be drilled.

Solution
1. From Figure 5.13 (a) the maximum anticipated pressure is 17.2 ppg (equivalent). What are the total imposed pressures?

   17.2 ppg (formation pressure-actual mud weight)

   0.3 ppg (trip margin-actual mud weight)

   0.3 ppg (surge pressure-equivalent mud weight)

   + 0.2 ppg (safety margin-equivalent mud weight)

   18.0 ppg

2. Determine those formations that cannot withstand a 18.0 ppg pressure. The deepest of these zones is the (minimum) setting depth.
3. From Fig 5.13 (b) the (minimum) depth would be 12,100 ft.

Most wells will require that this procedure be used to determine several setting depths. An example is a well that has an intermediate string and one or more liners. Example 5.13 was designed to show this step. The same procedure is followed as presented in Example 5.4 and a

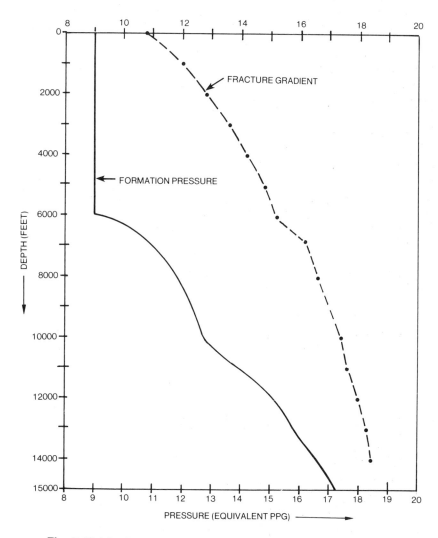

Fig. 5-13 (a) Projected formation pressures and fracture gradients

depth of 6,800 ft is selected as the setting depth for the shallower string (Fig. 5.13 c).

The intermediate string of casing is usually the shallowest string of pipe where other design criteria must also be taken into account. One of these is differential pipe sticking. As shown in Fig. 5.14, some initial designs may yield a solution where the actual drilling of the well may produce pipe sticking problems due to high differential pressures. In this case, the greatest tendency would be at 11,000 ft.

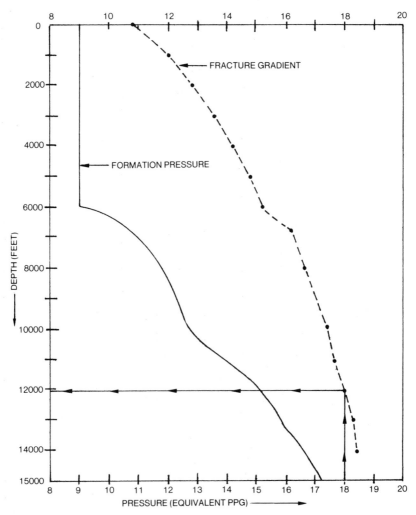

Fig. 5-13 (b)   Illustration of the procedures used to determine the setting depth for the deep string of casing for Example 5-4

Field studies have shown that an average of approximately 2,300 psi differential pressure would be the maximum that should be imposed on normal-pressure zones and still expect to remain free, assuming that optimum mud properties are maintained.[10] If this is the case in the initial design, the pipe must be set shallower to minimize the possibility of sticking either the drill string or casing.

Surface casing setting depth selection. At some depth in a well, kick-imposed equivalent mud weights will be greater than those result-

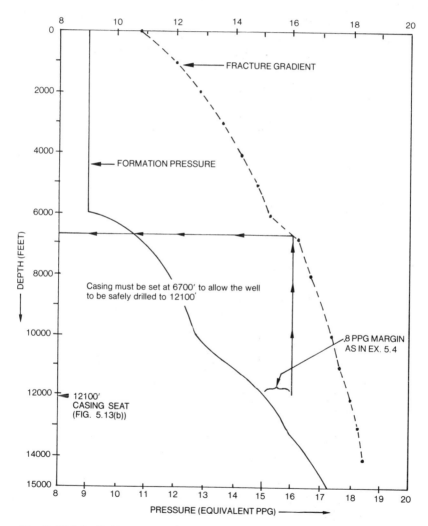

Fig. 5-13 (c) Subsequent procedures used to select the next setting depth

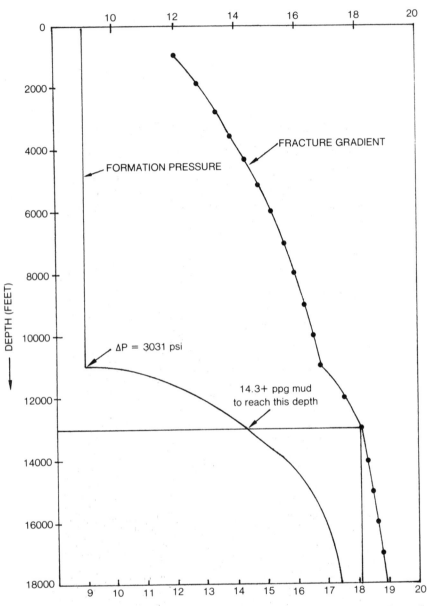

If pipe is set at 13000 ft. as indicated, a 14.3+ ppg mud weight will be required. This will impose a differential pressure of 3031 psi at the 11,000 ft. normal pressured zone. As a result, the operator has the option of (1) to set the casing shallower than 13000 ft. and run a drilling liner, or (2) use optimum mud properties in an effort to prevent differential sticking when the mud weight reaches 14.3 ppg.

Fig. 5-14 Pipe sticking tendencies may often require that alternate casing depths be used

ing from friction and surge pressures during normal drilling activities. When this occurs, the casing setting depth must be selected based on kick tolerance rather than the procedures used for intermediate or deeper strings.

The first string usually subject to these conditions will be the surface casing, although occasionally a deeper string must be designed based on this principle.

Kick-imposed equivalent mud weights are the cause for most underground blowouts. When a kick occurs, the shut-in casing pressure added to the drilling mud hydrostatic pressure exceeds the formation fracture pressure and results in an induced fracture. Therefore the objective of a seat selection procedure that avoids underground blowouts would be to choose a depth that exhibits sufficient competency to withstand the pressures imposed by reasonable kick conditions.

A simple relationship for kick pressures can be seen in Fig. 5.15.

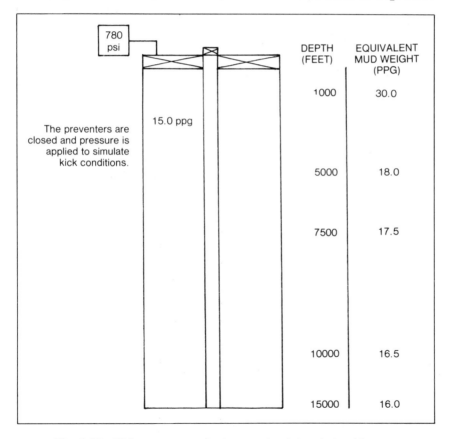

Fig. 5-15 Kick pressure-equivalent mud weight relationships

Under these conditions, the equation that describes the equivalent mud weights at any depth is:

$$\text{EMW} = (\text{Total depth/surface depth}) (0.5) + \text{original mud weight}$$

Eq. 5.9

Where: EMW = equivalent mud weight at any depth, ppg
Total depth = depth of deepest interval, ft
Surface depth = depth of interest, ft
(0.5) = incremental kill mud weight increase, ppg
Original mud weight = mud weight in use, ppg

This equation can be used in an iterative manner along with a suitable theoretical fracture gradient calculation to determine a surface pipe depth that will have sufficient strength to resist kick pressures. Initially a shallow depth is chosen for which the fracture gradient and equivalent mud weights are calculated. If the equivalent mud weight is greater than the fracture gradient, a deeper interval must be selected and the calculations repeated.

This procedure is followed until the fracture gradient exceeds the equivalent mud weight. When this occurs, a depth has been selected that will withstand the designed kick pressures. Example 5.5 illustrates the use of Eq. 5.9.

Example 5.5 An operator wishes to drill a well to 12,000 ft and use sufficient surface casing so that an intermediate string is not required. If the maximum anticipated mud weight at 12,000 ft is 11.9 ppg, where must the surface casing be set? Use Equation 5.9 and the Eaton fracture gradient chart.

Solution
1. Select a shallow depth and calculate the equivalent mud weight and fracture gradient.
   a. EMW = (Total depth/surface depth) (0.5 ppg) + 11.9 ppg
      = (12,000 ft/1,800 ft) (0.5 ppg) + 11.9 ppg
      = 15.2 ppg
   b. From Eaton's chart, the fracture gradient would be 12.7 ppg if normal formation pressures are assumed.
2. Since the EMW exceeds the fracture gradient, a deeper interval must be evaluated. Sample results are as follows:

Depth, ft	EMW, ppg	Fracture gradient, ppg
1,800	15.2	12.7
2,100	14.8	12.9
2,400	14.4	13.3
2,700	14.1	13.5
3,000	13.9	13.7
3,300	13.7	13.8
3,600	13.6	14.0

3. A depth of 3300–3600 feet would be selected as the surface casing setting depth.

The value of 0.5 ppg represents the average (maximum) mud weight increase necessary to kill a kick. Use of this variable in Equation 5.9 allows the operator to (inadvertently) drill into a formation in which the pressure was in excess of 0.5 ppg greater than the original calculated value and still safely control the kick that may occur.

In fact, if the original mud weight variable is 0.3–0.4 ppg greater than the anticipated formation pressure, the equation would account for formation pressure calculation errors of 0.8–0.9 ppg. If necessary, an operator may elect to alter the 0.5 ppg variable to whatever is deemed most suitable for his particular drilling environment.

A valid argument has been raised concerning Equation 5.9 and its representation of field circumstances. In actual kick situations, the equivalent mud weights are controlled to a certain degree by casing pressure which is not directly taken into account in the equation. However, an inspection of a casing pressure will show that the two components in the pressure are degree of underbalance between the original mud and the formation pressure and degree of underbalance between the influx fluid and formation pressure.

The first of these two components is in the equation in the form of the incremental mud weight increase term, while the latter is not taken into account. In most kick situations, the average value of the second component will range from 100–300 psi. If an operator believes that the second component is of significance worthy to alter the equation, this can easily be done by changing the incremental mud weight increase term to a higher value.

The setting depth selection procedure presented in this section obviously cannot serve under all possible drilling conditions. However, an evaluation of the existing drilling conditions can be used to develop procedures similar to the ones presented. Often, a minor modification to the techniques developed in these illustrations will serve the operator well in meeting his drilling requirements.

Casing Design

(Author's note: This section on casing design was authored by Charles M. Prentice of Prentice and Records Enterprises, Inc. The original paper was first published as SPE 2560 and later reprinted in the Journal of Petroleum Technology.[11] Special acknowledgements are given to Mr. Prentice and the Society of Petroleum Engineers for their permission to reprint this material.)

In keeping with the format of this text a complete example problem is provided for each type of casing design. To avoid a discontinuity within the text, the examples are printed in Appendix B.

Introduction

As long as we drill in areas of the lost returns, abnormal formation pressures, differential sticking, heaving shale, and crooked holes, the need for designed casing is recognized. The attainable, optimum condition is to design casing to withstand the maximum loads from the imposed drilling problems for the minimum cost.

To properly evaluate the maximum loads imposed on different types of designs, each type must be considered separately. To this end, six types of casing are defined.

1. Conductor casing
2. Surface casing
3. Intermediate casing
4. Intermediate casing with a drilling liner
5. The drilling liner
6. Production casing

The loading for burst should be considered first since it is found that burst will dictate a majority of "maximum load" casing strings. The weights, grades, and section lengths for the burst design can be determined. The collapse load can next be evaluated, and the string sections upgraded as necessary. Once the weights, grades, and section lengths have been determined to satisfy burst and collapse loadings, the tension load can be evaluated. The tube can be upgraded as necessary and the coupling types determined.

The final step is a check for biaxial reductions in burst strength and collapse resistance caused by compression and tension loads respectively. If these reductions show the strength of any part of the section to be less than the potential load, the section(s) affected can be upgraded to withstand this loading.

By initially choosing the least expensive weights and grades of casing which satisfy burst loading, and upgrading only as directed by the prescribed sequence, the resulting design will be the least expensive possible which fulfills all maximum loading requirements.

As previously stated, different applications of casing strings dictate different loading configurations. Since intermediate casing offers the most complex set of considerations, it will be discussed first.

The starting point in "maximum load" casing design is an evaluation of maximum loads pertinent to a geographical region or geological section. It would be difficult, if not impossible, to list the considerations for all possible combinations of regions and sections. With that in mind, a general procedure applicable to the United States Gulf Coast area will be used in this text. Actual conditions must be substituted for specific different applications.

Intermediate Casing

Burst. In order to evaluate the burst loading, the values of surface and bottom hole burst limits must first be established. The surface burst pressure limit is arbitrary and is set equal to the related working pressure of the surface equipment (well head, BOP's, etc.) to be used. The burst limitation at the bottom of the casing is calculated and is equal to the predicted fracture gradient of the formation immediately below the casing shoe plus a safety factor.

Since the value of the fracture gradient is generally expressed in terms of mud weight, the recommended safety factor is 1.0 ppg. Thus, the bottom hole burst pressure, defined as the injection pressure, is equal to the fracture gradient expressed as mud weight plus the safety factor of 1.0 ppg, converted to pressure.

With the end points determined, the maximum burst load line may be constructed. Since the maximum load will occur when the end points are satisfied simultaneously, the loading will necessarily be provided under kick conditions. A characteristic of kick loading is the existence of two or more fluids in the well bore which are the mud being drilled with at the time of the kick and the influx fluid(s).

Since consideration is given only to maximum loads, the fluids considered will be the heaviest mud weight projected for use below the casing string, and gas as the single influx fluid. The position of these fluids in the borehole is important as illustrated in Fig. 5.16. If the gas is considered to be on top, the load line will appear as the line labeled (1). If the positions are reversed so that mud is placed on top, the load line will be as illustrated by line (2). It is evident that (2) exerts the greater burst loading by the amount of shaded area between the lines. Therefore the configuration with the heaviest mud weight to be used on top, gas on the bottom, and the end points satisfied simultaneously will constitute the maximum load line.

Remaining is the determination of the lengths of the columns of mud and gas. By assuming a gradient for the gas (0.115 psi/ft), this may be accomplished by solving the following simultaneous equations.

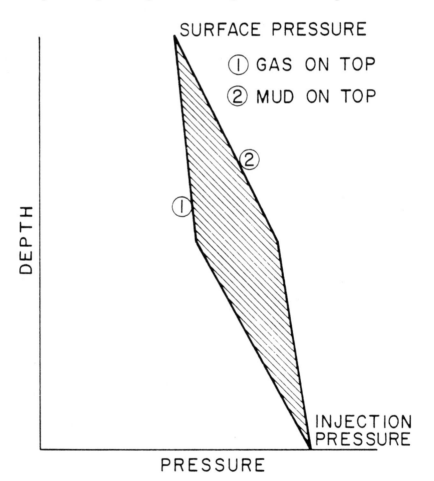

Fig. 5-16   Fluid positions

$$x + y = D \qquad \text{Eq. 5.10}$$
$$P_s + xG_m + yG_g = \text{injection pressure} \qquad \text{Eq. 5.11}$$
$$= 0.052 \text{ (F.G. + S.F.) D}$$

Where: x = Length of mud column, ft
y = Length of gas column, ft
D = Setting depth of the casing, ft
P_s = Surface pressure, psi

BURST DESIGN

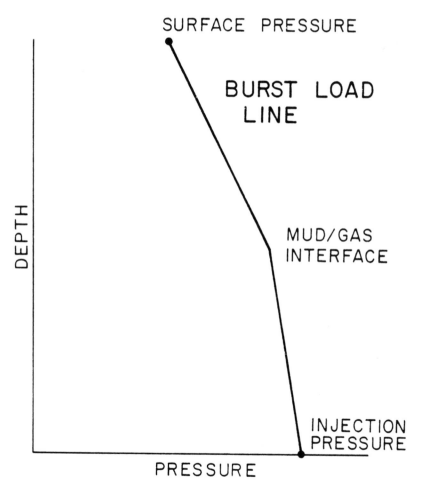

SURFACE PRESSURE

BURST LOAD LINE

MUD/GAS INTERFACE

DEPTH

INJECTION PRESSURE

PRESSURE

Fig. 5-17 Burst design-load line

G_m = Gradient of the heaviest mud weight to
be used, psi/ft.
G_g = Gradient of gas–assumed as 0.115 psi/ft.
F.G. = Calculated fracture gradient, ppg
S.F. = Safety factor–recommended as 1.0 ppg
0.052 = Conversion constant

BURST DESIGN

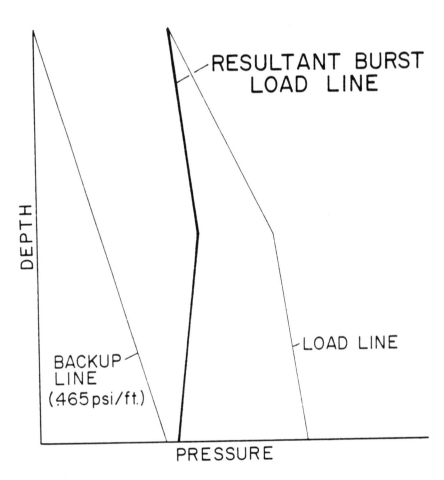

Fig. 5-18 Burst design-resultant line

With the simultaneous solution of the equations, the length of the respective columns of fluids becomes known, and the resulting load line is illustrated in Fig. 5.17.

The load acting to burst the casing is now known at every increment of depth by reference to the graphical representation. A load resisting this burst, imposed by the fluid occupying the annular space behind the casing, may also be calculated and applied.

BURST DESIGN

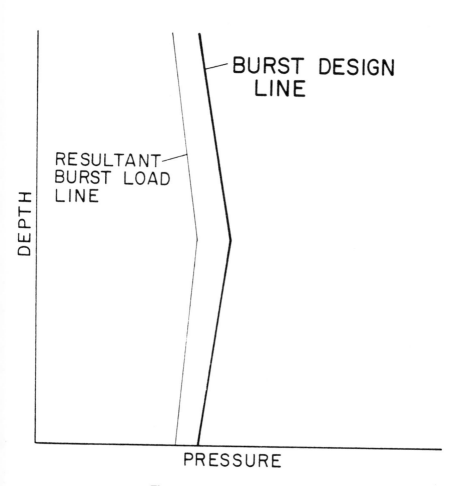

Fig. 5-19 (a)   Burst design line

Due to weight degradation of the fluid behind the casing in contact with the formations (backup fluid), and in keeping with the maximum load concept, it is assumed that backup is provided by a column of fluid equal in density to saltwater with a gradient of 0.465 psi/ft. By plotting this burst-resisting backup load and subtracting it from the burst load line, the actual pressure load at each depth increment can be determined. This load is labeled with the resultant burst load in Fig. 5.18.

If a design factor is deemed necessary to allow for wear, it may be applied to the resultant burst load to obtain the design line. For illustrative purposes, a design factor of 1.1 is applied and the design line shown in Fig. 5.19 (a) is created. If no design factor is necessary, the resultant burst load line becomes the design line.

BURST DESIGN

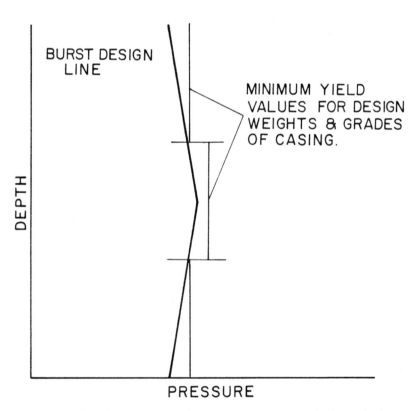

Fig. 5-19 (b)   Selection of casing weights and grades for burst design

Starting at either end of the design line, plot the published values of burst strength for the least expensive weight and grade of casing that exceeds the design load. The section length is determined by intersection with the design line. The strength of the next applicable weight and/or grade can be plotted to intersection, and this procedure repeated until the string is completely designed for burst (Fig. 5.19 b).

If this procedure is rigidly adhered to, it is possible to end up with more sections than would be practical to handle in the field, particularly for offshore operations. For this reason, a compromise may be made

COLLAPSE DESIGN

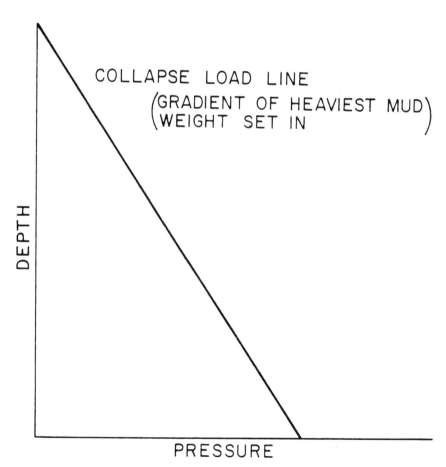

Fig. 5-20   Collapse design-load line

between cost and practicality by setting a restrictive number of sections which would be allowed.

Upon completion of this phase, the designer will have the weight, grades, and section lengths of the casing which will satisfy the burst loading. This tentative design is set aside pending evaluation of the collapse loading.

Collapse.   The collapse load for intermediate casing is imposed by the fluid in the annular space, assumed to be the heaviest mud weight the string is projected to be run in. An ambiguity lies in the fact that in the

COLLAPSE   DESIGN

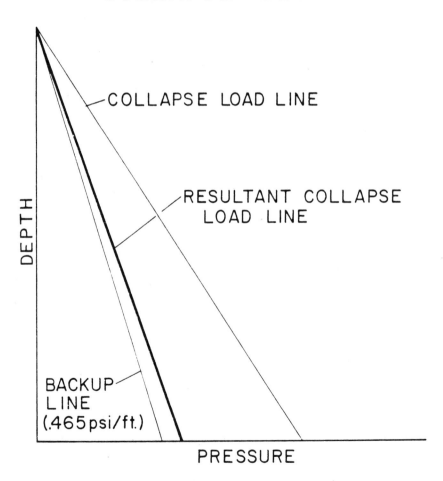

Fig. 5-21   Collapse design-resultant line

burst loading considerations, the density of the fluid behind the casing was taken as minimal (0.465 psi/ft saltwater). If the precepts of maximum load are to be followed, however, no other assumption is valid (Fig. 5.20). The backup fluid for collapse considerations also adheres to the maximum load concept.

Maximum collapse loading will occur when the mud level inside the casing drops, attendant to loss of circulation. At the intermediate casing shoe it is improbable that the hydrostatic pressure exerted by the reduced mud column would be less than that exerted by a full column of

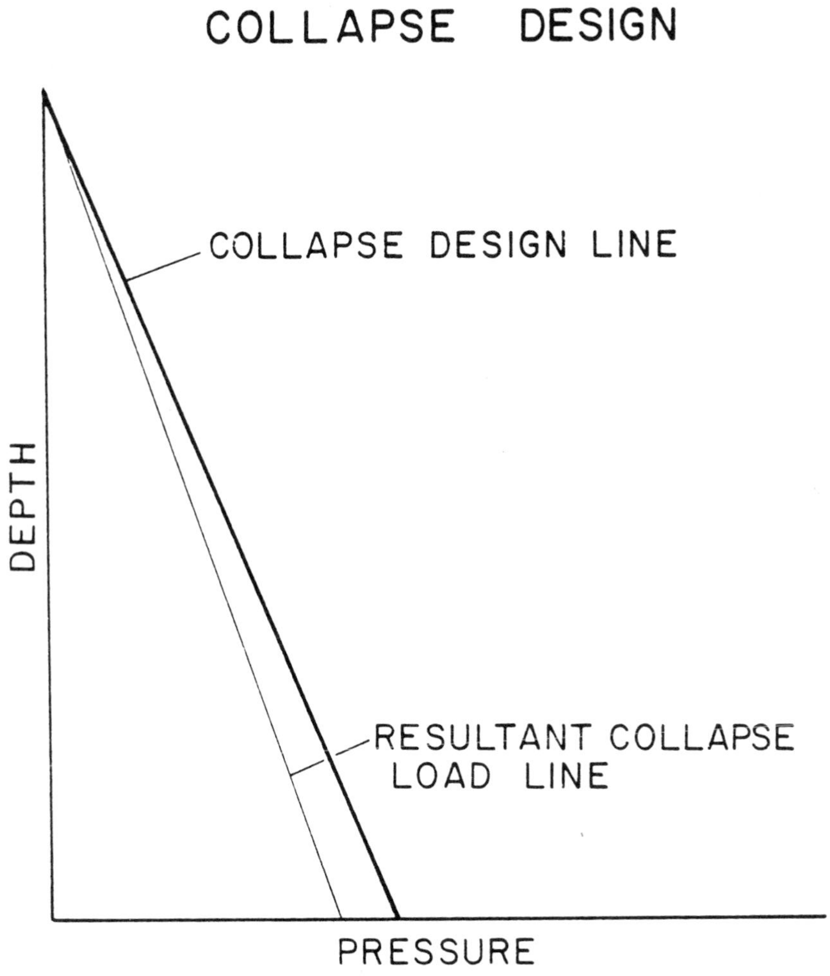

COLLAPSE   DESIGN

Fig. 5-22 (a)   Collapse design line

saltwater. Therefore, maximum collapse loading results when minimum backup is provided.

This occurs when circulation is lost while drilling below the string with the heaviest mud weight projected for use. The fluid level falls so that the column of heavy mud remaining exerts a pressure at the shoe equal to a full column of saltwater (Fig. 5.21). The resultant collapse loading is the backup line subtracted from the load line as shown in Fig. 5.22 (a). Application of a collapse design factor of 1.1 results in the collapse design line.

On the graphical representation of the collapse design line, the collapse resistances of the sections dictated by burst considerations should now be plotted and checked. Should the collapse resistances fall below the collapse design line, the section should be upgraded for collapse.

COLLAPSE DESIGN

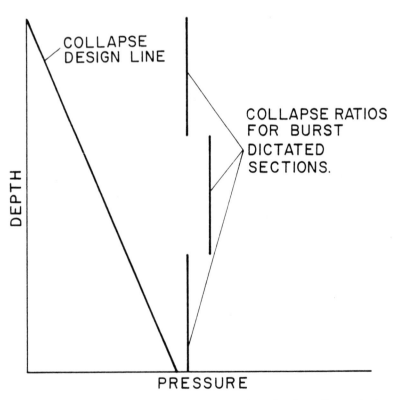

Fig. 5-22 (b)   Selection of casing weights and grades for collapse design

When the checking and necessary upgrading are completed, the result is a design of weights, grades, and section lengths which satisfy the burst and collapse maximum loads as shown in Fig. 5.22 (b).

Tension. Knowing the weights, grades and section lengths based on burst and collapse design, the tension load (both positive and negative) can be evaluated.

Buoyancy has been omitted in some casing designs in the past for various reasons. Due to the actual manner in which the buoyant force is

EFFECT OF BUOYANCY

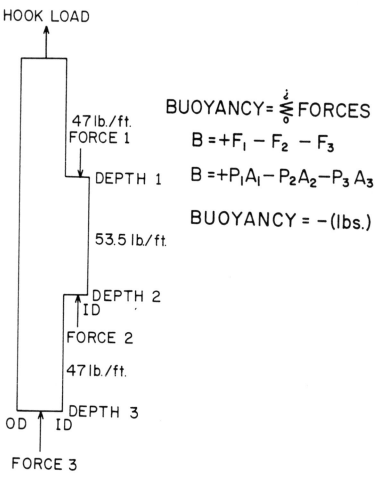

Fig. 5-23 Effect of buoyancy

applied to a casing string, burst and collapse resistances are altered by the effect of biaxial stresses. For this reason buoyancy cannot be overlooked in a maximum load design.

The effect of buoyancy is commonly thought of as the reduction in string weight when run in liquid as compared to the string in air. However, no allowance is made regarding the manner in which the buoyant force is applied to the casing. The buoyancy, or reduction in string weight, as noted on the surface is actually the resultant of forces acting on all the exposed, horizontally oriented areas of the casing string.

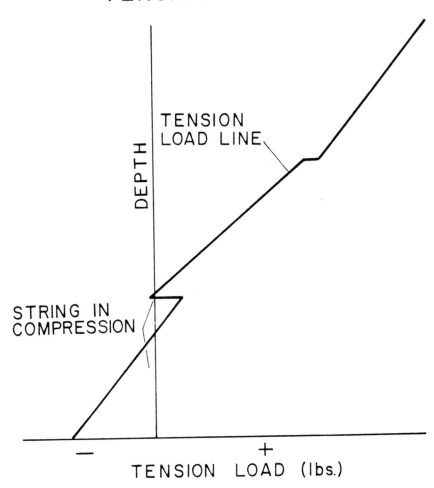

Fig. 5-24 Tension design-load line

The forces are equal to the hydrostatic pressures at each respective depth times the exposed areas, and defined as negative if acting upward. The areas in reference are the tube end areas, the shoulders at changing casing weights, and to a small degree the shoulders on collars. Fig. 5.23 shows the references forces acting at each exposed area of a casing string, with the resultant loading indicated as negative tension (compression). The forces acting on the areas of collar shoulders are sufficiently small as to be considered negligible in a practical casing design.

The reduction in hook load observed at the surface is the same as that calculated using either the "buoyancy factor" method, or by calculating the weight of the volume of known density liquid displaced and subtracting it from the dry weight. However, the tension loadings differ

TENSION DESIGN

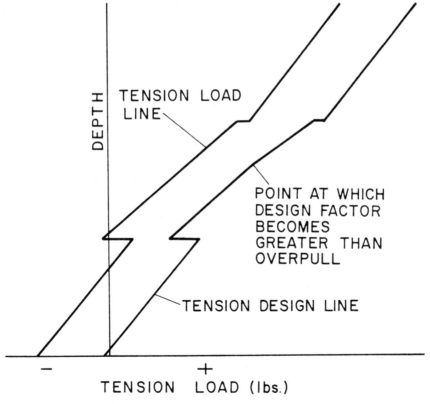

Fig. 5-25 Tension design line

greatly. Once the magnitude and location of the forces are determined, the tension load line may be graphically constructed as in Fig. 5.24. It is noteworthy that more than one section of the casing string may be loaded in compression.

In order to obtain a design line for tension, it is recommended that a design factor be used with a conditional minimum overpull value included. The recommended values for the safety factors are 1.6 for the design factor and/or 100,000-lb overpull, whichever is greater.

TENSION DESIGN

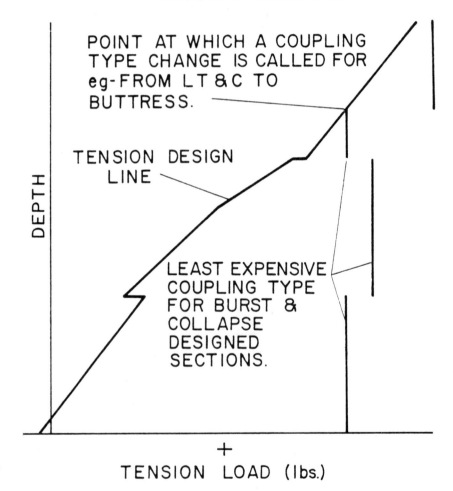

Fig. 5-26 Selection of casing couplings

This is a form of compromise with the Goins, Collings, and O'Brien "marginal loading" concept,[12] and allows for safely pulling on stuck casing to some definite predetermined value (100,000 lb in this case). The graphical representation of this combination of design factors is shown in Fig. 5.25, and labeled the "tension design line."

With few exceptions, the weakest part of a joint of casing in tension is the coupling. Therefore, the tension design line is used primarily for determination of coupling type. The least expensive coupling strengths which satisfy the design can be plotted, and the proper couplings determined as shown in Fig. 5.26.

At this point the entire string is designed for burst, collapse and tension, and the weights, grades, section lengths, and coupling types are known. Remaining is to check the reductions in burst and collapse resistances caused by biaxial loading.

The tension load line, Fig. 5.27, which shows tension loading versus depth is used to evaluate the effect of biaxial loading. By noting the magnitude of plus (tension) or minus (compression) loads at the top and bottom of each section, the strength reductions can be calculated using the Holmquist and Nadai ellipse. With the reduced values known at the

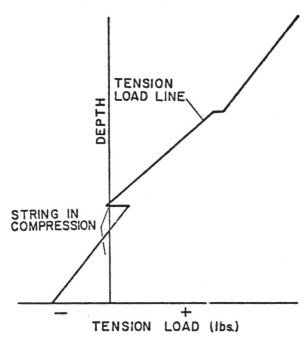

Fig. 5-27 Tension design line for biaxial loading consideration

ends of each of the sections, a new strength line can be constructed by connecting the end points with a straight line. Should the reduced values indicate an under design, the section should be upgraded.

In addition to the above discussed tension loading, another type of tension load must be considered in deviated holes or holes with severe doglegs. These conditions result in point loading which must be evaluated as additional tensions.[13] The equation used to calculate the magnitude of this load is an adaptation of the bending force calculation (Eulers Equation). The equation is of the following form:

$$F_b = 218 \ (d_p) \ (\phi) \ (A_s) \ \frac{6KJ}{Tanh \ (6KJ)} \qquad \qquad \text{Eq. 5.12}$$

Where: $F_b =$ Bending force (tension load additive), lb
                     $d_p =$ Pipe, OD, in.
                     $\phi =$ Dogleg angle, degrees/100 ft
                     $A_s =$ Pipe wall area, sq in.
                     $K = \sqrt{\text{Tension Load at Point (lb)}} \ / \ EI$
                     $E =$ Modulus of Elasticity, 30,000,000 psi
                     $I =$ Moment of Inertia (in.^{4}) $= (\ \pi \ / \ 64) \ (d_p^{\ 4} - d_{i_p}^{\ 4})$
                     $d_{i_p} =$ ID of Pipe, in.
                     $J =$ Joint length, 30 ft or 40 ft
            $Tanh =$ Hyperbolic tangent of value 6KJ

Surface Casing Design

Only the intermediate casing design follows the outlined procedure in its entirety. Surface casing is the first of the four types of variations of the procedures.

Burst. Due to the relatively low injection pressures associated with surface casing, it is recommended that a surface pressure limit be disregarded. Determine the injection pressure at the casing shoe, and assume a column of gas back to the surface. Figure 5.28 shows the burst load line.

Pressure at the surface will be equal to the injection pressure less the hydrostatic pressure of a column of gas (0.115 psi/ft), rather than a set limit. Backup is provided by a saltwater column and the section, weight, and grade determination procedure is the same as outlined for intermediate casing.

Collapse. Due to the possibility of lost returns allowing the fluid level to fall below the surface casing shoe, no backup load is applied to

SURFACE CASING
BURST DESIGN

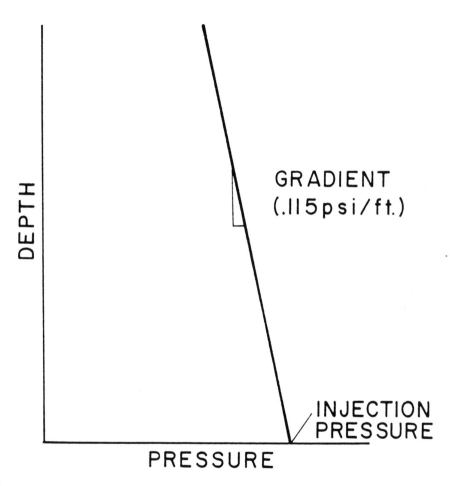

Fig. 5-28   Surface casing burst design

collapse loading. The load line with a design factor applied becomes the design line. Checking the design line for the burst-dictated sections is the same as for intermediate casing. The tension and biaxial reduction calculations are as outlined previously.

Intermediate Casing With a Drilling Liner and Liner

Burst.   If a drilling liner is to be included in the drilling of a well, the design of the intermediate casing string will be altered slightly. Since the injection pressure and heaviest mud weight used will be greater below the liner, these values are to be used to design the intermediate string as well as the liner.

The procedure for evaluation will remain the same as intermediate casing. A surface pressure limit will be decided upon. The injection pressure at the liner shoe will be calculated, and the load line developed just as in the intermediate string design. The backup fluid will remain saltwater, and the resultant loading defined as before. A design factor is applied to obtain the design line. This line is used to design both the intermediate casing and the liner for burst, as shown in Fig. 5.29.

Collapse.   Collapse, tension, and biaxial loading remain the same as previously discussed. However, the designer must give consideration to the discontinuity in the load line which results from the separate casing strings being set in two different mud weights.

Production Casing

Burst.   The burst design for production casing involves several assumptions which warrant discussion. One of the assumptions is that the packer fluid used is equal in density to the mud weight in the annular space behind the casing. This is not strictly in keeping with maximum loading considerations and makes a good case for utilizing lightweight packer fluids. The result of this assumption is that the effect of the load and backup fluids cancel out and, at this point, the casing has no burst load or backup.

The second assumption made is a tubing leak near the surface, resulting in bottom hole pressure less gas hydrostatic being introduced as a burst load over the entire length of the production casing. The bottom hole pressure used for this calculation can be predicted from offset data, measured pressures from tests, by log interpretation, or by assuving it equivalent to the maximum projected mud weight at total depth. As may have been noted, the fracture gradient and the injection pressure are not used to design production casing. A design factor is applied to the load

INTERMEDIATE CASING & DRILL IN LINER BURST DESIGN

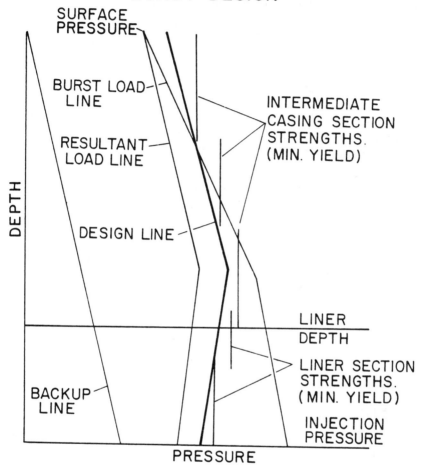

Fig. 5-29 Intermediate casing and drill in liner burst design

line resulting in a design line, as shown in Fig. 5.30, and the design for burst can proceed as previously discussed.

Collapse. Due to the possibility of tubing leaks, artificial lift, and plugged perforations, the collapse design for production casing incorporates no consideration for backup fluid. The string is designed dry inside.

The collapse load is supplied by the hydrostatic pressure of the heaviest mud weight the string is to be run in, and the design factor is applied directly to this load. The resulting design line is used as in the other types of design to check and upgrade as necessary the burst design. The tension and biaxial reductions are evaluated as previously outlined.

Hydrogen Sulfide

Drilling in a potential H_2S (hydrogen sulfide-bearing) zone requires that precautionary steps be taken to ensure the safety of the rig crew and the continued integrity of the control equipment. The dramatic and drastic effects of H_2S can cause death and metal failures in a relatively short time and can complicate an already serious problem like a kick or blowout. As a result, a working knowledge of the effects of H_2S and

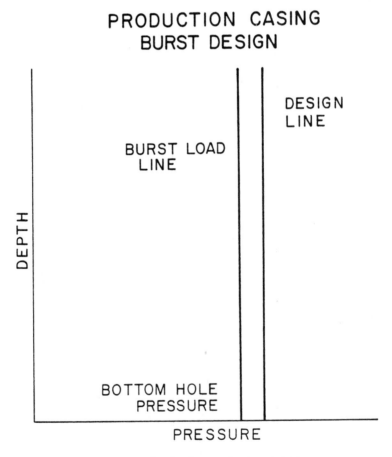

Fig. 5-30 Production casing burst design

proper control procedures is necessary to safe planning and drilling techniques.

Rules and regulations. National, state, and local regulations may govern drilling and production practices in potential H_2S areas. In the U.S., these agencies may include federal bodies such as O.S.H.A. (Occupational Safety and Health Administration), N.I.O.S.H. (National Institute for Occupational Safety and Health, U.S.G.S. (United States Geological Survey), or others. U.S. state agencies and some European nations require adherence to guidelines that may be more stringent than U.S. federal plans. Regardless of the body that has jurisdiction, criteria common to most of the rules and regulations must be followed.

Radius of Exposure. A procedure for determining the radius of a certain level of concentration of H_2S gas to public exposure is generally used to establish which rules and regulations if any apply to an area. Various concentration levels and radii may be used, and different plans must be implemented depending on the results of each radius of exposure calculation and the surrounding surface conditions.

One of the more common schemes used to determine radii of exposure is the Pasquill-Gifford equation. The concentration levels generally employed are 100 ppm and 500 ppm. The Pasquill-Gifford equation for each of these levels is as follows:

1. For determining the location of the 100 ppm radius of exposure,
$$x = [(1.589) (H_2S) (Q)]^{(0.6258)} \qquad \text{Eq. 5.13}$$

2. For determining the location of the 500 ppm radius of exposure,
$$x = [(0.4546) (H_2S) (Q)]^{(0.6258)} \qquad \text{Eq. 5.14}$$

Where: x = radius of exposure, ft
Q = maximum volume determined to be available for escape in cfd (at standard temperature and pressure)
H_2S = mole fraction of hydrogen sulfide in the gaseous mixture available for escape.

The volume used as the escape rate in determining the radius of exposure is generally based on the following guidelines:

1. The maximum daily rate of gas containing H_2S handled by that system element for which the radius of exposure is calculated.
2. For existing gas wells, the current adjusted open-flow rate, or operator's estimate of the well's capacity to flow against zero back-pressure at the wellhead.

3. For new wells drilled in development areas, the escape rate shall be determined by using the current adjusted open-flow rate of offset wells, or the field average current adjusted open-flow rate, whichever is larger.

4. For drilling a well in an area where insufficient data exist to calculate a radius of exposure, but where H_2S may be expected, a 100 ppm radius of exposure equal to 3,000 ft is usually assumed. A lesser assumed radius is often considered when a written request with adequate justification is given.

After the calculations for radius of exposure have been made, efforts must be made to determine if the well in question is subject to the rules and regulations. The guidelines most commonly used for this purpose are:

1. The 100 ppm radius of exposure is in excess of 50 ft and includes any part of a city, town, village, park, dwelling, school bus stop, or similar area that is expected to be populated.

2. The 500 ppm radius of exposure is greater than 50 ft and includes any part of a road owned by and maintained for public access or use.

3. The 100 ppm radius of exposure is greater than 3,000 ft.

A sample calculation may serve to illustrate the exposure equation and its application.

Example 5.6 Assume that a proposed drilling location for a development well in an area in which the maximum open hole rate of the best offset well or the rate carried by the production system was as follows:

$Q = 2$ MMcfd

$H_2S = 10\%$ or 0.1 mole fraction

For 100 ppm radius:

$x = [(1.589)\ (0.1)\ (2,000,000)]^{(0.6258)} = 2,775$ ft

For 500 ppm radius:

$x = [(0.4546)\ (0.1)\ (2,000,000)]^{(0.6258)} = 1,268$ ft

In this example, both the 100 and 500 ppm radii exceed 50 feet, and the requirements of this chapter would apply if a public area is within 2,775 ft, or if a public road is within 1,268 ft. If neither was the case, the operation would not be subject to the requirements of the regulations because the 100 ppm radius was calculated to be less than 3,000 ft.

If the escape rate (Q) of this same gas (10% H_2S) had been 2.5 MMcfd, the 100 ppm radius would be 3,190 ft. In this case, the operation would be subject to the regulations because the exposure exceeded 3,000 ft, even though no public road or dwellings were within the radius. However, a 10 MMcfd well or system with an H_2S concentration of 2% in the same geographical location (public not included) would not be under the rules because the radius of exposure is less than 3,000 ft (2,775 ft).

Characteristics and Effects of Hydrogen Sulfide

Characteristics of H_2S include its explosive nature, its toxicity and its ability to cause sudden metal failures. Characteristics of hydrogen sulfide taken from N.I.O.S.H. standards[14] show the gas is colorless and exhibits a rotten egg odor only when inhaled in small concentrations. Its specific gravity of 1.192 is higher than air which has a specific gravity of 1.000 at 60°F. H_2S burns with a blue flame and produces sulfur dioxide, which is also a toxic gas. H_2S forms an explosive mixture with air or oxygen. For comparison, hydrogen sulfide will burn when mixed in 4.3–45% air, while methane will burn only when mixed with 5–15% air. The gas ignites at 500°F (250°C). For comparison, methane autoignites at 1000°F. It is soluble in water.

One primary characteristic relative to detection that is most often misunderstood is the ability to smell hydrogen sulfide gas as a rotten egg odor. In low, safe concentrations, the gas does smell similar to rotten eggs. However, when the concentration reaches high, toxic levels, olfactory nerves are quickly deadened and can no longer be used as a detection device.

Since H_2S is heavier than air, the gas will settle and collect in low places near the rig such as the cellar and around the mud pits. Due to this, proper consideration must be given to the location of monitors to ensure that the gas will be detected before it reaches dangerous levels.

Effects on Personnel. Sour gas, or hydrogen sulfide, has serious effects on the health and safety of rig personnel (Table 5.1). Death is not the only serious consideration since unconsciousness may allow rig crewmen to fall from heights or into mud pits.

The long-term effects of hydrogen sulfide are not well understood. As an example, personnel that have been exposed to hydrogen sulfide tend to have a reduced resistance to its effects when again exposed to the gas. As a result of the lack of understanding concerning these effects, several government agencies are proposing that personnel records be maintained for 30 years on all workers exposed to the substance.

Effects of hydrogen sulfide on metal. It is generally known that hydrogen sulfide has ill effects on metals although the exact details are not well understood. Some of the terms that have been applied to the hydrogen sulfide-metal reaction are hydrogen blistering, hydrogen embrittlement, stress cracking, and sulfide stress cracking. Each of these

Table 5.1 Toxicity of hydrogen sulfide

PPM (Parts per Million)	0–2 Minutes	2–15 Minutes	15–30 Minutes	30 Minutes To One Hour	1–4 Hours	4–8 Hours	8–48 Hours
5–100				Mild conjunctivitis; respiratory tract irritation			
100–150		Coughing; Irritation of eyes; loss of sense of smell.	Disturbed respiration; pain in eyes; sleepiness	Throat irritation	Salivation and mucous discharge; sharp pain in eyes; coughing	Increased symptoms.*	Hemorrhage and death.*
150–200		Loss of sense of smell.	Throat and eye irritation.	Throat and eye irritation	Difficult breathing; blurred vision; light shy.	Serious irritating effect.*	Hemorrhage and death.*
250–350	Irritation of eyes, loss of sense of smell	Irritation of eyes	Painful secretion of tears, weariness	Light shy, nasal catarrh; pain in eyes, difficult breathing	Hemorrhage and death.*		
350–450		Irritation of eyes; loss of sense of smell.	Difficult respiration; coughing; irritation of eyes.	Increased irritation of eyes and nasal tract; dull pain in head; weariness; light shy.	Dizziness; weakness; increased irritation; death.	Death.*	

| 500–600 | Coughing; collapse and unconsciousness.* | Respiratory disturbances; Irritation of eyes; collapse.* | Serious eye irritation; light shy palpitation of heart; a few cases of death. | Severe pain in eyes and head; dizziness; trembling of extremities; great weakness and death.* |
| 600 or greater | Collapse;* unconsciousness;* death. | | | |

*Data secured from experiments on dogs which have a susceptibility similar to men.
Source: National Safety Council data sheet D-chem. 16

is similar in its effect and, in fact, is caused by the same phenomenon. When hydrogen atoms are formed on a metal surface by a corrosion reaction, they often combine to form gaseous molecular hydrogen which is released into the environment. However, some of the hydrogen atoms are absorbed by the metal. The atomic hydrogen migrates to the grain boundaries of the metal and recombines to form molecular hydrogen which occupies a greater volume than the hydrogen atoms. The formation of the molecular hydrogen causes internal stresses to increase, which in turn causes hydrogen blistering or embrittlement to occur. The blistering will occur with metals that have an average yield strength less than 90,000 psi, while embrittlement occurs with metals having a higher yield strength.

When hydrogen sulfide is present in the electrolyte, the sulfide ion reduces the rate at which the hydrogen atoms combine outside of the metal. This creates a larger concentration of atomic hydrogen on the metal surface. A greater portion of the hydrogen atoms entering the metal increases the tendency for blistering or embrittlement to occur.

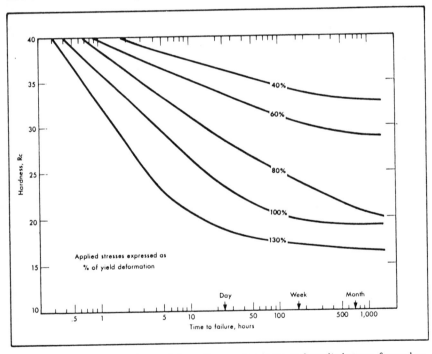

Approximate failure time vs. hardness and applied stress for carbon steel, 3,000 ppm H_2S in a 5% solution of NaCl

Fig. 5-31   Effect of stress level on time to failure

Failures due to hydrogen embrittlement often do not occur immediately after application of the load or exposure to the hydrogen-producing
environment. This is referred to as delayed failure. The time before
failure is referred to as the incubation period during which hydrogen is
diffusing to points of high axial stress. The time to failure decreases as
the amount of hydrogen absorbed, applied stress, and strength level
increases.

Spontaneous brittle failure that occurs in steel and other high strength
alloys when exposed to moist hydrogen sulfide and other sulfuric environments is frequently referred to as sulfide stress cracking. It is
generally thought to be a form of hydrogen embrittlement. Although the
mechanism of sulfide cracking is not completely understood, it is generally accepted that four conditions must be present before cracking can
occur: hydrogen sulfide, water, high-strength steel, and applied or residual stress.

Stress level, either applied or residual, affects sulfide cracking tendencies. The time to failure decreases as the stress level increases (Fig. 5.31).
In most cases stress results from a tensile load or from the application of

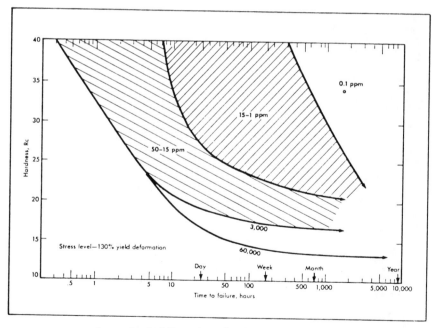

*Approximate failure time of carbon steel in 5% NaCl and various ppm
of H_2S*

Fig. 5-32   Effect of hydrogen sulfide concentration on time to failure.

pressure. However, residual stresses and hard spots can be created by welding or cold working the material.

The time to failure decreases as the hydrogen sulfide concentration increases, as shown in Fig. 5.32. Although delayed failure can occur at very low concentrations, the time to failure becomes great. There is evidence that cracking susceptibility decreases above 175–200°F regardless of the sulfide concentration. Although the exact temperatures are not yet defined, this principle can be used in well planning for tubular goods.

Hydrogen Sulfide, H₂S, Detection

Equipment used to detect hydrogen sulfide may include fixed-location monitors, personal detectors, mud monitors with electronic probes, or chemicals for analysis of the drilling fluid. The monitors may be qualitative or quantitative and may function with chemical or electronic sensors.

The most important concern with any H₂S air detector is proper

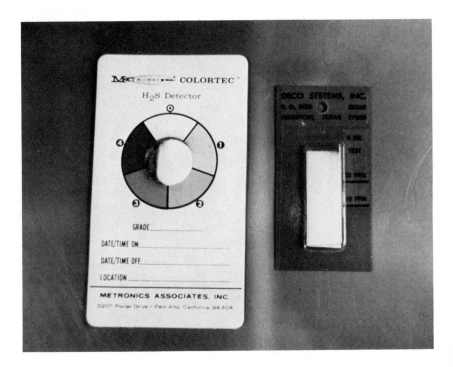

Fig. 5-33 Badge type paper detector

placement of the sensor units. Since H_2S is heavier than air, it will settle in low areas. The personal units should be attached to the clothing or carried level with the waist. The electronic rig monitors have portable sensor heads that should be placed in low areas such as the cellar and near the pits. A sensor should also be placed near the shale shaker since it is the first location where the mud will receive exposure to the air.

Lead acetate paper detectors. Several reasonably semi-quantitative detectors for hydrogen sulfide are based on lead acetate paper. As the gas contacts the paper, the lead acetate impregnated in the paper reacts with the gas to form lead sulfide which causes the paper to change color from white to various shades of brown or black. The degree of color change depends on the H_2S concentration which can be roughly estimated by comparing the observed color to a control chart or table.

The primary advantage of these detectors is that they are carried by each crewman, enabling him to detect the gas wherever he may be. This provides an additional measure of safety to each crewman as well as an atmosphere of security. When necessary, the paper can be changed to provide a new chemical surface for gas detection.

Fig. 5-34 Spot check paper detector

The reaction time required for the detector to function is a disadvantage of the tool. The total of 3–5 minutes necessary can be excessive and dangerous when large concentrations of hydrogen sulfide are encountered. Also, it is advisable to consider the lead acetate paper as a qualitative indicator rather than as a determination of the concentration.

Fig. 5-35 Capsule detector

The two paper detectors most often employed are the badge type and the spot check, which are shown in Figs. 5.33 and 5.34 respectively. The badge typa is clipped to the clothing, while the spot check can be carried in the hand or pocket. Note the color codes used to determine the H_2S concentration.

Capsule detector.   The capsule detector resembles an ammonia type capsule and is filled with chemical granules (Fig. 5.35). The capsule is broken and attached to the clothing with a string. If hydrogen sulfide contacts the granules, a brown discoloration will be observed.

This detector should be used only as an indicator of H_2S because of the limitations of the capsule. The life of the tube is approximately 6 days after it is broken. The maximum concentration of the gas that can be measured accurately is 20 ppm.

Draeger detector.   The Draeger unit is one of the most widely used tools for quantitative gas detection. It can be altered to measure almost any type of gas and, as a result, is used extensively in hydrogen sulfide detection.

The tool consists of a calibrated glass tube filled with lead acetate granules. A pump is used to draw gas samples into the tube, and the level of color change denotes the H_2S concentration. Several scales are usually

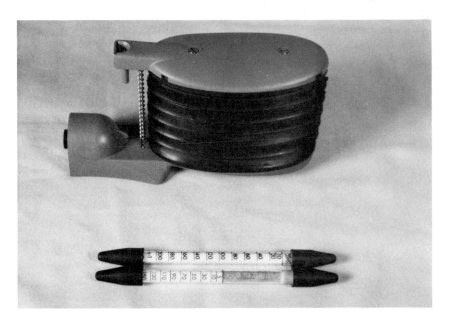

Fig. 5-36   Draeger detection unit with several tubes

presented on the glass tube to denote high and low concentrations. The pump is usually the bellows type as shown in Figs. 5.36–5.38.

The simple operating procedure increases the utility of the tool. The tips of the detector tube are broken and inserted into the suction outlet of the Draeger unit. Ten compressions of the bellows are required to ensure an accurate reading in low concentrations of hydrogen sulfide. As the gas is drawn into the tube by the bulb, the lead acetate granules become discolored denoting the quantitative measurement of the gas concentration.

The accuracy of the measurements depends on the training and practice of the personnel using the unit. As varying amounts of air are drawn into the unit, the measurements will be different than if ten compressions were used. In high concentrations of H_2S, only one compression is required to activate the high scale on the unit.

The measurements obtained with the Draeger unit are usually reliable. Since there are no electronic parts, the unit is not subject to electronic malfunction. The shelf life of an unbroken tube is approximately two years, and the tube can be used after the tips are broken as long as no indication of H_2S is present.

Belt detectors. The belt type of hydrogen sulfide detector is an electronic unit usually attached to the crewman's belt. The unit is oper-

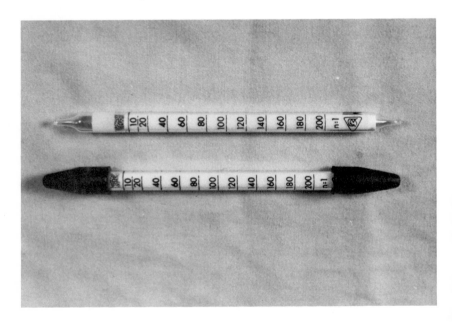

Fig. 5-37   High and low scales for quantitative detection

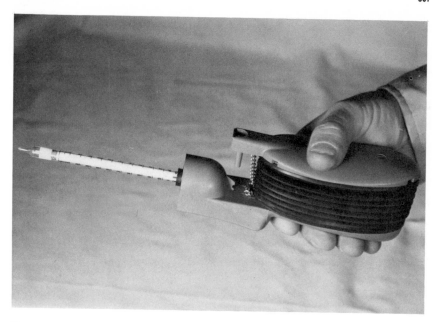

Fig. 5-38   Compression of the Draeger unit with a glass detection tube in position.

Fig. 5-39   Electronic belt detector with an alternate power source

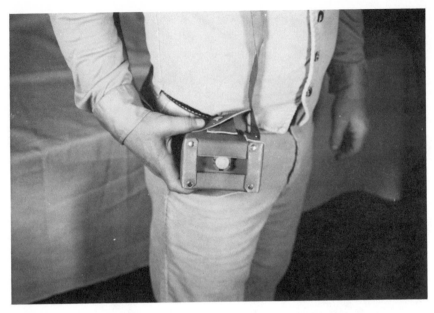

Fig. 5-40   Sensor head for the belt type detector

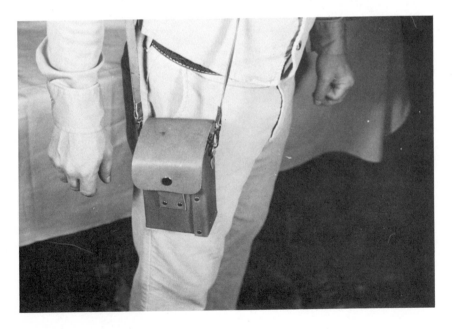

Fig. 5-41   The portable electronic detector can be worn on the belt or carried over the shoulder.

ated by rechargeable and/or replaceable batteries. The detector has a sensor head that will monitor hydrogen sulfide gas and report in a visible readout for concentrations of 5–10 ppm. An audible alarm can be used and is usually pre-set to respond at 20 ppm. The response time for the unit is approximately 35 seconds. The belt type detector is shown in Fig. 5.39–5.41.

Fixed-location monitors.   The rig monitor is a fixed-location, quantitative, electronic device designed for permanent, full-time operation. Sensor heads are placed at various locations on the rig and attached to the detection unit which is housed in a hard plastic or metal case. On the monitor, a readout in ppm concentration will be shown on a needle type indicator.

A rotating beacon or strobe light attached to the unit will activate automatically when a specified amount of gas has been detected. An audible alarm can be used to denote a higher level of gas concentration. The response time for the monitor is approximately 35 seconds for concentrations of 0–10 ppm.

The detection unit, depending on brand name and model, can have from 1 to 12 channels to which are attached the sensors. The most common units have 4 to 6 channels. The unit shown in Figs 5.42–5.44 has 3 channels.

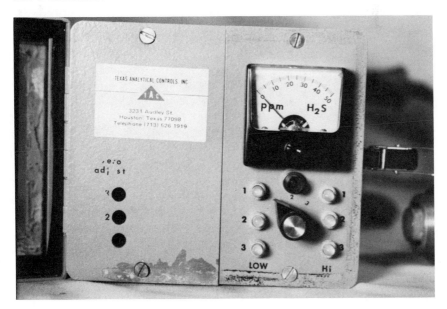

Fig. 5-42   Electronic rig monitor with multiple sensors. The unit has audible and visible alarms.

Fig. 5-43 The sensor head as it would appear at the rig.

Fig. 5-44 The high and low alarms activate a flashing light and a horn when various H_2S concentrations are detected.

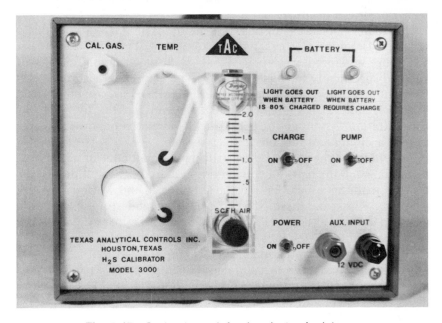

Fig. 5-45 Calibration unit for the electronic detectors

The rig monitor or the belt type detector must be calibrated and tested periodically to insure that it is functioning properly. A calibration instrument similar to that shown in Fig. 5.45 is used. A known concentration sample is placed in the machine and fed into the monitor. If the monitor does not respond accurately, it is adjusted accordingly. The calibration unit is fully self-contained with rechargeable batteries.

Mud analysis for hydrogen sulfide. Several testing procedures are available for evaluating hydrogen sulfide, sulfide, and sulfide scavenger concentrations in the drilling fluid.[15] Some of these include the Hach test, Garrett gas train, iodine test for the determination of hydrogen sulfide in water, quantitative copper carbonate concentration, quantitative Ironite Sponge concentration, and electronic probes such as the Mud Duck (Delphian Corp.). The most commonly used methods during drilling are the Garrett gas train and the electronic Mud Duck.

Chemical means to analyze sulfides in the drilling muds and filtrate should be quick, accurate, reliable, and simple. The Garrett gas train, a recent innovation with A.P.I. approval, meets all these requirements. The equipment and procedures do not respond falsely to other fluid components, primarily because this method eliminates most of the interference problems by using a gas train to separate H_2S gas from the liquid phase.[16] Self-contained Garrett gas train analysis kits are available from many major mud companies.

Electronic analysis of sulfides in the mud system can be accomplished with the Delphian Corp. patented Mud Duck. The tool uses special electrodes and processes their signals with electronic circuiting developed specifically for this purpose.[17] The equipment also provides a constant monitoring of the drilling fluid pH since the form of dissolved hydrogen sulfide in the mud system depends on the hydrogen ion concentration. Although the Mud Duck is a relatively new tool, it has proved accurate and reliable in field service.

Breathing apparatus. When drilling is conducted in an environment that contains harmful concentration of hydrogen sulfide gas, protective breathing apparatus must be supplied to and worn by the crew members. The apparatus must be a pressure demand, supplied air respirator; and it can be either a self-contained breathing apparatus (SCBA) or a hose line, supplied air respirator. Strict rules and regulations govern the use of breathing apparatus and ensure that proper, certified equipment is used, the equipment is worn and maintained properly, each individual receives personal fitting and use instruction, and that the individual meets all medical and physical requirements for respirator use.

The self-contained breathing apparatus (SCBA) can be seen in Fig. 5.46. It has a limited supply of air that may range from 5 to 20 minutes

Fig. 5-46 Self-contained breathing apparatus (SCBA)

depending on the size of the bottle, the amount of air in the bottle, and the user's physical activity while wearing the unit. The SCBA has an audible alarm to notify the user when the air supply is almost depleted. A pressure regulator maintains a slight, positive pressure within the facepiece to ensure that the user has a constant supply of air upon his demand.

In drilling situations where the worker must perform his routine duties in a contaminated atmosphere, the SCBA is not ideal because it must be refilled at frequent intervals. When this is the case, a supplied air respirator is usually employed. The respirator uses the same facepieces and regulator as on the SCBA, but draws its air supply from a hose line connected either to an air compressor or to a series of large volume, compressed-air bottles. The hose line unit, or work unit as it is termed, will usually have a 5-minute capacity bottle to be worn by the worker so that he may excape from the area should the hose line unit fail for any reason (Fig. 5.47).

Proper fitting of the facepiece is important when wearing breathing apparatus. If the facepiece does not fit the user's face pressure-tight, it will seriously diminishes the effectiveness of the entire breathing apparatus. Some of the many physical conditions that prohibit a proper face seal are growth of beard, sideburns, a skullcap that projects under the facepiece, temple pieces on glasses, and the absence of one or more dentures. Wearing contact lenses is prohibited because they may "float" on the eye when the breathing apparatus pressurizes the facepiece.

Mud System and H_2S Corrosion

The corrosivity of any mud system will depend to a large degree on the conductivity of the electrolyte. Corrosion rates increase as the conductivity of the mud, or electrolyte, increases. Conversely, the rates decrease as the electrolyte conductivity decreases. If a mud has essentially no conductivity, corrosion will be very low. This is the case when using an oil mud as a drilling fluid since the oil mud has a very low conductivity. Table 5.2 shows the effect of H_2S corrosion rates in various common drilling fluids.[18]

Corrosion control is a greater concern in water base fluids than oil base fluids. As was seen in Table 5.2, the primary corrosion effects were in the water base muds with the lignite/lignosulfonate additives while significant levels of corrosion were noted in the nondispersed system. Since these two systems are being used extensively in the industry for reasons not related to corrosion, special efforts must be made to make the systems more resistant to corrosion.

The rates at which corrosion occurs are influenced by the amount of dissolved gases in the electrolyte. As can be seen in Table 5.3, the corrosion of steel in brine shows a significant increase as a gas is dissolved in the fluid.[19] The increase is even greater when several gases are dissolved in the electrolyte.

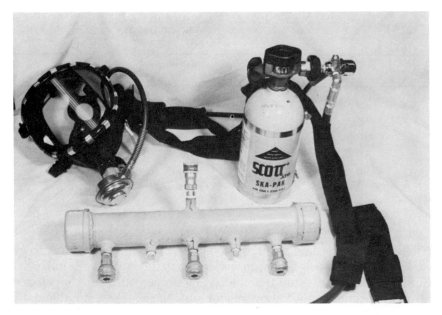

Fig. 5-47 Supplied air respirator

Table 5.2 Effect of mud type on corrosion
(Courtesy of NL/Baroid)*

Mud types	H₂S presence on coupons	Hydrogen embrittlement	Corrosion rates, MPY
Invermul (3 lb/bbl lime)	No	No	5.30
Invermul (8 lb/bbl lime)	No	No	3.99
Low lime	No	No	3.23
High lime	No	No	3.42
Nondispersed—low lime with saturated salt, polymer, starch	Yes	Yes	26.60
Lignite/lignosulfonate (starting pH 9–11)	Yes	Yes	107.47
Lignite/lignosulfonate (starting pH 11)	Yes	Yes	70.02

* Series of tests using mild steel coupons and prestressed bearings contaminated with 2,400 ppm H_2S rolled 16 hr at 150°F. MPY = mils/year.

Table 5.3 Corrosion fatigue of steel in brine
(Courtesy of Oil and Gas Journal)

Dissolved gas	% decrease from air-endurance limit
H_2S	20
CO_2	41
CO_2 + Air	41
H_2S + Air	48
H_2S + CO_2	62
Air	65

Reduction in the amount of dissolved gases in the mud system can be attained in several ways. The most effective methods are to minimize the entrance of the gases into the mud (1) by proper utilization of surface equipment such as mud hoppers and guns to reduce the mud and air mixing, and (2) by prevention of formation fluid entries into the well that may contain dissolved gases as CO_2 or H_2S. Another important method for gas content reduction is through the use of chemical agents that remove the gases.

pH control.   The pH of fluid is a measure of the hydrogen ion concentration. The pH values are presented on a scale of 1 to 14 with 1−7 considered to be acidic and 7−14 considered as basic. A value of 7 is neutral (Fig. 5.48).

The pH of the mud systems must be monitored and controlled since most mud additives function more effectively in a pH range of 9−11. Also, corrosion rates are affected by the hydrogen ion concentration, and personnel safety problems are increased in certain pH ranges. Although most mud systems function more effectively in higher pH ranges, the individual additive pH values are often low. As an example, chrome

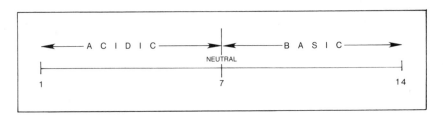

Fig. 5-48   pH ranges

lignosulfonate solution has a pH of about 3.6. A pH adjuster, such as sodium hydroxide (caustic soda), can raise the level to the desired range.

The rate of corrosion is affected significantly by the pH or hydrogen ion concentration of the electrolyte. As the pH decreases, the embrittlement tendencies increase due to the larger concentration of hydrogen ions. Fig. 5.49 shows the effect of pH on the time to failure of a sample metal ring. Note that the time to failure is high at a pH value of 9.5.

Another effect of pH relative to hydrogen sulfide is due to the solubility of the gas at higher pH levels. In Fig. 5.50, this relationship can be seen. The total concentration of dissolved gas exists as hydrogen sulfide at pH ranges of 3−6. It begins to convert to hydrogen sulfide and sulfide ions at pH ranges of 6−14. In the range of 6−9, a mixture of hydrogen sulfide and sulfide (monovalent and divalent) ions are present.

The presence of the sulfide ion has only a small corrosion effect itself, but it does increase the tendencies for hydrogen embrittlement and sulfide stress cracking by retarding the rate at which atomic hydrogen is allowed to escape from any corrosion point on the metal surface.

Caution must be exercised when using pH control as a corrosion

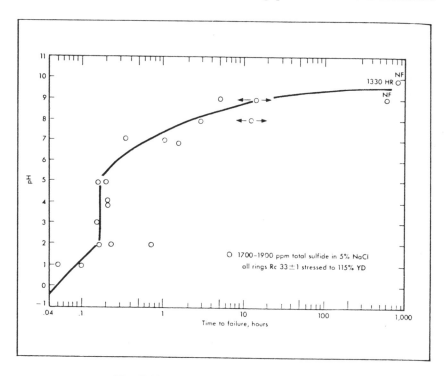

Fig. 5-49   Effect of pH on time to failure

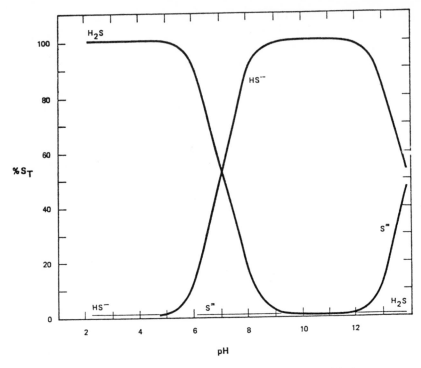

Fig. 5-50 Effect of pH on hydrogen sulfide and sulfide ions

preventive measure. As noted in Fig. 5.50, high pH converts the hydrogen sulfide molecule to ionic sulfides, which are not as directly dangerous to human life. This tends to mask the problem by hiding the hydrogen sulfide. If the pH is reduced for any reason, the sulfides may revert to H_2S and cause severe problems.

Another problem area is the use of chemical control agents such as scavengers to remove the hydrogen sulfide while the pH of the system is not conducive for the use of the particular additive employed. Certain agents require extremely high pH values to function while others need significantly lower ranges. A high pH additive generally will not function in a low pH system while the converse is true for the low pH additive. Therefore, additives and pH ranges must be properly chosen for effective hydrogen sulfide control.

Corrosion inhibiting fluids.   One of the most widely used methods of protecting downhole equipment is chemical inhibition. Since corrosion is a reaction on metal surfaces, any modification of the steel-electrolyte interface will affect the corrosion rate. Inhibitors are chemicals that when added in small quantities to a corrosive system will alter the steel-electrolyte interface and reduce the corrosion rate. The action of these materials may be described as oil wetting the steel surface.

Most inhibitors are amine-fatty acid salts formulated into either oil soluble or oil soluble-water dispersible materials. The inhibitor molecule can be imagined as having a polar end and an oil-soluble end. The polar end can adhere to solid surfaces such as the steel tubular goods. The oil soluble end absorbs a film of oil that protects the surface.[20] The dual nature of the inhibitor molecule also gives detergent properties to the chemical.

Inhibitors are used in continuous or periodic treatments. The main requirement for treatment is to contact the metal surface with sufficient inhibitor for a long time to obtain a film coating. New treatments should be made at regular intervals to maintain the film.

Hydrogen Sulfide Scavengers

Scavengers are additives designed to remove a contaminant from the mud system. The additive generally does not prevent corrosion such as hydrogen embrittlement but does reduce the severity of the problem by sequestering the hydrogen sulfide or sulfide ion which would have increased the embrittlement tendencies. There are many types of additives with different properties.

Most scavengers function in either a surface adsorption manner or through ionic precipitation. If the scavenger is based on the surface adsorption technique, the mud must be thoroughly mixed to insure that a sufficient number of collisions occur between the hydrogen sulfide and the scavenger for a completion of the process. In the ionic reactions, the solution characteristics of the scavengers must be studied to insure that such variables as fluid pH and salinity are conducive to use of the additive.

The primary hydrogen sulfide scavengers used in the industry are metallic compounds based on copper, zinc, or iron.

Copper products. An effective copper derivative used in hydrogen sulfide scavengers is copper carbonate. This product was the first material to be widely used because of its effeciency as a scavenger. Even though basic copper carbonate is essentially insoluble in the drilling fluid, it has been found to be sufficiently reactive to precipitate hydrogen sulfide as copper sulfide, which is insoluble. Copper carbonate does not adversely affect the drilling fluid properties.

Copper derivatives have side effects that limit their application. Copper compounds may react with iron (the drill string) causing the iron to corrode or copper plating to occur. Therefore, when copper products are used as hydrogen sulfide scavengers, one corrosion problem is simply replaced with another.

Zinc products. In an effort to avoid the corrosion problems posed by copper, metallic compounds with oxidation potentials closer to iron have been used. The most common metal that will meet this requirement and still readily form a sulfide is zinc.

Zinc carbonate is one of the most widely used sulfide scavengers in the industry. It utilizes an ionic reaction as well as a surface reaction for scavenging, and it functions most effectively at pH values above 9. The compound is relatively safe and poses no immediate hazards.

Although zinc carbonate is an effective scavenger, there are problems associated with the compound. Concentrations above 3 lb/bbl in weighted, high solids systems cause high gel strength. The zinc ion may cause a reaction similar to calcium contamination in some muds. The compound does not prevent hydrogen embrittlement. A high pH is required to solubilize the product, and since its specific gravity is roughly the same as barite, it may settle in muds and brines that have a low carrying capacity.[21]

Organically chelated zinc compounds may be used in brines where the zinc carbonate would settle. These compounds are efficient but contain only one fourth as much zinc as conventional zinc carbonate. The product may also function as a thinner in non-dispersed muds.

Zinc chromate is an efficient scavenger that also minimizes hydrogen embrittlement and maintains low corrosion rates. The compound uses an oxidation reaction to form a sulfate when the pH is greater than 9. The problem with zinc chromate is the environmental restriction on the use of any chromate product.

Iron products. Ironite Sponge, a tradename for iron oxide, is a hydrogen sulfide scavenger. The surface reaction is not restricted by temperature or time and the product does not degrade mud properties. The iron oxide, or Sponge, reacts with the hydrogen sulfide to form the stable iron sulfide, pyrite.

Sponge has a specific gravity of 4.4 which allows it to replace barite on an even basis as a density additive. The use of a high specific gravity material is advantageous because it limits the buildup of low gravity solids. The particle size averages 6−8 microns and ranges between 1.5−50 microns which is comparable to barite. The uniform spherical shape and size of the Sponge particle creates a low abrasion level.

Ironite Sponge is ferromagnetic. It is strongly attracted to a magnet but will not retain magnetism. High saturation magnetism is the basis for the simple field test to determine Sponge concentration. The low remanent magnetism prevents it from being attracted to drill pipe or casing.

The reaction between the hydrogen sulfide and iron oxide appears to be stable and irreversible under most conditions. One pound of Ironite Sponge will react with 0.7 pounds of hydrogen sulfide. The reactions occur most effectively in a pH range of 6–9. Values above 9 restrict the scavenging of H_2S because it is in the sulfide form (Fig. 5.50). This fact must be approached with caution because it may lead rig personnel to assume that an excess of Sponge is present when in reality the hydrogen sulfide is in a form that cannot readily be scavenged by the additive. If the pH drops, the gas or liquid may revert to hydrogen sulfide and react

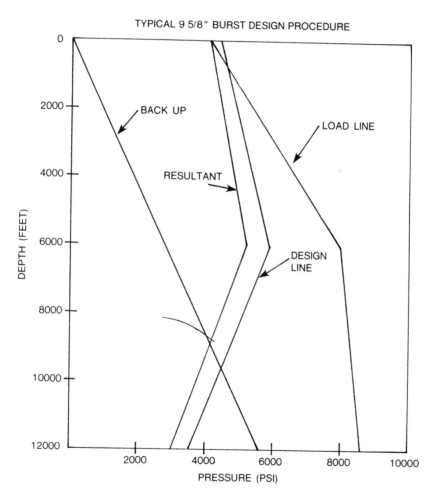

Fig. 5-51 (a) Typical casing burst design

out the existing Sponge, while leaving an excess of unreacted hydrogen sulfide.

Tubular Design

In the design of tubulars for use in hydrogen sulfide environments, several considerations must be made. Among these are the required strengths of the pipe and the temperature involved. Each of these variables must be established before the string of pipe can be designed.

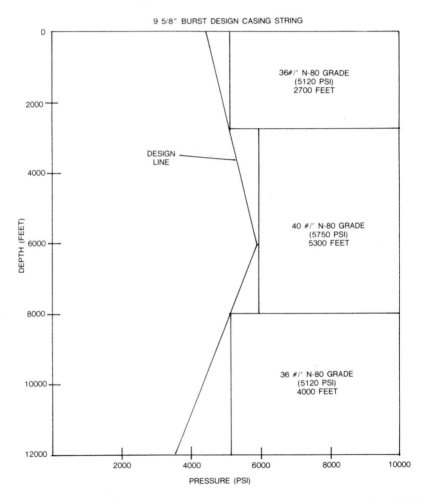

Fig. 5-51 (b)   Selection of casing weights and grades for the design in Fig. 5-51 (a)

Casing design. The initial step in designing casing tubulars for a well is to determine the maximum pressures that could be imposed on the string. After these pressures have been established, casing weights and grades are chosen that will contain these pressures (Fig. 5.51 a and b). Casing strengths can be determined from sources similar to those presented in the Appendix.

Once the initial design has been established as shown in Fig. 5.51 (b), consideration must be given to the effect of sour gas on the design. Hydrogen embrittlement appears to occur in metals with a hardness

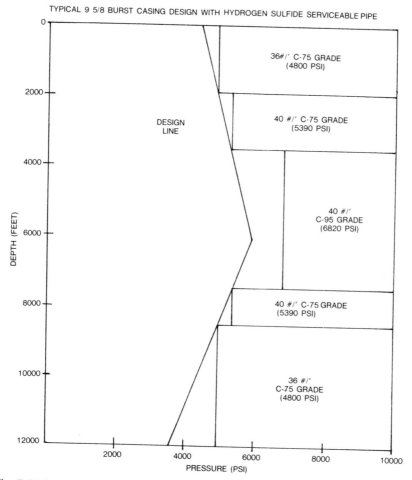

TYPICAL 9 5/8 BURST CASING DESIGN WITH HYDROGEN SULFIDE SERVICEABLE PIPE

Fig. 5-51 (c)   Selection of casing weights and grades for the design in Fig. 5-51 (a) when hydrogen sulfide is to be encountered.

greater than 22 on the Rockwell scale. This hardness corresponds to casing that exhibits yield strengths greater than 90,000 psi.

All casing grades above this strength or hardness must be considered susceptible to embrittlement under normal conditions. Table 5.4 lists several casing grades and denotes their acceptability for hydrogen sulfide service. In this table, "C" grades are special corrosion resistant pipe and are suitable for usage in H_2S. L-80 and MN-80 are also acceptable for

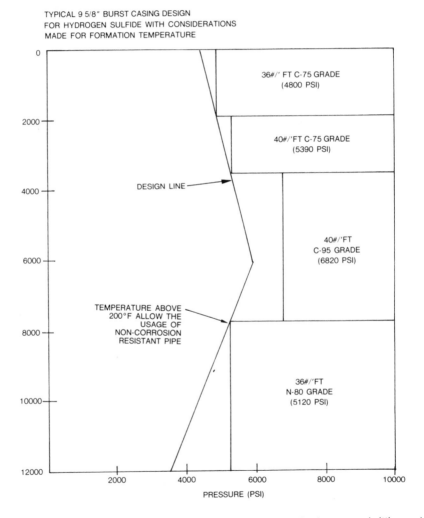

TYPICAL 9 5/8" BURST CASING DESIGN
FOR HYDROGEN SULFIDE WITH CONSIDERATIONS
MADE FOR FORMATION TEMPERATURE

Fig. 5-51 (d)   Application of the effect of temperature on hydrogen embrittlement for the casing design in Fig. 5-51 (c).

Table 5.4 Casing grades acceptable for H_2S service

Casing grade	H_2S service*	Comments**
H-40	Yes	—
K-55	Yes	—
C-75	Yes	—
N-80	No	Above 200°F
L-80	Yes	—
MN-80	Yes	—
C-90	Yes	—
SOO-90	Yes	—
S-95	No	Above 200°F
SOO-95	No	Above 200°F
S-105	No	Above 200°F
P-110	No	Above 200°F

* Service conditions for any H_2S environment.
** Denotes grades usable above 200°F. It should be noted that many operators have successfully employed a 150°F design constant. The constant of 200°F presented here is used as a conservative value since there is still some disagreement upon the exact value.

H_2S service although N-80 is not usable under normal conditions. The design in Fig. 5.51 (b) would be altered to that shown in Fig. 5.51 (c).

Temperature must also be considered in a casing design. Field experience has shown that temperatures above 175−200°F tend to minimize hydrogen embrittlement. Using the formation temperature gradient in the area in question, all casing that will exist permanently in a temperature above 200°F can be used as designed in the initial step of the procedure (Fig. 5.51 d).

High-pressured wells that are subject to temperature less than 200°F will often require casing strengths greater than 90,000 psi. In these cases, the tubular strings must be specially designed with a larger wall thickness of softer metal. Although this process is expensive, it is the only approach to use without relying exclusively on scavengers to remove the hydrogen sulfide or sulfide ion.

Drill pipe design. The design of a drill string is similar to the approach used in casing design. After the strength requirements have been established, H_2S considerations are made with respect to pipe strength and embrittlement susceptibility. Table 5.5 lists several grades of drill pipe available. Drill collar design is not often considered with

Table 5.5 Grades of drill pipe for H_2S service

Grade	H_2S service
D	Yes
E	Yes
X-95	Yes
G-105	No
S-135	No

respect to H_2S since it will generally be used in temperatures that will minimize embrittlement.

H_2S Protection for Blowout Prevention Equipment

After the blowout prevention equipment has been designed to meet the pressure requirements necessary to drill the well, efforts must be made to insure that the equipment is resistant to hydrogen sulfide corrosion or sulfide stress cracking (SSC). Certain equipment will meet these requirements when new but will fall below minimum standards as the equipment is subjected to normal wear. For this reason, when used in a potential H_2S environment, all well control equipment must be tested on a regular basis for corrosion resistance.

Field and laboratory studies have been conducted to establish minimum acceptable metal hardness standards that would be allowable for use in hydrogen sulfide areas. The results from the study showed that metals with a hardness less than 22 on the Rockwell scale could be used satisfactorily without sulfide stress cracking failure. Metals with a hardness greater than 22 showed a tendency towards stress cracking with the problem becoming severe in the high strength metals. All blowout prevention equipment must either conform to these hardness requirements or be comprised of metal alloys resistant to hydrogen sulfide at higher hardness values.

Many blowout preventers will conform initially to this hardness requirement. However, as normal drilling mechanics progress, stresses due to case hardening or other reasons may render the equipment unusable due to the development of localized areas having a hardness greater than 22. This equipment can be restored to a satisfactory condition by heat treating in a manner that will relieve the stresses. This procedure is applicable only to equipment that would have a suitable hardness under normal conditions.

Table 5.6 H$_2$S modification for common ram type preventers
(Courtesy of SPE-AIME)

Component	Standard service	H$_2$S service
Cameron type U Ram assembly		
(1) Ram	4340 R$_C$ = 20–23	4340 R$_C$ = 20–24
(2) Front packer	Specially compounded gas-resistant Hycor	Same
(3) Top seal	Rubber	Same
Bonnet bolts	4142 bolt class 354 (R$_C$ = 35)	A 286
Operating piston	4140 R$_C$ = 22–30	4140 R$_C$ = 24
Connecting rod seal ring	Rubber	Specially compounded gas-resistant Hycor
Bonnet seal	Rubber-impregnated asbestos	Hycor-impregnated asbestos
Bonnet bolt retainer O-ring	Rubber	Hycor
Int. flange to bonnet cap screw	4142 bolt class 354 (R$_C$ = 35)	4142 bolt class 354 (R$_C$ = 22)
Shaffer type LWS Door cap screws	4140 BHN = 285–341	4140 BHN = 225–235
Cylinder cap screws	4340	4140 R$_C$ ≤ 22
Cylinder head cap screws	4340	4140 R$_C$ ≤ 22
Ram shaft	410 SS	4130 R$_C$ ≤ 22
Hinge bracket cap screw	4340	4140 R$_C$ ≤ 22
Secondary seal plug	Heat treated 4140	304 SS
Thread protector plug	Heat treated 4140	304 SS
Ram rubber retain- ing screws	Heat treated 4140 (BHN = 385)	4140 R$_C$ ≤ 22
Ram block retain- ing screws	Heat treated 4140	K-Monel 500
Studs	4140 B-7	A-286 or soft B-7
Nuts	2H	2H R$_C$ ≤ 22

Another procedure to protect equipment from hydrogen sulfide is to use special trim that is acceptable in this environment. As an example, special synthetic H_2S resistant seals and elastometers should be used in pressure control valves to prevent their degradation from exposure to H_2S. Table 5.6 shows the modifications made by Shell Oil Co. to certain preventers used to drill in potential H_2S wells.[22]

Contingency Planning

As in all drilling situations, events will occur which do not follow the regular mode of operation. These events should be considered before they occur so they can be either avoided or mitigated. Contingency plans are designed to account for these abnormal occurences by providing an operating plan that can be implemented as necessary. Although contingency plans can cover any drilling problem, they are generally designed for well control or hydrogen sulfide drilling.

Contingency plans must be specific in assigning areas of responsibilities and duties so that each crew member knows what is expected of him during an emergency. The plans must also include provisions for authoritative succession or alternates for the assigned duties in the event that the accident incapacitates the person primarily responsible.

Sample contingency plans have been provided in the Appendix. These plans cover basic well control actions and hydrogen sulfide drilling. They must be considered only as guidelines since a specific plan must encompass the actual circumstances surrounding the drilling event.

Problems

5.1   Determine fracture gradients for the following sets of conditions. Use the methods of (1) Hubbert and Willis, (2) Matthews and Kelly, and (3) Eaton. Assume "Louisiana" conditions for the Matthews and Kelly calculations.

Depth, ft	Formation pressure, ppg
3,000	Normal
13,000	13.1
9,000	9.6
6,500	9.0

8,000	10.2
11,000	15.1
17,000	18.0
4,500	9.9
10,500	Normal
15,000	15.6

Solution: 3,000 Ft

Hubbert and Willis —12.2–14.0 ppg
Matthews and Kelly —13.8 ppg
Eaton —13.9 ppg

5.2 Prepare a graph of fracture gradients versus depth for the methods used in Problem 5.1 Assume normal formation pressures.

5.3* Calculate the fracture gradient for the following set of deepwater conditions. Use the Prentice approach.
Freeboard = 50 ft
Water depth = 1,700 ft
Casing depth below seafloor = 6,000 ft

5.4 Use the following leak off data to determine the formation fracture gradient. Casing is set at 12,000 ft and the mud weight is 13.9 ppg.

Volume pumped, bbl	Pressure, psi
0	0
1	175
2	400
2½	590
3	680
3½	760
4	650
4½	740
5	830
5½	920
6	1,010
6½	1,100
7	1,190
7½	1,260
8	1,280
8½	1,300

5.5 Calculate the formation fracture gradient for the following conditions. Use the pressure data from Problem 5.4.

Depth of casing, ft	Mud weight in use, ppg
9,300	9.0
16,000	13.9
13,100	15.1
6,400	9.5

Solution: 9,300 ft–11.6 ppg

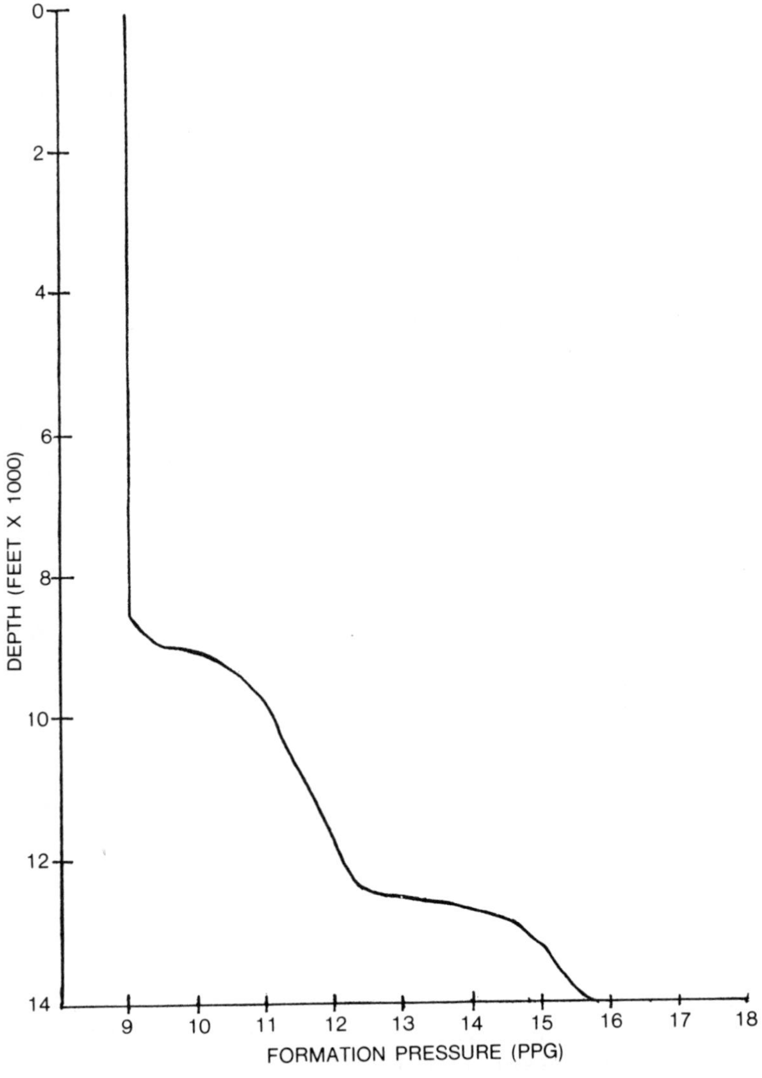

Fig. 5-52 Formation pressures for Problem 5-6

5.6 Calculate the casing setting depths for the formation pressure plot shown in Fig. 5.52. Use the Eaton method (chart) to determine the fracture gradients.

Solution: Intermediate—9,600 ft
Surface —2,400 ft

5.7 Select casing setting depths for the pressure plot in Fig. 5.12.
5.8 Select casing setting depths for the pressure plot in Figure 5.53. Use the Eaton method to determine the fracture gradients.
5.9 Use the following data to determine the proper setting depth for a surface casing string.
Intermediate casing depth = 11,000 ft
Mud weight in use at 11,000 ft = 12.1 ppg
5.10 Design the following casing string. Use the data and solution from Problem 5.6.

Casing string type:	Intermediate
Casing size =	9⅝ in.
Surface equipment rating =	5,000 psi
Minimum drift acceptable =	8½ in.
Limit on number of sections =	4

5.11 Redesign the casing string in Problem 5.10 and use 7⅝-in. casing with a minimum acceptable drift of 6.5 in.
5.12 Design the intermediate casing and liner (if necessary) in Problem 5.8. Use the following criteria.

Intermediate casing

Casing size =	7 in.
Surface equipment rating =	5,000 psi
Minimum drift acceptable =	5½ in.
Limit on number of sections =	4

Liner

Casing size =	5 in.
Length of overlap =	500 ft
Surface equipment rating =	5,000 psi
Minimum drift acceptable =	4.0 in.
Limit on number of sections =	1
Special considerations:	Couplings must be stream-lined (flush joint or extreme line)

5.13 Redesign the intermediate string in Problem 5.12 assuming that no liner will be used. Compare the results.

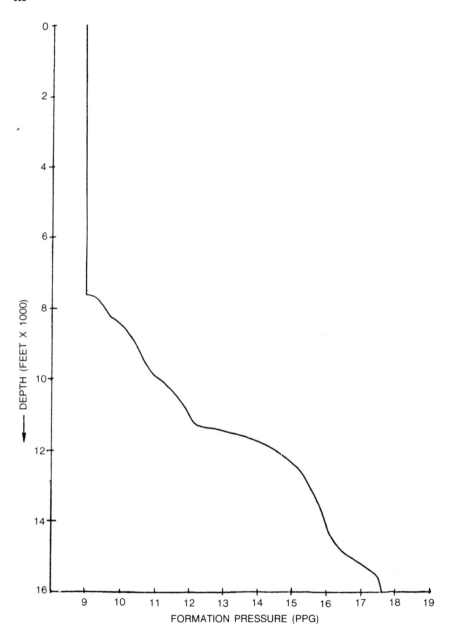

Fig. 5-53 Formation pressures for Problem 5-8

5.14 Design the surface casing for Problem 5.6.
      Casing size =   13⅜ in.
      Minimum drift acceptable =   12.0 in.
      Limit on number of sections = 3
5.15 Design the surface casing for Problem 5.8.
      Casing size =   9⅝ in.
      Minimum drift acceptable =   7⅞ in.
      Limit on number of sections = 2

5.16 Redesign the surface casing for Problem 5.8 and use 2 sections of
      10¾-in. casing.
5.17 Design production casing for Problem 5.6.
      Casing size =   5 in.
      Setting depth =   Hole bottom
      Producing zone pressure =   Maximum mud weight
      Type of production =   Gas
      Packer fluid weight =   10.0 ppg
5.18 Design production casing for Problem 5.8.
      Casing size =   4½ in.
      Setting depth =   Hole bottom
      Producing zone pressure =   Maximum mud weight
      Type of production =   Gas
      Packer fluid weight =   14.0 ppg
5.19 Use the Pasquill-Gifford equation to determine radii of exposure
      for the following data.

      $Q = 3$ MMcfd
      $H_2S = 12\%$

                              Solution: 100 ppm—4,008 ft
                                        500 ppm—1,831 ft

5.20 Prepare a plot of radii of exposure versus flow rate (Q) for a 5%
      H_2S well. Use a maximum flow rate of 10 MMcfd for 100 ppm and
      the 500 ppm plots.
5.21 A well is to be drilled that will contain 3% H_2S. If a resident lives
      2,400 ft from the well, what is the maximum flow rate that can be
      tolerated before exceeding 100 ppm concentration at the resi-
      dence.
5.22 Redesign the casing string in Problem 5.10 assuming that it is to be
      used in a hydrogen sulfide environment. The formation tempera-
      ture gradient for the area is 0.015°F/ft.

References

1.   Fertl, W.H. and Timko, D.J.: "Application of Well Logs to Geopressure Problems in the Search, Drilling, and Production of Hydrocarbons," French Petroleum Institute Paper No. 4, (June, 1971).

2.   Pilkington, P.E.: "Fracture Gradient Estimates in Tertiary Basins," *Petroleum Engineer International* (May, 1978).

3.   Hubbert, M. King and Willis, D.G.: "Mechanics of Hydraulic Fracturing," Trans., AIME (1957) 210, 153–166.

4.   Matthews, W.R. and Kelly, John: "How to Predict Formation Pressure and Fracture Gradient," *Oil and Gas Journal* (Feb. 20, 1967).

5.   Eaton, Ben A.: "Fracture Gradient Prediction and its Application in Oilfield Operations," *Journal of Petroleum Technology,* (October, 1969), 1353–1360.

6.   Fertl, W.H., "Predicting Fracture Pressure Gradients for More Efficient Drilling," *Petroleum Engineer* (December, 1976).

7.   Christman, Stan A.: "Offshore Fracture Gradients," *Journal of Petroleum Technology* (August, 1973), 910–914.

8.   Personal communications with Charles M. Prentice of Prentice and Records Enterprises, Inc., Lafayette, Louisiana.

9.   MacPherson, L.A. and Berry, L.N.: "Prediction of Fracture Gradients from Log-derived Elastic Moduli," *Log Analyst,* 13 (5); 12–19.

10. Adams, N.J.: "A Field Case Analysis of Differential Pressure Pipe Sticking," SPE 6716 presented at the 1977 Fall Technical Conference, Denver, Colo.

11. Prentice, C.M.: "Maximum Load Casing Design," *Journal of Petroleum Technology* (July, 1970) 805–811.

12. Goins, W.C., Collings, B.J., and O'Brien, T.B.: "A New Approach to Tubular String Design," *World Oil*, November, December 1965, January, February, 1966.

13. Higdon, A., Ohlson, E.H., Steles, W.B., Weese, John, A., and Riley, W.F.: "Mechanics of Materials," John Wiley and Sons, Inc., 1976.

14. NIOSH Publication # 77–158: "Occupational Exposure to Hydrogen Sulfide," May, 1977.

15. Adams, N.J. and Carter, D.: "Hydrogen Sulfide in the Drilling Industry," paper presented at the 1979 Deep Drilling Symposium, Amarillo, Tex.

16. Garrett, R.L.: "A New Field Method for the Quantitative Determination of Sulfides in Water-Base Drilling Fluids," *Journal of Petroleum Technology* (September, 1977) 1195–1202.

17. Hadden, David M.: "A System for Continuous On-Site Measurement of Sulfides in Water-Base Drilling Muds," SPE 6664 presented at the 1977 Sour Gas Symposium in Tyler, Texas.

18. N.L. Baroid BTC Laboratory Report # STB–461.

19. Patton, C.C.: "Corrosion Control in Drilling," *Oil and Gas Journal*, (July 29, 1974).

20.   Amoco Chemicals Corp. Bulletin WT–1A, "Sour Corrosion and its Prevention."

21.   Reynolds, D.: "How to Cope with Sulfide Corrosion," Drilling-DCW (June, 1976).

22.   Bruist, E.H.: "A New Approach in Relief Well Drilling," 1972 Transactions Volume 253.

6

Abnormal Pressure Detection

ABNORMAL PRESSURE DETECTION IS ESSENTIAL to proper well planning and drilling practices. Some more common applications of pressure detection principles are to lessen the frequency and severity of kicks and blowouts, to reduce differential pipe sticking tendencies by using the minimum required mud weights, to maximize penetration rates, and to reduce formation damage problems resulting from excessive mud weights. When applied in an orderly fashion, these principles and techniques can be used to drill safely with minimum time and cost.

By definition, an abnormal pressure is any pressure that is different than the established normal trend for the given conditions. The variations may be (1) less than normal which are usually termed subnormal, or (2) greater than normal which have been called geopressure, superpressure, or simply abnormal pressure.

Subnormal pressures present few direct well control problems. The reader should be aware, however, that subnormal pressures do cause many drilling problems and indirectly may cause well control problems. Pressures greater than normal do cause well control problems and will therefore be presented in this discussion. For clarity, the term abnormal pressure will be used to identify the pressures greater than normal.

Origin of Abnormal Formation Pressure

Normal formation pressure in an area is equal to the hydrostatic pressure of the native formation fluids. In most cases, the fluids will vary from fresh water with a density of 8.33 ppg (0.433 psi/ft) to 9.0 ppg salt water (0.465 psi/ft). (It should be noted, however, that some field reports have indicated instances where the normal formation fluid density was greater than 9.0 ppg.) Regardless of the fluid density, the normal pressured formation can be considered as an open hydraulic system where pressure can easily be communicated throughout.

Abnormal pressured formations do not have the freedom of pressure communications exhibited by the normal zones. If in fact they did, the high pressures would rapidly dissipate and revert to normal pressures. It can therefore be concluded that a pressure entrapment mechanism of some type must be present before abnormal pressures can be generated and maintained. Fertl and Timko[1] listed several of the more common entrapment seals found thoughout the world (Table 6.1).

Assuming that a pressure seal is present, the causes or origins of pressure depend on such items as lithology, mineralogy, tectonic action, and rates of sedimentation. Fertl [2] has listed many of the field reported

Table 6.1
Suggested types of formation pressure seals
(after Fertl and Timko)

Type of seal	Nature of trap	Examples
Vertical	Massive shales and siltstones	Gulf Coast, U.S.A.
	Massive salts	Zechstein, North Germany
	Anhydrite	North Sea, Middle East
	Gypsum Limestone, marl, chalk Dolomite	U.S.A., U.S.S.R.
Transverse	Faults Salt and shale diapirs	worldwide
Combination of vertical and transverse		worldwide

causes of high pressures (Table 6.2). The most common cause found throughout the world is that related to abnormal compaction of sediments. Due to the frequency with which this type of pressure generation has been observed, it will be presented in detail in this section. (For further information concerning the other sources of abnormal pressure generation, see Reference 2.)

Compaction of sediments. The normal sedimentation process involves the deposition of layers of various rock particles. As these layers continue to build and increase the overburden (total rock) pressure, the underlying sediments are forced downward to allow for further surface deposition. The overburden pressure in this case is defined as the total of the rock matrix pressure and the formation fluid pressure. Under normal drilling conditions, the formation fluid pressure is the only concern due to its ability to cause fluid flow into the wellbore under certain geological conditions, and the general inability of the rock matrix to move into the wellbore due to its semirigid structure.

Table 6.2
Origins for the generation of abnormal fluid pressure
(after Fertl[2])

Piezometric fluid level (artesian water system)
Reservoir structure
Repressuring of reservoir rock
Rate of sedimentation and depositional environment
Paleopressures
Tectonic activities
    Faults
    Shale diapirism
    (mud volcanoes)
    Salt diapirism
    Sandstone dikes
    Earthquakes
Osmotic phenomena
Diagenesis phenomena
    Diagenesis of clay sediments
    Diagenesis of sulfates
    Diagenesis of volcanic ash
Massive areal rock salt deposition
Permafrost environment
Thermodynamic and biochemical causes

The manner in which the rock matrix accepts the increasing overburden load will serve to explain the generation of abnormal pressures in this environment. As both the surface deposition and the resultant total overburden increase, the underlying rock must accept the load.

The primary manner in which the rock matrix can increase its strength is to increase the grain-to-grain contact of the individual rock particles. This implies that the resultant porosity must decrease with depth under normal sedimentary conditions. This relationship can be seen in Fig. 6.1 where overburden stress is equated to depth. If the normal porosity compaction process is prohibited by disallowing the fluids in the pore spaces to escape, the rock matrix cannot increase its grain-to-grain contact or its overburden support capabilities.

Since the total overburden load continues to increase and the rock matrix can no longer carry its burden, the fluids in the pores of the rock must therefore begin to support an abnormally large amount of the

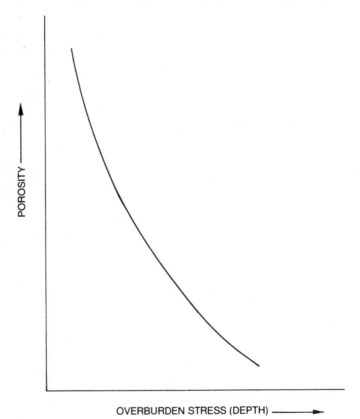

Fig. 6-1   Effect of overburden stress on formation porosity during normal compaction

overburden resulting in higher-than-normal fluid pressures (Fig. 6.2). The depth at which the abnormally high porosities are encountered is usually assumed to be the initial depth at which abnormal pressures are encountered.

A sealing mechanism must be present to entrap the abnormal pressures in their environment. The most common sealing mechanism found in continuous depositional basins is rapid deposition of a low permeability layer of rock such as a clean shale section. The shale reduces normal fluid escape causing undercompaction and abnormal fluid pressures.

An illustration of this phenomenon can be seen from the electric log show in Fig. 6.3. In this field case, the shale section under the massive sands entrapped the fluids in the lower intervals and created pressure gradients greater than 0.465 psi/ft.

Pressure Detection Methods

There are many methods used to predict, detect, and evaluate formation fluid pressures. Fertl[2] has developed a representative listing of these

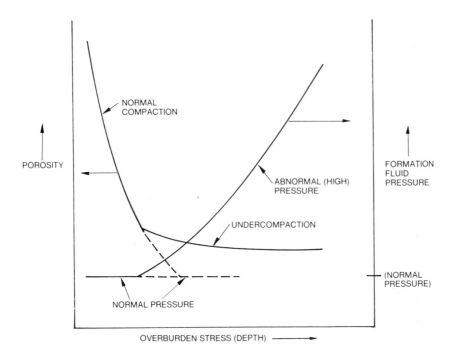

Fig. 6-2 The formation fluid pressure will increase above the normal value at the depth where undercompaction begins to occur.

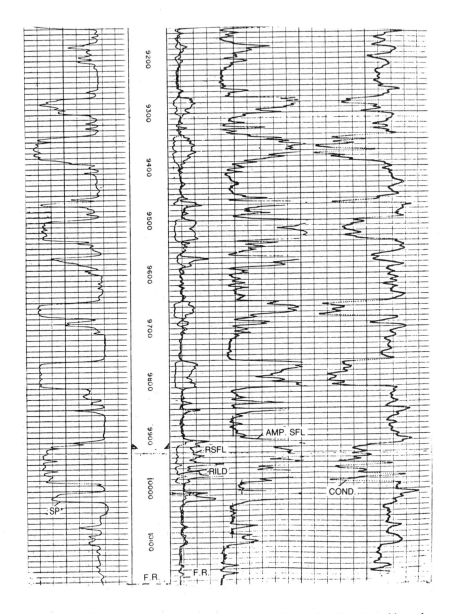

Fig. 6-3 Illustration of an electric log from a well in which the deposition of an impermeable shale barrier generated abnormal pressures in the lower intervals. In the well, the barrier is from 10,100 ft.

Table 6.3 Techniques available to predict, detect, and evaluate overpressures (after Fertl)

Source of data	Pressure indicators	Time of recording
Geophysical methods	Seismic (formation velocity) Gravity Magnetics Electrical prospecting methods	Prior to spudding well
Drilling parameters	Drilling rate d-exponent Modified d-exponent Drilling rate equations Drilling porosity and formation pressure logs Logging while drilling Torque Drag	While drilling (no delay time)
Drilling mud parameters	Mud-gas cutting Flow-line mud weight Pressure kicks Flow-line temperature Resistivity, chloride ion, and other novel concepts Pit level and total pit volume Hole fill-up Mud flow rate	While drilling (delayed by the time required for mud return)
Shale cuttings parameters	Bulk density Shale factor Volume, shape, and size Novel, miscellaneous methods	While drilling (delayed by the time required for mud return)

Well logging	Electrical surveys
	  resistivity
	  conductivity
	  shale formation factor
	  salinity variations
	Interval transit time
	Bulk density
	Hydrogen index
	Thermal neutron capture cross
	  section
	Nuclear magnetic resonance
	Downhole gravity data
		After drilling

Direct pressure measuring devices	Pressure bombs Drill-stem test Wire-line formation test	When well is tested or completed

344   Well Control Problems and Solutions

Table 6.4 Pressure detection methods while drilling

Qualitative methods	Quantitative methods
Paleontology Offset well correlation Temperature anomalies Gas counting Mud and/or cuttings resistivity Delta chlorides Cuttings character Hole condition Cuttings content (shale factor)	Log analysis (porosity detection) resistivity (conductivity) sonic Bulk density Drilling equations "d_c" exponent computerized drilling models Kicks

methods (Table 6.3). Most of the methods in this listing can be subdivided into qualitative or quantitative detection methods as shown in Table 6.4. Table 6.4 contains only those methods that are applicable during the drilling phase of a well.

Qualitative Methods of Abnormal Pressure Detection

Many techniques can be used during drilling to indicate the presence of abnormal fluid pressures. Although they may not provide a quantitative estimate of the pressures, they are useful in forewarning the operator to use caution in further drilling operations. Some of the qualitative indicators are currently undergoing analysis to determine if there is a manner in which they can be used quantitatively.

Paleontology. The science dealing with the life of past geologic periods especially as known from fossil remains is often used as a qualitative pressure indicator. The cuttings from the well in question are observed to determine if they contain fossils known to be associated with overpressured rock sections in the area or in an offset well. The disadvantage of using paleontology as a primary pressure indicator is that it requires the presence of a paleontologist at the rig site during the drilling operations.

Offsets. Offset well correlation has been used in most areas of the world to correlate between known pressure environments in offset wells and the unknown pressures in the well being drilled. The most common type of correlation is that of lithology identification between the two wells and subsequently assuming that the pressure environments will be similar in common geologic zones.

Fig. 6.4 shows the manner in which the correlation was successfully applied in a Gulf Coast well. Although Fig. 6.4 used a correlation between penetration rate and the S.P. curve from the offset well log, many other parameters can be used for correlation. These might include drill-

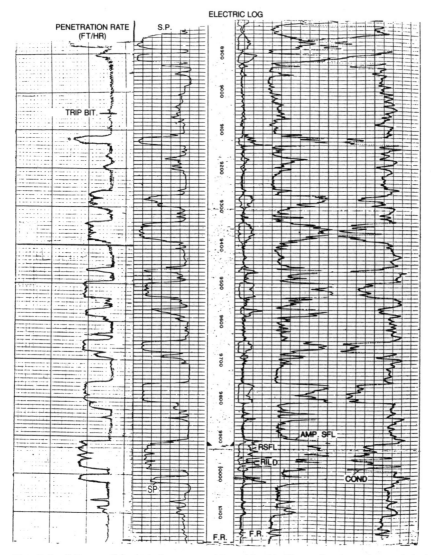

Fig. 6-4   Offset well correlation was achieved in this illustration by using a plot of penetration rates as a lithology indication. The sand and shale sequences were clearly defined in this example.

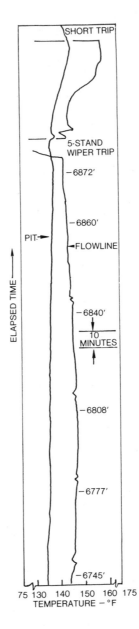

Fig. 6-5 Typical results from flowline temperature monitoring

ing rates, cuttings comparison, gas content, hole problems, and fluid content in marker zones.

Temperature. Temperature anomalies have been proposed by many authors[3,4] to be an effective qualitative indicator. Wilson and Bush[5] presented field results from several wells that illustrated the application. Although difficult to monitor, temperature anomalies do provide an additional monitoring device to detect transition zones into high pressures.

The development of temperature anomalies in abnormally pressured zones is due to the high fluid-filled porosity. As the earth radiates heat in a constant manner, any thermal conductivity change in the rock will result in a temperature anomaly. Since water conducts heat approximately 60% as effectively as clay, any zone with a high water content will act as a resistance to heat flow. Therefore, when greater-than-normal temperatures are encountered, high porosities are inferred which indicates abnormal pressures.

The flow line temperature is usually monitored as an indicator of formation temperature. A probe is placed in the mud flow line from which circulating temperatures are recorded. Typical results are shown in Fig. 6.5. The readings from the probe are used to calculate a temperature gradient using an equation similar to Equation 6.1.

$$G = 100 \ (T_2 - T_1)/D_2 - D_1 \qquad \text{Eq. 6.1}$$

where: D = depth, ft
$\qquad$ T = flowline temperature, °F
$\qquad$ G = geothermal gradient, °F/100 ft
$\qquad$ 1 = subscript denoting shallow point
$\qquad$ 2 = subscript denoting deeper point

By plotting these temperature gradients, or in some cases, the absolute temperatures versus depth, entry into abnormal pressures can be detected by noting unusual temperature changes. Fig. 6.6 (a)–(e) illustrate worldwide field results from this method.

Mud gas. Gas counting in the return mud has often been listed as an indicator of abnormal pressures. In recent years, however, experience has shown that this is not a reliable indicator due to the various sources of complications that must be considered when measuring the gas content, in addition to a variety of causes for gas in a formation.

Resistivity. Mud and/or cuttings resistivity in addition to the delta-chlorides concept are receiving interest as pressure indicators. As the high porosities of the abnormally pressured formations are encountered

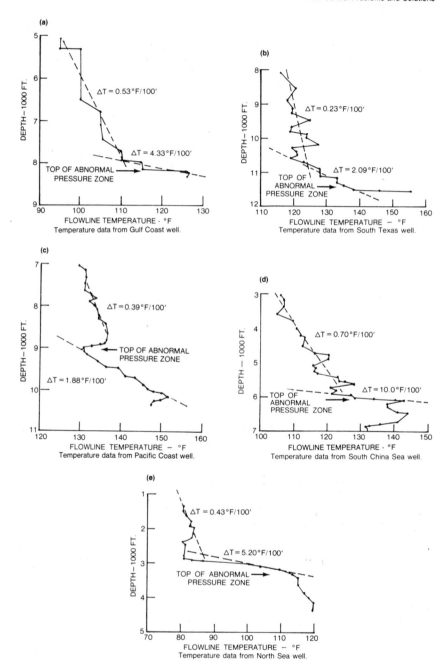

Fig. 6-6   Illustration of temperature monitoring that was used to detect pressure transition zones

while drilling, the crushed rock will liberate its formation fluid to the mud stream. This should alter the mud resistivity and chloride content of the drilling fluid, assuming that the salinity of the formation waters is different than the drilling mud. In addition, the resistivity of the cuttings should vary with the increased porosity. Fig. 6.7 illustrates the expected variations, and Fig. 6.8 shows a typical delta-chloride plot.

The primary difficulty with delta-chlorides is detecting the relatively minute changes in chloride content within the short transition intervals often found in areas such as the Louisiana and Texas Gulf Coast. In addition, the changing mud resistivities are affected by factors such as make-up water and mud additives as well as salinity of the formation waters. As a result, the reliability of these methods suggests that they be used as secondary indicators when attempting to monitor possible transition zones.

Cuttings. The character of the cuttings may serve to indicate entry into pressured zones. Observing the cuttings is perhaps the simplest as well as oldest method used in the industry.

During the drilling process, a bit tooth will penetrate the formation and gouge a piece of rock which becomes the cutting. Although the cutting has been created, its removal from below the bit will depend on the drilling fluid type and hydraulics, and on differential pressure.

The controlling variable when discussing pressure detection is the amount of differential pressure. When large amounts of differential pressure are present, the cuttings are held in place and therefore must be reground until their reduced size allows removal. This is termed the "chip hold down effect."

When small amounts of differential pressure are present, however, the cuttings are removed from below the bit before they can be reground. This is indicated by large cuttings at the shaker. It is inferred from the larger cuttings that the differential pressure must be decreasing. If the

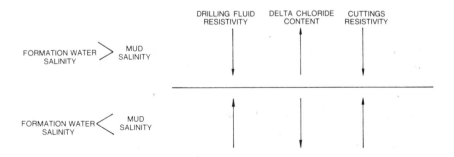

Fig. 6-7   Expected results from (1) mud and cuttings resistivities and (2) delta chloride monitoring when used as a pressure monitoring device.

mud weight were held constant while drilling the interval, it is assumed
that the formation pressure must be increasing.

Tight hole.   The hole condition, as measured by torque and drag, will
often indicate entry in pressured zones. As the formation pressure
approaches and exceeds the mud hydrostatic pressure, the walls of the
borehole may exhibit tendencies to heave into the wellbore or neck-
down due to the pressure differential. When this occurs, the force required to rotate and move the pipe may increase.

Shale.   The cuttings content (shale factor) method is a new approach
of detecting abnormal pressures. The method is based on analyzing the
montmorillonite content of the cuttings by using the common methylene
blue test or the more sophisticated X-ray diffraction technique. It has

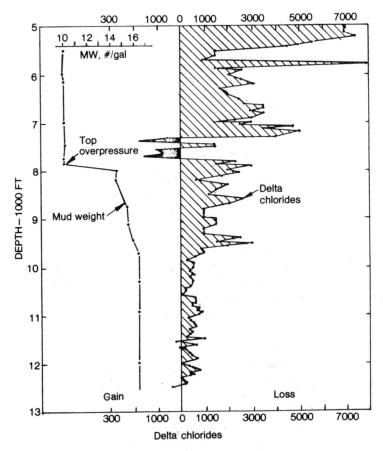

Fig. 6-8 Typical delta chloride plot

been determined that certain clay states can be associated with either the cap rock or the clays in the overpressured intervals.

Quantitative Methods

Proper well planning procedures such as mud weight selection and casing design require that a reasonably accurate estimate of the magnitude of formation pressures is obtained. The most common methods used during the drilling phase are presented in Table 6.4. These quantitative methods can be subdivided into log analysis which requires that a log is run at several key depths in the well, and drilling equations or bulk density analyses, both of which can provide engineering estimates while the drilling is in progress.

Log analysis. Log analysis is a common procedure for pore pressure estimation in both offset wells and the actual well drilling. The analysis techniques use the effect of the abnormally high porosities on rock properties such as electrical conductivity, sonic travel time, and bulk density. Each of the logs that are to be presented here are based on one of these principles. The logs are the resistivity (or reciprocated conductivity) log and the sonic log. It should be noted, however, that any log dependent primarily on porosity for its responses can be used in a quantitative evaluation of formation pressures.

The log originally used in pressure detection was the resistivity log. The response of the log is based on the resistivity of the total sample which includes the rock matrix and the fluid-filled porosity. If a zone is penetrated that has abnormally high porosities (and associated high pressures), the resistivity of the rock will be reduced, due to the greater conductivity of water than rock matrix. The expected response can be seen in Fig. 6.9.

Fig. 6.9 illustrates several important points. Since the high formation pressures were originally developed in shale sections and later equalized the sand zone pressures, only the clean shale sections are used as plotting points. This excludes sand resistivities, silty shale, lime or limey shale, or any other types of rock that may be encountered. As the shale resistivities are selected and plotted, a normal trend line should develop prior to entry into the pressured zone.

Upon penetrating an abnormally pressured zone, a deviation or divergence will be noted. The degree of divergence is used to estimate the magnitude of the formation pressures. This concept of the development of the normal trend and noting any divergence will be used with most pressure detection techniques.

An actual field case can be seen in Fig. 6.10. The impermeable shale

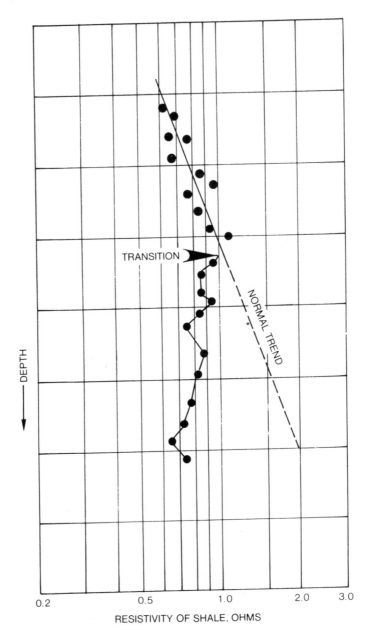

Fig. 6-9 Generalized shale resistivity plot

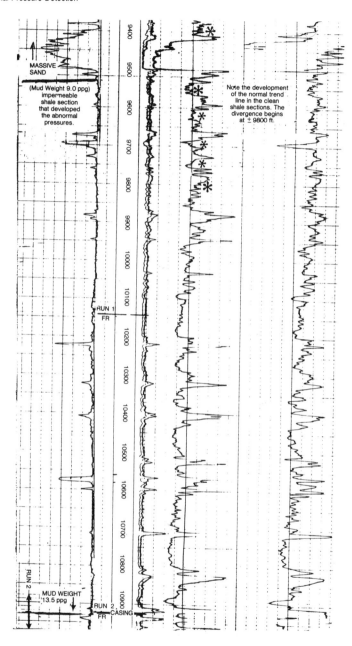

Fig. 6-10 Illustration of an electric log from a well in which the deposition of an impermeable shale barrier generated abnormal pressures in the lower intervals. In this well, the barrier is from 9,500-9,700 ft.

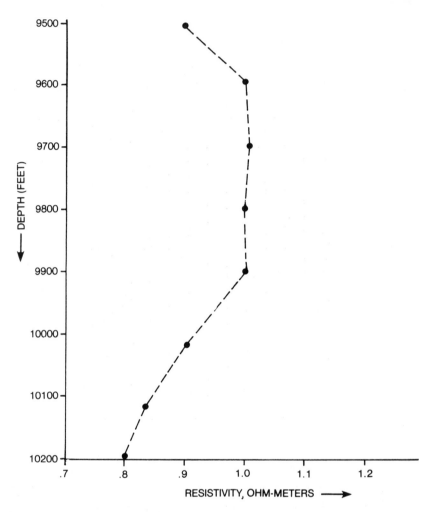

Fig. 6-11   Shale resistivities from the log shown in Fig. 6-10 are plotted vs. depth. Note the departure from the normal trend line at 10,000 ft.

section was entered at about 9,500 ft. Although this section contained normal pressure from 9,500–9,800 ft, as evidenced by the increasing resistivity of the normal trend, the reversal can be seen from 9,800–10,900 ft. The mud weight was 9.0 ppg at 9,500 ft but was increased to 13.5 ppg at 10,900 ft. A plot of the key shale resistivities points is shown in Fig. 6.11.

Hottman and Johnson[6] developed a technique based on empirical relationships whereby an estimate of formation pressures could be made by noting the ratio between the observed and normal rock resistivities. Their data points, shown in Table 6.5, was used to construct the curve in Fig. 6.12. As explained by the authors, the following steps are necessary to estimate the formation pressure.

1. The normal trend is established by plotting the logarithm of shale resistivity versus depth.
2. The top of the pressured interval is found by noting the depth at which the plotted points diverge from the trend.

Table 6.5 Formation pressures and shale resistivity ratios in overpressured Miocene/Oligocene formations, U.S. Gulf Coast area (after Hottman and Johnson, 1965)

Parish or County and State	Well	Depth ft	Pressure psi	FPG* psi/ft	Shale resistivity ratio** Ωm
St. Martin, La.	A	12,400	10,240	0.83	2.60
Cameron, La.	B	10,070	7,500	0.74	1.70
Cameron, La.	B	10,150	8,000	0.79	1.95
	C	13,100	11,600	0.89	4.20
	D	9,370	5,000	0.53	1.15
Offshore	E	12,300	6,350	0.52	1.15
St. Mary, La.	F	12,500	6,440	0.52	1.30
		14,000	11,500	0.82	2.40
Jefferson Davis,	G	10,948	7,970	0.73	1.78
La.	H	10,800	7,600	0.70	1.92
		10,750	7,600	0.71	1.77
Cameron, La.	I	12,900	11,000	0.85	3.30
Iberia, La.	J	13,844	7,200	0.52	1.10
		15,353	12,100	0.79	2.30
Lafayette, La.	K	12,600	9,000	0.71	1.60
		12,900	9,000	0.70	1.70
	L	11,750	8,700	0.74	1.60
	M	14,550	10,800	0.74	1.85
Cameron, La.	N	11,070	9,400	0.85	3.90
Terrebonne, La.	O	11,900	8,100	0.68	1.70
		13,600	10,900	0.80	2.35
Jefferson, Tex.	P	10,000	8,750	0.88	3.20
St. Martin, La.	Q	10,800	7,680	0.71	1.60
Cameron, La.	R	12,700	11,150	0.88	2.80
		13,500	11,600	0.86	2.50
		13,950	12,500	0.90	2.75

*Formation fluid pressure gradient.
**Ratio of resistivity of normally pressured shale to observed resistivity of overpressured shale: $R_{sh(n)}/R_{sh(ob)}$.

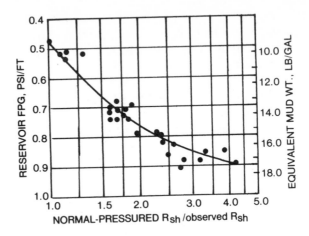

Fig. 6-12   Empirical correlation of formation pressure gradients vs. a ratio of normal
to observed shale resistivities

3. The pressure gradient at any depth is found as follows:
   a. The ratio of the extrapolated normal shale resistivity to the
      observed shale resistivity is determined.
   b. The formation pressure corresponding to the calculated ratio is
      found from Fig. 6.12.

Example 6.1 will illustrate the application of the Hottman and Johnson
method.

Example 6.1 Using the following data, plot the shale resistivities on
semi-log paper. Where does the entry into abnormal pressure occur?
Using the Hottman and Johnson method, calculate the formation
pressures at the bottom of the well.

Depth, ft	Shale resistivity, ohm m^2/m
5,000	0.70
5,500	0.79
6,000	0.74
6,500	0.76
7,000	0.78
7,500	0.79
8,000	0.80
8,500	0.84
9,000	0.9

9,500	0.96
10,000	1.04
10,100	1.10
10,200	1.02
10,300	1.06
10,400	1.10
10,500	1.12
10,600	1.10
10,700	1.10
10,800	·1.08
10,900	1.05
11,000	1.02
11,100	1.00
11,200	1.00
11,300	1.00
11,400	0.96
11,500	0.94
11,600	0.92
11,700	0.84
11,800	0.78
11,900	0.70
12,000	0.64

Solution

1. Plot the data on semi-logarithmic paper as shown in Fig. 6.13.
2. Divergence from the normal trend denotes entry into abnormal pressures. In Fig. 6.13, this occurs at 10,500 ft.
3. To calculate the formation pressure at 12,000 ft, extrapolate the normal trend line to 12,000 ft.
4. Read the values of R_{normal} and $R_{observed}$ and calculate $R_{normal}/R_{observed}$.
5. From Fig. 6.13, R_{normal} is 1.4 and $R_{observed}$ is 0.64.
   $$R = 1.4/0.64$$
   $$= 2.18$$
6. Using Fig. 6.12 and a ratio of 2.18, the formation pressure at 12,000 ft is calculated to be 14.8 ppg.

Overlays. Subsequent to the Hottman and Johnson approach, unpublished techniques were developed that used an overlay or underlay for a quick evaluation of formation pressures. The overlay (and underlay) contains a series of parallel lines that represent formation pressure expressed as mud weight (Fig. 6.14).

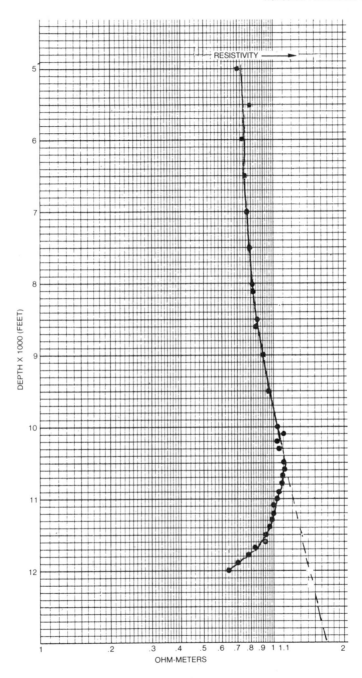

Fig. 6-13 Shale resistivity plot for Example 6-1

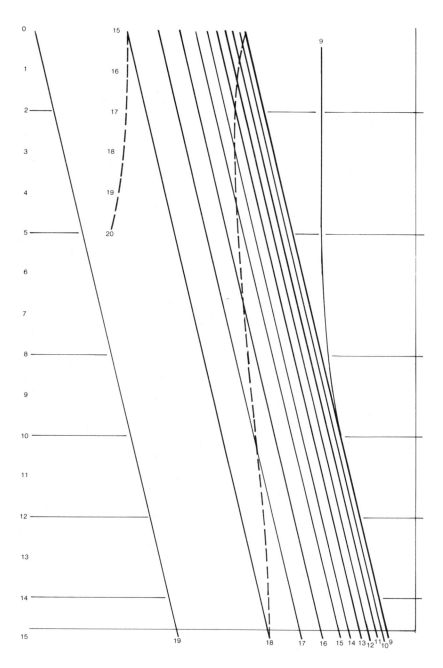

Fig. 6-14 Shale resistivity overlay (after Prentice et al.)

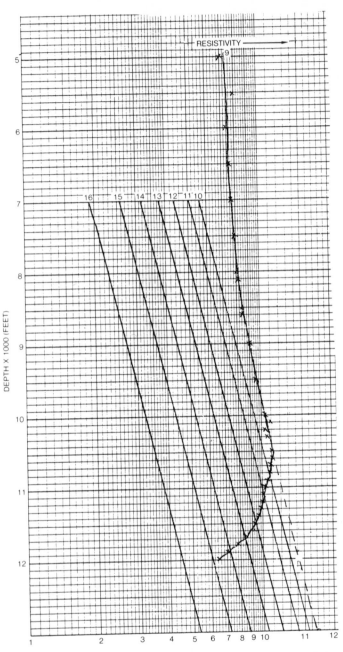

Fig. 6-15 The overlay can be used with the data from Example 6-1 to rapidly estimate the formation pressure. Note that the results are the same as those obtained from the Hottman-Johnson method.

The overlay, developed by Prentice et al.,[7] is shifted left and right until the normal pressure line is aligned with the normal trend. Formation pressures are read directly from a visual inspection of the location of the resistivity plots within the framework of the parallel lines. As an example, the data from Example 6.1 was plotted in Fig. 6.15 and the overlay was used to estimate the formation pressure.

Different types of overlays have been developed for pressure determinations. Some are for use with resistivity or conductivity curves while others are used with the sonic logs. In addition, overlays have been developed for the various geological ages for each log type.

There are many pitfalls to avoid when using an overlay. Most can be shifted left or right but are depth fixed and therefore cannot be moved vertically. Overlays are generally developed for one type of semi-log paper and cannot be interchanged. This necessitates the design of a different overlay if paper sizes must be changed.

Another common mistake when using the resistivity overlay is an attempt to use it for conductivity values by turning the overlay to its reverse side. In addition, overlays do not account for abnormal water

Table 6.6 Formation pressure and shale acoustic log data in overpressured Miocene/Oligocene formations, U.S. Gulf Coast area (after Hottman and Johnson, 1965)

Parish or County and State	Well	Depth, ft	Pressure, psi	FPG* psi/ft	$\Delta t_{ob(sh)} - \Delta t_{n(sh)}$, μsec/ft
Terrebonne, La.	1	13,387	11,647	0.87	22
Offshore Lafourche, La.	2	11,000	6,820	0.62	9
Assumption, La.	3	10,820	8,872	0.82	21
Offshore Vermilion, La.	4	11,900	9,996	0.84	27
Offshore Terrebonne, La.	5	13,118	11,281	0.86	27
East Baton Rouge, La.	6	10,980	8,015	0.73	13
St. Martin, La.	7	11,500	6,210	0.54	4
Offshore St. Mary, La.	8	13,350	11,481	0.86	30
Calcasieu, La.	9	11,800	6,608	0.56	7
Offshore St. Mary, La.	10	13,010	10,928	0.84	23
Offshore St. Mary, La.	11	13,825	12,719	0.92	33
Offshore Plaquemines, La.	12	8,874	5,324	0.60	5
Cameron, La.	13	11,115	9,781	0.88	32
Cameron, La.	14	11,435	11,292	0.90	38
Jefferson, Tex.	15	10,890	9,910	0.91	39
Terrebonne, La.	16	11,050	8,951	0.81	21
Offshore Galveston, Tex.	17	11,750	11,398	0.97	56
Chambers, Tex.	18	12,080	9,422	0.78	18

*Formation fluid pressure gradient.

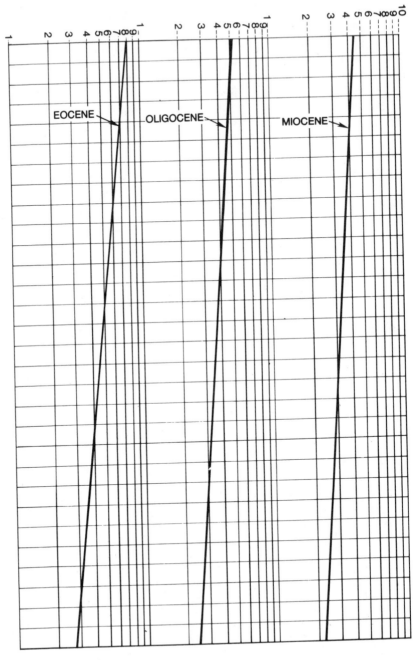

Fig. 6-16 Normal trend for travel time plots for several geologic conditions

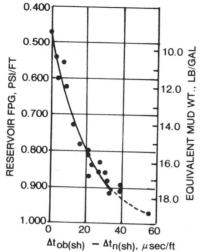

Fig. 6-17 Empirical correlation of formation pressure gradients vs. a difference between observed and normal travel times

salinity changes. When these changes are encountered, different techniques must be employed that normalize the salinity effect.[8]

The sonic log has been successfully used as a pressure evaluation tool. The technique utilizes the difference in travel times between high porosity overpressured zones and the low porosity, normally pressured zones. The basic relationship between travel times can be seen in Fig. 6.16.

Hottman and Johnson studied the wells shown in Table 6.6 and developed the pressure relationship shown in Fig. 6.17. The manner in which formation pressures are calculated using the Hottman and Johnson approach is similar to their method for resistivity plots as illustrated in Example 6.1.

Observed transit times are plotted and the normal trend line is established and extrapolated throughout the pressured region. At the depth of interest, the difference between the observed and normal travel times is established. This difference is used with Fig. 6.17 to estimate the formation pressure. This procedure is illustrated in Example 6.2.

Example 6.2 Use the following sonic travel time data to determine the formation pressure at 13,000 ft. Where does entry in abnormal pressures occur?

Depth, ft	Travel time μ sec/ft
6,000	160
6,500	153
7,000	145
7,500	138

8,000	130
8,500	120
9,000	120
9,200	113
9,400	115
9,600	110
9,800	108
10,000	116
10,200	115
10,400	116
10,600	118
10,800	115
11,000	120
11,200	122
11,400	118
11,500	119
11,600	120
11,700	120
11,800	119
11,900	118
12,000	121
12,100	120
12,200	119
12,300	118
12,400	120
12,500	118
12,600	120
12,700	120
12,800	121
12,900	121
13,000	121

Solution:
1. Plot the data on semi-log paper as shown in Fig. 6.18.
2. The divergence from the normal trend at 9,800 ft denotes entry into the pressured zone.
3. At 13,000 ft, the difference between the extrapolated normal trend and observed values is 44 μsec/ft.
4. Enter Fig. 6.17 with a value of 44 μsec/ft, and read the formation pressure as 18.2 ppg.

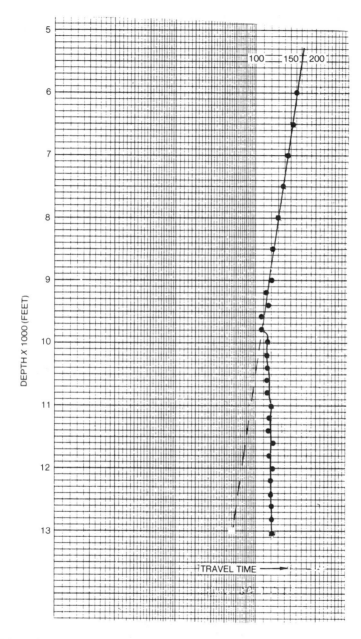

Fig. 6-18 Illustration of the sonic travel time data used in Example 6-2

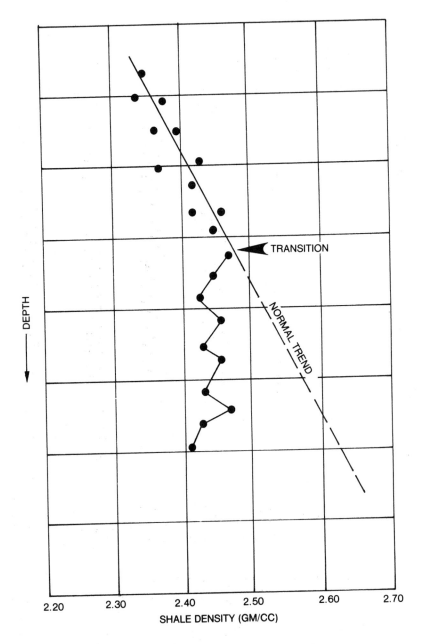

Fig. 6-19 Generalized shale density plot

Bulk density. When drilling in normally pressured zones, the bulk density of the drilled rock should increase due to compaction, or porosity reduction.[9] As high formation pressures are encountered, the associated high porosities will cause a deviation in the expected bulk density trend. A typical plot of bulk densities is seen in Fig. 6.19. The transition from normal to abnormal pressures occurs at the depth where divergence from the normal trend is observed.

The results from a typical field case are seen in Fig. 6.24. The resistivity plot shows transition zones at 10,700 ft and 12,500 ft. The density log detected the lower transition zone but was unable to define the upper transition zone due to the lack of an established trend line.

The most common field method for measuring bulk densities is the variable density column. A graduated cylinder is filled with a mixture of halogenated hydrocarbons such that the fluid density increases with depth. Several known-density, colored, glass beads are dropped into the liquid and their final depths are recorded. A graph that relates density to depth is prepared (Fig. 6.21).

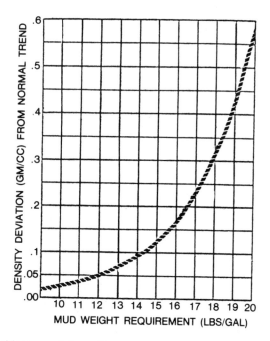

Fig. 6-20 Empirical correlation of formation pressures vs. the difference between normal and observed shale densities

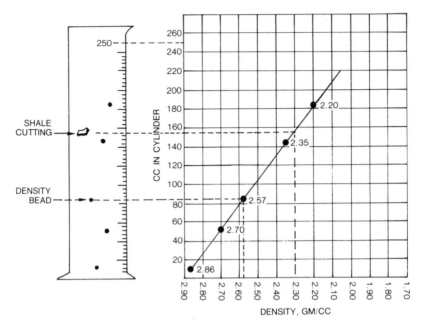

Fig. 6-21   Illustration of the manner in which the variable density column is used to determine shale densities (after Bourgoyne).

Several clean cuttings are dropped into the cylinder and their final depths are recorded and averaged. The average density is plotted versus well depth. A normal trend line is established and any divergence can be noted. Example 6.3 will illustrate this method.

Example 6.3 The following cuttings data were collected on a North Sea well. Average and plot the data. Use the graph in Fig. 6.21 to determine the shale densities. Graph the densities and determine where the transition zone is encountered.

Depth, ft	Cuttings depth in density column, cc		
	#1	#2	#3
6,000	191	193	191
6,500	185	184	186
7,000	181	182	183
7,500	170	173	190
8,000	167	166	169
8,500	158	159	159
9,000	155	154	156
9,200	152	153	153

9,400	170	168	172
9,600	153	150	151
9,800	151	151	152
10,000	151	151	151
10,100	150	151	151
10,200	147	147	149
10,300	148	148	170
10,400	147	149	147
10,500	147	150	149
10,600	149	148	149
10,700	155	154	180
10,800	155	158	156
10,900	160	159	161
11,000	153	152	190

Solution
1. The data were averaged* and the results are as follows:

Depth, ft	Average column depth, cc	Density, gm/cc
6,000	192	2.16
6,500	185	2.18
7,000	182	2.22
7,500	172	2.24
8,000	168	2.26
8,500	159	2.29
9,000	155	2.31
9,200	153	2.32
9,400	170	2.25
9,600	151	2.33
9,800	151	2.33
10,000	151	2.33
10,100	151	2.33
10,200	148	2.34
10,300	148	2.34
10,400	148	2.34
10,500	149	2.33
10,600	149	2.33
10,700	155	2.31
10,800	156	2.30
10,900	160	2.28
11,000	153	2.32

*Note that certain values were not used in the averaging.

2. The shale densities are plotted in Fig. 6.22. The transition zone
   begins at 9,400 ft.

Another common tool for determining bulk densities is with the mud
balance. The cup is filled with clean, shale cuttings until a weight of 8.33
ppg is recorded. The cup is subsequently filled with fresh water and the
mixture is weighed. The shale density is calculated as follows:

$$\text{Density} = 8.33/(16.67 - W_2) \qquad \text{Eq. 6.2}$$

where: Density = gm/cc
        W_2 = final weight, ppg

Example 6.4 gives a sample set of calculations using this procedure.

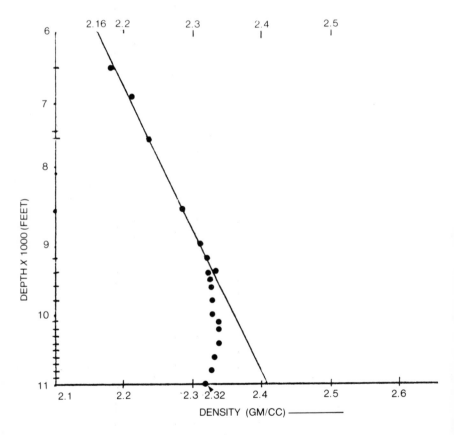

Fig. 6-22 Shale density plot of the values given in Example 6-3

Example 6.4 On a given well in Oklahoma, the following data were obtained by using the mud balance method. Calculate and plot the densities. Where is entry into abnormal pressures observed?

Depth, ft	Average of 5 samples at each depth, ppg
6,000	12.88
6,500	12.91
7,000	12.93
7,500	12.98
8,000	13.00
8,200	13.00
8,400	13.01
8,600	13.03
8,800	13.03
9,000	13.04
9,200	13.06
9,400	13.06
9,600	13.08
9,800	13.09
10,000	13.08
10,200	13.09
10,400	13.09
10,600	13.09
10,800	13.11
11,000	13.09
11,200	13.11
11,400	13.11
11,600	13.12
11,800	13.11
12,000	13.11

Solution
1. The density for each depth interval was calculated using Eq. 6.2. The results are as follows:

Depth, ft	Density, gm/cc
6,000	2.20
6,500	2.22
7,000	2.23
7,500	2.26

8,000	2.27
8,200	2.27
8,400	2.28
8,600	2.29
8,800	2.29
9,000	2.30
9,200	2.31
9,400	2.31
9,600	2.32
9,800	2.33
10,000	2.32
10,200	2.33
10,400	2.33
10,600	2.33
10,800	2.34
11,000	2.33
11,200	2.34
11,400	2.34
11,600	2.35
11,800	2.34
12,000	2.34

2. The densities are plotted in Figure 6.23. The transition zone is at 9,800 ft.

After a sufficient amount of data has been gathered in a particular area, an empirical relationship between normal trend divergence and formation pressure can be established. An example prepared by Boatman is shown in Fig. 6.20. In addition, overlays such as that shown in Fig. 6.25 are available and can be used to provide a quick interpretation from the density plot.[10]

Drilling equations.   Many mathematical models have been proposed in an effort to describe the relationship of several drilling variables on penetration rate. Several of the models are designed for easy application in the field while others require computerization. When conscientously applied, most of the available models can accurately detect and quantify abnormal formation pressures.

An attempt to quantify differential pressure is the basis of most drilling models. If this value is known, the formation pressure can readily be calculated. Garnier and van Lingen[11] showed that differential pressure has a definite effect on penetration (Fig. 6.26). In field studies, Benit and

Vidrine[12] found evidence that the range in differential pressure of 0–500 psi has the greatest effect in reducing penetration rates (Fig. 6.27).

The most common model used in the industry perhaps is the "d_c" exponent. The basis of the model is found in Bingham's equation to define the drilling process.[13] The equation is as follows:

$$\frac{R}{60N} = a \left[\frac{12W}{d_B} \right]^b \qquad \text{Eq. 6.3}$$

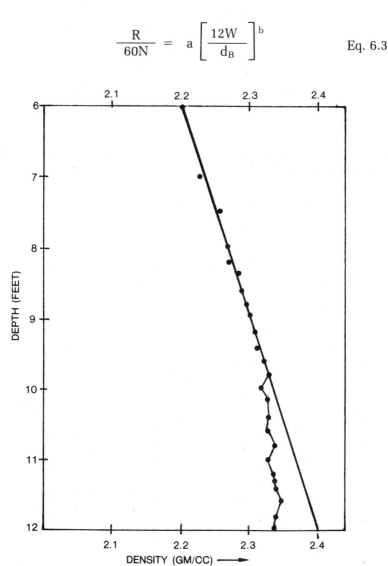

Fig. 6-23 Plot of the mud balanced-derived shale densities in Example 6-4

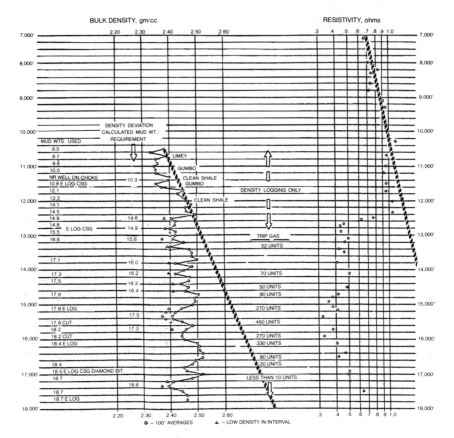

Fig. 6-24 Illustration of an actual case in which shale densities are used as a pressure monitoring device. Note that a resistivity plot is also shown.

Where: $R = $ bit penetration rate, ft/hr

$N = $ rotary speed, rpm

$W = $ bit weight, 1,000 lb

$d_B = $ bit diameter, in.

$b = $ bit weight exponent, dimensionless

$a = $ formation drillability constant, dimensionless

Jordan and Shirley[14] modified Bingham's equation to the form as follows:

$$d = \log (R/60N)/\log (12W/1000 \ d_B) \qquad \text{Eq. 6.4}$$

where "d" replaces "b" in Bingham's model. In Equation 6.4, the authors introduced several scaling constants and assigned a value of unity to the

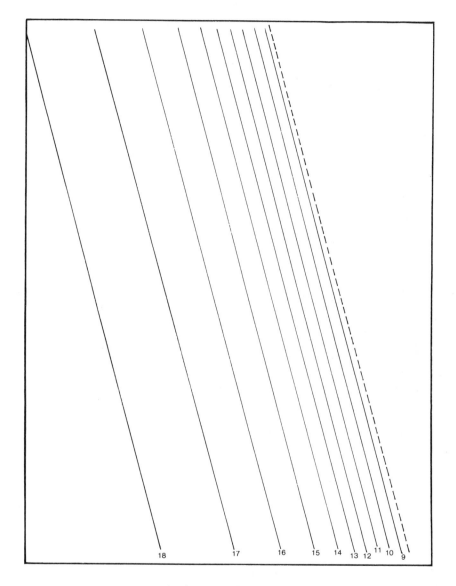

Fig. 6-25 Shale density overlay

drillability constant, "a". In field applications the "d" exponent should respond to the effect of differential pressure as shown in Fig. 6.28.

Rehm and McClendon[15] brought the equation to its final form by realizing that mud weight increases would mask the difference between

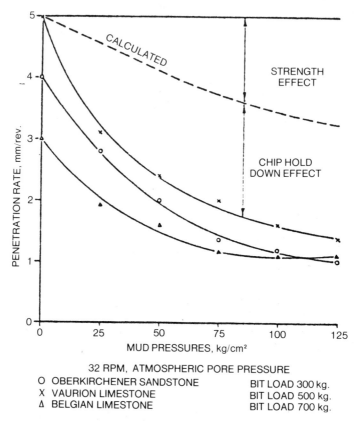

32 RPM, ATMOSPHERIC PORE PRESSURE

O OBERKIRCHENER SANDSTONE	BIT LOAD 300 kg.
X VAURION LIMESTONE	BIT LOAD 500 kg.
Δ BELGIAN LIMESTONE	BIT LOAD 700 kg.

Fig. 6-26 Differential pressure causes a reduction in penetration rates

normal and actual formation pressures. They proposed the normalizing ratio in Equation 6.5 to account for the effect of mud weight increases.

$$d_c = d(\text{normal formation pressure})/(\text{actual mud weight})$$

Eq. 6.5

Where: d_c = corrected d-exponent
       d = original value from Eq. 6.4
       Normal formation pressure = ppg
       Actual mud weight = ppg

Overlays have been developed that aid in the evaluation of formation pressures using the "d_c" exponent (Fig. 6.29).[16] The "d_c" exponents are calculated using Equation 6.4 and 6.5 and plotted on semi-logarithmic

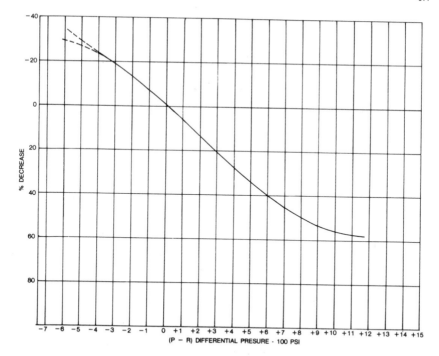

Fig. 6-27 Effect of differential pressure on penetration rates in field cases

paper. The most common scales are 2 cycle paper and 1 inch = 500 ft for the vertical scale. The overlay is used for a rapid estimation of the pressures. Fig. 6.30 shows four examples of the application of the overlay in actual field cases.

Bourgoyne and Young[17] proposed a more complex mathematical model for relating the effect of several variables on penetration rate. Their equation is as follows:

$$R = Exp. [2.303(A_1 + A_2(10{,}000 - D) + A_3D(G_p - \rho)A_4 D^{0.69}(G_p - 9)]$$
$$\times [(W/4d_B)^{A_5}(N/100)^{A_6} Exp. (-A_7h)(F_j/1000)^{A_8})]$$

Eq. 6.6

Where: R = penetration rate, ft/hr
D = well depth, ft
G_p = pore pressure gradient, ppg
ρ = equivalent drilling fluid density, ppg
W = weight on bit, 1,000 lbs

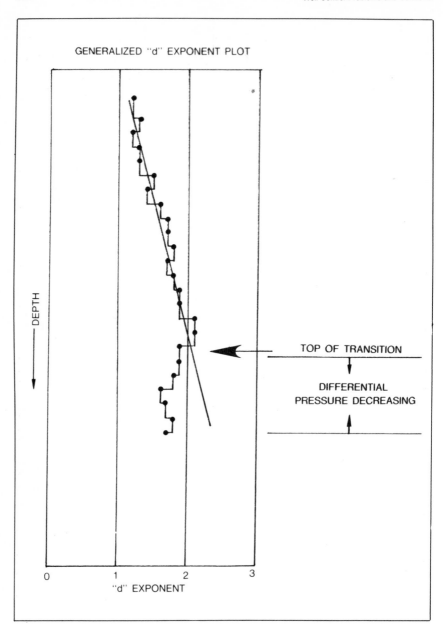

Fig. 6-28 Generalized "d" exponent plot

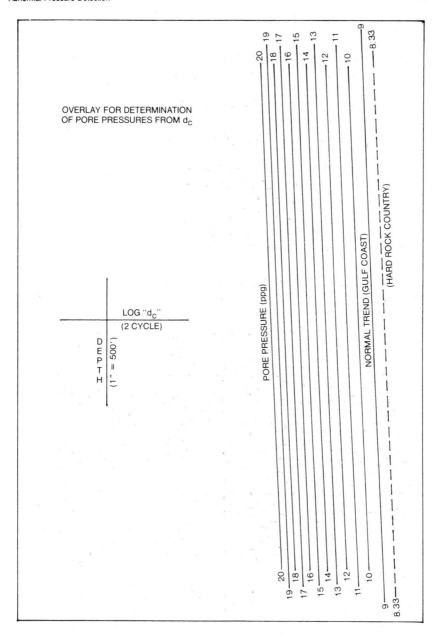

Fig. 6-29 "d_c" exponent overlay (after Zamora).

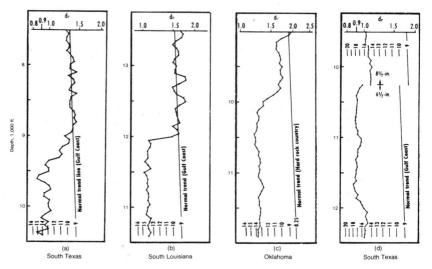

Fig. 6-30 Results of field cases where "d_c" exponent was used to detect transition zones.

N   = rotary speed, rpm
d_B = bit diameter, in.
h   = fractional tooth dullness
F_j = hydraulic impact force beneath the bit
$A_1 - A_8$ = constants which must be chosen based on local drilling conditions, evaluated by regressive analysis
Exp (x)   = symbolic representation of exponential function e^x

The parameters plotted from the Bourgoyne and Young model are similar to those seen in the Jordan and Shirley equation. They are given by the following equations.

$$K_p = Log \frac{R \, Exp \, (A_7h)}{\left(\dfrac{W}{4d}\right)^{A_5} \left(\dfrac{N}{100}\right)^{A_6} \left(\dfrac{F_j}{1000}\right)^{A_8}} \qquad Eq. \ 6.7$$

where $K_p{'}$ is the plotting parameter.

$$K_p{'} = K_p + A_2(D - 10,000) + A_3D \, (\rho - 9)$$

The formation pressure is determined by Equation 6.9.

$$G_p = 9.0 + \frac{K_p - A_1}{A_3D + A_4D^{0.69}}$$ Eq. 6.9

The coefficients $A_1 - A_8$ are based on local drilling conditions and therefore must be established on the initial wells in an area.

Sample results of the Bourgoyne and Young equation are given in Table 6.7. A plot of the parameter K_p' is shown in Fig. 6.31. The overlay shown in Fig. 6.31 was constructed using the coefficients $A_1 - A_8$.

Pressure kicks have been shown in previous chapters to be an accurate means of estimating formation pressures. Using Equation 6.10, the pressures can readily be calculated.

$$BHP = SIDPP + DPHP$$ Eq. 6.10

Where: BHP = formation pressure, psi
SIDPP = shut-in drill pipe pressure, psi
DPHP = drill pipe hydrostatic pressure, psi
It should be obvious that it is seldom desired to use a kick to establish formation pressures.

Problems

6.1 Use the following data to construct a plot of absolute temperature versus depth. In addition, calculate and plot the temperature gradient (°F/ft) versus depth. Where does entry into abnormal pressures occur?

Depth, ft	Flowline temperature, °F
6,000	96
6,200	100
6,400	99
6,600	103
6,800	105
7,000	106
7,200	110
7,400	111
7,600	112
7,800	118
8,000	120
8,200	121

Table 6.7 Example pore pressure evaluation using the normalized drillability parameter (after Bourgoyne and Young[7])

Depth, ft	Drilling rate, ft/hr	Bit weight, 1,000 lb/in	Rotary speed, rpm	Tooth dullness, fraction	Jet impact, Lbs	Equivalent circ. den., ppg	K_p	K_p'	G_p, ppg
9,515	23.0	2.58	113	0.77	964	9.5	1.57	1.66	9.1
9,830	22.0	1.15	126	0.38	964	9.5	1.62	1.73	9.3
10,130	14.0	0.81	129	0.74	827	9.6	1.57	1.72	9.3
10,390	16.0	1.02	78	0.24	984	9.7	1.53	1.72	9.3
10,500	19.0	1.69	81	0.61	984	9.7	1.57	1.77	9.4
10,575	13.0	1.56	81	0.73	984	9.7	1.44	1.65	9.0
10,840	16.6	1.63	67	0.38	932	9.8	1.50	1.75	9.4
10,960	15.9	1.83	65	0.57	878	9.8	1.50	1.76	9.4
11,060	15.7	2.03	69	0.72	878	9.8	1.50	1.76	9.4
11,475	14.0	1.69	77	0.20	887	10.3	1.38	1.80	9.5
11,775	13.5	2.31	53	0.12	852	11.8	1.32	2.12	10.6
11,940	6.2	2.26	67	0.20	976	15.3	0.97	2.63	12.2
12,070	9.6	2.07	84	0.08	993	15.7	1.14	2.90	13.1
12,315	15.5	3.11	60	0.40	1,185	16.3	1.33	3.25	14.3
12,900	31.4	2.82	85	0.42	1,150	16.7	1.64	3.71	15.7
12,975	42.7	3.48	77	0.17	1,221	16.7	1.70	3.77	15.9
13,055	38.6	3.29	75	0.29	1,161	16.8	1.69	3.79	16.0
13,255	43.4	2.82	76	0.43	1,161	16.8	1.79	3.91	16.3

$A_1 = 1.64$ $A_2 = 0.000074$ $A_3D = 0.24$ $A_4 = 0.0001$ $A_5 = 0.43$ $A_6 = 0.21$ $A_7 = 0.41$ $A_8 = 0.16$

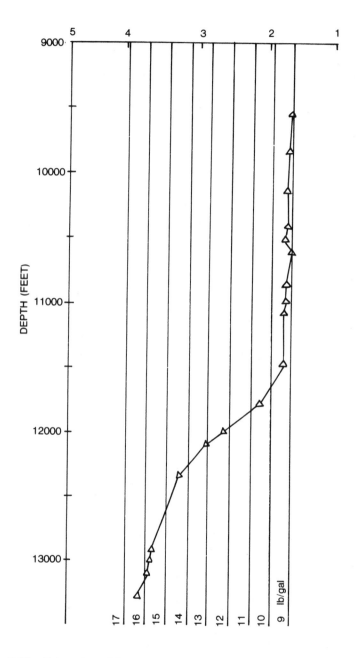

Fig. 6-31 Estimation of formation pore pressure using the drillability plot

8,400	126
8,600	129
8,800	131
9,000	133
9,200	135
9,400	138
9,600	139
9,800	141
10,000	145
10,200	136
10,400	141
10,600	150
10,800	163
11,000	

6.2 Construct a graph of temperature gradients versus depth for the following dala. Where does the increase in formation pressure occur?

Depth, ft	Formation temperature, °F
12,000	262
12,100	264
12,200	263
12,300	267
12,400	271
12,500	269
12,600	269
12,700	273
12,800	274
12,900	273
13,000	272
13,100	278
13,200	281
13,300	283
13,400	284
13,500	286
13,600	280
13,700	282
13,800	290
13,900	299
14,000	308

6.3 Using the following data, plot the shale resistivities on semi-logarithmic paper. Where is the top of the abnormal pressured zone? Construct a normal trend line and extrapolate it through the abnormal region. Use the Hottman and Johnson approach to calculate the formation pressures at each 500-ft interval beginning at 6,000 ft.

Depth, ft	Resistivity, ohm m^2/m
6,000	0.70
6,500	0.73
7,000	0.74
7,500	0.78
8,000	0.80
8,500	0.85
9,000	0.90
9,500	0.96
10,000	1.06
10,200	1.10
10,400	1.15
10,600	1.10
10,800	0.98
11,000	0.88
11,200	0.80
11,400	0.74
11,600	0.70
11,800	0.64
12,000	0.60

Solution: 12,000 ft = 15.1 ppg

6.4 A well is to be drilled to 10,000 ft. The shale resistivities for the controlling offset well are given below. Use a trip margin of 0.3 ppg mud weight. What would be the minimum required mud density in the well that is to be drilled? Use the Hottman and Johnson method.

Depth, ft	Resistivity, ohm m^2/m
6,000	0.62
6,500	0.64
7,000	0.65
7,200	0.64
7,400	0.68
7,600	0.68
7,800	0.70
8,000	0.71

8,200	0.72
8,400	0.74
8,600	0.78
8,800	0.75
9,000	0.72
9,200	0.72
9,400	0.74
9,600	0.71
9,800	0.70
10,000	0.70

6.5 Use the following data to determine if a well can be drilled safely to 15,000 ft without incurring a kick or stuck pipe. Assume that the formation pressures in the well that is to be drilled will be the same as those in the offset well.

New Well

Casing setting depth = 11,500 ft
Mud weight to be used = 17.0 ppg
Maximum allowable differential pressure = 3,000 psi

Offset Well

Depth, ft	Shale resistivities, ohm m^2/m
5,000	0.38
5,500	0.39
6,000	0.40
6,500	0.42
7,000	0.50
7,500	0.40
8,000	0.45
8,500	0.45
9,000	0.50
9,100	0.48
9,200	0.48
9,300	0.50
9,400	0.50
9,500	0.54
9,600	0.50
9,700	0.52
9,800	0.54
9,900	0.55
10,000	0.56

10,100	0.56
10,200	0.58
10,300	0.58
10,400	0.62
10,500	0.62
10,600	0.65
10,700	0.64
10,800	0.66
10,900	0.66
11,000	0.66
11,100	0.64
11,200	0.60
11,300	0.52
11,400	0.45
11,500	0.40
11,600	0.35
11,700	0.32
11,800	0.30
11,900	0.28
12,000	0.26
12,100	0.28
12,200	0.28
12,300	0.27
12,400	0.27
12,500	0.27
12,600	0.26
12,700	0.25
12,800	0.22
12,900	0.23
13,000	0.30
13,100	0.38
13,200	0.42
13,300	0.48
13,400	0.50
13,500	0.60
13,600	0.70
13,700	0.74
13,800	0.76
13,900	0.80
14,000	0.80
14,100	0.78
14,200	0.75

14,300	0.70
14,400	0.65
14,500	0.60
14,600	0.58
14,700	0.50
14,800	0.48
14,900	0.46
15,000	0.40

6.6 Use the following data to calculate the formation pressure at each 500-ft interval in the well.

Depth, ft	Travel time, μsec/ft
6,000	130
6,250	128
6,500	126
6,750	121
7,000	122
7,250	116
7,500	117
7,750	115
8,000	112
8,250	110
8,500	105
8,750	106
9,000	105
9,250	103
9,500	100
9,750	105
10,000	110
10,250	113
10,500	116
10,750	118
11,000	120

Solution: 11,000 ft = 15.0 ppg

6.7 Can the following data be used for formation pressure evaluation? If so, what is the pressure at 12,000 ft? If not useable, give the reasons.

Depth, ft	Travel time, μsec/ft
10,000	96
10,200	95

10,400	98
10,600	101
10,800	101
11,000	103
11,200	106
11,400	109
11,600	110
11,800	111
12,000	113

6.8 Calculate the required mud weights to drill each section of a well using offset well data for control. Use a 0.3 ppg trip margin of mud weight.

Section	Interval, ft
1	3,500–10,400
2	10,400–12,900
3	12,900–14,500

Offset well sonic survey

Depth, ft	Travel time, μsec/ft
3,500	156
4,000	152
4,500	148
5,000	143
5,500	140
6,000	136
6,500	132
7,000	129
7,500	125
8,000	121
8,200	120
8,400	119
8,600	117
8,800	116
9,000	114
9,200	113
9,400	111
9,600	110
9,800	110
10,000	108
10,100	111
10,200	115

10,300	120
10,400	120
10,500	120
10,600	120
10,700	121
10,800	122
10,900	123
11,000	124
11,100	125
11,200	126
11,300	125
11,400	125
11,500	130
11,600	129
11,700	129
11,800	128
11,900	125
12,000	123
12,100	122
12,200	123
12,300	124
12,400	122
12,500	121
12,600	120
12,700	120
12,800	119
12,900	115
13,000	108
13,100	107
13,200	106
13,300	106
13,400	105
13,500	106
13,600	105
13,700	104
13,800	103
13,900	102
14,000	101
14,100	100
14,200	100
14,300	99
14,400	99
14,500	98

6.9 In Example 6.3, calculate the formation pressure at every 200-ft interval. Use the Boatman correlation shown in Fig. 6.24.

Solution: 11,000 ft = 13.8 ppg

6.10 Calculate the formation pressure at every 500-ft interval in Example 6.4. Use Boatman's correlation.

6.11 The following values for the "d" exponent were calculated on a South Louisiana well. Use the associated mud weights to calculate the "d_c" exponent. Plot the "d" and "d_c" exponent on linear coordinate paper. Where is the top of abnormal pressures encountered? Can the "d" exponent be used to detect the top of the pressures? Give the reasons for your answer. Assume the normal formation pressure to be equivalent to 9 ppg.

Depth, ft	"d"	Mud weight, ppg
7,000	1.40	9.0
7,250	1.43	9.0
7,500	1.44	9.0
7,750	1.44	9.1
8,000	1.45	9.3
8,250	1.46	9.3
8,500	1.47	9.3
8,750	1.48	9.3
9,000	1.46	9.8
9,250	1.51	9.8
9,500	1.50	10.5
9,750	1.55	11.1
10,000	1.49	11.1
10,250	1.55	13.0
10,500	1.55	13.5
10,750	1.55	13.5
11,000	1.54	13.5
11,250	1.57	14.1
11,500	1.60	14.1
11,750	1.60	14.4
12,000	1.65	14.9

6.12 Calculate and plot the "d" and "d_c" exponents for the following problem. Assume a 9.0 ppg normal formation pressure.

Depth, ft	Penetration rate ft/hr	Bit weight 1,000 lb
6,000	106	35
6,500	103	35

7,000	76.9	35
7,500	66	35
8,000	44.6	30
8,500	46	30
9,000	39.4	30
9,500	35	30
10,000	30.8	30
10,200	26.3	30
10,400	24.7	30
10,600	23.2	30
10,800	21.8	30
11,000	19.1	30
11,200	17.9	30
11,400	16.8	30
11,600	21.9	35
11,800	20.6	35
12,000	20.6	35
12,200	20	35
12,400	18	35
12,600	18	35
12,800	17	35

Depth, ft	Rotary speed, rpm	Bit diameter, in.	Mud weight, ppg
6,000	120	8.5	9.0
6,500	120	8.5	9.0
7,000	110	8.5	9.0
7,500	110	8.5	9.0
8,000	110	8.5	9.4
8,500	110	7.87	9.4
9,000	110	7.87	9.4
9,500	110	7.87	9.8
10,000	110	7.87	10.1
10,200	100	7.87	10.1
10,400	100	7.87	10.1
10,600	100	7.87	10.5
10,800	90	7.87	11.1
11,000	90	7.87	11.1
11,200	90	7.87	11.3
11,400	90	7.87	11.6
11,600	90	7.87	11.6
11,800	90	7.87	11.8

12,000	90	7.87	13.1
12,200	90	7.87	13.4
12,400	90	7.87	13.6
12,600	90	7.87	14.2
12,800	90	7.87	14.5

References

1. Fertl, W.H., and Timko, D.J.: "How Downhole Temperatures, Pressures Affect Drilling" Part 1, "Origin of Abnormal Formation Pressures." World Oil (June, 1972).

2. Fertl, W.H.: "Abnormal Formation Pressures, Implications to Exploration, Drilling, and Production of Oil and Gas Resources." Elsevier Scientific Publishing Co., (1976).

3. Lewis, C.R., and Rose, S.C.: "A Theory Relating High Temperatures and Overpressures." Journal of Petroleum Technology (1970).

4. Fertl, W.H., and Timko, D.J.: "How Abnormal Pressure Detection Techniques are Applied," Oil and Gas Journal, 1970).

5. Wilson, G.J., and Bush, R.E.: "Pressure Prediction with Flowline Temperature Gradient," Journal of Petroleum Technology. (1973).

6. Hottman, C.E., and Johnson, R.K.: "Estimation of Formation Pressures from Log-Derived Shale Properties," Journal of Petroleum Technology, (June, 1965).

7. Personal communications with Charles M. Prentice, Prentice and Records Enterprises, Inc., Lafayette, La.

8.   Foster, J.B., and Whalen, H.E.: "Estimation of Formation Pressures from Electrical Surveys—Offshore Louisiana," *Journal of Petroleum Technology*, (February, 1965).

9.   Boatman, W.A.: "Shale Densities Key to Safer, Faster Drilling," *World Oil*, (August, 1967).

10. Baroid Division, NL/Petroleum Services, Houston, Tex.

11. Garnier, A.J., and van Lingen, N.H.: "Phenomena Affecting Drilling Rates of Depth," *Trans.*, AIME (1959) 216, 232.

12. Vidrine, D.J., and Benit, E.J.: "Field Verification of the Effect of Differential Pressure on Drilling Rate," *Journal of Petroleum Technology*, (July, 1968).

13. Bingham, M.G.: "A New Approach to Interpreting Rock Drillability," *Oil and Gas Journal* (November 2, 1964–April 5, 1965).

14. Jordan, J.R., and Shirley, O.J.: "Application of Drilling Performance Data to Overpressure Detection," *Journal of Petroleum Technology* (November, 1966).

15. Rehm, B., and McClendon, R.: "Measurement of Formation Pressure from Drilling Data," SPE 3601 (1971).

16. Zamora, M.: "Slide-rule Correlation Aids "d" Exponent Use." *Oil and Gas Journal*, (December 18, 1972).

17. Bourgoyne, A.T., and Young, F.S.: "A Multiple Regression Approach to Optimal Drilling and Abnormal Pressure Detection," *Journal of Petroleum Technology*, (August, 1974).

APPENDIX A

Table of Contents

Displacement of Drill Pipe*

O.D. Size (in.)	Drill Pipe				Displacement	
	I.U.	Weight (Lbs/Ft) E.U.	I.U. & E.U.	I.D. (in.)	Bbl/Ft	Bbl per 93′ Stand
2⅜	4.85	4.85		1.995	.0019	.174
	6.65	6.65		1.815	.0025	.237
2⅞	6.45			2.469	.002	.225
	6.85	6.85		2.441	.0027	.248
	8.35			2.323	.0032	.299
	10.40	10.40	10.40	2.151	.0041	.378
3½	8.50			3.063	.0032	.304
	9.50	9.50		2.992	.0038	.352
	11.20			2.900	.0043	.399
	13.30	13.30	13.30	2.764	.0054	.505
	15.50	15.50	15.50	2.602	.0061	.571
4	11.85	11.85		3.476	.0044	.414
	14.00	14.00	14.00	3.340	.0056	.526
	15.70		15.30	3.24	.0063	.589
4½	12.75			4.00	.0048	.448
	13.75	13.75		3.958	.0055	.515
	16.60	16.60	16.60	3.826	.00648	.602
		18.15	18.15	3.754	.0068	.635
	20.00	20.00	20.00	3.640	.0081	.754
5	16.25		16.25	4.408	.0064	.592
	19.50		19.50	4.276	.0075	.707
	20.50	20.50		4.214	.0078	.727
5½	21.90		21.90	4.778	.0094	.874
	24.70		24.70	4.670	.0105	.977
5⁹⁄₁₆	19.00			4.975	.0084	.777
	22.20		22.20	4.859	.0090	.845
	25.25			4.733	.0112	1.047
6⅝**	22.20			6.065	.008	.750
	25.20		25.20	5.965	.009	.852
	31.90			5.761	.012	1.07

Drill Pipe					Displacement	
O.D. Size (in.)	I.U.	Weight (Lbs/Ft) E.U.	I.U. & E.U.	I.D. (in.)	Bbl/Ft	Bbl per 93' Stand
7⅝**	29.25			6.969	.011	.988
8⅝**	40.00			7.825	.014	1.352

* -These figures do not include the effects of upsets or tool joints
**-Conventionally used in mining operations.

Capacity and Displacement of Drill Collars*

Drill Collar			Displacement		Capacity	
O.D. Size (in.)	Weight (Lbs/Ft)	I.D. (in.)	Bbl/Ft	Bbl per 93′ Stand	Bbl/Ft	Bbl per 93′ Stand
4½	51	1	.0187	1.74	.0009	.0837
	48	1½	.0175	1.62	.0022	.2046
	43	2	.0158	1.47	.0039	.3627
4¾	54	1½	.0197	1.83	.0022	.2046
	52	1¾	.0189	1.757	.0030	.279
	50	2	.018	1.67	.0039	.3627
5	61	1½	.0221	2.055	.0022	.2046
	59	1¾	.0213	1.98	.0030	.279
	56	2	.0204	1.89	.0039	.3627
5¼	68	1½	.0246	2.29	.0022	.2046
	65	1¾	.0238	2.21	.0030	.279
	63	2	.0229	2.13	.0039	.3627
5½	75	1½	.0272	2.53	.0022	.2046
	73	1¾	.0264	2.46	.0030	.279
	70	2	.0255	2.37	.0039	.3627
5¾	82	1½	.0299	2.78	.0022	.2046
	80	1¾	.0291	2.71	.0030	.279
	78	2	.0282	2.62	.0039	.3627
6	88	1¾	.032	2.98	.0030	.279
	85	2	.0311	2.89	.0039	.3627
	83	2¼	.0301	2.80	.0049	.4557
6¼	96	1¾	.0349	3.24	.0030	.279
	94	2	.034	3.16	.0039	.3627
	91	2¼	.033	3.07	.0049	.4557
6½	105	1¾	.038	3.53	.0030	.279
	102	2	.0371	3.45	.0039	.3627
	99	2¼	.0361	3.36	.0049	.4557
6¾	114	1¾	.0413	3.84	.0030	.279
	111	2	.0404	3.76	.0039	.3627
	108	2¼	.0394	3.66	.0049	.4557

Drill Collar			Displacement		Capacity	
O.D. Size (in.)	Weight (Lbs/Ft)	I.D. (in.)	Bbl/Ft	Bbl per 93' Stand	Bbl/Ft	Bbl per 93' Stand
7	120	2	.0437	4.06	.0039	.3627
	114	2½	.0415	3.86	.0061	.5673
	107	3	.0388	3.61	.0088	.8184
7¼	130	2	.0472	4.39	.0039	.3627
	124	2½	.045	4.18	.0061	.5673
	116	3	.0423	3.93	.0088	.8184
7½	139	2	.0507	4.72	.0039	.3627
	133	2½	.0485	4.51	.0061	.5673
	126	3	.0458	4.26	.0088	.8184
7¾	144	2½	.0522	4.85	.0061	.5673
	136	3	.0495	4.6	.0088	.8184
	128	3½	.0464	4.32	.0119	1.107
8	147	3	.0534	4.97	.0088	.8184
	143	3¼	.0519	4.83	.0103	.958
	138	3½	.0503	4.68	.0119	1.107

Capacities of Pipe and Hole

CAPACITIES OF VARIOUS DIAMETERS

HOLE DIAM (IN)	CU FT PER LIN FT	LIN FT PER CU FT	BARRELS PER LIN FT	LIN FT PER BARREL	GALLONS PER LIN FT
1	0.0055	183.3465	0.0010	1029.4142	0.0408
1/8	0.0069	144.8664	0.0012	813.3643	0.0516
1/4	0.0085	117.3418	0.0015	658.8251	0.0637
3/8	0.0103	96.9767	0.0018	544.4835	0.0771
1/2	0.0123	81.4873	0.0022	457.5174	0.0918
5/8	0.0144	69.4330	0.0026	389.8373	0.1077
3/4	0.0167	59.8682	0.0030	336.1352	0.1249
7/8	0.0192	52.1519	0.0034	292.8111	0.1434
2	0.0218	45.8366	0.0039	257.3535	0.1632
1/8	0.0246	40.6027	0.0044	227.9672	0.1842
1/4	0.0276	36.2166	0.0049	203.3411	0.2065
3/8	0.0308	32.5046	0.0055	182.5000	0.2301
1/2	0.0341	29.3354	0.0061	164.7063	0.2550
5/8	0.0376	26.6081	0.0067	149.3934	0.2811
3/4	0.0412	24.2442	0.0073	136.1209	0.3085
7/8	0.0451	22.1818	0.0080	124.5416	0.3372
3	0.0491	20.3718	0.0087	114.3794	0.3672
1/8	0.0533	18.7747	0.0095	105.4120	0.3984
1/4	0.0576	17.3582	0.0103	97.4593	0.4309
3/8	0.0621	16.0963	0.0111	90.3738	0.4647
1/2	0.0668	14.9671	0.0119	8.40338	0.4998
5/8	0.0717	13.9526	0.0128	78.3383	0.5361
3/4	0.0767	13.0380	0.0137	73.2028	0.5737
7/8	0.0819	12.2104	0.0146	68.5562	0.6126
4	0.0873	11.4592	0.0155	64.3384	0.6528
1/8	0.0928	10.7752	0.0165	60.4982	0.6942
1/4	0.0985	10.1507	0.0175	56.9918	0.7369
3/8	0.1044	9.5789	0.0186	53.7816	0.7809
1/2	0.1104	9.0541	0.0197	50.8353	0.8262
5/8	0.1167	8.5713	0.0208	48.1245	0.8727
3/4	0.1231	8.1262	0.0219	45.6250	0.9205
7/8	0.1296	7.7148	0.0231	43.3153	0.9696
5	0.1364	7.3339	0.0243	41.1766	1.0200
1/8	0.1433	6.9805	0.0255	39.1924	1.0716
1/4	0.1503	6.6520	0.0268	37.3484	1.1245
3/8	0.1576	6.3462	0.0281	35.6314	1.1787
1/2	0.1650	6.0610	0.0294	34.0302	1.2342
5/8	0.1726	5.7947	0.0307	32.5346	1.2909
3/4	0.1803	5.5455	0.0321	31.1354	1.3489
7/8	0.1833	5.3120	0.0335	29.8246	1.4082

NOTE: Some diameters and weights are non-API.
No allowance has been made for couplings or upsets.
*Plain end weights are indicated by asterisk.

CAPACITIES OF VARIOUS DIAMETERS

HOLE DIAM (IN)	CU FT PER LIN FT	LIN FT PER CU FT	BARRELS PER LIN FT	LIN FT PER BARREL	GALLONS PER LIN FT
6	0.1963	5.0930	0.0350	28.5948	1.4688
⅛	0.2046	4.8872	0.0364	27.4396	1.5306
¼	0.2131	4.6937	0.0379	26.3530	1.5937
⅜	0.2217	4.5114	0.0395	25.3297	1.6581
½	0.2304	4.3396	0.0410	24.3648	1.7238
⅝	0.2394	4.1773	0.0426	23.4541	1.7907
¾	0.2485	4.0241	0.0443	22.5935	1.8589
⅞	0.2578	3.8791	0.0459	21.7793	1.9284
7	0.2673	3.7418	0.0476	21.0085	1.9992
⅛	0.2769	3.6116	0.0493	20.2778	2.0712
¼	0.2867	3.4882	0.0511	19.5846	2.1445
⅜	0.2967	3.3709	0.0528	18.9263	2.2191
½	0.3068	3.2595	0.0546	18.3007	2.2950
⅝	0.3171	3.1535	0.0565	17.7056	2.3721
¾	0.3276	3.0526	0.0583	17.1390	2.4505
⅞	0.3382	2.9565	0.0602	16.5993	2.5302
8	0.3491	2.8648	0.0622	16.0846	2.6112
⅛	0.3601	2.7773	0.0641	15.5935	2.6934
¼	0.3712	2.6938	0.0661	15.1245	2.7769
⅜	0.3826	2.6140	0.0681	14.6764	2.8617
½	0.3941	2.5377	0.0702	14.2479	2.9478
⅝	0.4057	2.4646	0.0723	13.8380	3.0351
¾	0.4176	2.3947	0.0744	13.4454	3.1237
⅞	0.4296	2.3277	0.0765	13.0693	3.2136
9	0.4418	2.2635	0.0787	12.7088	3.3048
⅛	0.4541	2.2019	0.0809	12.3630	3.3972
¼	0.4667	2.1428	0.0831	12.0311	3.4909
⅜	0.4794	2.0861	0.0854	11.7124	3.5859
½	0.4922	2.0315	0.0877	11.4063	3.6822
⅝	0.5053	1.9791	0.0900	11.1119	3.7797
¾	0.5185	1.9287	0.0923	10.8288	3.8785
⅞	0.5319	1.8802	0.0947	10.5564	3.9786
10	0.5454	1.8335	0.0971	10.2941	4.0800
⅛	0.5591	1.7885	0.0996	10.0415	4.1826
¼	0.5730	1.7451	0.1021	9.7981	4.2865
⅜	0.5871	1.7033	0.1046	9.5634	4.3917
½	0.6013	1.6630	0.1071	9.3371	4.4982
⅝	0.6157	1.6241	0.1097	9.1187	4.6059
¾	0.6303	1.5866	0.1123	8.9079	4.7149
⅞	0.6450	1.5503	0.1149	8.7043	4.8252
11	0.6600	1.5153	0.1175	8.5076	4.9368
⅛	0.6750	1.4814	0.1202	8.3174	5.0496
¼	0.6903	1.4487	0.1229	8.1336	5.1637
⅜	0.7057	1.4170	0.1257	7.9559	5.2791
½	0.7213	1.3864	0.1285	7.7839	5.3958
⅝	0.7371	1.3567	0.1313	7.6174	5.5137
¾	0.7530	1.3280	0.1341	7.4561	5.6329
⅞	0.7691	1.3002	0.1370	7.3000	5.7534

Well Control Problems and Solutions

CAPACITIES OF VARIOUS DIAMETERS

HOLE DIAM (IN)	CU FT PER LIN FT	LIN FT PER CU FT	BARRELS PER LIN FT	LIN FT PER BARREL	GALLONS PER LIN FT
12	0.7854	1.2732	0.1399	7.1487	5.8752
1/8	0.8018	1.2471	0.1428	7.0021	5.9982
1/4	0.8185	1.2218	0.1458	6.8599	6.1225
3/8	0.8353	1.1972	0.1488	6.7220	6.2481
1/2	0.8522	1.1734	0.1518	6.5883	6.3750
5/8	0.8693	1.1503	0.1548	6.4564	6.5031
3/4	0.8866	1.1279	0.1579	6.3324	6.6325
7/8	0.9041	1.1061	0.1610	6.2101	6.7632
13	0.9218	1.0849	0.1642	6.0912	6.8952
1/8	0.9396	1.0643	0.1673	5.9757	7.0284
1/4	0.9575	1.0443	0.1705	5.8635	7.1629
3/8	0.9757	1.0249	0.1738	5.7544	7.2987
1/2	0.9940	1.0060	0.1770	5.6484	7.4358
5/8	1.0125	0.9876	0.1803	5.5452	7.5741
3/4	1.0312	0.9698	0.1837	5.4448	7.7137
7/8	1.0500	0.9524	0.1870	5.3472	7.8546
14	1.0690	0.9354	0.1904	5.2521	7.9968
1/8	1.0882	0.9190	0.1938	5.1596	8.1402
1/4	1.1075	0.9029	0.1973	5.0694	8.2849
3/8	1.1270	0.8873	0.2007	4.9817	8.4309
1/2	1.1467	0.8720	0.2042	4.8961	8.5782
5/8	1.1666	0.8572	0.2078	4.8128	8.7267
3/4	1.1866	0.8427	0.2113	4.7316	8.8765
7/8	1.2068	0.8286	0.2149	4.6524	9.0276
15	1.2272	0.8149	0.2186	4.5752	9.1800
1/8	1.2477	0.8015	0.2222	4.4999	9.3336
1/4	1.2684	0.7884	0.2259	4.4264	9.4885
3/8	1.2893	0.7756	0.2296	4.3547	9.6447
1/2	1.3104	0.7631	0.2334	4.2848	9.8022
5/8	1.3316	0.7510	0.2372	4.2165	9.9609
3/4	1.3530	0.7391	0.2410	4.1498	10.1209
7/8	1.3745	0.7275	0.2448	4.0847	10.2822
16	1.3963	0.7162	0.2487	4.0211	10.4448
1/4	1.4402	0.6943	0.2565	3.8984	10.7737
1/2	1.4849	0.6734	0.2645	3.7811	11.1078
3/4	1.5302	0.6535	0.2725	3.6691	11.4469
17	1.5763	0.6344	0.2807	3.5620	11.7912
1/4	1.6230	0.6162	0.2891	3.4595	12.1405
1/2	1.6703	0.5987	0.2975	3.3614	12.4950
3/4	1.7184	0.5819	0.3061	3.2673	12.8545
18	1.7671	0.5659	0.3147	3.1772	13.2192
1/4	1.8166	0.5505	0.3235	3.0908	13.5889
1/2	1.8667	0.5357	0.3325	3.0078	13.9638
3/4	1.9175	0.5215	0.3415	2.9281	14.3437
19	1.9689	0.5079	0.3507	2.8516	14.7288
1/4	2.0211	0.4948	0.3600	2.7780	15.1189
1/2	2.0739	0.4822	0.3694	2.7072	15.5142
3/4	2.1275	0.4700	0.3789	2.6391	15.9145

CAPACITIES OF VARIOUS DIAMETERS

HOLE DIAM (IN)	CU FT PER LIN FT	LIN FT PER CU FT	BARRELS PER LIN FT	LIN FT PER BARREL	GALLONS PER LIN FT
20	2.1817	0.4584	0.3886	2.5735	16.3200
¼	2.2365	0.4471	0.3983	2.5104	16.7305
½	2.2921	0.4363	0.4082	2.4495	17.1462
¾	2.3484	0.4258	0.4183	2.3909	17.5669
21	2.4053	0.4158	0.4284	2.3343	17.9928
¼	2.4629	0.4060	0.4387	2.2797	18.4237
½	2.5212	0.3966	0.4490	2.2270	18.8598
¾	2.5802	0.3876	0.4595	2.1761	19.3009
22	2.6398	0.3788	0.4702	2.1269	19.7472
¼	2.7001	0.3704	0.4809	2.0794	20.1985
½	2.7612	0.3622	0.4918	2.0334	20.6550
¾	2.8229	0.3542	0.5028	1.9890	21.1165
23	2.8852	0.3466	0.5139	1.9460	21.5831
¼	2.9483	0.3392	0.5251	1.9043	22.0549
½	3.0121	0.3320	0.5365	1.8640	22.5317
¾	3.0765	0.3250	0.5479	1.8250	23.0137
24	3.1416	0.3183	0.5595	1.7872	23.5007
¼	3.2074	0.3118	0.5713	1.7505	23.9929
½	3.2739	0.3055	0.5831	1.7150	24.4901
¾	3.3410	0.2993	0.5951	1.6805	24.9925
25	3.4088	0.2934	0.6071	1.6471	25.4999
¼	3.4774	0.2876	0.6193	1.6146	26.0125
½	3.5466	0.2820	0.6317	1.5831	26.5301
¾	3.6164	0.2765	0.6441	1.5525	27.0529
26	3.6870	0.2712	0.6567	1.5228	27.5807
¼	3.7583	0.2661	0.6694	1.4939	28.1137
½	3.8302	0.2611	0.6822	1.4659	28.6517
¾	3.9028	0.2562	0.6951	1.4386	29.1949
27	3.9761	0.2515	0.7082	1.4121	29.7431
¼	4.0501	0.2469	0.7213	1.3863	30.2965
½	4.1247	0.2424	0.7346	1.3612	30.8549
¾	4.2000	0.2381	0.7481	1.3368	31.4185
28	4.2761	0.2339	0.7616	1.3130	31.9871
¼	4.3528	0.2297	0.7753	1.2899	32.5609
½	4.4301	0.2257	0.7890	1.2674	33.1397
¾	4.5082	0.2218	0.8029	1.2454	33.7237
29	4.5869	0.2180	0.8170	1.2240	34.3127
¼	4.6664	0.2143	0.8311	1.2032	34.9069
½	4.7465	0.2107	0.8454	1.1829	35.5061
¾	4.8273	0.2072	0.8598	1.1631	36.1105
30	4.9087	0.2037	0.8743	1.1438	36.7199
¼	4.9909	0.2004	0.8889	1.1250	37.3345
½	5.0737	0.1971	0.9037	1.1066	37.9541
¾	5.1572	0.1939	0.9185	1.0887	38.5789

CAPACITIES OF VARIOUS DIAMETERS

HOLE DIAM (IN)	CU FT PER LIN FT	LIN FT PER CU FT	BARRELS PER LIN FT	LIN FT PER BARREL	GALLONS PER LIN FT
31	5.2414	0.1908	0.9335	1.0712	39.2087
¼	5.3263	0.1877	0.9487	1.0541	39.8437
½	5.4119	0.1848	0.9639	1.0375	40.4837
¾	5.4981	0.1819	0.9793	1.0212	41.1289
32	5.5851	0.1790	0.9947	1.0053	41.7791
¼	5.6727	0.1763	1.0103	0.9898	42.4345
½	5.7610	0.1736	1.0261	0.9746	43.0949
¾	5.8499	0.1709	1.0419	0.9598	43.7604
33	5.9396	0.1684	1.0579	0.9453	44.4311
¼	6.0299	0.1658	1.0740	0.9311	45.1068
½	6.1209	0.1634	1.0902	0.9173	45.7877
¾	6.2126	0.1610	1.1065	0.9037	46.4736
34	6.3050	0.1586	1.1230	0.8905	47.1647
¼	6.3981	0.1563	1.1395	0.8775	47.8608
½	6.4918	0.1540	1.1562	0.8649	48.5621
¾	6.5862	0.1518	1.1731	0.8525	49.2684
35	6.6813	0.1497	1.1900	0.8403	49.9799
¼	6.7771	0.1476	1.2071	0.8285	50.6964
½	6.8736	0.1455	1.2242	0.8168	51.4181
¾	6.9707	0.1435	1.2415	0.8054	52.1448
36	7.0686	0.1415	1.2590	0.7943	52.8767
¼	7.1671	0.1395	1.2765	0.7834	53.6136
½	7.2663	0.1376	1.2942	0.7727	54.3557
¾	7.3662	0.1358	1.3120	0.7622	55.1028
37	7.4667	0.1339	1.3299	0.7519	55.8551
¼	7.5680	0.1321	1.3479	0.7419	56.6124
½	7.6699	0.1304	1.3661	0.7320	57.3749
¾	7.7725	0.1287	1.3843	0.7224	58.1424
38	7.8758	0.1270	1.4027	0.7129	58.9151
¼	7.9798	0.1253	1.4213	0.7036	59.6928
½	8.0844	0.1237	1.4399	0.6945	60.4757
¾	8.1898	0.1221	1.4587	0.6856	61.2636
39	8.2958	0.1205	1.4775	0.6768	62.0567
¼	8.4025	0.1190	1.4965	0.6682	62.8548
½	8.5098	0.1175	1.5157	0.6598	63.6581
¾	8.6179	0.1160	1.5349	0.6515	64.4664
40	8.7266	0.1146	1.5543	0.6434	65.2798
¼	8.8361	0.1132	1.5738	0.6354	66.0984
½	8.9462	0.1118	1.5934	0.6276	66.9220
¾	9.0570	0.1104	1.6131	0.6199	67.7508

CAPACITIES OF VARIOUS DIAMETERS

HOLE DIAM (IN)	CU FT PER LIN FT	LIN FT PER CU FT	BARRELS PER LIN FT	LIN FT PER BARREL	GALLONS PER LIN FT
41	9.1684	0.1091	1.6330	0.6124	68.5846
¼	9.2806	0.1078	1.6529	0.6050	69.4236
½	9.3934	0.1065	1.6730	0.5977	70.2676
¾	9.5069	0.1052	1.6933	0.5906	71.1168
42	9.6211	0.1039	1.7136	0.5836	71.9710
¼	9.7360	0.1027	1.7341	0.5767	72.8304
½	9.8516	0.1015	1.7546	0.5699	73.6948
¾	9.9678	0.1003	1.7753	0.5633	74.5644
43	10.0847	0.0992	1.7962	0.5567	75.4390
¼	10.2023	0.0980	1.8171	0.5503	76.3188
½	10.3206	0.0969	1.8382	0.5440	77.2036
¾	10.4396	0.0958	1.8594	0.5378	78.0936
44	10.5592	0.0947	1.8807	0.5317	78.9886
¼	10.6796	0.0936	1.9021	0.5257	79.8888
½	10.8006	0.0926	1.9237	0.5198	80.7940
¾	10.9223	0.0916	1.9453	0.5140	81.7044
45	11.0447	0.0905	1.9671	0.5084	82.6198
¼	11.1677	0.0895	1.9891	0.5028	83.5404
½	11.2915	0.0886	2.0111	0.4972	84.4660
¾	11.4159	0.0876	2.0333	0.4918	85.3968
46	11.5410	0.0866	2.0555	0.4865	86.3326
¼	11.6668	0.0857	2.0779	0.4812	87.2735
½	11.7932	0.0848	2.1005	0.4761	88.2196
¾	11.9204	0.0839	2.1231	0.4710	89.1707
47	12.0482	0.0830	2.1459	0.4660	90.1270
¼	12.1767	0.0821	2.1688	0.4611	91.0883
½	12.3059	0.0813	2.1918	0.4563	92.0548
¾	12.4358	0.0804	2.2149	0.4515	93.0263
48	12.5664	0.0796	2.2382	0.4468	94.0030
¼	12.6976	0.0788	2.2615	0.4422	94.9847
½	12.8295	0.0779	2.2850	0.4376	95.9716
¾	12.9621	0.0771	2.3087	0.4332	96.9635

TUBING SIZES AND CAPACITIES

SIZE OD INCHES	WEIGHT PER LIN FT	ID	CU FT PER LIN FT	LIN FT PER CU FT	BARRELS PER LIN FT	LIN FT PER BARREL	GALLONS PER LIN FT
0.750	0.42	0.636	.0022	453.2715	.0004	2544.9304	.0165
1.000	0.67	0.866	.0041	244.4763	.0007	1372.6328	.0306
1.050	1.14	0.824	.0037	270.0338	.0007	1516.1275	.0277
	1.20	0.824	.0037	270.0338	.0007	1516.1275	.0277
	1.55	0.742	.0030	333.0158	.0005	1869.7448	.0225
1.315	1.30	1.125	.0069	144.8664	.0012	813.3643	.0516
	1.43	1.097	.0066	152.3559	.0012	855.4150	.0491
	1.63	1.065	.0062	161.6491	.0011	907.5926	.0463
	1.70	1.049	.0060	166.6179	.0011	935.4900	.0449
	1.72	1.049	.0060	166.6179	.0011	935.4900	.0449
	1.80	1.049	.0060	166.6179	.0011	935.4900	.0449
	1.90	1.049	.0060	166.6179	.0011	935.4900	.0449
	2.25	0.957	.0050	200.1929	.0009	1123.9999	.0374
	2.30	0.957	.0050	200.1929	.0009	1123.9999	.0374
1.660	2.10	1.410	.0108	92.2220	.0019	517.7879	.0811
	2.30	1.380	.0104	96.2752	.0018	540.5451	.0777
	2.40	1.380	.0104	96.2752	.0018	540.5451	.0777
	3.02	1.278	.0089	112.2563	.0016	630.2726	.0666
	3.24	1.264	.0087	114.7568	.0016	644.3117	.0652
	3.29	1.264	.0087	114.7568	.0016	644.3117	.0652
1.900	2.40	1.650	.0148	67.3449	.0026	378.1136	.1111
	2.75	1.610	.0141	70.7328	.0025	397.1352	.1058
	2.90	1.610	.0141	70.7328	.0025	397.1352	.1058
	3.64	1.500	.0123	81.4873	.0022	457.5174	.0918
	4.19	1.462	.0117	85.7784	.0021	481.6099	.0872
2	3.30	1.670	.0152	65.7415	.0027	369.1112	.1138
	3.40	1.670	.0152	65.7415	.0027	369.1112	.1138
2¹⁄₁₆	2.66	1.813	.0179	55.7798	.0032	313.1804	.1341
	3.25	1.750	.0167	59.8682	.0030	336.1352	.1249
	4.50	1.613	.0142	70.4699	.0025	395.6593	.1062
2⅜	3.10	2.125	.0246	40.6027	.0044	227.9672	.1842
	3.32	2.107	.0242	41.2994	.0043	231.8788	.1811
	4.00	2.041	.0227	44.0136	.0040	247.1179	.1700
	4.60	1.995	.0217	46.0667	.0039	258.6451	.1624
	4.70	1.995	.0217	46.0667	.0039	258.6451	.1624
	5.00	1.947	.0207	48.3661	.0037	271.5553	.1547
	5.30	1.939	.0205	48.7660	0037	273.8007	.1534
	5.80	1.867	.0190	52.5998	.0034	295.3259	.1422
	5.95	1.867	.0190	52.5998	.0034	295.3259	.1422
	6.20	1.853	.0187	53.3976	.0033	299.8053	.1401
	7.70	1.073	.0158	63.2184	.0028	354.9448	.1183

TUBING SIZES AND CAPACITIES

SIZE OD INCHES	WEIGHT PER LIN FT	ID	CU FT PER LIN FT	LIN FT PER CU FT	BARRELS PER LIN FT	LIN FT PER BARREL	GALLONS PER LIN FT
	4.36	2.579	.0363	27.5658	.0065	154.7702	.2714
	4.64	2.563	.0358	27.9110	.0064	156.7086	.2680
	5.90	2.469	.0332	30.0767	.0059	168.8682	.2487
	6.40	2.441	.0325	30.7707	.0058	172.7645	.2431
	6.50	2.441	.0325	30.7707	.0058	172.7645	.2431
	7.90	2.323	.0294	33.9762	.0052	190.7619	.2202
2⅞	8.60	2.259	.0278	35.9286	.0050	201.7240	.2082
	8.70	2.259	.0278	35.9286	.0050	201.7240	.2082
	9.50	2.195	.0263	38.0543	.0047	213.6590	.1966
	10.40	2.151	.0252	39.6271	.0045	222.4894	.1888
	10.70	2.091	.0238	41.9338	.0042	235.4410	.1784
	11.00	2.065	.0233	4299964	.0041	241.4071	.1740
	11.65	1.995	.0217	46.0667	.0039	258.6451	.1624
	5.63	3.188	.0554	18.0400	.0099	101.2870	.4147
	7.70	3.068	.0513	19.4788	.0091	109.3653	.3840
	8.50	3.018	.0497	20.1296	.0088	113.0191	.3716
	8.90	3.018	.0497	20.1296	.0088	113.0191	.3716
	9.20	2.992	.0488	20.4809	.0087	114.9918	.3652
	9.30	2.992	.0488	20.4809	.0087	114.9918	.3652
	10.20	2.992	.0488	20.4809	.0087	114.9918	.3652
	10.30	2.922	.0466	21.4740	.0083	120.5674	.3484
3½	11.20	2.900	.0459	21.8010	.0082	122.4036	.3431
	12.70	2.750	.0412	24.2442	.0073	136.1209	.3085
	12.80	2.764	.0417	23.9992	.0074	134.7454	.3117
	12.95	2.750	.0412	24.2442	.0073	136.1209	.3085
	13.30	2.764	.0417	23.9992	.0074	134.7454	.3117
	15.50	2.602	.0369	27.0806	.0066	152.0462	.2762
	15.80	2.548	.0354	28.2406	.0063	158.5591	.2649
	16.70	2.480	.0335	29.8105	.0060	167.3735	.2509
	17.05	2.440	.0325	30.7959	.0058	172.9062	.2429
	9.25	3.548	.0687	14.5648	.0122	81.7754	.5136
	9.40	3.548	.0687	14.5648	.0122	81.7754	.5136
	9.50	3.548	.0687	14.5648	.0122	81.7754	.5136
	10.80	3.476	.0659	15.1745	.0117	85.1982	.4930
	10.90	3.476	.0659	15.1745	.0117	85.1982	.4930
4	11.00	3.476	.0659	15.1745	.0117	85.1982	.4930
	11.60	3.428	.0641	15.6024	.0114	87.6009	.4794
	13.30	3.340	.0608	16.4354	.0108	92.2778	.4551
	13.40	3.340	.0608	16.4354	.0108	92.2778	.4551
	19.00	3.000	.0491	20.3718	.0087	114.3794	.3672
	22.50	2.780	.0422	23.7237	.0075	133.1989	.3153
	11.00	4.026	.0884	11.3116	.0157	63.5101	.6613
	11.80	3.990	.0868	11.5167	.0155	64.6613	.6495
	12.60	3.958	.0854	11.7036	.0152	65.7111	.6392
	12.75	3.958	.0854	11.7036	.0152	65.7111	.6392
	13.50	3.920	.0838	11.9316	.0149	66.9912	.6269
	15.40	3.826	.0798	12.5251	.0142	70.3235	.5972
4½	15.50	3.826	.0798	12.5251	.0142	70.3235	.5972
	16.90	3.754	.0769	13.0102	.0137	73.0469	.5750
	19.20	3.640	.0723	13.8379	.0129	77.6940	.5406
	21.60	3.500	.0668	14.9671	.0119	84.0338	.4998
	24.60	3.380	.0623	16.0487	.0111	90.1066	.4661
	26.50	3.240	.0573	17.4656	.0102	98.0619	.4283

CASING SIZES AND CAPACITIES

SIZE OD INCHES	WEIGHT PER LIN FT	ID	CU FT PER LIN FT	LIN FT PER CU FT	BARRELS PER LIN FT	LIN FT PER BARREL	GALLONS PER LIN FT
	9.26	3.550	0.0687	14.5484	.0122	81.6833	0.5142
	9.50	3.550	0.0687	14.5484	.0122	81.6833	0.5142
4	11.00	3.480	0.0661	15.1396	.0118	85.0025	0.4941
	11.60	3.430	0.0642	15.5842	.0114	87.4988	0.4800
	12.60	3.364	0.0617	16.2017	.0110	90.9658	0.4617
	9.50	4.090	0.0912	10.9604	.0163	61.5380	0.6825
	10.50	4.052	0.0896	11.1669	.0159	62.6977	0.6699
	10.98	4.030	0.0886	11.2892	.0158	63.3841	0.6626
	11.00	4.026	0.0884	11.3116	.0157	63.5101	0.6613
	11.60	4.000	0.0873	11.4592	.0155	64.3384	0.6528
	11.75	3.990	0.0868	11.5167	.0155	64.6613	0.6495
	12.60	3.958	0.0854	11.7036	.0152	65.7111	0.6392
4½	12.75	3.960	0.0855	11.6918	.0152	65.6447	0.6398
	13.50	3.920	0.0838	11.9316	.0149	66.9912	0.6269
	15.10	3.826	0.0798	12.5251	.0142	70.3235	0.5972
	16.60	3.826	0.0798	12.5251	.0142	70.3235	0.5972
	18.80	3.640	0.0723	13.8379	.0129	77.6940	0.5406
	21.60	3.500	0.0668	14.9671	.0119	84.0338	0.4998
	24.60	3.380	0.0623	16.0487	.0111	90.1066	0.4661
	26.50	3.240	0.0573	17.4656	.0102	98.0619	0.4283
	16.00	4.082	0.0909	11.0034	.0162	61.7795	0.6798
	16.50	4.070	0.0903	11.0684	.0161	62.1443	0.6758
4¾	18.00	4.000	0.0873	11.4592	.0155	64.3384	0.6528
	20.00	3.910	0.0834	11.9928	.0149	67.3343	0.6238
	21.00	3.850	0.0808	12.3695	.0144	69.4494	0.6048
	11.50	4.560	0.1134	8.8174	.0202	49.5063	0.8484
	12.85	4.500	0.1104	9.0541	.0197	50.8353	0.8262
	13.00	4.494	0.1102	9.0783	.0196	50.9711	0.8240
	14.00	4.450	0.1080	9.2588	.0192	51.9841	0.8079
	15.00	4.408	0.1060	9.4360	.0189	52.9794	0.7928
5	18.00	4.276	0.0997	10.0276	.0178	56.3008	0.7460
	20.30	4.184	0.0955	10.4734	.0170	58.8040	0.7142
	21.00	4.154	0.0941	10.6253	.0168	59.6564	0.7040
	23.20	4.044	0.0892	11.2112	.0159	62.9460	0.6672
	24.20	4.000	0.0873	11.4592	.0155	64.3384	0.6528
5¼	16.00	4.650	0.1179	8.4794	.0210	47.6085	0.8822
	13.00	5.044	0.1388	7.2065	.0247	40.4613	1.0380
	14.00	5.012	0.1370	7.2988	.0244	40.9796	1.0249
	15.00	4.974	0.1349	7.4107	.0240	41.6082	1.0094
	15.50	4.950	0.1336	7.4828	.0238	42.0126	0.9997
	17.00	4.892	0.1305	7.6613	.0232	43.0147	0.9764
5½	20.00	4.778	0.1245	8.0312	.0222	45.0918	0.9314
	23.00	4.670	0.1189	8.4070	.0212	47.2016	0.8898
	25.00	4.580	0.1144	8.7406	.0204	49.0749	0.8558
	26.00	4.548	0.1128	8.8640	.0201	49.7679	0.8439
	32.30	4.276	0.0997	10.0276	.0178	56.3008	0.7460
	36.40	4.090	0.0912	10.9604	.0163	61.5380	0.6825

CASING SIZES AND CAPACITIES

SIZE OD INCHES	WEIGHT PER LIN FT	ID	CU FT PER LIN FT	LIN FT PER CU FT	BARRELS PER LIN FT	LIN FT PER BARREL	GALLONS PER LIN FT
	14.00	5.290	0.1526	6.5518	.0272	36.7857	1.1417
	17.00	5.190	0.1469	6.8067	.0262	38.2169	1.0990
	19.50	5.090	0.1413	7.0768	.0252	39.7333	1.0570
5¾	20.00	5.090	0.1413	7.0768	.0252	39.7333	1.0570
	22.50	4.990	0.1358	7.3633	.0242	41.3418	1.0159
	23.00	4.990	0.1358	7.3633	.0242	41.3418	1.0159
	25.20	4.890	0.1304	7.6675	.0232	43.0499	0.9756
	15.00	5.524	0.1664	6.0085	.0296	33.7352	1.2450
	16.00	5.500	0.1650	6.0610	.0294	34.0302	1.2342
	18.00	5.424	0.1605	6.2321	.0286	34.9906	1.2003
6	20.00	5.352	0.1562	6.4009	.0278	35.9383	1.1687
	23.00	5.240	0.1498	6.6774	.0267	37.4910	1.1203
	26.00	5.140	0.1441	6.9398	.0257	38.9640	1.0779
	13.00	6.260	0.2137	4.6787	.0381	26.2689	1.5989
	17.00	6.135	0.2053	4.8713	.0366	27.3502	1.5356
	20.00	6.049	0.1996	5.0108	.0355	28.1334	1.4929
	22.00	5.980	0.1950	5.1271	.0347	28.7864	1.4590
	24.00	5.921	0.1912	5.2298	.0341	29.3630	1.4304
	25.00	5.880	0.1886	5.3030	.0336	29.7739	1.4106
6⅝	26.00	5.855	0.1870	5.3483	.0333	30.0287	1.3987
	26.80	5.837	0.1858	5.3814	.0331	30.2142	1.3901
	28.00	5.791	0.1829	5.4672	.0326	30.6961	1.3683
	29.00	5.761	0.1810	5.5243	.0322	31.0166	1.3541
	31.80	5.675	0.1757	5.6930	.0313	31.9638	1.3140
	32.00	5.675	0.1757	5.6930	.0313	31.9638	1.3140
	34.00	5.595	0.1707	5.8570	.0304	32.8844	1.2772
	17.00	6.538	0.2331	4.2893	.0415	24.0824	1.7440
	20.00	6.456	0.2273	4.3989	.0405	24.6981	1.7005
	22.00	6.398	0.2233	4.4790	.0398	25.1479	1.6701
	23.00	6.366	0.2210	4.5242	.0394	25.4014	1.6535
	24.00	6.336	0.2190	4.5671	.0390	25.6425	1.6379
	26.00	6.276	0.2148	4.6549	.0383	26.1351	1.6070
	28.00	6.214	0.2106	4.7482	.0375	26.6592	1.5754
	29.00	6.184	0.2086	4.7944	.0371	26.9185	1.5603
	29.80	6.168	0.2075	4.8193	.0370	27.0584	1.5522
7	30.00	6.154	0.2066	4.8413	.0368	27.1816	1.5452
	32.00	6.094	0.2026	4.9371	.0361	27.7195	1.5152
	35.00	6.004	0.1966	5.0862	.0350	28.5567	1.4708
	38.00	5.920	0.1911	5.2315	.0340	29.3729	1.4299
	40.20	5.836	0.1858	5.3832	.0331	30.2245	1.3896
	41.00	5.820	0.1847	5.4129	.0329	30.3909	1.3820
	43.00	5.736	0.1795	5.5726	.0320	31.2876	1.3424
	44.00	5.720	0.1785	5.6038	.0318	31.4629	1.3349
	49.50	5.540	0.1674	5.9738	.0298	33.5406	1.2522

CASING SIZES AND CAPACITIES

SIZE OD INCHES	WEIGHT PER LIN FT	ID	CU FT PER LIN FT	LIN FT PER CU FT	BARRELS PER LIN FT	LIN FT PER BARREL	GALLONS PER LIN FT
	20.00	7.125	0.2769	3.6116	.0493	20.2778	2.0712
	24.00	7.025	0.2692	3.7152	.0479	20.8592	2.0135
	26.40	6.969	0.2649	3.7751	.0472	21.1958	1.9815
	29.70	6.875	0.2578	3.8791	.0459	21.7793	1.9284
	33.70	6.765	0.2496	4.0062	.0445	22.4934	1.8672
7⅝	34.00	6.760	0.2492	4.0122	.0444	22.5267	1.8645
	35.50	6.710	0.2456	4.0722	.0437	22.8636	1.8370
	38.00	6.655	0.2416	4.1398	.0430	23.2431	1.8070
	39.00	6.625	0.2394	4.1773	.0426	23.4541	1.7907
	45.30	6.435	0.2259	4.4277	.0402	24.8595	1.6895
7¾	46.10	6.560	0.2347	4.2605	.0418	23.9212	1.7558
8	26.00	7.386	0.2975	3.3609	.0530	18.8700	2.2258
	28.00	7.485	0.3056	3.2726	.0544	18.3741	2.2858
	32.00	7.385	0.2975	3.3618	.0530	18.8751	2.2252
	35.50	7.285	0.2895	3.4547	.0516	19.3968	2.1653
8⅛	36.00	7.285	0.2895	3.4547	.0516	19.3968	2.1653
	39.50	7.185	0.2816	3.5516	.0501	19.9405	2.1063
	40.00	7.185	0.2816	3.5516	.0501	19.9405	2.1063
	42.00	7.125	0.2769	3.6116	.0493	20.2778	2.0712
	24.00	8.097	0.3576	2.7966	.0637	15.7015	2.6749
	28.00	8.017	0.3506	2.8527	.0624	16.0165	2.6223
	32.00	7.921	0.3422	2.9222	.0609	16.4070	2.5599
	36.00	7.825	0.3340	2.9944	.0595	16.8121	2.4982
	38.00	7.775	0.3297	3.0330	.0587	17.0290	2.4664
8⅝	40.00	7.725	0.3255	3.0724	.0580	17.2502	2.4348
	43.00	7.651	0.3193	3.1321	.0569	17.5855	2.3883
	44.00	7.625	0.3171	3.1535	.0565	17.7056	2.3721
	49.00	7.511	0.3077	392500	.0548	18.2471	2.3017
	52.00	7.435	0.3015	3.3167	.0537	18.6221	2.2554
	34.00	8.290	0.3748	2.6679	.0668	14.9789	2.8039
	38.00	8.196	0.3664	2.7294	.0653	15.3245	2.7407
	40.00	8.150	0.3623	2.7603	.0645	15.4980	2.7100
	41.20	8.150	0.3623	2.7603	.0645	15.4980	2.7100
9	45.00	8.032	0.3519	2.8420	.0627	15.9567	2.6321
	46.10	8.032	0.3519	2.8420	.0627	15.9567	2.6321
	54.00	7.810	0.3327	3.0059	.0593	16.8767	2.4886
	55.20	7.812	0.3329	3.0043	.0593	16.8681	2.4899

CASING SIZES AND CAPACITIES

SIZE OD INCHES	WEIGHT PER LIN FT	ID	CU FT PER LIN FT	LIN FT PER CU FT	BARRELS PER LIN FT	LIN FT PER BARREL	GALLONS PER LIN FT
	29.30	9.063	0.4480	2.2322	.0798	12.5327	3.3512
	32.30	9.001	0.4419	2.2630	.0787	12.7060	3.3055
	36.00	8.921	0.4341	2.3038	.0773	12.9349	3.2470
	38.00	8.885	0.4306	2.3225	.0767	13.0399	3.2209
	40.00	8.835	0.4257	2.3489	.0758	13.1879	3.1847
	42.00	8.799	0.4223	2.3681	.0752	13.2961	3.1588
	43.50	8.755	0.4181	2.3920	.0745	13.4301	3.1273
9⅝	44.30	8.750	0.4176	2.3947	.0744	13.4454	3.1237
	47.00	8.681	0.4110	2.4329	.0732	13.6600	3.0747
	47.20	8.680	0.4109	2.4335	.0732	13.6631	3.0740
	53.50	8.535	0.3973	2.5169	.0708	14.1313	2.9721
	57.40	8.450	0.3894	2.5678	.0694	14.4171	2.9132
	58.40	8.435	0.3881	2.5769	.0691	14.4684	2.9029
	61.10	8.375	0.3826	2.6140	.0681	14.6764	2.8617
10	33.00	9.384	0.4803	2.0821	.0855	11.6900	3.5928
	60.00	8.780	0.4205	2.3784	.0749	13.3537	3.1452
	32.75	10.192	0.5666	1.7650	.1009	9.9099	4.2382
	35.75	10.140	0.5608	1.7832	.0999	10.0118	4.1950
	40.50	10.050	0.5509	1.8153	.0981	10.1920	4.1209
	45.50	9.950	0.5400	1.8519	.0962	10.3979	4.0393
	46.20	9.950	0.5400	1.8519	.0962	10.3979	4.0393
	48.00	9.902	0.5348	1.8699	.0952	10.4989	4.0004
10¾	49.50	9.850	0.5292	1.8897	.0943	10.6101	3.9585
	51.00	9.850	0.5292	1.8897	.0943	10.6101	3.9585
	54.00	9.784	0.5221	1.9153	.0930	10.7537	3.9056
	55.50	9.760	0.5195	1.9247	.0925	10.8066	3.8865
	60.70	9.660	0.5090	1.9648	.0906	11.0315	3.8073
	65.70	9.560	0.4985	2.0061	.0888	11.2635	3.7289
	71.10	9.450	0.4871	2.0531	.0868	11.5273	3.6435
	76.00	9.350	0.4768	2.0972	.0849	11.7752	3.5668
	81.00	9.250	0.4667	2.1428	.0831	12.0311	3.4909
	38.00	11.150	0.6781	1.4748	.1208	8.2802	5.0723
	42.00	11.084	0.6701	1.4924	.1193	8.3791	5.0125
	47.00	11.000	0.6600	1.5153	.1175	8.5076	4.9368
	50.00	10.950	0.6540	1.5291	.1165	8.5854	4.8920
11¾	54.00	10.880	0.6456	1.5489	.1150	8.6963	4.8297
	60.00	10.772	0.6329	1.5801	.1127	8.8715	4.7343
	61.00	10.770	0.6326	1.5807	.1127	8.8748	4.7325
	65.00	10.682	0.6223	1.6068	.1108	9.0216	4.6555
12	40.00	11.384	0.7068	1.4148	.1259	7.9433	5.2875
	33.38*	12.250	0.8185	1.2218	.1458	6.8599	6.1225
	37.42*	12.188	0.8102	1.2343	.1443	6.9299	6.0607
	41.45*	12.126	0.8020	1.2469	.1428	7.0009	5.9992
12¾	43.77*	12.090	0.7972	1.2544	.1420	7.0427	5.9636
	45.58*	12.062	0.7935	1.2602	.1413	7.0754	5.9361
	49.56*	12.000	0.7854	1.2732	.1399	7.1487	5.8752
	53.00	11.970	0.7815	1.2796	.1392	7.1846	5.8458

CASING SIZES AND CAPACITIES

SIZE OD INCHES	WEIGHT PER LIN FT	ID	CU FT PER LIN FT	LIN FT PER CU FT	BARRELS PER LIN FT	LIN FT PER BARREL	GALLONS PER LIN FT
13	40.00	12.438	0.8438	1.1851	.1503	6.6541	6.3119
	45.00	12.360	0.8332	1.2002	.1484	6.7383	6.2330
	50.00	12.282	0.8227	1.2154	.1465	6.8242	6.1546
	54.00	12.200	0.8118	1.2318	.1446	6.9162	6.0727
13⅜	48.00	12.715	0.8818	1.1341	.1571	6.3673	6.5962
	54.50	12.615	0.8680	1.1521	.1546	6.4687	6.4928
	61.00	12.515	0.8543	1.1706	.1521	6.5725	6.3903
	68.00	12.415	0.8407	1.1895	.1497	6.6788	6.2886
	72.00	12.347	0.8315	1.2027	.1481	6.7525	6.2199
	77.00	12.275	0.8218	1.2168	.1464	6.8320	6.1476
	83.50	12.175	0.8085	1.2369	.1440	6.9447	6.0478
	85.00	12.159	0.8063	1.2402	.1436	6.9630	6.0319
	92.00	12.031	0.7895	1.2667	.1406	7.1119	5.9056
	98.00	11.937	0.7772	1.2867	.1384	7.2244	5.8137
14	50.00	13.344	0.9712	1.0297	.1730	5.7812	7.2649
16	55.00	15.375	1.2893	0.7756	.2296	4.3547	9.6447
	65.00	15.250	1.2684	0.7884	.2259	4.4264	9.4885
	70.00	15.198	1.2598	0.7938	.2244	4.4567	9.4239
	75.00	15.124	1.2476	0.8016	.2222	4.5005	9.3324
	84.00	15.010	1.2288	0.8138	.2189	4.5691	9.1922
	109.00	14.688	1.1767	0.8499	.2096	4.7716	8.8021
	118.00	14.570	1.1578	0.8637	.2062	4.8492	8.6612
18	80.00	17.180	1.6098	0.6212	.2867	3.4877	12.0422
18⅝	78.00	17.855	1.7388	0.5751	.3097	3.2290	13.0071
	87.50	17.755	1.7194	0.5816	.3062	3.2655	12.8618
	96.50	17.655	1.7001	0.5882	.3028	3.3026	12.7173
20	90.00	19.190	2.0085	0.4979	.3577	2.7954	15.0248
	94.00	19.124	1.9947	0.5013	.3553	2.8147	14.9216
	106.50	19.000	1.9689	0.5079	.3507	2.8516	14.7288
	133.00	18.730	1.9134	0.5226	.3408	2.9344	14.3131
	169.00	18.376	1.8417	0.5430	.3280	3.0485	13.7772
21½	92.50	20.710	2.3393	0.4275	.4166	2.4001	17.4992
	103.00	20.610	2.3168	0.4316	.4126	2.4235	17.3307
	114.00	20.510	2.2943	0.4359	.4086	2.4471	17.1629
24½	88.00	23.850	3.1024	0.3223	.5526	1.8097	23.2079
	100.50	23.750	3.0765	0.3250	.5479	1.8250	23.0137
	113.00	23.650	3.0506	0.3278	.5433	1.8405	22.8203
30	98.93*	29.376	4.7067	0.2125	.8383	1.1929	35.2083
	118.65*	29.250	4.6664	0.2143	.8311	1.2032	34.9069
	157.53*	29.000	4.5869	0.2180	.8170	1.2240	34.3127
	196.08*	28.750	4.5082	0.2218	.8029	1.2454	33.7237
	234.29*	28.500	4.4301	0.2257	.7890	1.2674	33.1397
	309.72*	28.000	4.2761	0.2339	.7616	1.3130	31.9871
	346.93*	27.750	4.2000	0.2381	.7481	1.3368	31.4185
	383.81*	27.500	4.1247	0.2424	.7346	1.3612	30.8549
	546.57*	27.000	3.9761	0.2515	.7082	1.4121	29.7431

EXTREME LINE CASING
SIZES AND CAPACITIES

SIZE OD INCHES	WEIGHT PER LIN FT	ID	CU FT PER LIN FT	LIN FT PER CU FT	BARRELS PER LIN FT	LIN FT PER BARREL	GALLONS PER LIN FT
4½	11.60	4.000	.0873	11.4592	.0155	64.3384	0.6528
	13.50	3.920	.0838	11.9316	.0149	66.9912	0.6269
	15.10	3.826	.0798	12.5251	.0142	70.3235	0.5972
4¾	16.00	4.082	.0909	11.0034	.0162	61.7795	0.6798
	18.00	4.000	.0873	11.4592	.0155	64.3384	0.6528
5	15.00	4.408	.1060	9.4360	.0189	52.9794	0.7928
	18.00	4.276	.0997	10.0276	.0178	56.3008	0.7460
	21.00	4.154	.0941	10.6253	.0168	59.6564	0.7040
5½	15.50	4.950	.1336	7.4828	.0238	42.0126	0.9997
	17.00	4.892	.1305	7.6613	.0232	43.0147	0.9764
	20.00	4.778	.1245	8.0312	.0222	45.0918	0.9314
	23.00	4.670	.1189	8.4070	.0212	47.2016	0.8898
	25.00	4.580	.1144	8.7406	.0204	49.0749	0.8558
5¾	19.50	5.090	.1413	7.0768	.0252	39.7333	1.0570
	22.50	4.990	.1358	7.3633	.0242	41.3418	1.0159
	25.20	4.890	.1304	7.6675	.0232	43.0499	0.9756
6	18.00	5.424	.1605	6.2321	.0286	34.9906	1.2003
	20.00	5.352	.1562	6.4009	.0278	35.9383	1.1687
	23.00	5.240	.1498	6.6774	.0267	37.4910	1.1203
	26.00	5.140	.1441	6.9398	.0257	38.9640	1.0779
6⅝	24.00	5.921	.1912	5.2298	.0341	29.3630	1.4304
	26.00	5.855	.1870	5.3483	.0333	30.0287	1.3987
	28.00	5.791	.1829	5.4672	.0326	30.6961	1.3683
	29.00	5.791	.1829	5.4672	.0326	30.6961	1.3683
	32.00	5.675	.1757	5.6930	.0313	31.9638	1.3140
	34.00	5.595	.1707	5.8570	.0304	32.8844	1.2772
7	23.00	6.366	.2210	4.5242	.0394	25.4014	1.6535
	24.00	6.336	.2190	4.5671	.0390	25.6425	1.6379
	26.00	6.276	.2148	4.6549	.0383	26.1351	1.6070
	28.00	6.214	.2106	4.7482	.0375	26.6592	1.5754
	29.00	6.184	.2086	4.7944	.0371	26.9185	1.5603
	30.00	6.154	.2066	4.8413	.0368	27.1816	1.5452
	32.00	6.094	.2026	4.9371	.0361	27.7195	1.5152
	33.70	6.048	.1995	5.0124	.0355	28.1428	1.4924
	35.00	6.004	.1966	5.0862	.0350	28.5567	1.4708
	35.30	6.000	.1963	5.0930	.0350	28.5948	1.4688
	38.00	5.920	.1911	5.2315	.0340	29.3729	1.4299
	40.00	5.836	.1858	5.3832	.0331	30.2245	1.3896

EXTREME LINE CASING
SIZES AND CAPACITIES

SIZE OD INCHES	WEIGHT PER LIN FT	ID	CU FT PER LIN FT	LIN FT PER CU FT	BARRELS PER LIN FT	LIN FT PER BARREL	GALLONS PER LIN FT
	26.40	6.969	.2649	3.7751	.0472	21.1958	1.9815
	29.70	6.875	.2578	3.8791	.0459	21.7793	1.9284
	33.70	6.765	.2496	4.0062	.0445	22.4934	1.8672
	36.00	6.705	.2452	4.0783	.0437	22.8977	1.8342
7⅝	38.70	6.625	.2394	4.1773	.0426	23.4541	1.7907
	39.00	6.625	.2394	4.1773	.0426	23.4541	1.7907
	45.00	6.445	.2266	4.4139	.0404	24.7825	1.6947
	45.30	6.435	.2259	4.4277	.0402	24.8595	1.6895
	32.00	7.921	.3422	2.9222	.0609	16.4070	2.5599
	36.00	7.825	.3340	2.9944	.0595	16.8121	2.4982
	38.00	7.775	.3297	3.0330	.0587	17.0290	2.4664
	40.00	7.725	.3255	3.0724	.0580	17.2502	2.4348
8⅝	43.00	7.651	.3193	3.1321	.0569	17.5855	2.3883
	44.00	7.625	.3171	3.1535	.0565	17.7056	2.3721
	48.00	7.537	.3098	3.2276	.0552	18.1215	2.3177
	49.00	7.511	.3077	3.2500	.0548	18.2471	2.3017
	34.00	8.290	.3748	2.6679	.0668	14.9789	2.8039
	38.00	8.196	.3664	2.7294	.0653	15.3245	2.7407
9	40.00	8.150	.3623	2.7603	.0645	15.4980	2.7100
	45.00	8.032	.3519	2.8420	.0627	15.9567	2.6321
	50.20	7.910	.3413	2.9304	.0608	16.4527	2.5528
	40.00	8.835	.4257	2.3489	.0758	13.1879	3.1849
	43.50	8.755	.4181	2.3920	.0745	13.4301	3.1273
	47.00	8.681	.4110	2.4329	.0732	13.6600	3.0747
9⅝	53.50	8.535	.3973	2.5169	.0708	14.1313	2.9721
	58.00	8.435	.3881	2.5769	.0691	14.4684	2.9029
	58.40	8.435	.3881	2.5769	.0691	14.4684	2.9029
	61.10	8.375	.3826	2.6140	.0681	14.6764	2.8617
	41.50	9.200	.4616	2.1662	.0822	12.1623	3.4533
	45.50	9.120	.4536	2.2044	.0808	12.3766	3.3935
10	50.50	9.016	.4434	2.2555	.0790	12.6638	3.3166
	55.50	8.908	.4328	2.3105	.0771	12.9727	3.2376
	61.20	8.790	.4214	2.3730	.0751	13.3233	3.1524
	45.50	9.950	.5400	1.8519	.0962	10.3979	4.0393
	51.00	9.850	.5292	1.8897	.0943	10.6101	3.9585
	55.50	9.760	.5195	1.9247	.0925	10.8066	3.8865
	60.70	9.660	.5090	1.9648	.0906	11.0315	3.8073
10¾	65.70	9.560	.4985	2.0061	.0888	11.2635	3.7289
	71.10	9.450	.4871	2.0531	.0868	11.5273	3.6435
	76.00	9.350	.4768	2.0972	.0849	11.7752	3.5668
	81.00	9.250	.4667	2.1428	.0831	12.0311	3.4909

ALUMINUM LINER
SIZES AND CAPACITIES

SIZE OD INCHES	WEIGHT PER LIN FT	ID	CU FT PER LIN FT	LIN FT PER CU FT	BARRELS PER LIN FT	LIN FT PER BARREL	GALLONS PER LIN FT
2⅞	2.65	2.323	.0294	33.9762	.0052	190.7619	0.2202
3½	3.55	2.900	.0459	21.8010	.0082	122.4036	0.3431
4	4.35	3.364	.0617	16.2017	.0110	90.9658	0.4617
4½	5.18	3.826	.0798	12.5251	.0142	70.3235	0.5972
4¾	5.41	4.082	.0909	11.0034	.0162	61.7795	0.6798
5	6.08	4.290	.1004	9.9623	.0179	55.9340	0.7509
5½	6.80	4.778	.1245	8.0312	.0222	45.0918	0.9314
5¾	7.43	5.000	.1364	7.3339	.0243	41.1766	1.0200
6⅝	10.00	5.761	.1810	5.5243	.0322	31.0166	1.3541
7	10.25	6.154	.2066	4.8413	.0368	27.1816	1.5452
8⅝	14.50	7.651	.3193	3.1321	.0569	17.5855	2.3883

API PLAIN END LINER
SIZES AND CAPACITIES

SIZE OD INCHES	WEIGHT PER LIN FT	ID	CU FT PER LIN FT	LIN FT PER CU FT	BARRELS PER LIN FT	LIN FT PER BARREL	GALLONS PER LIN FT
3½	9.91	2.922	.0466	21.4740	.0083	120.5674	0.3484
4	11.34	3.428	.0641	15.6024	.0114	87.6009	0.4794
4½	13.04	3.920	.0838	11.9316	.0149	66.9912	0.6269
5	17.93	4.276	.0997	10.0276	.0178	56.3008	0.7460
5½	19.81	4.778	.1245	8.0312	.0222	45.0918	0.9314
6	22.81	5.241	.1498	6.6749	.0267	37.4767	1.1207
6⅝	27.65	5.791	.1829	5.4672	.0326	30.6961	1.3683

DRILL PIPE SIZES AND CAPACITIES

SIZE OD INCHES	WEIGHT PER LIN FT	ID	CU FT PER LIN FT	LIN FT PER CU FT	BARRELS PER LIN FT	LIN FT PER BARREL	GALLONS PER LIN FT
1.900	3.75	1.500	.0123	81.4873	.0022	457.5174	.0918
	4.80	2.000	.0218	45.8366	.0039	257.3535	.1632
2⅜	4.85	1.995	.0217	46.0667	.0039	258.6451	.1624
	6.65	1.815	.0180	55.6569	.0032	312.4905	.1344
	6.45	2.469	.0332	30.0767	.0059	168.8682	.2487
	6.85	2.441	.0325	30.7707	.0058	172.7645	.2431
2⅞	8.35	2.323	.0294	33.9762	.0052	190.7619	.2202
	10.40	2.151	.0252	39.6271	.0045	222.4894	.1888
	8.50	3.063	.0512	19.5424	.0091	109.7226	.3828
	9.50	2.992	.0488	20.4809	.0087	114.9918	.3652
3½	11.20	2.900	.0459	21.8010	.0082	122.4036	.3431
	13.30	2.764	.0417	23.9992	.0074	134.7454	.3117
	15.50	2.602	.0369	27.0806	.0066	152.0462	.2762
3⅞	14.50	3.181	.0552	18.1195	.0098	101.7332	.4128
	11.00	3.500	.0668	14.9671	.0119	84.0338	.4998
	11.85	3.476	.0659	15.1745	.0117	85.1982	.4930
4	14.00	3.340	.0608	16.4354	.0108	92.2778	.4551
	15.30	3.244	.0574	17.4225	.0102	97.8202	.4294
	12.75	4.000	.0873	11.4592	.0155	64.3384	.6528
	13.75	3.958	.0854	11.7036	.0152	65.7111	.6392
4½	16.60	3.826	.0798	12.5251	.0142	70.3235	.5972
	18.16	3.754	.0769	13.0102	.0137	73.0469	.5750
	20.00	3.640	.0723	13.8379	.0129	77.6940	.5406
4¾	19.08	4.000	.0873	11.4592	.0155	64.3384	.6528
	14.20	4.500	.1104	9.0541	.0197	50.8353	.8262
	15.00	4.408	.1060	9.4360	.0189	52.9794	.7928
	16.25	4.408	.1060	9.4360	.0189	52.9794	.7928
5	19.50	4.276	.0997	10.0276	.0178	56.3008	.7460
	20.50	4.214	.0969	10.3248	.0173	57.9697	.7245
	25.60	4.000	.0873	11.4592	.0155	64.3384	.6528
	21.90	4.778	.1245	8.0312	.0222	45.0918	.9314
5½	24.70	4.670	.1189	8.4070	.0212	47.2016	.8898
	19.00	4.975	.1350	7.4078	.0240	41.5914	.0098
5⁹⁄₁₆	22.20	4.859	.1288	7.7657	.0229	43.6010	.9633
	25.25	4.733	.1222	8.1846	.0218	45.9533	.9140
5¾	23.40	5.000	.1364	7.3339	.0243	41.1766	.0200
	22.20	6.065	.2006	4.9844	.0357	27.9852	.5008
6⅝	25.20	5.965	.1941	5.1529	.0346	28.9314	.4517
	31.90	5.761	.1810	5.5243	.0322	31.0166	.3541
	28.75	6.965	.2646	3.7795	.0471	21.2201	.9793
7⅝	29.25	6.969	.2649	3.7751	.0472	21.1958	.9815
	40.00	7.825	.3340	2.9944	.0595	16.8121	.4982
8⅝	46.50	7.625	.3171	3.1535	.0565	17.7056	.3721

Volume and Fill between Pipe and Hole

1	ONE STRING OF 1.050″ OD TUBING	VOLUME AND FILL BETWEEN PIPE AND HOLE			
HOLE DIAM (IN)	**CU FT PER LIN FT**	**LIN FT PER CU FT**	**BARRELS PER LIN FT**	**LIN FT PER BARREL**	**GALLONS PER LIN FT**
2	0.0158	63.2775	0.0028	355.2767	0.1182
⅛	0.0186	53.7181	0.0033	301.6046	0.1393
¼	0.0216	46.2996	0.0038	259.9531	0.1616
⅜	0.0248	40.4014	0.0044	226.8369	0.1852
½	0.0281	35.6186	0.0050	199.9833	0.2100
⅝	0.0316	31.6763	0.0056	177.8493	0.2362
¾	0.0352	28.3818	0.0063	159.3520	0.2636
⅞	0.0391	25.5959	0.0070	143.7102	0.2923
3	0.0431	23.2158	0.0077	130.3468	0.3222
⅛	0.0473	21.1640	0.0084	118.8271	0.3535
¼	0.0516	19.3812	0.0092	108.8176	0.3860
⅜	0.0561	17.8212	0.0100	100.0585	0.4198
½	0.0608	16.4473	0.0108	92.3448	0.4548
⅝	0.0657	15.2305	0.0117	85.5128	0.4912
¾	0.0707	14.1471	0.0126	79.4301	0.5288
⅞	0.0759	13.1780	0.0135	73.9887	0.5677
4	0.0813	12.3072	0.0145	69.0998	0.6078
⅛	0.0868	11.5217	0.0155	64.6896	0.6493
¼	0.0925	10.8105	0.0165	60.6966	0.6920
⅜	0.0984	10.1644	0.0175	57.0688	0.7360
½	0.1044	9.5755	0.0186	53.7623	0.7812
⅝	0.1107	9.0371	0.0197	50.7397	0.8278
¾	0.1170	8.5436	0.0208	47.9690	0.8756
⅞	0.1236	8.0901	0.0220	45.4224	0.9247
5	0.1303	7.6722	0.0232	43.0762	0.9750
⅛	0.1372	7.2863	0.0244	40.9096	1.0267
¼	0.1443	6.9292	0.0257	38.9045	1.0796
⅜	0.1516	6.5980	0.0270	37.0451	1.1338
½	0.1590	6.2903	0.0283	35.3174	1.1892
⅝	0.1666	6.0039	0.0297	33.7091	1.2460
¾	0.1743	5.7367	0.0310	32.2095	1.3040
⅞	0.1822	5.4873	0.0325	30.8087	1.3663
6	0.1903	5.2539	0.0339	29.4982	1.4238
⅛	0.1986	5.0352	0.0354	28.2704	1.4857
¼	0.2070	4.8300	0.0369	27.1184	1.5488
⅜	0.2156	4.6372	0.0384	26.0360	1.6132
½	0.2244	4.4558	0.0400	25.0177	1.6788
⅝	0.2334	4.2850	0.0416	24.0584	1.7458
¾	0.2425	4.1239	0.0432	23.1537	1.8140
⅞	0.2518	3.9717	0.0448	22.2995	1.8835
7	0.2612	3.8279	0.0465	21.4920	1.9542
⅛	0.2709	3.6918	0.0482	20.7279	2.0263
¼	0.2807	3.5629	0.0500	20.0042	2.0996
⅜	0.2906	3.4407	0.0518	19.3179	2.1742
½	0.3008	3.3247	0.0536	18.6666	2.2500
⅝	0.3111	3.2145	0.0554	18.0478	2.3272
¾	0.3216	3.1097	0.0573	17.4595	2.4056
⅞	0.3322	3.0100	0.0592	16.8997	2.4852

NOTE: No allowance has been made for couplings or upsets.

2

TWO STRINGS OF 1.050″ OD TUBING

VOLUME AND FILL BETWEEN PIPE AND HOLE

HOLE DIAM (IN)	CU FT PER LIN FT	LIN FT PER CU FT	BARRELS PER LIN FT	LIN FT PER BARREL	GALLONS PER LIN FT
3	0.0371	26.9826	0.0066	151.4958	0.2772
⅛	0.0412	24.2502	0.0073	136.1546	0.3085
¼	0.0456	21.9380	0.0081	123.1725	0.3410
⅜	0.0501	19.9602	0.0089	112.0679	0.3748
½	0.0548	18.2525	0.0098	102.4803	0.4098
⅝	0.0596	16.7660	0.0106	94.1340	0.4462
¾	0.0647	15.4625	0.0115	86.8154	0.4838
⅞	0.0699	14.3121	0.0124	80.3563	0.5227
4	0.0752	13.2908	0.0134	74.6223	0.5628
⅛	0.0808	12.3794	0.0144	69.5051	0.6043
¼	0.0865	11.5621	0.0154	64.9165	0.6470
⅜	0.0924	10.8261	0.01.5	60.7839	0.6910
½	0.0984	10.1605	0.0175	57.0471	0.7362
⅝	0.1046	9.5565	0.0186	53.6555	0.7828
¾	0.1110	9.0063	0.0198	50.5668	0.8306
⅞	0.1176	8.5038	0.0209	47.7451	0.8797
5	0.1243	8.0433	0.0221	45.1596	0.9300
⅛	0.1312	7.6202	0.0234	42.7842	0.9817
¼	0.1383	7.2305	0.0246	40.5960	1.0346
⅜	0.1455	6.8706	0.0259	38.5756	1.0888
½	0.1530	6.5376	0.0272	36.7058	1.1442
⅝	0.1605	6.2287	0.0286	34.9717	1.2010
¾	0.1683	5.9417	0.0300	33.3603	1.2590
⅞	0.1762	5.6745	0.0314	31.8599	1.3183
6	0.1843	5.4253	0.0328	30.4605	1.3788
⅛	0.1926	5.1924	0.0343	29.1531	1.4407
¼	0.2010	4.9745	0.0358	27.9296	1.5038
⅜	0.2096	4.7702	0.0373	26.7828	1.5682
½	0.2184	4.5785	0.0389	25.7064	1.6338
⅝	0.2274	4.3983	0.0405	24.6947	1.7008
¾	0.2365	4.2287	0.0421	23.7425	1.7690
⅞	0.2458	4.0689	0.0438	22.8451	1.8385
7	0.2552	3.9181	0.0455	21.9984	1.9092
⅛	0.2649	3.7756	0.0472	21.1985	1.9813
¼	0.2747	3.6409	0.0489	20.4421	2.0546
⅜	0.2846	3.5134	0.0507	19.7260	2.1292
½	0.2948	3.3925	0.0525	19.0474	2.2050
⅝	0.3051	3.2778	0.0543	18.4036	2.2822
¾	0.3156	3.1689	0.0562	17.7922	2.3606
⅞	0.3262	3.0655	0.0581	17.2112	2.4403
8	0.3370	2.9670	0.0600	16.6585	2.5212
⅛	0.3480	2.8733	0.0620	16.1323	2.6035
¼	0.3592	2.7840	0.0640	15.6309	2.6870
⅜	0.3705	2.6988	0.0660	15.1528	2.7718
½	0.3820	2.6176	0.0680	14.6965	2.8578
⅝	0.3937	2.5399	0.0701	14.2607	2.9452
¾	0.4056	2.4657	0.0722	13.8441	3.0338
⅞	0.4176	2.3948	0.0744	13.4457	3.1237

3 THREE STRINGS OF 1.050″ OD TUBING — VOLUME AND FILL BETWEEN PIPE AND HOLE

HOLE DIAM (IN)	CU FT PER LIN FT	LIN FT PER CU FT	BARRELS PER LIN FT	LIN FT PER BARREL	GALLONS PER LIN FT
3	0.0310	32.2084	0.0055	180.8369	0.2323
⅛	0.0352	28.3901	0.0063	159.3983	0.2635
¼	0.0396	25.2717	0.0070	141.8903	0.2960
⅜	0.0441	22.6826	0.0079	127.3535	0.3298
½	0.0488	20.5028	0.0087	115.1148	0.3649
⅝	0.0536	18.6458	0.0096	104.6884	0.4012
¾	0.0587	17.0476	0.0104	95.7149	0.4388
⅞	0.0639	15.6598	0.0114	87.9231	0.4777
4	0.0692	14.4453	0.0123	81.1041	0.5179
⅛	0.0748	13.3750	0.0133	75.0952	0.5593
¼	0.0805	12.4261	0.0143	69.7671	0.6020
⅜	0.0864	11.5799	0.0154	65.0165	0.6460
½	0.0924	10.8217	0.0165	60.7593	0.6913
⅝	0.0986	10.1391	0.0176	56.9268	0.7378
¾	0.1050	9.5220	0.0187	53.4622	0.7856
⅞	0.1116	8.9620	0.0199	50.3181	0.8347
5	0.1183	8.4521	0.0211	47.4548	0.8851
⅛	0.1252	7.9861	0.0223	44.8388	0.9367
¼	0.1323	7.5591	0.0236	42.4413	0.9896
⅜	0.1395	7.1667	0.0249	40.2380	1.0438
½	0.1469	6.8051	0.0262	38.2078	1.0993
⅝	0.1545	6.4711	0.0275	36.3325	1.1560
¾	0.1623	6.1619	0.0289	34.5963	1.2140
⅞	0.1702	5.8750	0.0303	32.9855	1.2733
6	0.1783	5.6082	0.0318	31.4878	1.3339
⅛	0.1866	5.3597	0.0332	30.0927	1.3957
¼	0.1950	5.1279	0.0347	28.7908	1.4588
⅜	0.2036	4.9111	0.0363	27.5737	1.5232
½	0.2124	4.7081	0.0378	26.4342	1.5889
⅝	0.2213	4.5178	0.0394	25.3656	1.6558
¾	0.2305	4.3390	0.0410	24.3619	1.7240
⅞	0.2398	4.1709	0.0427	23.4181	1.7935
7	0.2492	4.0126	0.0444	22.5292	1.8642
⅛	0.2588	3.8633	0.0461	21.6910	1.9363
¼	0.2686	3.7224	0.0478	20.8997	2.0096
⅜	0.2786	3.5892	0.0496	20.1517	2.0842
½	0.2888	3.4631	0.0514	19.4440	2.1600
⅝	0.2991	3.3437	0.0533	18.7736	2.2372
¾	0.3096	3.2305	0.0551	18.1379	2.3156
⅞	0.3202	3.1230	0.0570	17.5344	2.3953
8	0.3310	3.0209	0.0590	16.9611	2.4762
⅛	0.3420	2.9238	0.0609	16.4160	2.5585
¼	0.3532	2.8314	0.0629	15.8971	2.6420
⅜	0.3645	2.7433	0.0649	15.4028	2.7268
½	0.3760	2.6594	0.0670	14.9315	2.8128
⅝	0.3877	2.5793	0.0691	14.4818	2.9002
¾	0.3995	2.5029	0.0712	14.0525	2.9888
⅞	0.4116	2.4298	0.0733	13.6422	3.0787

4

| FOUR STRINGS OF 1.050" OD TUBING | | VOLUME AND FILL BETWEEN PIPE AND HOLE | | |

HOLE DIAM (IN)	CU FT PER LIN FT	LIN FT PER CU FT	BARRELS PER LIN FT	LIN FT PER BARREL	GALLONS PER LIN FT
3	0.0250	39.9448	0.0045	224.2732	0.1873
1/8	0.0292	34.2344	0.0052	192.2118	0.2185
1/4	0.0336	29.8003	0.0060	167.3164	0.2510
3/8	0.0381	26.2651	0.0068	147.4673	0.2848
1/2	0.0428	23.3860	0.0076	131.3028	0.3199
5/8	0.0476	21.0004	0.0085	117.9084	0.3562
3/4	0.0526	18.9947	0.0094	106.6474	0.3938
7/8	0.0578	17.2877	0.0103	97.0630	0.4327
4	0.0632	15.8194	0.0113	88.8192	0.4729
1/8	0.0688	14.5448	0.0122	81.6631	0.5143
1/4	0.0745	13.4295	0.0133	75.4011	0.5570
3/8	0.0803	12.4466	0.0143	69.8826	0.6010
1/2	0.0864	11.5749	0.0154	64.9883	0.6463
5/8	0.0926	10.7974	0.0165	60.6229	0.6928
3/4	0.0990	10.1003	0.0176	56.7092	0.7406
7/8	0.1056	9.4725	0.0188	53.1842	0.7897
5	0.1123	8.9046	0.0200	49.9958	0.8401
1/8	0.1192	8.3890	0.0212	47.1007	0.8917
1/4	0.1263	7.9191	0.0225	44.4623	0.9446
3/8	0.1335	7.4895	0.0238	42.0502	0.9988
1/2	0.1409	7.0955	0.0251	39.8380	1.0543
5/8	0.1485	6.7331	0.0265	37.8035	1.1110
3/4	0.1563	6.3990	0.0278	35.9276	1.1690
7/8	0.1642	6.0901	0.0292	34.1934	1.2283
6	0.1723	5.8039	0.0307	32.5867	1.2889
1/8	0.1806	5.5382	0.0322	31.0948	1.3507
1/4	0.1890	5.2910	0.0337	29.7068	1.4138
3/8	0.1976	5.0605	0.0352	28.4128	1.4782
1/2	0.2064	4.8453	0.0368	27.2044	1.5439
5/8	0.2153	4.6440	0.0384	26.0739	1.6108
3/4	0.2245	4.4553	0.0440	25.0146	1.6790
7/8	0.2337	4.2782	0.0416	24.0205	1.7485
7	0.2432	4.1118	0.0433	23.0862	1.8193
1/8	0.2528	3.9552	0.0450	22.2069	1.8913
1/4	0.2626	3.8076	0.0468	21.3782	1.9646
3/8	0.2726	3.6684	0.0486	20.5963	2.0392
1/2	0.2827	3.5368	0.0504	19.8575	2.1151
5/8	0.2931	3.4123	0.0522	19.1588	2.1922
3/4	0.3035	3.2945	0.0541	18.4972	2.2706
7/8	0.3142	3.1828	0.0560	17.8700	2.3503
8	0.3250	3.0768	0.0579	17.2749	2.4313
1/8	0.3360	2.9761	0.0598	16.7097	2.5135
1/4	0.3472	2.8804	0.0618	16.1724	2.5970
3/8	0.3585	2.7894	0.0639	15.6611	2.6818
1/2	0.3700	2.7026	0.0659	15.1741	2.7679
5/8	0.3817	2.6200	0.0680	14.7100	2.8552
3/4	0.3935	2.5411	0.0701	14.2672	2.9438
7/8	0.4055	2.4658	0.0722	13.8445	3.0337

1

ONE STRING OF
1.315" OD
TUBING

VOLUME AND FILL BETWEEN PIPE AND HOLE

HOLE DIAM (IN)		CU FT PER LIN FT	LIN FT PER CU FT	BARRELS PER LIN FT	LIN FT PER BARREL	GALLONS PER LIN FT
2		0.0124	80.7418	0.0022	453.3316	0.0926
	⅛	0.0152	65.8005	0.0027	369.4424	0.1137
	¼	0.0182	55.0049	0.0032	308.8297	0.1360
	⅜	0.0213	46.8749	0.0038	263.1830	0.1596
	½	0.0247	40.5564	0.0044	227.7075	0.1844
	⅝	0.0282	35.5226	0.0050	199.4448	0.2106
	¾	0.0318	31.4311	0.0057	176.4728	0.2380
	⅞	0.0357	28.0501	0.0063	157.4895	0.2667
3		0.0397	25.2169	0.0071	141.5825	0.2966
	⅛	0.0438	22.8145	0.0078	128.0939	0.3279
	¼	0.0482	20.7563	0.0086	116.5382	0.3604
	⅜	0.0527	18.9772	0.0094	106.5492	0.3942
	½	0.0574	17.4271	0.0102	97.8458	0.4292
	⅝	0.0622	16.0670	0.0111	90.2093	0.4656
	¾	0.0673	14.8660	0.0120	83.4664	0.5032
	⅞	0.0725	13.7996	0.0129	77.4788	0.5421
4		0.0778	12.8477	0.0139	72.1344	0.5822
	⅛	0.0834	11.9941	0.0148	67.3418	0.6237
	¼	0.0891	11.2253	0.0159	63.0256	0.6664
	⅜	0.0950	10.5303	0.0169	59.1230	0.7104
	½	0.1010	9.8995	0.0180	55.5816	0.7556
	⅝	0.1072	9.3252	0.0191	52.3571	0.8022
	¾	0.1136	8.8007	0.0202	49.4120	0.8500
	⅞	0.1202	8.3202	0.0214	46.7143	0.8991
5		0.1269	7.8788	0.0226	44.2364	0.9494
	⅛	0.1338	7.4724	0.0238	41.9546	1.0011
	¼	0.1409	7.0973	0.0251	39.8484	1.0540
	⅜	0.1481	6.7503	0.0264	37.8999	1.1082
	½	0.1556	6.4285	0.0277	36.0935	1.1636
	⅝	0.1631	6.1297	0.0291	34.4154	1.2204
	¾	0.1709	5.8515	0.0304	32.8537	1.2784
	⅞	0.1788	5.5922	0.0318	31.3976	1.3377
6		0.1869	5.3499	0.0333	30.0377	1.3982
	⅛	0.1952	5.1234	0.0348	28.7655	1.4601
	¼	0.2036	4.9111	0.0363	27.5736	1.5232
	⅜	0.2122	4.7119	0.0378	26.4553	1.5876
	½	0.2210	4.5248	0.0394	25.4046	1.6532
	⅝	0.2300	4.3487	0.0410	24.4160	1.7202
	¾	0.2391	4.1828	0.0426	23.4848	1.7884
	⅞	0.2484	4.0264	0.0442	22.6064	1.8579
7		0.2578	3.8786	0.0459	21.7770	1.9286
	⅛	0.2675	3.7390	0.0476	20.9929	2.0007
	¼	0.2773	3.6068	0.0494	20.2508	2.0740
	⅜	0.2872	3.4816	0.0512	19.5478	2.1486
	½	0.2974	3.3629	0.0530	18.8811	2.2244
	⅝	0.3077	3.2502	0.0548	18.2483	2.3016
	¾	0.3182	3.1431	0.0567	17.6471	2.3800
	⅞	0.3288	3.0413	0.0586	17.0754	2.4597

2 VOLUME AND FILL BETWEEN PIPE AND HOLE

TWO STRINGS OF 1.315" OD TUBING

HOLE DIAM (IN)	CU FT PER LIN FT	LIN FT PER CU FT	BARRELS PER LIN FT	LIN FT PER BARREL	GALLONS PER LIN FT
3	0.0302	33.0858	0.0054	185.7629	0.2261
1/8	0.0344	29.0695	0.0061	163.2132	0.2573
1/4	0.0387	25.8087	0.0069	144.9053	0.2898
3/8	0.0433	23.1143	0.0077	129.7770	0.3236
1/2	0.0480	20.8549	0.0085	117.0913	0.3587
5/8	0.0528	18.9365	0.0094	106.3205	0.3950
3/4	0.0578	17.2902	0.0103	97.0775	0.4326
7/8	0.0630	15.8643	0.0112	89.0714	0.4715
4	0.0684	14.6191	0.0122	82.0803	0.5117
1/8	0.0739	13.5239	0.0132	75.9313	0.5531
1/4	0.0797	12.5545	0.0102	70.4883	0.5958
3/8	0.0855	11.6914	0.0152	65.6423	0.6398
1/2	0.0916	10.9190	0.0163	61.3055	0.6851
5/8	0.0978	10.2244	0.0174	57.4060	0.7316
3/4	0.1042	9.5973	0.0186	53.8846	0.7794
7/8	0.1108	9.0287	0.0197	50.6921	0.8285
5	0.1175	8.5113	0.0209	47.7874	0.8789
1/8	0.1244	8.0390	0.0222	45.1355	0.9305
1/4	0.1315	7.6065	0.0234	42.7071	0.9834
3/8	0.1387	7.2092	0.0247	40.4768	1.0376
1/2	0.1461	6.8434	0.0260	38.4231	1.0931
5/8	0.1537	6.5058	0.0274	36.5271	1.1498
3/4	0.1615	6.1933	0.0288	34.7727	1.2078
7/8	0.1694	5.9035	0.0302	33.1458	1.2671
6	0.1775	5.6342	0.0316	31.6338	1.3277
1/8	0.1858	5.3835	0.0331	30.2261	1.3895
1/4	0.1942	5.1496	0.0346	28.9128	1.4526
3/8	0.2028	4.9310	0.0361	27.6857	1.5170
1/2	0.2116	4.7265	0.0377	26.5371	1.5827
5/8	0.2205	4.5347	0.0393	25.4603	1.6496
3/4	0.2296	4.3546	0.0409	24.4493	1.7178
7/8	0.2389	4.1853	0.0426	23.4988	1.7873
7	0.2484	4.0259	0.0442	22.6038	1.8581
1/8	0.2580	3.8757	0.0460	21.7602	1.9301
1/4	0.2678	3.7338	0.0477	20.9639	2.0034
3/8	0.2778	3.5998	0.0495	20.2115	2.0780
1/2	0.2879	3.4730	0.0513	19.4996	2.1539
5/8	0.2982	3.3529	0.0531	18.8254	2.2310
3/4	0.3087	3.2391	0.0550	18.1862	2.3094
7/8	0.3194	3.1311	0.0569	17.5796	2.3891
8	0.3302	3.0284	0.0588	17.0034	2.4701
1/8	0.3412	2.9309	0.0608	16.4556	2.5523
1/4	0.3524	2.8380	0.0628	15.9342	2.6358
3/8	0.3637	2.7496	0.0648	15.4376	2.7206
1/2	0.3752	2.6652	0.0668	14.9643	2.8067
5/8	0.3869	2.5848	0.0689	14.5127	2.8940
3/4	0.3987	2.5080	0.0710	14.0815	2.9826
7/8	0.4107	2.4346	0.0732	13.6695	3.0725

3

THREE STRINGS OF 1.315" OD TUBING

VOLUME AND FILL BETWEEN PIPE AND HOLE

HOLE DIAM (IN)	CU FT PER LIN FT	LIN FT PER CU FT	BARRELS PER LIN FT	LIN FT PER BARREL	GALLONS PER LIN FT
3	0.0208	48.0931	0.0037	270.0227	0.1555
1/8	0.0250	40.0499	0.0044	224.8636	0.1868
1/4	0.0293	34.1121	0.0052	191.5252	0.2193
3/8	0.0338	29.5580	0.0060	165.9556	0.2531
1/2	0.0385	25.9612	0.0069	145.7614	0.2881
5/8	0.0434	23.0539	0.0077	129.4380	0.3245
3/4	0.0484	20.6592	0.0086	115.9926	0.3621
7/8	0.0536	18.6556	0.0095	104.7435	0.4010
4	0.0590	16.9572	0.0105	95.2075	0.4411
1/8	0.0645	15.5011	0.0115	87.0323	0.4826
1/4	0.0702	14.2407	0.0125	79.9556	0.5253
3/8	0.0761	13.1403	0.0136	73.7775	0.5693
1/2	0.0822	12.1725	0.0146	68.3436	0.6145
5/8	0.0884	11.3156	0.0157	63.5325	0.6611
3/4	0.0948	10.5524	0.0169	59.2475	0.7089
7/8	0.1013	9.8690	0.0180	55.4105	0.7580
5	0.1081	9.2542	0.0192	51.9583	0.8083
1/8	0.1150	8.6985	0.0205	48.8384	0.8600
1/4	0.1220	8.1943	0.0217	46.0077	0.9129
3/8	0.1293	7.7352	0.0230	43.4298	0.9671
1/2	0.1367	7.3156	0.0243	41.0742	1.0225
5/8	0.1443	6.9310	0.0257	38.9149	1.0793
3/4	0.1520	6.5775	0.0271	36.9299	1.1373
7/8	0.1600	6.2516	0.0285	35.1001	1.1966
6	0.1681	5.9504	0.0299	33.4092	1.2571
1/8	0.1763	5.6715	0.0314	31.8429	1.3190
1/4	0.1848	5.4125	0.0329	30.3888	1.3821
3/8	0.1934	5.1715	0.0344	29.0361	1.4465
1/2	0.2021	4.9470	0.0360	27.7752	1.5121
5/8	0.2111	4.7373	0.0376	26.5978	1.5791
3/4	0.2202	4.5411	0.0392	25.4964	1.6473
7/8	0.2295	4.3573	0.0409	24.4645	1.7168
7	0.2390	4.1848	0.0426	23.4960	1.7875
1/8	0.2486	4.0227	0.0443	22.5858	1.8596
1/4	0.2548	3.8701	0.0460	21.7291	1.9239
3/8	0.2684	3.7263	0.0478	20.9218	2.0075
1/2	0.2785	3.5906	0.0496	20.1600	2.0833
5/8	0.2888	3.4624	0.0514	19.4402	2.1605
3/4	0.2993	3.3412	0.0533	18.7593	2.2389
7/8	0.3099	3.2263	0.0552	18.1146	2.3186
8	0.3208	3.1175	0.0571	17.5034	2.3995
1/8	0.3318	3.0142	0.0591	16.9234	2.4818
1/4	0.3429	2.9161	0.0611	16.3724	2.5653
3/8	0.3543	2.8228	0.0631	15.8486	2.6501
1/2	0.3658	2.7340	0.0651	15.3501	2.7361
5/8	0.3774	2.6494	0.0672	14.8753	2.8235
3/4	0.3893	2.5688	0.0693	14.4227	2.9121
7/8	0.4013	2.4919	0.0715	13.9908	3.0020

4

FOUR STRINGS OF 1.315" OD TUBING	**VOLUME AND FILL BETWEEN PIPE AND HOLE**

HOLE DIAM (IN)	CU FT PER LIN FT	LIN FT PER CU FT	BARRELS PER LIN FT	LIN FT PER BARREL	GALLONS PER LIN FT
3	0.0114	88.0162	0.0020	494.1741	0.0850
1/8	0.0155	64.3609	0.0028	361.3596	0.1162
1/4	0.0199	50.2925	0.0035	282.3717	0.1487
3/8	0.0244	40.9830	0.0043	230.1022	0.1825
1/2	0.0291	34.3790	0.0052	193.0236	0.2176
5/8	0.0339	29.4593	0.0060	165.4016	0.2539
3/4	0.0390	25.6587	0.0069	144.0627	0.2915
7/8	0.0442	22.6389	0.0079	127.1082	0.3304
4	0.0495	20.1855	0.0088	113.3329	0.3706
1/8	0.0551	18.1554	0.0098	101.9351	0.4120
1/4	0.0608	16.4501	0.0108	92.3606	0.4547
3/8	0.0667	14.9992	0.0119	84.2144	0.4987
1/2	0.0727	13.7512	0.0130	77.2074	0.5440
5/8	0.0789	12.6675	0.0141	71.1230	0.5905
3/4	0.0853	11.7187	0.0152	65.7958	0.6383
7/8	0.0919	10.8819	0.0164	61.0975	0.6874
5	0.0986	10.1391	0.0176	56.9269	0.7378
1/8	0.1055	9.4759	0.0188	53.2032	0.7894
1/4	0.1126	8.8807	0.0201	49.8612	0.8423
3/8	0.1198	8.3439	0.0213	46.8475	0.8965
1/2	0.1273	7.8578	0.0227	44.1182	0.9520
5/8	0.1348	7.4158	0.0240	41.6367	1.0087
3/4	0.1426	7.0125	0.0254	39.3724	1.0667
7/8	0.1505	6.6433	0.0268	37.2993	1.1260
6	0.1586	6.3042	0.0283	35.3956	1.1866
1/8	0.1669	5.9920	0.0297	33.6424	1.2484
1/4	0.1753	5.7036	0.0312	32.0235	1.3115
3/8	0.1839	5.4367	0.0328	30.5249	1.3759
1/2	0.1927	5.1891	0.0343	29.1346	1.4416
5/8	0.2017	4.9588	0.0359	27.8418	1.5085
3/4	0.2108	4.7443	0.0375	26.6373	1.5767
7/8	0.2201	4.5440	0.0392	25.5129	1.6462
7	0.2295	4.3568	0.0409	24.4615	1.7170
1/8	0.2392	4.1813	0.0426	23.4765	1.7890
1/4	0.2490	4.0167	0.0443	22.5523	1.8623
3/8	0.2589	3.8621	0.0461	21.6839	1.9369
1/2	0.2691	3.7165	0.0479	20.8666	2.0128
5/8	0.2794	3.5793	0.0498	20.0964	2.0899
3/4	0.2899	3.4499	0.0516	19.3697	2.1683
7/8	0.3005	3.3276	0.0535	18.6831	2.2480
8	0.3113	3.2119	0.0555	18.0336	2.3290
1/8	0.3223	3.1024	0.0574	17.4186	2.4112
1/4	0.3335	2.9985	0.0594	16.8355	2.4947
3/8	0.3448	2.9000	0.0614	16.2821	2.5795
1/2	0.3563	2.8063	0.0635	15.7564	2.6656
5/8	0.3680	2.7173	0.0655	15.2565	2.7529
3/4	0.3799	2.6326	0.0677	14.7807	2.8415
7/8	0.3919	2.5518	0.0698	14.3275	2.9314

1 ONE STRING OF 1.660" OD TUBING

VOLUME AND FILL BETWEEN PIPE AND HOLE

HOLE DIAM (IN)	CU FT PER LIN FT	LIN FT PER CU FT	BARRELS PER LIN FT	LIN FT PER BARREL	GALLONS PER LIN FT
2	0.0068	147.3373	0.0012	827.2373	0.0508
1/8	0.0096	104.1727	0.0017	584.8861	0.0718
1/4	0.0126	79.4774	0.0022	446.2327	0.0941
3/8	0.0157	63.5511	0.0028	356.8129	0.1177
1/2	0.0191	52.4687	0.0034	294.5897	0.1426
5/8	0.0226	44.3399	0.0040	248.9499	0.1687
3/4	0.0262	38.1424	0.0047	214.1534	0.1961
7/8	0.0301	33.2751	0.0054	186.8257	0.2248
3	0.0341	29.3617	0.0061	164.8540	0.2548
1/8	0.0382	26.1549	0.0068	146.8489	0.2860
1/4	0.0426	23.4852	0.0076	131.8595	0.3185
3/8	0.0471	21.2329	0.0084	119.2138	0.3523
1/2	0.0518	19.3110	0.0092	108.4233	0.3874
5/8	0.0566	17.6549	0.0101	99.1249	0.4237
3/4	0.0617	16.2155	0.0110	91.0430	0.4613
7/8	0.0669	14.9548	0.0119	83.9651	0.5002
4	0.0722	13.8433	0.0129	77.7245	0.5404
1/8	0.0778	12.8574	0.0139	72.1888	0.5818
1/4	0.0835	11.9780	0.0149	67.2516	0.6245
3/8	0.0894	11.1899	0.0159	62.8265	0.6685
1/2	0.0954	10.4803	0.0170	58.8425	0.7138
5/8	0.1016	9.8388	0.0181	55.2408	0.7603
3/4	0.1080	9.2567	0.0192	51.9725	0.8081
7/8	0.1146	8.7266	0.0204	48.9963	0.8572
5	0.1213	8.2424	0.0216	46.2775	0.9076
1/8	0.1282	7.7987	0.0228	43.7862	0.9592
1/4	0.1353	7.3909	0.0241	41.4971	1.0121
3/8	0.1425	7.0154	0.0254	39.3883	1.0663
1/2	0.1500	6.6685	0.0267	37.4409	1.1218
5/8	0.1575	6.3475	0.0281	35.6383	1.1785
3/4	0.1653	6.0497	0.0294	33.9663	1.2365
7/8	0.1732	5.7729	0.0309	32.4123	1.2958
6	0.1813	5.5151	0.0323	30.9650	1.3564
1/8	0.1896	5.2746	0.0338	29.6149	1.4182
1/4	0.1980	5.0499	0.0353	28.3531	1.4813
3/8	0.2066	4.8396	0.0368	27.1721	1.5457
1/2	0.2154	4.6423	0.0384	26.0648	1.6114
5/8	0.2244	4.4572	0.0400	25.0252	1.6783
3/4	0.2335	4.2831	0.0416	24.0479	1.7465
7/8	0.2428	4.1192	0.0432	23.1277	1.8160
7	0.2522	3.9647	0.0449	22.2603	1.8868
1/8	0.2619	3.8189	0.0466	21.4417	1.9588
1/4	0.2717	3.6811	0.0484	20.6681	2.0321
3/8	0.2816	3.5508	0.0502	19.9364	2.1067
1/2	0.2918	3.4274	0.0520	19.2434	2.1826
5/8	0.3021	3.3104	0.0538	18.5865	2.2597
3/4	0.3126	3.1994	0.0557	17.9632	2.3381
7/8	0.3232	3.0939	0.0576	17.3711	2.4178

2 — TWO STRINGS OF 1.660" OD TUBING — VOLUME AND FILL BETWEEN PIPE AND HOLE

HOLE DIAM (IN)	CU FT PER LIN FT	LIN FT PER CU FT	BARRELS PER LIN FT	LIN FT PER BARREL	GALLONS PER LIN FT
4	0.0572	17.4802	0.0102	98.1441	0.4279
1/8	0.0627	15.9370	0.0112	89.4798	0.4694
1/4	0.0685	14.6078	0.0122	82.0165	0.5121
3/8	0.0743	13.4523	0.0132	75.5288	0.5561
1/2	0.0804	12.4397	0.0143	69.8438	0.6013
5/8	0.0866	11.5462	0.0154	64.8269	0.6479
3/4	0.0930	10.7526	0.0166	60.3716	0.6957
7/8	0.0996	10.0439	0.0177	56.3926	0.7448
5	0.1063	9.4078	0.0189	52.8208	0.7951
1/8	0.1132	8.8341	0.0202	49.5997	0.8468
1/4	0.1203	8.3145	0.0214	46.6827	0.8997
3/8	0.1275	7.8422	0.0227	44.0308	0.9539
1/2	0.1349	7.4113	0.0240	41.6113	1.0093
5/8	0.1425	7.0169	0.0254	39.3967	1.0661
3/4	0.1503	6.6547	0.0268	37.3635	1.1241
7/8	0.1582	6.3213	0.0282	35.4916	1.1834
6	0.1663	6.0136	0.0296	33.7637	1.2439
1/8	0.1746	5.7288	0.0311	32.1647	1.3058
1/4	0.1830	5.4647	0.0326	30.6818	1.3689
3/8	0.1916	5.2192	0.0341	29.3035	1.4333
1/2	0.2004	4.9905	0.0357	28.0198	1.4989
5/8	0.2093	4.7772	0.0373	26.8220	1.5659
3/4	0.2184	4.5778	0.0389	25.7024	1.6341
7/8	0.2277	4.3911	0.0406	24.6540	1.7036
7	0.2372	4.2159	0.0422	23.6708	1.7743
1/8	0.2468	4.0515	0.0440	22.7473	1.8464
1/4	0.2566	3.8967	0.0457	21.8785	1.9197
3/8	0.2666	3.7510	0.0475	21.0603	1.9943
1/2	0.2767	3.6135	0.0493	20.2885	2.0701
5/8	0.2870	3.4837	0.0511	19.5597	2.1473
3/4	0.2975	3.3610	0.0530	18.8706	2.2257
7/8	0.3082	3.2448	0.0549	18.2183	2.3054
8	0.3190	3.1347	0.0568	17.6002	2.3863
1/8	0.3300	3.0303	0.0588	17.0139	2.4686
1/4	0.3412	2.9311	0.0608	16.4571	2.5521
3/8	0.3525	2.8369	0.0628	15.9279	2.6369
1/2	0.3640	2.7472	0.0648	15.4245	2.7229
5/8	0.3757	2.6618	0.0669	14.9452	2.8103
3/4	0.3875	2.5805	0.0690	14.4883	2.8989
7/8	0.3995	2.5029	0.0712	14.0526	2.9888
9	0.4117	2.4288	0.0733	13.6366	3.0799
1/8	0.4241	2.3580	0.0755	13.2393	3.1724
1/4	0.4366	2.2904	0.0778	12.8594	3.2661
3/8	0.4493	2.2256	0.0800	12.4960	3.3611
1/2	0.4622	2.1637	0.0823	12.1481	3.4573
5/8	0.4752	2.1043	0.0846	11.8148	3.5549
3/4	0.4884	2.0474	0.0870	11.4952	3.6537
7/8	0.5018	1.9928	0.0894	11.1887	3.7538

3

THREE STRINGS	VOLUME AND FILL BETWEEN
OF	PIPE AND HOLE
1.660" OD	
TUBING	

HOLE DIAM (IN)	CU FT PER LIN FT	LIN FT PER CU FT	BARRELS PER LIN FT	LIN FT PER BARREL	GALLONS PER LIN FT
4	0.0422	23.7090	0.0075	133.1162	0.3155
1/8	0.0477	20.9567	0.0085	117.6631	0.3570
1/4	0.0534	18.7170	0.0095	105.0884	0.3997
3/8	0.0593	16.8613	0.0106	94.6690	0.4437
1/2	0.0654	15.3003	0.0116	85.9048	0.4889
5/8	0.0716	13.9705	0.0127	78.4386	0.5355
3/4	0.0780	12.8253	0.0139	72.0087	0.5833
7/8	0.0845	11.8297	0.0151	66.4189	0.6324
5	0.0913	10.9570	0.0163	61.5193	0.6827
1/8	0.0982	10.1866	0.0175	57.1934	0.7344
1/4	0.1052	9.5019	0.0187	53.3494	0.7873
3/8	0.1125	8.8900	0.0200	49.9138	0.8415
1/2	0.1199	8.3403	0.0214	46.8273	0.8969
5/8	0.1275	7.8441	0.0227	44.0413	0.9536
3/4	0.1352	7.3943	0.0241	41.5158	1.0117
7/8	0.1432	6.9849	0.0255	39.2175	1.0709
6	0.1513	6.6111	0.0269	37.1185	1.1315
1/8	0.1595	6.2685	0.0284	35.1951	1.1933
1/4	0.1680	5.9536	0.0299	33.4272	1.2565
3/8	0.1766	5.6634	0.0314	31.7977	1.3208
1/2	0.1853	5.3952	0.0330	30.2919	1.3865
5/8	0.1943	5.1467	0.0346	28.8968	1.4534
3/4	0.2034	4.9160	0.0362	27.6014	1.5217
7/8	0.2127	4.7013	0.0379	26.3960	1.5911
7	0.2222	4.5012	0.0396	25.2721	1.6619
1/8	0.2318	4.3142	0.0413	24.2222	1.7339
1/4	0.2416	4.1391	0.0430	23.2396	1.8073
3/8	0.2516	3.9751	0.0448	22.3185	1.8818
1/2	0.2617	3.8211	0.0466	21.4536	1.9577
5/8	0.2720	3.6762	0.0484	20.6404	2.0348
3/4	0.2825	3.5398	0.0503	19.8745	2.1133
7/8	0.2932	3.4112	0.0522	19.1523	2.1929
8	0.3040	3.2897	0.0541	18.4704	2.2739
1/8	0.3150	3.1749	0.0561	17.8257	2.3561
1/4	0.3261	3.0662	0.0581	17.2155	2.4397
3/8	0.3375	2.9632	0.0601	16.6373	2.5244
1/2	0.3490	2.8655	0.0622	16.0888	2.6105
5/8	0.3606	2.7728	0.0642	15.5680	2.6978
3/4	0.3725	2.6846	0.0663	15.0729	2.7865
7/8	0.3845	2.6007	0.0685	14.6019	2.8763
9	0.3967	2.5208	0.0707	14.1533	2.9675
1/8	0.4091	2.4447	0.0729	13.7257	3.0599
1/4	0.4216	2.3720	0.0751	13.3179	3.1537
3/8	0.4343	2.3027	0.0773	12.9285	3.2486
1/2	0.4471	2.2364	0.0796	12.5564	3.3449
5/8	0.4602	2.1730	0.0820	12.2006	3.4424
3/4	0.4734	2.1124	0.0843	11.8602	3.5413
7/8	0.4868	2.0543	0.0867	11.5342	3.6413

4   FOUR STRINGS OF 1.660″ OD TUBING   VOLUME AND FILL BETWEEN PIPE AND HOLE

HOLE DIAM (IN)	CU FT PER LIN FT	LIN FT PER CU FT	BARRELS PER LIN FT	LIN FT PER BARREL	GALLONS PER LIN FT
4	0.0271	36.8343	0.0048	206.8093	0.2031
⅛	0.0327	30.5923	0.0058	171.7630	0.2445
¼	0.0384	26.0432	0.0068	146.2215	0.2872
⅜	0.0443	22.5846	0.0079	126.8029	0.3312
½	0.0503	19.8694	0.0090	111.5582	0.3765
⅝	0.0565	17.6835	0.0101	99.2855	0.4230
¾	0.0629	15.8878	0.0112	89.2032	0.4708
⅞	0.0695	14.3878	0.0124	80.7813	0.5199
5	0.0762	13.1172	0.0136	73.6474	0.5703
⅛	0.0831	12.0281	0.0148	67.5326	0.6219
¼	0.0902	11.0850	0.0161	62.2375	0.6748
⅜	0.0975	10.2610	0.0174	57.6114	0.7290
½	0.1049	9.5356	0.0187	53.5384	0.7845
⅝	0.1125	8.8924	0.0200	49.9274	0.8412
¾	0.1202	8.3188	0.0214	46.7064	0.8992
⅞	0.1281	7.8042	0.0228	43.8175	0.9585
6	0.1362	7.3404	0.0243	41.2135	1.0191
⅛	0.1445	6.9205	0.0257	38.8558	1.0809
¼	0.1529	6.5387	0.0272	36.7122	1.1440
⅜	0.1615	6.1903	0.0288	34.7561	1.2084
½	0.1703	5.8713	0.0303	32.9649	1.2741
⅝	0.1793	5.5782	0.0319	31.3194	1.3410
¾	0.1884	5.3082	0.0336	29.8035	1.4092
⅞	0.1977	5.0588	0.0352	28.4029	1.4787
7	0.2071	4.8278	0.0369	27.1058	1.5495
⅛	0.2168	4.6133	0.0386	25.9016	1.6215
¼	0.2266	4.4137	0.0404	24.7812	1.6948
⅜	0.2365	4.2277	0.0421	23.7366	1.7694
½	0.2467	4.0539	0.0439	22.7608	1.8453
⅝	0.2570	3.8912	0.0458	21.8475	1.9224
¾	0.2675	3.7387	0.0476	20.9913	2.0008
⅞	0.2781	3.5955	0.0495	20.1873	2.0805
8	0.2889	3.4608	0.0515	19.4311	2.1615
⅛	0.2999	3.3340	0.0534	18.7189	2.2437
¼	0.3111	3.2143	0.0554	18.0472	2.3272
⅜	0.3224	3.1014	0.0574	17.4128	2.4120
½	0.3339	2.9945	0.0595	16.8129	2.4981
⅝	0.3456	2.8934	0.0616	16.2450	2.5854
¾	0.3575	2.7975	0.0637	15.7066	2.6740
⅞	0.3695	2.7065	0.0658	15.1958	2.7639
9	0.3817	2.6201	0.0680	14.7106	2.8551
⅛	0.3940	2.5379	0.0702	14.2493	2.9475
¼	0.4066	2.4597	0.0724	13.8102	3.0412
⅜	0.4193	2.3852	0.0747	13.3919	3.1362
½	0.4321	2.3142	0.0770	12.9931	3.2325
⅝	0.4452	2.2464	0.0793	12.6126	3.3300
¾	0.4584	2.1817	0.0816	12.2491	3.4288
⅞	0.4717	2.1198	0.0840	11.9017	3.5289

1

| ONE STRING OF 1.900″ OD TUBING | VOLUME AND FILL BETWEEN PIPE AND HOLE |

HOLE DIAM (IN)	CU FT PER LIN FT	LIN FT PER CU FT	BARRELS PER LIN FT	LIN FT PER BARREL	GALLONS PER LIN FT
3	0.0294	34.0160	0.0052	190.9859	0.2199
1/8	0.0336	29.7852	0.0060	167.2315	0.2511
1/4	0.0379	26.3713	0.0068	148.0639	0.2837
3/8	0.0424	23.5645	0.0076	132.3048	0.3174
1/2	0.0471	21.2207	0.0084	119.1452	0.3525
5/8	0.0520	19.2376	0.0093	108.0112	0.3888
3/4	0.0570	17.5409	0.0102	98.4850	0.4265
7/8	0.0622	16.0751	0.0111	90.2550	0.4653
4	0.0676	14.7979	0.0120	83.0843	0.5055
1/8	0.0731	13.6768	0.0130	76.7897	0.5469
1/4	0.0788	12.6861	0.0140	71.2274	0.5897
3/8	0.0847	11.8055	090151	66.2829	0.6336
1/2	0.0908	11.0184	0.0162	61.8638	0.6789
5/8	0.0970	10.3116	0.0173	57.8953	0.7254
3/4	0.1034	9.6740	0.0184	54.3155	0.7733
7/8	0.1099	9.0965	0.0196	51.0733	0.8223
5	0.1167	8.5716	0.0208	48.1260	0.8727
1/8	0.1236	8.0928	0.0220	45.4375	0.9243
1/4	0.1306	7.6546	0.0233	42.9773	0.9773
3/8	0.1379	7.2525	0.0246	40.7195	1.0314
1/2	0.1453	6.8824	0.0259	38.6417	1.0869
5/8	0.1529	6.5409	0.0272	36.7246	1.1436
3/4	0.1606	6.2252	0.0286	34.9517	1.2017
7/8	0.1686	5.9325	0.0300	33.3083	1.2609
6	0.1767	5.6606	0.0315	31.7819	1.3215
1/8	0.1849	5.4076	0.0329	30.3612	1.3833
1/4	0.1934	5.1716	0.0344	29.0364	1.4465
3/8	0.2020	4.9512	0.0360	27.7990	1.5108
1/2	0.2107	4.7450	0.0375	26.6412	1.5765
5/8	0.2197	4.5517	0.0391	25.5561	1.6434
3/4	0.2288	4.3703	0.0408	24.5376	1.7117
7/8	0.2381	4.1998	0.0424	23.5803	1.7811
7	0.2476	4.0394	0.0441	22.6793	1.8519
1/8	0.2572	3.8881	0.0458	21.8301	1.9239
1/4	0.2670	3.7454	0.0476	21.0288	1.9973
3/8	0.2770	3.6106	0.0493	20.2718	2.0718
1/2	0.2871	3.4830	0.0511	19.5557	2.1477
5/8	0.2974	3.3623	0.0530	18.8777	2.2248
3/4	0.3079	3.2478	0.0548	18.2351	2.3033
7/8	0.3186	3.1392	0.0567	17.6253	2.3829
8	0.3294	3.0360	0.0587	17.0461	2.4639
1/8	0.3404	2.9380	0.0606	16.4955	2.5461
1/4	0.3515	2.8447	0.0626	15.9717	2.6297
3/8	0.3629	2.7558	0.0646	15.4728	2.7144
1/2	0.3744	2.6711	0.0667	14.9973	2.8005
5/8	0.3860	2.5903	0.0688	14.5437	2.8878
3/4	0.3979	2.5132	0.0709	14.1107	2.9765
7/8	0.4099	2.4396	0.0730	13.6971	3.0663

2 — TWO STRINGS OF 1.900″ OD TUBING — VOLUME AND FILL BETWEEN PIPE AND HOLE

HOLE DIAM (IN)	CU FT PER LIN FT	LIN FT PER CU FT	BARRELS PER LIN FT	LIN FT PER BARREL	GALLONS PER LIN FT
5	0.0970	10.3120	0.0173	57.8973	0.7254
1/8	0.1039	9.6267	0.0185	54.0499	0.7771
1/4	0.1110	9.0130	0.0198	50.6041	0.8300
3/8	0.1182	8.4606	0.0211	47.5027	0.8842
1/2	0.1256	7.9612	0.0224	44.6988	0.9396
5/8	0.1332	7.5079	0.0237	42.1535	0.9964
3/4	0.1409	7.0948	0.0251	39.8342	1.0544
7/8	0.1489	6.7171	0.0265	37.7135	1.1137
6	0.1570	6.3706	0.0280	35.7684	1.1742
1/8	0.1652	6.0519	0.0294	33.9790	1.2361
1/4	0.1737	5.7579	0.0309	32.3283	1.2992
3/8	0.1823	5.4860	0.0325	30.8018	1.3636
1/2	0.1911	5.2340	0.0340	29.3866	1.4292
5/8	0.2000	4.9998	0.0356	28.0719	1.4962
3/4	0.2091	4.7818	0.0372	26.8479	1.5644
7/8	0.2184	4.5784	0.0389	25.7060	1.6339
7	0.2279	4.3884	0.0406	24.6389	1.7046
1/8	0.2375	4.2104	0.0423	23.6399	1.7767
1/4	0.2473	4.0436	0.0440	22.7031	1.8500
3/8	0.2573	3.8869	0.0458	21.8232	1.9246
1/2	0.2674	3.7395	0.0476	20.9956	2.0004
5/8	0.2777	3.6006	0.0495	20.2161	2.0776
3/4	0.2882	3.4697	0.0513	19.4808	2.1560
7/8	0.2989	3.3460	0.0532	18.7864	2.2357
8	0.3097	3.2291	0.0552	18.1299	2.3166
1/8	0.3207	3.1184	0.0571	17.5083	2.3989
1/4	0.3318	3.0135	0.0591	16.9193	2.4824
3/8	0.3432	2.9139	0.0611	16.3605	2.5672
1/2	0.3547	2.8194	0.0632	15.8298	2.6532
5/8	0.3664	2.7296	0.0653	15.3254	2.7406
3/4	0.3782	2.6441	0.0674	14.8454	2.8292
7/8	0.3902	2.5627	0.0695	14.3882	2.9191
9	0.4024	2.4850	0.0717	13.9525	3.0102
1/8	0.4148	2.4110	0.0739	13.5368	3.1027
1/4	0.4273	2.3403	0.0761	13.1399	3.1964
3/8	0.4400	2.2728	0.0784	12.7607	3.2914
1/2	0.4529	2.2082	0.0807	12.3981	3.3876
5/8	0.4659	2.1464	0.0830	12.0511	3.4852
3/4	0.4791	2.0872	0.0853	11.7189	3.5840
7/8	0.4925	2.0305	0.0877	11.4005	3.6841
10	0.5060	1.9761	0.0901	11.0952	3.7854
1/8	0.5198	1.9240	0.0926	10.8023	3.8881
1/4	0.5336	1.8739	0.0950	10.5211	3.9920
3/8	0.5477	1.8258	0.0976	10.2510	4.0972
1/2	0.5619	1.7795	0.1001	9.9914	4.2036
5/8	0.5763	1.7351	0.1027	9.7417	4.3114
3/4	0.5909	1.6923	0.1052	9.5015	4.4204
7/8	0.6057	1.6511	0.1079	9.2702	4.5307

3

THREE STRINGS OF 1.900″ OD TUBING

VOLUME AND FILL BETWEEN PIPE AND HOLE

HOLE DIAM (IN)	CU FT PER LIN FT	LIN FT PER CU FT	BARRELS PER LIN FT	LIN FT PER BARREL	GALLONS PER LIN FT
5	0.0773	12.9391	0.0138	72.6474	0.5781
⅛	0.0842	11.8781	0.0150	66.6908	0.6298
¼	0.0913	10.9575	0.0163	61.5218	0.6827
⅜	0.0985	10.1517	0.0175	56.9977	0.7369
½	0.1059	9.4411	0.0189	53.0079	0.7923
⅝	0.1135	8.8102	0.0202	49.4658	0.8491
¾	0.1213	8.2468	0.0216	46.3022	0.9071
⅞	0.1292	7.7408	0.0230	43.4616	0.9664
6	0.1373	7.2843	0.0245	40.8985	1.0269
⅛	0.1455	6.8706	0.0259	38.5756	1.0888
¼	0.1540	6.4942	0.0274	36.4620	1.1519
⅜	0.1626	6.1504	0.0290	34.5318	1.2163
½	0.1714	5.8353	0.0305	32.7630	1.2819
⅝	0.1803	5.5458	0.0321	31.1372	1.3489
¾	0.1894	5.2788	0.0337	29.6384	1.4171
⅞	0.1987	5.0321	0.0354	28.2530	1.4866
7	0.2082	4.8034	0.0371	26.9692	1.5573
⅛	0.2178	4.5911	0.0388	25.7768	1.6294
¼	0.2276	4.3934	0.0405	24.6670	1.7027
⅜	0.2376	4.2090	0.0423	23.6318	1.7773
½	0.2477	4.0367	0.0441	22.6643	1.8531
⅝	0.2580	3.8754	0.0460	21.7586	1.9303
¾	0.2685	3.7241	0.0478	20.9092	2.0087
⅞	0.2792	3.5820	0.0497	20.1114	2.0884
8	0.2900	3.4483	0.0517	19.3608	2.1693
⅛	0.3010	3.3224	0.0536	18.6537	2.2516
¼	0.3122	3.2035	0.0556	17.9865	2.3351
⅜	0.3235	3.0913	0.0576	17.3563	2.4199
½	0.3350	2.9851	0.0597	16.7602	2.5059
⅝	0.3467	2.8846	0.0617	16.1958	2.5933
¾	0.3585	2.7893	0.0639	15.6607	2.6819
⅞	0.3705	2.6988	0.0660	15.1528	2.7718
9	0.3827	2.6129	0.0682	14.6703	2.8629
⅛	0.3951	2.5312	0.0704	14.2114	2.9554
¼	0.4076	2.4534	0.0726	13.7747	3.0491
⅜	0.4203	2.3793	0.0749	13.3585	3.1441
½	0.4332	2.3086	0.0772	12.9616	3.2403
⅝	0.4462	2.2411	0.0795	12.5829	3.3379
¾	0.4594	2.1767	0.0818	12.2211	3.4367
⅞	0.4728	2.1151	0.0842	11.8753	3.5368
10	0.4863	2.0561	0.0866	11.5444	3.6381
⅛	0.5001	1.9997	0.0891	11.2277	3.7408
¼	0.5140	1.9457	0.0915	10.9242	3.8447
⅜	0.5280	1.8939	0.0940	10.6333	3.9499
½	0.5423	1.8442	0.0966	10.3542	4.0563
⅝	0.5567	1.7964	0.0991	10.0863	4.1641
¾	0.5712	1.7506	0.1017	9.8290	4.2731
⅞	0.5860	1.7066	0.1044	9.5817	4.3834

| 4 | FOUR STRINGS OF 1.900" OD TUBING | **VOLUME AND FILL BETWEEN PIPE AND HOLE** | | | |

HOLE DIAM (IN)	CU FT PER LIN FT	LIN FT PER CU FT	BARRELS PER LIN FT	LIN FT PER BARREL	GALLONS PER LIN FT
5	0.0576	17.3624	0.0103	97.4824	0.4308
⅛	0.0645	15.5042	0.0115	87.0495	0.4825
¼	0.0716	13.9719	0.0127	78.4465	0.5354
⅜	0.0788	12.6878	0.0140	71.2367	0.5896
½	0.0862	11.5969	0.0154	65.1116	0.6450
⅝	0.0938	10.6593	0.0167	59.8475	0.7018
¾	0.1016	9.8454	0.0181	55.2780	0.7598
⅞	0.1095	9.1328	0.0195	51.2768	0.8191
6	0.1176	8.5040	0.0209	47.7465	0.8796
⅛	0.1259	7.9455	0.0224	44.6105	0.9415
¼	0.1343	7.4463	0.0239	41.8079	1.0046
⅜	0.1429	6.9978	0.0255	39.2897	1.0690
½	0.1517	6.5928	0.0270	37.0160	1.1346
⅝	0.1606	6.2256	0.0286	34.9539	1.2016
¾	0.1697	5.8911	0.0302	33.0762	1.2698
⅞	0.1790	5.5855	0.0319	31.3601	1.3393
7	0.1885	5.3052	0.0336	29.7863	1.4100
⅛	0.1981	5.0473	0.0353	28.3385	1.4821
¼	0.2079	4.8094	0.0370	27.0028	1.5554
⅜	0.2179	4.5893	0.0388	25.7672	1.6300
½	0.2280	4.3852	0.0406	24.6212	1.7058
⅝	0.2383	4.1955	0.0425	23.5561	1.7830
¾	0.2488	4.0188	0.0443	22.5637	1.8614
⅞	0.2595	3.8538	0.0462	21.6374	1.9411
8	0.2703	3.6995	0.0481	20.7711	2.0220
⅛	0.2813	3.5549	0.0501	19.9593	2.1043
¼	0.2925	3.4192	0.0521	19.1974	2.1878
⅜	0.3038	3.2916	0.0541	18.4812	2.2726
½	0.3153	3.1715	0.0562	17.8069	2.3586
⅝	0.3270	3.0583	0.0582	17.1710	2.4460
¾	0.3388	2.9514	0.0603	16.5707	2.5346
⅞	0.3508	2.8503	0.0625	16.0032	2.6245
9	0.3630	2.7546	0.0647	15.4660	2.7156
⅛	0.3754	2.6639	0.0669	14.9568	2.8081
¼	0.3879	2.5779	0.0691	14.4738	2.9018
⅜	0.4006	2.4962	0.0714	14.0150	2.9968
½	0.4135	2.4185	0.0736	13.5789	3.0930
⅝	0.4265	2.3446	0.0760	13.1638	3.1906
¾	0.4397	2.2741	0.0783	12.7683	3.2894
⅞	0.4531	2.2070	0.0807	12.3913	3.3895
10	0.4667	2.1429	0.0831	12.0315	3.4908
⅛	0.4804	2.0817	0.0856	11.6878	3.5935
¼	0.4943	2.0232	0.0880	11.3594	3.6974
⅜	0.5083	1.9672	0.0905	11.0451	3.8026
½	0.5226	1.9136	0.0931	10.7443	3.9090
⅝	0.5370	1.8623	0.0956	10.4561	4.0168
¾	0.5515	1.8131	0.0982	10.1799	4.1258
⅞	0.5663	1.7659	0.1009	9.9148	4.2361

1

ONE STRING OF
2.000" OD
TUBING

VOLUME AND FILL BETWEEN PIPE AND HOLE

HOLE DIAM (IN)	CU FT PER LIN FT	LIN FT PER CU FT	BARRELS PER LIN FT	LIN FT PER BARREL	GALLONS PER LIN FT
3	0.0273	36.6693	0.0049	205.8828	0.2040
⅛	0.0314	31.7999	0.0056	178.5434	0.2352
¼	0.0358	27.9385	0.0064	156.8631	0.2677
⅜	0.0403	24.8080	0.0072	139.2865	0.3015
½	0.0450	22.2238	0.0080	124.7775	0.3366
⅝	0.0499	20.0584	0.0089	112.6197	0.3729
¾	0.0549	18.2208	0.0098	102.3020	0.4105
⅞	0.0601	16.6442	0.0107	93.4504	0.4494
4	0.0654	15.2789	0.0117	85.7845	0.4896
⅛	0.0710	14.0866	0.0126	79.0906	0.5310
¼	0.0767	13.0380	0.0137	73.2028	0.5737
⅜	0.0826	12.1096	0.0147	67.9902	0.6177
½	0.0886	11.2829	0.0158	63.3486	0.6630
⅝	0.0949	10.5428	0.0169	59.1936	0.7095
¾	0.1012	9.8773	0.0180	55.4567	0.7573
⅞	0.1078	9.2760	0.0192	52.0810	0.8064
5	0.1145	8.7308	0.0204	49.0197	0.8568
⅛	0.1214	8.2345	0.0216	46.2333	0.9084
¼	0.1285	7.7813	0.0229	43.6887	0.9613
⅜	0.1358	7.3661	0.0242	41.3575	1.0155
½	0.1432	6.9846	0.0255	39.2158	1.0710
⅝	0.1508	6.6332	0.0269	37.2428	1.1277
¾	0.1585	6.3087	0.0282	35.4207	1.1857
⅞	0.1664	6.0083	0.0296	33.7340	1.2450
6	0.1745	5.7296	0.0311	32.1692	1.3056
⅛	0.1828	5.4705	0.0326	30.7145	1.3674
¼	0.1912	5.2291	0.0341	29.3594	1.4305
⅜	0.1998	5.0039	0.0356	28.0949	1.4949
½	0.2086	4.7934	0.0372	26.9128	1.5606
⅝	0.2176	4.5962	0.0388	25.8059	1.6275
¾	0.2267	4.4113	0.0404	24.7679	1.6957
⅞	0.2360	4.2377	0.0420	23.7929	1.7652
7	0.2454	4.0744	0.0437	22.8759	1.8360
⅛	0.2551	3.9205	0.0454	22.0122	1.9080
¼	0.2649	3.7755	0.0472	21.1977	1.9813
⅜	0.2748	3.6385	0.0490	20.4287	2.0559
½	0.2850	3.5090	0.0508	19.7017	2.1318
⅝	0.2953	3.3865	0.0526	19.0137	2.2089
¾	0.3058	3.2704	0.0545	18.3619	2.2873
⅞	0.3164	3.1603	0.0564	17.7437	2.3670
8	0.3272	3.0558	0.0583	17.1569	2.4480
⅛	0.3382	2.9565	0.0602	16.5993	2.5302
¼	0.3494	2.8620	0.0622	16.0689	2.6137
⅜	0.3607	2.7721	0.0643	15.5640	2.6985
½	0.3722	2.6864	0.0663	15.0830	2.7846
⅝	0.3839	2.6047	0.0684	14.6243	2.8719
¾	0.3958	2.5267	0.0705	14.1866	2.9605
⅞	0.4078	2.4523	0.0726	13.7685	3.0504

2

TWO STRINGS OF 2.000" OD TUBING

VOLUME AND FILL BETWEEN PIPE AND HOLE

HOLE DIAM (IN)	CU FT PER LIN FT	LIN FT PER CU FT	BARRELS PER LIN FT	LIN FT PER BARREL	GALLONS PER LIN FT
5	0.0927	10.7851	0.0165	60.5538	0.6936
1/8	0.0996	10.0378	0.0177	56.3580	0.7452
1/4	0.1067	9.3723	0.0190	52.6218	0.7981
3/8	0.1139	8.7765	0.0203	49.2764	0.8523
1/2	0.1214	8.2403	0.0216	46.2658	0.9078
5/8	0.1289	7.7556	0.0230	43.5443	0.9645
3/4	0.1367	7.3156	0.0243	41.0739	1.0225
7/8	0.1446	6.9147	0.0258	38.8229	1.0818
6	0.1527	6.5481	0.0272	36.7648	1.1424
1/8	0.1610	6.2118	0.0287	34.8769	1.2042
1/4	0.1694	5.9025	0.0302	33.1401	1.2673
3/8	0.1780	5.6171	0.0317	31.5378	1.3317
1/2	0.1868	5.3532	0.0333	30.0559	1.3974
5/8	0.1958	5.1085	0.0349	28.6820	1.4643
3/4	0.2049	4.8811	0.0365	27.4054	1.5325
7/8	0.2142	4.6694	0.0381	26.2167	1.6020
7	0.2236	4.4719	0.0398	25.1077	1.6728
1/8	0.2333	4.2872	0.0415	24.0711	1.7448
1/4	0.2431	4.1144	0.0433	23.1005	1.8181
3/8	0.2530	3.9522	0.0451	22.1901	1.8927
1/2	0.2632	3.7999	0.0469	21.3350	1.9686
5/8	0.2735	3.6566	0.0487	20.5305	2.0457
3/4	0.2840	3.5217	0.0506	19.7727	2.1241
7/8	0.2946	3.3943	0.0525	19.0577	2.2038
8	0.3054	3.2740	0.0544	18.3824	2.2848
1/8	0.3164	3.1603	0.0564	17.7437	2.3670
1/4	0.3276	3.0526	0.0583	17.1390	2.4505
3/8	0.3389	2.9505	0.0604	16.5659	2.5353
1/2	0.3504	2.8536	0.0624	16.0220	2.6214
5/8	0.3621	2.7616	0.0645	15.5054	2.7087
3/4	0.3740	2.6742	0.0666	15.0142	2.7973
7/8	0.3860	2.5909	0.0687	14.5468	2.8872
9	0.3982	2.5116	0.0709	14.1016	2.9784
1/8	0.4105	2.4360	0.0731	13.6771	3.0708
1/4	0.4230	2.3639	0.0753	13.2721	3.1645
3/8	0.4357	2.2950	0.0776	12.8853	3.2595
1/2	0.4486	2.2291	0.0799	12.5157	3.3558
5/8	0.4616	2.1662	0.0822	12.1622	3.4533
3/4	0.4749	2.1059	0.0846	11.8239	3.5521
7/8	0.4882	2.0482	0.0870	11.4998	3.6522
10	0.5018	1.9929	0.0894	11.1893	3.7536
1/8	0.5155	1.9399	0.0918	10.8915	3.8562
1/4	0.5294	1.8890	0.0943	10.6057	3.9601
3/8	0.5435	1.8401	0.0968	10.3313	4.0653
1/2	0.5577	1.7931	0.0993	10.0676	4.1718
5/8	0.5721	1.7480	0.1019	9.8142	4.2795
3/4	0.5867	1.7046	0.1045	9.5704	4.3885
7/8	0.6014	1.6628	0.1071	9.3358	4.4988

3

THREE STRINGS OF 2.000" OD TUBING	**VOLUME AND FILL BETWEEN PIPE AND HOLE**

HOLE DIAM (IN)	CU FT PER LIN FT	LIN FT PER CU FT	BARRELS PER LIN FT	LIN FT PER BARREL	GALLONS PER LIN FT
5	0.0709	14.1036	0.0126	79.1857	0.5304
⅛	0.0778	12.8523	0.0139	72.1605	0.5820
¼	0.0849	11.7813	0.0151	66.1471	0.6349
⅜	0.0921	10.8549	0.0164	60.9459	0.6891
½	0.0995	10.0464	0.0177	56.4063	0.7446
⅝	0.1071	9.3351	0.0191	52.4125	0.8013
¾	0.1149	8.7049	0.0205	48.8743	0.8593
⅞	0.1228	8.1431	0.0219	45.7200	0.9186
6	0.1309	7.6394	0.0233	42.8923	0.9792
⅛	0.1392	7.1857	0.0248	40.3445	1.0410
¼	0.1476	6.7749	0.0263	38.0384	1.1041
⅜	0.1562	6.4016	0.0278	35.9424	1.1685
½	0.1650	6.0610	0.0294	34.0302	1.2342
⅝	0.1739	5.7492	0.0310	32.2795	1.3011
¾	0.1831	5.4628	0.0326	30.6716	1.3693
⅞	0.1923	5.1990	0.0343	29.1903	1.4388
7	0.2018	4.9553	0.0359	27.8220	1.5096
⅛	0.2114	4.7296	0.0377	26.5548	1.5816
¼	0.2212	4.5201	0.0394	25.3785	1.6549
⅜	0.2312	4.3252	0.0412	24.2840	1.7295
½	0.2413	4.1434	0.0430	23.2636	1.8054
⅝	0.2517	3.9736	0.0448	22.3104	1.8825
¾	0.2621	3.8148	0.0467	21.4182	1.9609
⅞	0.2728	3.6658	0.0486	20.5819	2.0406
8	0.2836	3.5259	0.0505	19.7964	2.1216
⅛	0.2946	3.3943	0.0525	19.0577	2.2038
¼	0.3058	3.2704	0.0545	18.3619	2.2873
⅜	0.3171	3.1535	0.0565	17.7056	2.3721
½	0.3286	3.0431	0.0585	17.0857	2.4582
⅝	0.3403	2.9387	0.0606	16.4995	2.5455
¾	0.3521	2.8398	0.0627	15.9445	2.6341
⅞	0.3641	2.7461	0.0649	15.4183	2.7240
9	0.3763	2.6572	0.0670	14.9190	2.8152
⅛	0.3887	2.5727	0.0692	14.4448	2.9076
¼	0.4012	2.4924	0.0715	13.9937	3.0013
⅜	0.4139	2.4159	0.0737	13.5644	3.0963
½	0.4268	2.3431	0.0760	13.1555	3.1926
⅝	0.4398	2.2736	0.0783	12.7655	3.2901
¾	0.4530	2.2073	0.0807	12.3932	3.3889
⅞	0.4664	2.1440	0.0831	12.0377	3.4890
10	0.4800	2.0835	0.0855	11.6979	3.5904
⅛	0.4937	2.0256	0.0879	11.3728	3.6930
¼	0.5076	1.9701	0.0904	11.0615	3.7969
⅜	0.5216	1.9170	0.0929	10.7634	3.9021
½	0.5359	1.8661	0.0954	10.4775	4.0086
⅝	0.5503	1.8173	0.0980	10.2033	4.1163
¾	0.5648	1.7704	0.1006	9.9400	4.2253
⅞	0.5796	1.7254	0.1032	9.6872	4.3356

4

FOUR STRINGS OF
2.000" OD
TUBING

VOLUME AND FILL BETWEEN
PIPE AND HOLE

HOLE DIAM (IN)	CU FT PER LIN FT	LIN FT PER CU FT	BARRELS PER LIN FT	LIN FT PER BARREL	GALLONS PER LIN FT
5	0.0491	20.3718	0.0087	114.3794	0.3672
⅛	0.0560	17.8602	0.0100	100.2778	0.4188
¼	0.0631	15.8570	0.0112	89.0304	0.4717
⅜	0.0703	14.2232	0.0125	79.8576	0.5259
½	0.0777	12.8664	0.0138	72.2396	0.5814
⅝	0.0853	11.7225	0.0152	65.8167	0.6381
¾	0.0931	10.7456	0.0166	60.3320	0.6961
⅞	0.1010	9.9023	0.0180	55.5971	0.7554
6	0.1091	9.1673	0.0194	51.4707	0.8160
⅛	0.1173	8.5216	0.0209	47.8450	0.8778
¼	0.1258	7.9500	0.0224	44.6358	0.9409
⅜	0.1344	7.4408	0.0239	41.7771	1.0053
½	0.1432	6.9846	0.0255	39.2158	1.0710
⅝	0.1521	6.5738	0.0271	36.9090	1.1379
¾	0.1612	6.2020	0.0287	34.8216	1.2061
⅞	0.1705	5.8642	0.0304	32.9248	1.2756
7	0.1800	5.5560	0.0321	31.1944	1.3464
⅛	0.1896	5.2738	0.0338	29.6101	1.4184
¼	0.1994	5.0146	0.0355	28.1549	1.4917
⅜	0.2094	4.7758	0.0373	26.8142	1.5663
½	0.2195	4.5552	0.0391	25.5755	1.6422
⅝	0.2298	4.3508	0.0409	24.4281	1.7193
¾	0.2403	4.1611	0.0428	23.3626	1.7977
⅞	0.2510	3.9844	0.0447	22.3710	1.8774
8	0.2618	3.8197	0.0466	21.4461	1.9584
⅛	0.2728	3.6658	0.0486	20.5819	2.0406
¼	0.2840	3.5217	0.0506	19.7727	2.1241
⅜	0.2953	3.3865	0.0526	19.0137	2.2089
½	0.3068	3.2595	0.0546	18.3007	2.2950
⅝	0.3185	3.1400	0.0567	17.6298	2.3823
¾	0.3303	3.0274	0.0588	16.9976	2.4709
⅞	0.3423	2.9211	0.0610	16.4009	2.5608
9	0.3545	2.8207	0.0631	15.8371	2.6520
⅛	0.3669	2.7257	0.0653	15.3037	2.7444
¼	0.3794	2.6357	0.0676	14.7984	2.8381
⅜	0.3921	2.5504	0.0698	14.3192	2.9331
½	0.4050	2.4693	0.0721	13.8642	3.0294
⅝	0.4180	2.3923	0.0745	13.4317	3.1269
¾	0.4312	2.3190	0.0768	13.0203	3.2257
⅞	0.4446	2.2492	0.0792	12.6284	3.3258
10	0.4581	2.1827	0.0816	12.2549	3.4272
⅛	0.4719	2.1192	0.0840	11.8986	3.5298
¼	0.4858	2.0586	0.0865	11.5583	3.6337
⅜	0.4998	2.0007	0.0890	11.2332	3.7389
½	0.5141	1.9453	0.0916	10.9222	3.8454
⅝	0.5285	1.8923	0.0941	10.6245	3.9531
¾	0.5430	1.8415	0.0967	10.3394	4.0621
⅞	0.5578	1.7928	0.0993	10.0661	4.1724

1

ONE STRING OF
2-1/16″ OD
TUBING

VOLUME AND FILL BETWEEN
PIPE AND HOLE

HOLE DIAM (IN)	CU FT PER LIN FT	LIN FT PER CU FT	BARRELS PER LIN FT	LIN FT PER BARREL	GALLONS PER LIN FT
3	0.0259	38.6478	0.0046	216.9914	0.1936
1/8	0.0301	33.2773	0.0054	186.8382	0.2248
1/4	0.0344	29.0725	0.0061	163.2299	0.2573
3/8	0.0389	25.6980	0.0069	144.2836	0.2911
1/2	0.0436	22.9354	0.0078	128.7729	0.3262
5/8	0.0485	20.6363	0.0086	115.8643	0.3625
3/4	0.0535	18.6964	0.0095	104.9723	0.4001
7/8	0.0587	17.0402	0.0105	95.6735	0.4390
4	0.0641	15.6119	0.0114	87.6542	0.4792
1/8	0.0696	14.3692	0.0124	80.6773	0.5206
1/4	0.0753	13.2797	0.0134	74.5599	0.5633
3/8	0.0812	12.3178	0.0145	69.1594	0.6073
1/2	0.0872	11.4634	0.0155	64.3624	0.6526
5/8	0.0935	10.7003	0.0166	60.0779	0.6991
3/4	0.0998	10.0154	0.0178	56.2321	0.7469
7/8	0.1064	9.3977	0.0190	52.7643	0.7960
5	0.1131	8.8385	0.0202	49.6246	0.8464
1/8	0.1200	8.3303	0.0214	46.7710	0.8980
1/4	0.1271	7.8667	0.0226	44.1685	0.9509
3/8	0.1344	7.4426	0.0239	41.7872	1.0051
1/2	0.1418	7.0534	0.0253	39.6019	1.0606
5/8	0.1494	6.6952	0.0266	37.5909	1.1173
3/4	0.1571	6.3648	0.0280	35.7354	1.1753
7/8	0.1650	6.0591	0.0294	34.0194	1.2346
6	0.1731	5.7758	0.0308	32.4286	1.2952
1/8	0.1814	5.5126	0.0323	30.9508	1.3570
1/4	0.1898	5.2676	0.0338	29.5753	1.4201
3/8	0.1984	5.0391	0.0353	28.2925	1.4845
1/2	0.2072	4.8257	0.0369	27.0941	1.5502
5/8	0.2162	4.6259	0.0385	25.9726	1.6171
3/4	0.2253	4.4387	0.0401	24.9213	1.6853
7/8	0.2346	4.2629	0.0418	23.9345	1.7548
7	0.2440	4.0977	0.0435	23.0067	1.8256
1/8	0.2537	3.9421	0.0452	22.1333	1.8976
1/4	0.2635	3.7955	0.0469	21.3100	1.9709
3/8	0.2734	3.6571	0.0487	20.5330	2.0455
1/2	0.2836	3.5263	0.0505	19.7987	2.1214
5/8	0.2939	3.4026	0.0523	19.1040	2.1985
3/4	0.3044	3.2854	0.0542	18.4461	2.2769
7/8	0.3150	3.1743	0.0561	17.8224	2.3566
8	0.3259	3.0689	0.0580	17.2304	2.4376
1/8	0.3368	2.9687	0.0600	16.6681	2.5198
1/4	0.3480	2.8735	0.0620	16.1334	2.6033
3/8	0.3593	2.7828	0.0640	15.6245	2.6881
1/2	0.3708	2.6965	0.0661	15.1398	2.7741
5/8	0.3825	2.6142	0.0681	14.6777	2.8615
3/4	0.3944	2.5357	0.0702	14.2368	2.9501
7/8	0.4064	2.4607	0.0724	13.8158	3.0400

2 — TWO STRINGS OF 2-1/16" OD TUBING — VOLUME AND FILL BETWEEN PIPE AND HOLE

HOLE DIAM (IN)	CU FT PER LIN FT	LIN FT PER CU FT	BARRELS PER LIN FT	LIN FT PER BARREL	GALLONS PER LIN FT
5	0.0899	11.1200	0.0160	62.4339	0.6727
1/8	0.0968	10.3272	0.0172	57.9831	0.7243
1/4	0.1039	9.6242	0.0185	54.0359	0.7773
3/8	0.1111	8.9970	0.0198	50.5143	0.8314
1/2	0.1186	9.4344	0.0211	47.3554	0.8869
5/8	0.1261	7.9272	0.0225	44.5081	0.9436
3/4	0.1339	7.4681	0.0238	41.9304	1.0017
7/8	0.1418	7.0508	0.0253	39.5872	1.0609
6	0.1499	6.6700	0.0267	37.4495	1.1215
1/8	0.1582	6.3215	0.0282	35.4925	1.1833
1/4	0.1666	6.0014	0.0297	33.6954	1.2465
3/8	0.1752	5.7066	0.0312	32.0403	1.3108
1/2	0.1840	5.4344	0.0328	30.5120	1.3765
5/8	0.1930	5.1824	0.0344	29.0970	1.4434
3/4	0.2021	4.9485	0.0360	27.7840	1.5117
7/8	0.2114	4.7311	0.0376	26.5630	1.5811
7	0.2208	4.5284	0.0393	25.4251	1.6519
1/8	0.2305	4.3392	0.0410	24.3627	1.7239
1/4	0.2403	4.1622	0.0428	23.3689	1.7973
1/4	0.2403	4.1622	0.0428	23.3689	1.7973
3/8	0.2502	3.9963	0.0446	22.4377	1.8718
1/2	0.2604	3.8407	0.0464	21.5638	1.9477
5/8	0.2707	3.6944	0.0482	20.7423	2.0248
3/4	0.2812	3.5566	0.0501	19.9690	2.1033
7/8	0.2918	3.4268	0.0520	19.2401	2.1829
8	0.3026	3.3043	0.0539	18.5520	2.2639
1/8	0.3136	3.1884	0.0559	17.9017	2.3461
1/4	0.3248	3.0788	0.0578	17.2864	2.4297
3/8	0.3361	2.9750	0.0599	16.7035	2.5144
1/2	0.3476	2.8766	0.0619	16.1507	2.6005
5/8	0.3593	2.7831	0.0640	15.6259	2.6878
3/4	0.3712	2.6943	0.0661	15.1272	2.7765
7/8	0.3832	2.6098	0.0682	14.6528	2.8663
9	0.3954	2.5293	0.0704	14.2012	2.9575
1/8	0.4077	2.4527	0.0726	13.7707	3.0499
1/4	0.4202	2.3796	0.0748	13.3602	3.1437
3/8	0.4329	2.3098	0.0771	12.9684	3.2386
1/2	0.4458	2.2431	0.0794	12.5941	3.3349
5/8	0.4589	2.1794	0.0817	12.2362	3.4324
3/4	0.4721	2.1184	0.0841	11.8938	3.5313
7/8	0.4854	2.0600	0.0865	11.5660	3.6313
10	0.4990	2.0040	0.0889	11.2519	3.7327
1/8	0.5127	1.9504	0.0913	10.9508	3.8353
1/4	0.5266	1.8990	0.0938	10.6619	3.9393
3/8	0.5407	1.8496	0.0963	10.3846	4.0444
1/2	0.5549	1.8021	0.0988	10.1183	4.1509
5/8	0.5693	1.7566	0.1014	9.8623	4.2586
3/4	0.5839	1.7127	0.1040	9.6161	4.3677
7/8	0.5986	1.6705	0.1066	9.3793	4.4779

3 — THREE STRINGS OF 2-¹⁄₁₆″ OD TUBING — VOLUME AND FILL BETWEEN PIPE AND HOLE

HOLE DIAM (IN)	CU FT PER LIN FT	LIN FT PER CU FT	BARRELS PER LIN FT	LIN FT PER BARREL	GALLONS PER LIN FT
5	0.0667	14.9890	0.0119	84.1568	0.4991
⅛	0.0736	13.5835	0.0131	76.2658	0.5507
¼	0.0807	12.3928	0.0144	69.5804	0.6036
⅜	0.0879	11.3719	0.0157	63.8487	0.6578
½	0.0954	10.4877	0.0170	58.8839	0.7133
⅝	0.1029	9.7149	0.0183	54.5451	0.7700
¾	0.1107	9.0343	0.0197	50.7236	0.8280
⅞	0.1186	8.4306	0.0211	47.3344	0.8873
6	0.1267	7.8919	0.0226	44.3100	0.9479
⅛	0.1350	7.4086	0.0240	41.5963	1.0097
¼	0.1434	6.9728	0.0255	39.1493	1.0728
⅜	0.1520	6.5780	0.0271	36.9327	1.1372
½	0.1608	6.2189	0.0286	34.9166	1.2029
⅝	0.1697	5.8911	0.0302	33.0760	1.2698
¾	0.1789	5.5908	0.0319	31.3898	1.3380
⅞	0.1882	5.3147	0.0335	29.8401	1.4075
7	0.1976	5.0603	0.0352	28.4117	1.4783
⅛	0.2072	4.8252	0.0369	27.0915	1.5503
¼	0.2170	4.6073	0.0387	25.8682	1.6236
⅜	0.2270	4.4050	0.0404	24.7320	1.6982
½	0.2372	4.2166	0.0422	23.6744	1.7741
⅝	0.2475	4.0409	0.0441	22.6880	1.8512
¾	0.2580	3.8767	0.0459	21.7660	1.9296
⅞	0.2686	3.7229	0.0478	20.9028	2.0093
8	0.2794	3.5787	0.0498	20.0932	2.0903
⅛	0.2904	3.4433	0.0517	19.3325	2.1725
¼	0.3016	3.3158	0.0537	18.6169	2.2560
⅜	0.3129	3.1957	0.0557	17.9426	2.3408
½	0.3244	3.0824	0.0578	17.3063	2.4269
⅝	0.3361	2.9753	0.0599	16.7051	2.5142
¾	0.3479	2.8740	0.0620	16.1364	2.6028
⅞	0.3600	2.7781	0.0641	15.5977	2.6927
9	0.3721	2.6871	0.0663	15.0869	2.7839
⅛	0.3845	2.6007	0.0685	14.6021	2.8763
¼	0.3970	2.5187	0.0707	14.1414	2.9700
⅜	0.4097	2.4406	0.0730	13.7031	3.0650
½	0.4226	2.3663	0.0753	13.2858	3.1613
⅝	0.4356	2.2955	0.0776	12.8882	3.2588
¾	0.4488	2.2279	0.0799	12.5089	3.3576
⅞	0.4622	2.1634	0.0823	12.1468	3.4577
10	0.4758	2.1018	0.0847	11.8009	3.5591
⅛	0.4895	2.0429	0.0872	11.4701	3.6617
¼	0.5034	1.9865	0.0897	11.1536	3.7656
⅜	0.5175	1.9326	0.0922	10.8505	3.8708
½	0.5317	1.8808	0.0947	10.5600	3.9773
⅝	0.5461	1.8312	0.0973	10.2815	4.0850
¾	0.5607	1.7836	0.0999	10.0143	4.1940
⅞	0.5754	1.7379	0.1025	9.7577	4.3043

4

FOUR STRINGS OF
2-¹⁄₁₆″ OD
TUBING

VOLUME AND FILL BETWEEN
PIPE AND HOLE

HOLE DIAM (IN)	CU FT PER LIN FT	LIN FT PER CU FT	BARRELS PER LIN FT	LIN FT PER BARREL	GALLONS PER LIN FT
5	0.0435	22.9869	0.0077	129.0620	0.3254
⅛	0.0504	19.8389	0.0090	111.3874	0.3771
¼	0.0575	17.3976	0.0102	97.6801	0.4300
⅜	0.0647	15.4504	0.0115	86.7478	0.4842
½	0.0721	13.8625	0.0128	77.8319	0.5396
⅝	0.0797	12.5436	0.0142	70.4270	0.5964
¾	0.0875	11.4316	0.0156	64.1834	0.6544
⅞	0.0954	10.4819	0.0170	58.8514	0.7137
6	0.1035	9.6620	0.0184	54.2479	0.7742
⅛	0.1118	8.9473	0.0199	50.2355	0.8361
¼	0.1202	8.3193	0.0214	46.7095	0.8992
⅜	0.1288	7.7634	0.0229	43.5883	0.9636
½	0.1376	7.2681	0.0245	40.8075	1.0292
⅝	0.1465	6.8243	0.0261	38.3155	1.0962
¾	0.1557	6.4245	0.0277	36.0709	1.1644
⅞	0.1649	6.0627	0.0294	34.0395	1.2339
7	0.1744	5.7339	0.0311	32.1932	1.3046
⅛	0.1840	5.4338	0.0328	30.5086	1.3767
¼	0.1938	5.1591	0.0345	28.9661	1.4500
⅜	0.2038	4.9067	0.0363	27.5489	1.5246
½	0.2139	4.6741	0.0381	26.2431	1.6004
⅝	0.2243	4.4592	0.0399	25.0364	1.6776
¾	0.2347	4.2600	0.0418	23.9184	1.7560
⅞	0.2454	4.0751	0.0437	22.8801	1.8357
8	0.2562	3.9030	0.0456	21.9136	1.9166
⅛	0.2672	3.7424	0.0476	21.0120	1.9989
¼	0.2784	3.5923	0.0496	20.1693	2.0824
⅜	0.2897	3.4518	0.0516	19.3802	2.1672
½	0.3012	3.3199	0.0536	18.6400	2.2532
⅝	0.3129	3.1960	0.0557	17.9444	2.3406
¾	0.3247	3.0795	0.0578	17.2899	2.4292
⅞	0.3367	2.9696	0.0600	16.6729	2.5191
9	0.3489	2.8659	0.0621	16.0906	2.6102
⅛	0.3613	2.7678	0.0643	15.5403	2.7027
¼	0.3738	2.6751	0.0666	15.0195	2.7964
⅜	0.3865	2.5872	0.0688	14.5261	2.8914
½	0.3994	2.5038	0.0711	14.0580	2.9876
⅝	0.4124	2.4247	0.0735	13.6136	3.0852
¾	0.4256	2.3494	0.0758	13.1911	3.1840
⅞	0.4390	2.2778	0.0782	12.7891	3.2841
10	0.4526	2.2096	0.0806	12.4061	3.3854
⅛	0.4663	2.1446	0.0830	12.0411	3.4881
¼	0.4802	2.0826	0.0855	11.6928	3.5920
⅜	0.4942	2.0233	0.0880	11.3601	3.6972
½	0.5085	1.9667	0.0906	11.0421	3.8036
⅝	0.5229	1.9125	0.0931	10.7380	3.9114
¾	0.5374	1.8607	0.0957	10.4468	4.0204
⅞	0.5522	1.8110	0.0983	10.1679	4.1307

1

ONE STRING OF 2-⅜" OD TUBING

VOLUME AND FILL BETWEEN PIPE AND HOLE

HOLE DIAM (IN)	CU FT PER LIN FT	LIN FT PER CU FT	BARRELS PER LIN FT	LIN FT PER BARREL	GALLONS PER LIN FT
3	0.0183	54.5776	0.0033	306.4303	0.1371
⅛	0.0225	44.4476	0.0040	249.5549	0.1683
¼	0.0268	37.2514	0.0048	209.1508	0.2008
⅜	0.0314	31.8863	0.0056	179.0286	0.2346
½	0.0360	27.7404	0.0064	155.7506	0.2697
⅝	0.0409	24.4462	0.0073	137.2552	0.3060
¾	0.0459	21.7703	0.0082	122.2310	0.3436
⅞	0.0511	19.5570	0.0091	109.8042	0.3825
4	0.0565	17.6986	0.0101	99.3703	0.4227
⅛	0.0620	16.1184	0.0110	90.4979	0.4641
¼	0.0678	14.7600	0.0121	82.8711	0.5068
⅜	0.0736	13.5812	0.0131	76.2529	0.5508
½	0.0797	12.5499	0.0142	70.4626	0.5961
⅝	0.0859	11.6410	0.0153	65.3596	0.6426
¾	0.0923	10.8349	0.0164	60.8333	0.6904
⅞	0.0989	10.1157	0.0176	56.7953	0.7395
5	0.1056	9.4707	0.0188	53.1739	0.7899
⅛	0.1125	8.8895	0.0200	49.9110	0.8415
¼	0.1196	8.3636	0.0213	46.9583	0.8944
⅜	0.1268	7.8859	0.0226	44.2759	0.9486
½	0.1342	7.4503	0.0239	41.8302	1.0041
⅝	0.1418	7.0518	0.0253	39.5929	1.0608
¾	0.1496	6.6861	0.0266	37.5399	1.1188
⅞	0.1575	6.3497	0.0280	35.6507	1.1781
6	0.1656	6.0392	0.0295	33.9076	1.2387
⅛	0.1739	5.7520	0.0310	32.2953	1.3005
¼	0.1823	5.4858	0.0325	30.8006	1.3636
⅜	0.1909	5.2385	0.0340	29.4118	1.4280
½	0.1997	5.0082	0.0356	28.1189	1.4937
⅝	0.2086	4.7934	0.0372	26.9128	1.5606
¾	0.2177	4.5926	0.0388	25.7857	1.6288
⅞	0.2270	4.4047	0.0404	24.7307	1.6983
7	0.2365	4.2285	0.0421	23.7414	1.7691
⅛	0.2461	4.0631	0.0438	22.8125	1.8411
¼	0.2559	3.9075	0.0456	21.9389	1.9144
⅜	0.2659	3.7610	0.0474	21.1162	1.9890
½	0.2760	3.6228	0.0492	20.3404	2.0649
⅝	0.2863	3.4923	0.0510	19.6079	2.1420
¾	0.2968	3.3690	0.0529	18.9154	2.2204
⅞	0.3075	3.2523	0.0548	18.2601	2.3001
8	0.3183	3.1417	0.0567	17.6392	2.3811
⅛	0.3293	3.0368	0.0586	17.0503	2.4633
¼	0.3405	2.9372	0.0606	16.4912	2.5468
⅜	0.3518	2.8426	0.0627	15.9599	2.6316
½	0.3633	2.7526	0.0647	15.4545	2.7177
⅝	0.3750	2.6669	0.0668	14.9733	2.8050
¾	0.3868	2.5852	0.0689	14.5148	2.8936
⅞	0.3988	2.5073	0.0710	14.0775	2.9835

2

TWO STRINGS OF
2-⅜" OD
TUBING

VOLUME AND FILL BETWEEN
PIPE AND HOLE

HOLE DIAM (IN)	CU FT PER LIN FT	LIN FT PER CU FT	BARRELS PER LIN FT	LIN FT PER BARREL	GALLONS PER LIN FT
6	0.1348	7.4173	0.0240	41.6451	1.0085
⅛	0.1431	6.9888	0.0255	39.2391	1.0704
¼	0.1515	6.5996	0.0270	37.0543	1.1335
⅜	0.1601	6.2449	0.0285	35.0625	1.1979
½	0.1689	5.9204	0.0301	33.2404	1.2635
⅝	0.1779	5.6225	0.0317	31.5680	1.3305
¾	0.1870	5.3483	0.0333	30.0285	1.3987
⅞	0.1963	5.0952	0.0350	28.6073	1.4682
7	0.2057	4.8609	0.0366	27.2918	1.5389
⅛	0.2154	4.6435	0.0384	26.0714	1.6110
¼	0.2252	4.4414	0.0401	24.9366	1.6843
⅜	0.2351	4.2531	0.0419	23.8791	1.7589
½	0.2453	4.0772	0.0437	22.8918	1.8347
⅝	0.2556	3.9127	0.0455	21.9682	1.9119
¾	0.2661	3.7585	0.0474	21.1027	1.9903
⅞	0.2767	3.6139	0.0493	20.2903	2.0700
8	0.2875	3.4778	0.0512	19.5265	2.1509
⅛	0.2985	3.3498	0.0532	18.8075	2.2332
¼	0.3097	3.2290	0.0552	18.1295	2.3167
⅜	0.3210	3.1150	0.0572	17.4894	2.4015
½	0.3325	3.0072	0.0592	16.8843	2.4875
⅝	0.3442	2.9052	0.0613	16.3116	2.5749
¾	0.3561	2.8086	0.0634	15.7689	2.6635
⅞	0.3681	2.7169	0.0656	15.2541	2.7534
9	0.3803	2.6298	0.0677	14.7652	2.8445
⅛	0.3926	2.5470	0.0699	14.3005	2.9379
¼	0.4051	2.4683	0.0722	13.8583	3.0307
⅜	0.4178	2.3933	0.0744	13.4372	3.1257
½	0.4307	2.3218	0.0767	13.0357	3.2219
⅝	0.4437	2.2535	0.0790	12.6527	3.3195
¾	0.4570	2.1884	0.0814	12.2869	3.4183
⅞	0.4703	2.1261	0.0838	11.9374	3.5184
10	0.4839	2.0666	0.0862	11.6031	3.6197
⅛	0.4976	2.0096	0.0886	11.2832	3.7224
¼	0.5115	1.9550	0.0911	10.9768	3.8263
⅜	0.5256	1.9027	0.0936	10.6831	3.9315
½	0.5398	1.8526	0.0961	10.4014	4.0379
⅝	0.5542	1.8044	0.0987	10.1311	4.1457
¾	0.5688	1.7582	0.1013	9.8715	4.2547
⅞	0.5835	1.7138	0.1039	9.6221	4.3650
11	0.5984	1.6711	0.1066	9.3823	4.4765
⅛	0.6135	1.6300	0.1093	9.1516	4.5894
¼	0.6288	1.5904	0.1120	8.9296	4.7035
⅜	0.6442	1.5523	0.1147	8.7158	4.8189
½	0.6598	1.5157	0.1175	8.5098	4.9355
⅝	0.6755	1.4803	0.1203	8.3112	5.0535
¾	0.6915	1.4462	0.1232	8.1196	5.1727
⅞	0.7076	1.4132	0.1260	7.9348	5.2932

	THREE STRINGS	VOLUME AND FILL BETWEEN

3 THREE STRINGS OF 2-⅜″ OD TUBING **VOLUME AND FILL BETWEEN PIPE AND HOLE**

HOLE DIAM (IN)	CU FT PER LIN FT	LIN FT PER CU FT	BARRELS PER LIN FT	LIN FT PER BARREL	GALLONS PER LIN FT
6	0.1041	9.6103	0.0185	53.9578	0.7784
⅛	0.1123	8.9030	0.0200	49.9867	0.8402
¼	0.1208	8.2810	0.0215	46.4944	0.9033
⅜	0.1294	7.7300	0.0230	43.4009	0.9677
½	0.1381	7.2388	0.0246	40.6431	1.0334
⅝	0.1471	6.7985	0.0262	38.1706	1.1003
¾	0.1562	6.4016	0.0278	35.9424	1.1685
⅞	0.1655	6.0423	0.0295	33.9251	1.2380
7	0.1750	5.7156	0.0312	32.0908	1.3088
⅛	0.1846	5.4174	0.0329	30.4167	1.3808
¼	0.1944	5.1443	0.0346	28.8832	1.4541
⅜	0.2044	4.8933	0.0364	27.4739	1.5287
½	0.2145	4.6620	0.0382	26.1750	1.6046
⅝	0.2248	4.4481	0.0400	24.9744	1.6817
¾	0.2353	4.2500	0.0419	23.8618	1.7601
⅞	0.2459	4.0659	0.0438	22.8283	1.8398
8	0.2568	3.8945	0.0457	21.8661	1.9208
⅛	0.2678	3.7346	0.0477	20.9683	2.0030
¼	0.2789	3.5851	0.0497	20.1291	2.0865
⅜	0.2903	3.4451	0.0517	19.3431	2.1713
½	0.3018	3.3148	0.0537	18.6056	2.2574
⅝	0.3134	3.1904	0.0558	17.9126	2.3447
¾	0.3253	3.0742	0.0579	17.2603	2.4333
⅞	0.3373	2.9647	0.0601	16.6454	2.5232
9	0.3495	2.8613	0.0622	16.0650	2.6144
⅛	0.3618	2.7636	0.0644	15.5164	2.7068
¼	0.3744	2.6711	0.0667	14.9972	2.8005
⅜	0.3871	2.5835	0.0689	14.5052	2.8955
½	0.3999	2.5004	0.0712	14.0385	2.9918
⅝	0.4130	2.4214	0.0736	13.5952	3.0893
¾	0.4262	2.3464	0.0759	13.1739	3.1881
⅞	0.4396	2.2749	0.0783	12.7729	3.2882
10	0.4531	2.2069	0.0807	12.3909	3.3896
⅛	0.4668	2.1421	0.0831	12.0267	3.4922
¼	0.4807	2.0802	0.0856	11.6792	3.5961
⅜	0.4948	2.0210	0.0881	11.3473	3.7013
½	0.5090	1.9645	0.0907	11.0301	3.8078
⅝	0.5234	1.9105	0.0932	10.7266	3.9155
¾	0.5380	1.8587	0.0958	10.4360	4.0245
⅞	0.5527	1.8092	0.0984	10.1576	4.1348
11	0.5677	1.7616	0.1011	9.8908	4.2464
⅛	0.5827	1.7160	0.1038	9.6348	4.3592
¼	0.5980	1.6722	0.1065	9.3890	4.4733
⅜	0.6134	1.6302	0.1093	9.1529	4.5887
½	0.6290	1.5898	0.1120	8.9260	4.70»
⅝	0.6448	1.5509	0.1148	8.7077	4.8233
¾	0.6607	1.5135	0.1177	8.4977	4.9425
⅞	0.6768	1.4775	0.1205	8.2955	5.0630

4

FOUR STRINGS OF
2-⅜" OD
TUBING

VOLUME AND FILL BETWEEN PIPE AND HOLE

HOLE DIAM (IN)	CU FT PER LIN FT	LIN FT PER CU FT	BARRELS PER LIN FT	LIN FT PER BARREL	GALLONS PER LIN FT
6	0.0733	13.6444	0.0131	76.6076	0.5482
⅛	0.0816	12.2614	0.0145	68.8427	0.6101
¼	0.0900	11.1119	0.0160	62.3887	0.6732
⅜	0.0986	10.1419	0.0176	56.9425	0.7376
½	0.1074	9.3128	0.0191	52.2877	0.8032
⅝	0.1163	8.5965	0.0207	48.2656	0.8702
¾	0.1254	7.9716	0.0223	44.7571	0.9384
⅞	0.1347	7.4220	0.0240	41.6714	1.0079
7	0.1442	6.9351	0.0257	38.9377	1.0786
⅛	0.1538	6.5009	0.0274	36.5000	1.1507
¼	0.1636	6.1115	0.0291	34.3138	1.2240
⅜	0.1736	5.7605	0.0309	32.3429	1.2986
½	0.1837	5.4426	0.0327	30.5577	1.3744
⅝	0.1940	5.1533	0.0346	28.9339	1.4516
¾	0.2045	4.8892	0.0364	27.4510	1.5300
⅞	0.2152	4.6472	0.0383	26.0921	1.6097
8	0.2260	4.4247	0.0403	24.8426	1.6906
⅛	0.2370	4.2194	0.0422	23.6902	1.7729
¼	0.2482	4.0296	0.0442	22.6245	1.8564
⅜	0.2595	3.8536	0.0462	21.6363	1.9412
½	0.2710	3.6900	0.0483	20.7178	2.0272
⅝	0.2827	3.5376	0.0503	19.8621	2.1146
¾	0.2945	3.3953	0.0525	19.0632	2.2032
⅞	0.3065	3.2622	0.0546	18.3160	2.2931
9	0.3187	3.1375	0.0568	17.6156	2.3842
⅛	0.3311	3.0204	0.0590	16.9582	2.4767
¼	0.3436	2.9103	0.0612	16.3399	2.5704
⅜	0.3563	2.8065	0.0635	15.7576	2.6654
½	0.3692	2.7087	0.0658	15.2083	2.7616
⅝	0.3822	2.6163	0.0681	14.6895	2.8592
¾	0.3954	2.5289	0.0704	14.1988	2.9580
⅞	0.4088	2.4461	0.0728	13.7341	3.0581
10	0.4224	2.3677	0.0752	13.2935	3.1594
⅛	0.4361	2.2932	0.0777	12.8752	3.2621
¼	0.4500	2.2224	0.0801	12.4777	3.3660
⅜	0.4640	2.1550	0.0826	12.0996	3.4712
½	0.4783	2.0909	0.0852	11.7396	3.5776
⅝	0.4927	2.0298	0.0877	11.3964	3.6854
¾	0.5072	1.9715	0.0903	11.0690	3.7944
⅞	0.5220	1.9158	0.0930	10.7563	3.9047
11	0.5369	1.8626	0.0956	10.4575	4.0162
⅛	0.5520	1.8117	0.0983	10.1718	4.1291
¼	0.5672	1.7629	0.1010	9.8982	4.2432
⅜	0.5827	1.7163	0.1038	9.6362	4.3586
½	0.5983	1.6715	0.1066	9.3850	4.4752
⅝	0.6140	1.6286	0.1094	9.1440	4.5932
¾	0.6300	1.5874	0.1122	8.9127	4.7124
⅞	0.6461	1.5478	0.1151	8.6905	4.8329

1
ONE STRING OF 2-⅞" OD TUBING
VOLUME AND FILL BETWEEN PIPE AND HOLE

HOLE DIAM (IN)	CU FT PER LIN FT	LIN FT PER CU FT	BARRELS PER LIN FT	LIN FT PER BARREL	GALLONS PER LIN FT
4	0.0422	23.7054	0.0075	133.0960	0.3156
⅛	0.0477	20.9539	0.0085	117.6473	0.3570
¼	0.0534	18.7148	0.0095	105.0758	0.3997
⅜	0.0593	16.8594	0.0106	94.6588	0.4437
½	0.0654	15.2988	0.0116	85.8964	0.4890
⅝	0.0716	13.9693	0.0127	78.4316	0.5355
¾	0.0780	12.8242	0.0139	72.0027	0.5833
⅞	0.0845	11.8288	0.0151	66.4138	0.6324
5	0.0913	10.9563	0.0163	61.5149	0.6828
⅛	0.0982	10.1859	0.0175	57.1897	0.7344
¼	0.1052	9.5014	0.0187	53.3462	0.7873
⅜	0.1125	8.8895	0.0200	49.9110	0.8415
½	0.1199	8.3399	0.0214	46.8248	0.8970
⅝	0.1275	7.8437	0.0227	44.0391	0.9537
¾	0.1352	7.3939	0.0241	41.5139	1.0117
⅞	0.1432	6.9846	0.0255	39.2158	1.0710
6	0.1513	6.6108	0.0269	37.1169	1.1316
⅛	0.1595	6.2683	0.0284	35.1936	1.1934
¼	0.1680	5.9534	0.0299	33.4259	1.2565
⅜	0.1766	5.6632	0.0314	31.7966	1.3209
½	0.1854	5.3950	0.0330	30.2908	1.3866
⅝	0.1943	5.1466	0.0346	28.8958	1.4535
¾	0.2034	4.9159	0.0362	27.6005	1.5217
⅞	0.2127	4.7012	0.0379	26.3952	1.5912
7	0.2222	4.5010	0.0396	25.2714	1.6620
⅛	0.2318	4.3140	0.0413	24.2215	1.7340
¼	0.2416	4.1390	0.0430	23.2390	1.8073
⅜	0.2516	3.9750	0.0448	22.3179	1.8819
½	0.2617	3.8210	0.0466	21.4531	1.9578
⅝	0.2720	3.6761	0.0484	20.6399	2.0349
¾	0.2825	3.5397	0.0503	19.8741	2.1133
⅞	0.2932	3.4111	0.0522	19.1519	2.1930
8	0.3040	3.2896	0.0541	18.4700	2.2740
⅛	0.3150	3.1748	0.0561	17.8254	2.3562
¼	0.3261	3.0662	0.0581	17.2152	2.4397
⅜	0.3375	2.9632	0.0601	16.6370	2.5245
½	0.3490	2.8655	0.0622	16.0885	2.6106
⅝	0.3607	2.7727	0.0642	15.5677	2.6979
¾	0.3725	2.6846	0.0663	15.0726	2.7865
⅞	0.3845	2.6007	0.0685	14.6016	2.8764
9	0.3967	2.5208	0.0707	14.1531	2.9676
⅛	0.4091	2.4446	0.0729	13.7255	3.0600
¼	0.4216	2.3720	0.0751	13.3177	3.1537
⅜	0.4343	2.3026	0.0773	12.9283	3.2487
½	0.4472	2.2364	0.0796	12.5562	3.3450
⅝	0.4602	2.1730	0.0820	12.2005	3.4425
¾	0.4734	2.1124	0.0843	11.8600	3.5413
⅞	0.4868	2.0543	0.0867	11.5341	3.6414

2

**TWO STRINGS OF
2-⅞″ OD
TUBING**

VOLUME AND FILL BETWEEN
PIPE AND HOLE

HOLE DIAM (IN)	CU FT PER LIN FT	LIN FT PER CU FT	BARRELS PER LIN FT	LIN FT PER BARREL	GALLONS PER LIN FT
7	0.1771	5.6469	0.0315	31.7048	1.3247
⅛	0.1867	5.3556	0.0333	30.0696	1.3968
¼	0.1965	5.0885	0.0350	28.5700	1.4701
⅜	0.2065	4.8428	0.0368	27.1905	1.5447
½	0.2166	4.6161	0.0386	25.9176	1.6205
⅝	0.2269	4.4064	0.0404	24.7400	1.6977
¾	0.2374	4.2118	0.0423	23.6477	1.7761
⅞	0.2481	4.0310	0.0442	22.6323	1.8558
8	0.2589	3.8625	0.0461	21.6861	1.9367
⅛	0.2699	3.7051	0.0481	20.8028	2.0190
¼	0.2811	3.5580	0.0501	19.9765	2.1025
⅜	0.2924	3.4200	0.0521	19.2021	2.1873
½	0.3039	3.2906	0.0541	18.4752	2.2733
⅝	0.3156	3.1688	0.0562	17.7917	2.3607
¾	0.3274	3.0542	0.0583	17.1480	2.4493
⅞	0.3394	2.9461	0.0605	16.5409	2.5392
9	0.3516	2.8440	0.0626	15.9676	2.6303
⅛	0.3640	2.7474	0.0648	15.4255	2.7228
¼	0.3765	2.6560	0.0671	14.9123	2.8165
⅜	0.3892	2.5693	0.0693	14.4258	2.9115
½	0.4021	2.4871	0.0716	13.9641	3.0077
⅝	0.4151	2.4090	0.0739	13.5255	3.1053
¾	0.4283	2.3347	0.0763	13.1083	3.2041
⅞	0.4417	2.2640	0.0787	12.7113	3.3042
10	0.4553	2.1966	0.0811	12.3329	3.4055
⅛	0.4690	2.1323	0.0835	11.9721	3.5082
¼	0.4829	2.0710	0.0860	11.6277	3.6121
⅜	0.4969	2.0124	0.0885	11.2987	3.7173
½	0.5112	1.9563	0.0910	10.9841	3.8237
⅝	0.5256	1.9027	0.0936	10.6831	3.9315
¾	0.5401	1.8514	0.0962	10.3948	4.0405
⅞	0.5549	1.8022	0.0988	10.1186	4.1508
11	0.5698	1.7550	0.1015	9.8538	4.2623
⅛	0.5849	1.7098	0.1042	9.5997	4.3752
¼	0.6001	1.6663	0.1069	9.3557	4.4893
⅜	0.6156	1.6246	0.1096	9.1212	4.6047
½	0.6311	1.5844	0.1124	8.8958	4.7213
⅝	0.6469	1.5458	0.1152	8.6790	4.8393
¾	0.6629	1.5086	0.1181	8.4704	4.9585
⅞	0.6790	1.4728	0.1209	8.2694	5.0790
12	0.6952	1.4384	0.1238	8.0758	5.2007
⅛	0.7117	1.4051	0.1268	7.8892	5.3238
¼	0.7283	1.3731	0.1297	7.7092	5.4481
⅜	0.7451	1.3421	0.1327	7.5355	5.5736
½	0.7620	1.3123	0.1357	7.3678	5.7005
⅝	0.7792	1.2834	0.1388	7.2058	5.8286
¾	0.7965	1.2555	0.1419	7.0493	5.9581
⅞	0.8139	1.2286	0.1450	6.8980	6.0887

3

THREE STRINGS OF 2-⅞″ OD TUBING	**VOLUME AND FILL BETWEEN PIPE AND HOLE**

HOLE DIAM (IN)	CU FT PER LIN FT	LIN FT PER CU FT	BARRELS PER LIN FT	LIN FT PER BARREL	GALLONS PER LIN FT
7	0.1320	7.5753	0.0235	42.5323	0.9875
⅛	0.1416	7.0603	0.0252	39.6405	1.0595
¼	0.1514	6.6034	0.0270	37.0751	1.1328
⅜	0.1614	6.1954	0.0287	34.7849	1.2074
½	0.1716	5.8292	0.0306	32.7285	1.2833
⅝	0.1819	5.4987	0.0324	30.8728	1.3604
¾	0.1923	5.1990	0.0343	29.1903	1.4388
⅞	0.2030	4.9262	0.0362	27.6585	1.5185
8	0.2138	4.6768	0.0381	26.2585	1.5995
⅛	0.2248	4.4481	0.0400	24.9744	1.6817
¼	0.2360	4.2377	0.0420	23.7929	1.7652
⅜	0.2473	4.0435	0.0440	22.7024	1.8500
½	0.2588	3.8637	0.0461	21.6933	1.9361
⅝	0.2705	3.6970	0.0482	20.7569	2.0234
¾	0.2823	3.5419	0.0503	19.8861	2.1120
⅞	0.2944	3.3973	0.0524	19.0743	2.2019
9	0.3065	3.2622	0.0546	18.3160	2.2931
⅛	0.3189	3.1358	0.0568	17.6062	2.3855
¼	0.3314	3.0173	0.0590	16.9407	2.4792
⅜	0.3441	2.9059	0.0613	16.3156	2.5742
½	0.3570	2.8012	0.0636	15.7275	2.6705
⅝	0.3700	2.7025	0.0659	15.1733	2.7680
¾	0.3832	2.6093	0.0683	14.6503	2.8668
⅞	0.3966	2.5213	0.0706	14.1561	2.9669
10	0.4102	2.4380	0.0731	13.6884	3.0683
⅛	0.4239	2.3591	0.0755	13.2454	3.1709
¼	0.4378	2.2842	0.0780	12.8251	3.2748
⅜	0.4518	2.2132	0.0805	12.4260	3.3800
½	0.4661	2.1456	0.0830	12.0465	3.4865
⅝	0.4805	2.0813	0.0856	11.6854	3.5942
¾	0.4950	2.0200	0.0882	11.3415	3.7032
⅞	0.5098	1.9616	0.0908	11.0135	3.8135
11	0.5247	1.9058	0.0935	10.7004	3.9251
⅛	0.5398	1.8526	0.0961	10.4014	4.0379
¼	0.5550	1.8017	0.0989	10.1155	4.1520
⅜	0.5705	1.7529	0.1016	9.8420	4.2674
½	0.5861	1.7063	0.1044	9.5801	4.3841
⅝	0.6018	1.6616	0.1072	9.3292	4.5020
¾	0.6178	1.6187	0.1100	9.0885	4.6212
⅞	0.6339	1.5776	0.1129	8.8576	4.7417
12	0.6502	1.5381	0.1158	8.6358	4.8635
⅛	0.6666	1.5002	0.1187	8.4227	4.9865
¼	0.6832	1.4637	0.1217	8.2179	5.1108
⅜	0.7000	1.4286	0.1247	8.0208	5.2364
½	0.7170	1.3948	0.1277	7.8310	5.3633
⅝	0.7341	1.3622	0.1307	7.6483	5.4914
¾	0.7514	1.3309	0.1338	7.4722	5.6208
⅞	0.7689	1.3006	0.1369	7.3024	5.7515

4

| FOUR STRINGS OF 2⅞" OD TUBING | VOLUME AND FILL BETWEEN PIPE AND HOLE |

HOLE DIAM (IN)	CU FT PER LIN FT	LIN FT PER CU FT	BARRELS PER LIN FT	LIN FT PER BARREL	GALLONS PER LIN FT
7	0.0869	11.5041	0.0155	64.5907	0.6502
⅛	0.0966	10.3567	0.0172	58.1487	0.7223
¼	0.1064	9.4024	0.0189	52.7905	0.7956
⅜	0.1163	8.5965	0.0207	48.2656	0.8702
½	0.1265	7.9071	0.0225	44.3952	0.9460
⅝	0.1368	7.3110	0.0244	41.0483	1.0232
¾	0.1473	6.7906	0.0262	38.1265	1.1016
⅞	0.1579	6.3325	0.0281	35.5545	1.1813
8	0.1687	5.9264	0.0301	33.2740	1.2622
⅛	0.1797	5.5639	0.0320	31.2387	1.3445
¼	0.1909	5.2385	0.0340	29.4118	1.4280
⅜	0.2022	4.9449	0.0360	27.7634	1.5128
½	0.2137	4.6787	0.0381	26.2689	1.5988
⅝	0.2254	4.4364	0.0401	24.9083	1.6862
¾	0.2373	4.2149	0.0423	23.6647	1.7748
⅞	0.2493	4.0117	0.0444	22.5239	1.8647
9	0.2615	3.8247	0.0466	21.4741	1.9558
⅛	0.2738	3.6521	0.0488	20.5050	2.0483
¼	0.2863	3.4923	0.0510	19.6079	2.1420
⅜	0.2990	3.3440	0.0533	18.7753	2.2370
½	0.3119	3.2061	0.0556	18.0007	2.3332
⅝	0.3249	3.0774	0.0579	17.2784	2.4308
¾	0.3382	2.9572	0.0602	16.6035	2.5296
⅞	0.3515	2.8446	0.0626	15.9715	2.6297
10	0.3651	2.7391	0.0650	15.3787	2.7310
⅛	0.3788	2.6399	0.0675	14.8217	2.8337
¼	0.3927	2.5465	0.0699	14.2974	2.9376
⅜	0.4068	2.4584	0.0724	13.8032	3.0428
½	0.4210	2.3753	0.0750	13.3365	3.1492
⅝	0.4354	2.2968	0.0775	12.8954	3.2570
¾	0.4500	2.2224	0.0801	12.4777	3.3660
⅞	0.4647	2.1519	0.0828	12.0819	3.4763
11	0.4796	2.0850	0.0854	11.7062	3.5878
⅛	0.4947	2.0214	0.0881	11.3493	3.7007
¼	0.5100	1.9609	0.0908	11.0098	3.8148
⅜	0.5254	1.9034	0.0936	10.6865	3.9302
½	0.5410	1.8485	0.0964	10.3785	4.0468
⅝	0.5567	1.7961	0.0992	10.0846	4.1648
¾	0.5727	1.7462	0.1020	9.8039	4.2840
⅞	0.5888	1.6984	0.1049	9.5358	4.4045
12	0.6051	1.6527	0.1078	9.2792	4.5262
⅛	0.6215	1.6090	0.1107	9.0337	4.6493
¼	0.6381	1.5671	0.1137	8.7984	4.7736
⅜	0.6549	1.5269	0.1166	8.5729	4.8992
½	0.6719	1.4884	0.1197	8.3565	5.0260
⅝	0.6890	1.4514	0.1227	8.1487	5.1542
¾	0.7063	1.4158	0.1258	7.9491	5.2836
⅞	0.7238	1.3816	0.1289	7.7573	5.4143

VOLUME AND FILL BETWEEN PIPE AND HOLE

3-½" OD PIPE

HOLE DIAM (IN)	CU FT PER LIN FT	LIN FT PER CU FT	BARRELS PER LIN FT	LIN FT PER BARREL	GALLONS PER LIN FT
4	0.0205	48.8924	0.0036	274.5104	0.1530
⅛	0.0260	38.4727	0.0046	216.0082	0.1944
¼	0.0317	31.5435	0.0056	177.1035	0.2371
⅜	0.0376	26.6081	0.0067	149.3934	0.2811
½	0.0436	22.9183	0.0078	128.6768	0.3264
⅝	0.0499	20.0584	0.0089	112.6197	0.3729
¾	0.0562	17.7791	0.0100	99.8220	0.4207
⅞	0.0628	15.9215	0.0112	89.3928	0.4698
5	0.0695	14.3801	0.0124	80.7384	0.5202
⅛	0.0764	13.0816	0.0136	73.4476	0.5718
¼	0.0835	11.9736	0.0149	67.2270	0.6247
⅜	0.0908	11.0180	0.0162	61.8615	0.6789
½	0.0982	10.1859	0.0175	57.1897	0.7344
⅝	0.1058	9.4554	0.0188	53.0882	0.7911
¾	0.1135	8.8094	0.0202	49.4613	0.8491
⅞	0.1214	8.2345	0.0216	46.2333	0.9084
6	0.1295	7.7199	0.0231	43.3438	0.9690
⅛	0.1378	7.2568	0.0245	40.7437	1.0308
¼	0.1462	6.8381	0.0260	38.3931	1.0939
⅜	0.1548	6.4580	0.0276	36.2589	1.1583
½	0.1636	6.1115	0.0291	34.3138	1.2240
⅝	0.1726	5.7947	0.0307	32.5346	1.2909
¾	0.1817	5.5038	0.0324	30.9017	1.3591
⅞	0.1910	5.2361	0.0340	29.3987	1.4286
7	0.2004	4.9890	0.0357	28.0113	1.4994
⅛	0.2101	4.7603	0.0374	26.7272	1.5714
¼	0.2199	4.5481	0.0392	25.5359	1.6447
⅜	0.2298	4.3508	0.0409	24.4281	1.7193
½	0.2400	4.1670	0.0427	23.3958	1.7952
⅝	0.2503	3.9953	0.0446	22.4319	1.8723
¾	0.2608	3.8347	0.0464	21.5302	1.9507
⅞	0.2714	3.6842	0.0483	20.6852	2.0304
8	0.2823	3.5429	0.0503	19.8921	2.1114
⅛	0.2932	3.4101	0.0522	19.1463	2.1936
¼	0.3044	3.2850	0.0542	18.4442	2.2771
⅜	0.3157	3.1671	0.0562	17.7821	2.3619
½	0.3272	3.0558	0.0583	17.1569	2.4480
⅝	0.3389	2.9505	0.0604	16.5659	2.5353
¾	0.3508	2.8509	0.0625	16.0064	2.6239
⅞	0.3628	2.7564	0.0646	15.4763	2.7138
9	0.3750	2.6669	0.0668	14.9733	2.8050
⅛	0.3873	2.5818	0.0690	14.4956	2.8974
¼	0.3999	2.5009	0.0712	14.0415	2.9911
⅜	0.4126	2.4239	0.0735	13.6093	3.0861
½	0.4254	2.3506	0.0758	13.1976	3.1824
⅝	0.4385	2.2807	0.0781	12.8052	3.2799
¾	0.4517	2.2140	0.0804	12.4307	3.3787
⅞	0.4651	2.1503	0.0828	12.0730	3.4788

3-½″ OD PIPE	VOLUME AND FILL BETWEEN PIPE AND HOLE				

HOLE DIAM (IN)	CU FT PER LIN FT	LIN FT PER CU FT	BARRELS PER LIN FT	LIN FT PER BARREL	GALLONS PER LIN FT
10	0.4786	2.0894	0.0852	11.7312	3.5802
⅛	0.4923	2.0312	0.0877	11.4043	3.6828
¼	0.5062	1.9755	0.0902	11.0913	3.7867
⅜	0.5203	1.9221	0.0927	10.7916	3.8919
½	0.5345	1.8709	0.0952	10.5042	3.9984
⅝	0.5489	1.8218	0.0978	10.2286	4.1061
¾	0.5635	1.7747	0.1004	9.9641	4.2151
⅞	0.5782	1.7294	0.1030	9.7100	4.3254
11	0.5931	1.6859	0.1056	9.4659	4.4370
⅛	0.6Ø2	1.6441	0.1083	9.2311	4.5498
¼	0.6235	1.6039	0.1110	9.0053	4.6639
⅜	0.6389	1.5652	0.1138	8.7878	4.7793
½	0.6545	1.5279	0.1166	8.5785	4.8960
⅝	0.6703	1.4919	0.1194	8.3767	5.0139
¾	0.6862	1.4573	0.1222	8.1821	5.1331
⅞	0.7023	1.4239	0.1251	7.9945	5.2536
12	0.7186	1.3916	0.1280	7.8134	5.3754
⅛	0.7350	1.3605	0.1309	7.6386	5.4984
¼	0.7517	1.3304	0.1339	7.4697	5.6227
⅜	0.7684	1.3013	0.1369	7.3065	5.7483
½	0.7854	1.2732	0.1399	7.1487	5.8752
⅝	0.8025	1.2461	0.1429	6.9961	6.0033
¾	0.8373	1.1943	0.1491	6.7056	6.2634
13	0.8549	1.1697	0.1523	6.5672	6.3954
⅛	0.8727	1.1458	0.1554	6.4332	6.5286
¼	0.8907	1.1227	0.1586	6.3033	6.6631
⅜	0.9089	1.1003	0.1619	6.1775	6.7989
½	0.9272	1.0785	0.1651	6.0554	6.9360
⅝	0.9457	1.0574	0.1684	5.9370	7.0743
¾	0.9644	1.0370	0.1718	5.8221	7.2139
⅞	0.9832	1.0171	0.1751	5.7105	7.3548
14	1.0022	0.9978	0.1785	5.6023	7.4970
⅛	1.0214	0.9791	0.1819	5.4971	7.6404
¼	1.0407	0.9609	0.1854	5.3949	7.7851
⅜	1.0602	0.9432	0.1888	5.2956	7.9311
½	1.0799	0.9260	0.1923	5.1991	8.0784
⅝	1.0998	0.9093	0.1959	5.1052	8.2269
¾	1.1198	0.8930	0.1994	5.0139	8.3767
⅞	1.1400	0.8772	0.2030	4.9251	8.5278

4" OD PIPE	VOLUME AND FILL BETWEEN PIPE AND HOLE

HOLE DIAM (IN)	CU FT PER LIN FT	LIN FT PER CU FT	BARRELS PER LIN FT	LIN FT PER BARREL	GALLONS PER LIN FT
5	0.0491	20.3718	0.0087	114.3794	0.3672
1/8	0.0560	17.8602	0.0100	100.2778	0.4188
1/4	0.0631	15.8570	0.0112	89.0304	0.4717
3/8	0.0703	14.2232	0.0125	79.8576	0.5259
1/2	0.0777	12.8664	0.0138	72.2396	0.5814
5/8	0.0853	11.7225	0.0152	65.8167	0.6381
3/4	0.0931	10.7456	0.0166	60.3320	0.6961
7/8	0.1010	9.9023	0.0180	55.5971	0.7554
6	0.1091	9.1673	0.0194	51.4707	0.8160
1/8	0.1173	8.5216	0.0209	47.8450	0.8778
1/4	0.1258	7.9500	0.0224	44.6358	0.9409
3/8	0.1344	7.4408	0.0239	41.7771	1.0053
1/2	0.1432	6.9846	0.0255	39.2158	1.0710
5/8	0.1521	6.5738	0.0271	36.9090	1.1379
3/4	0.1612	6.2020	0.0287	34.8216	1.2061
7/8	0.1705	5.8642	0.0304	32.9248	1.2756
7	0.1800	5.5560	0.0321	31.1944	1.3464
1/8	0.1896	5.2738	0.0338	29.6101	1.4184
1/4	0.1994	5.0146	0.0335	28.1549	1.4917
3/8	0.2094	4.7758	0.0373	26.8142	1.5663
1/2	0.2195	4.5552	0.0391	25.5755	1.6422
5/8	0.2298	4.3508	0.0409	24.4281	1.7193
3/4	0.2403	4.1611	0.0428	23.3626	1.7977
7/8	0.2510	3.9844	0.0447	22.3710	1.8774
8	0.2618	3.8197	0.0466	21.4461	1.9584
1/8	0.2728	3.6658	0.0486	20.5819	2.0406
1/4	0.2840	3.5217	0.0506	19.7727	2.1241
3/8	0.2953	3.3865	0.0526	19.0137	2.2089
1/2	0.3068	3.2595	0.0546	18.3007	2.2950
5/8	0.3185	3.1400	0.0567	17.6298	2.3823
3/4	0.3303	3.0274	0.0588	16.9976	2.4709
7/8	0.3423	2.9211	0.0610	16.4009	2.5608
9	0.3545	2.8207	0.0631	15.8371	2.6520
1/8	0.3669	2.7257	0.0653	15.3037	2.7444
1/4	0.3794	2.6357	0.0676	14.7984	2.8381
3/8	0.3921	2.5504	0.0698	14.3192	2.9331
1/2	0.4050	2.4693	0.0721	13.8642	3.0294
5/8	0.4180	2.3923	0.0745	13.4317	3.1269
3/4	0.4312	2.3190	0.0768	13.0203	3.2257
7/8	0.4446	2.2492	0.0792	12.6284	3.3258
10	0.4581	2.1827	0.0816	12.2549	3.4272
1/8	0.4719	2.1192	0.0840	11.8986	3.5298
1/4	0.4858	2.0586	0.0865	11.5583	3.6337
3/8	0.4998	2.0007	0.0890	11.2332	3.7389
1/2	0.5141	1.9453	0.0916	10.9222	3.8454
5/8	0.5285	1.8923	0.0941	10.6245	3.9531
3/4	0.5430	1.8415	0.0967	10.3394	4.0621
7/8	0.5578	1.7928	0.0993	10.0661	4.1724

4" OD
PIPE

VOLUME AND FILL BETWEEN
PIPE AND HOLE

HOLE DIAM (IN)	CU FT PER LIN FT	LIN FT PER CU FT	BARRELS PER LIN FT	LIN FT PER BARREL	GALLONS PER LIN FT
11	0.5727	1.7462	0.1020	9.8039	4.2840
⅛	0.5878	1.7013	0.1047	9.5523	4.3968
¼	0.6030	1.6583	0.1074	9.3107	4.5109
⅜	0.6184	1.6169	0.1102	9.0785	4.6263
½	0.6340	1.5772	0.1129	8.8552	4.7430
⅝	0.6498	1.5389	0.1157	8.6403	4.8609
¾	0.6657	1.5021	0.1186	8.4335	4.9801
⅞	0.6819	1.4666	0.1214	8.2343	5.1006
12	0.6981	1.4324	0.1243	8.0423	5.2224
⅛	0.7146	1.3994	0.1273	7.8572	5.3454
¼	0.7312	1.3676	0.1302	7.6786	5.4697
⅜	0.7480	1.3369	0.1332	7.5063	5.5953
½	0.7649	1.3073	0.1362	7.3399	5.7222
⅝	0.7821	1.2787	0.1393	7.1791	5.8503
¾	0.7994	1.2510	0.1424	7.0237	5.9797
⅞	0.8168	1.2242	0.1455	6.8735	6.1104
13	0.8345	1.1983	0.1486	6.7282	6.2424
⅛	0.8523	1.1733	0.1518	6.5876	6.3756
¼	0.8703	1.1491	0.1550	6.4515	6.5101
⅜	0.8884	1.1256	0.1582	6.3197	6.6459
½	0.9068	1.1028	0.1615	6.1920	6.7830
⅝	0.9252	1.0808	0.1648	6.0682	6.9213
¾	0.9439	1.0594	0.1681	5.9482	7.0609
⅞	0.9627	1.0387	0.1715	5.8319	7.2018
14	0.9817	1.0186	0.1749	5.7190	7.3440
⅛	1.0009	0.9991	0.1783	5.6094	7.4874
¼	1.0203	0.9801	0.1817	5.5030	7.6321
⅜	1.0398	0.9617	0.1852	5.3998	7.7781
½	1.0595	0.9439	0.1887	5.2994	7.9254
⅝	1.0793	0.9265	0.1922	5.2019	8.0739
¾	1.0994	0.9096	0.1958	5.1072	8.2237
⅞	1.1196	0.8932	0.1994	5.0150	8.3748
15	1.1399	0.8773	0.2030	4.9254	8.5272
⅛	1.1605	0.8617	0.2067	4.8383	8.6808
¼	1.1812	0.8466	0.2104	4.7534	8.8357
⅜	1.2020	0.8319	0.2141	4.6709	8.9919
½	1.2231	0.8176	0.2178	4.5905	9.1494
⅝	1.2443	0.8037	0.2216	4.5122	9.3081
¾	1.2657	0.7906	0.2254	4.4359	9.4681
⅞	1.2873	0.7768	0.2293	4.3616	9.6294

4-½" OD
PIPE

VOLUME AND FILL BETWEEN PIPE AND HOLE

HOLE DIAM (IN)	CU FT PER LIN FT	LIN FT PER CU FT	BARRELS PER LIN FT	LIN FT PER BARREL	GALLONS PER LIN FT
5	0.0259	38.5993	0.0046	216.7188	0.1938
⅛	0.0328	30.4784	0.0058	171.1234	0.2454
¼	0.0399	25.0730	0.0071	140.7746	0.2983
⅜	0.0471	21.2191	0.0084	119.1365	0.3525
½	0.0545	18.3346	0.0097	102.9414	0.4080
⅝	0.0621	16.0963	0.0111	90.3738	0.4647
¾	0.0699	14.3100	0.0124	80.3445	0.5227
⅞	0.0778	12.8523	0.0139	72.1605	0.5820
6	0.0859	11.6410	0.0153	65.3596	0.6426
⅛	0.0942	10.6192	0.0168	59.6222	0.7044
¼	0.1026	9.7460	0.0183	54.7197	0.7675
⅜	0.1112	8.9917	0.0198	50.4847	0.8319
½	0.1200	8.3339	0.0214	46.7916	0.8976
⅝	0.1289	7.7556	0.0230	43.5443	0.9645
¾	0.1381	7.2433	0.0246	40.6682	1.0327
⅞	0.1473	6.7867	0.0262	38.1044	1.1022
7	0.1568	6.3773	0.0279	35.8057	1.1730
⅛	0.1664	6.0083	0.0296	33.7340	1.2450
¼	0.1762	5.6742	0.0314	31.8581	1.3183
⅜	0.1862	5.3703	0.0332	30.1522	1.3929
½	0.1963	5.0930	0.0350	28.5948	1.4688
⅝	0.2067	4.8388	0.0368	27.1680	1.5459
¾	0.2171	4.6052	0.0387	25.8566	1.6243
⅞	0.2278	4.3899	0.0406	24.6474	1.7040
8	0.2386	4.1908	0.0425	23.5295	1.7850
⅛	0.2496	4.0062	0.0445	22.4932	1.8672
¼	0.2608	3.8347	0.0464	21.5302	1.9507
⅜	0.2721	3.6750	0.0485	20.6334	2.0355
½	0.2836	3.5259	0.0505	19.7964	2.1216
⅝	0.2953	3.3865	0.0526	19.0137	2.2089
¾	0.3071	3.2559	0.0547	18.2804	2.2975
⅞	0.3192	3.1333	0.0568	17.5921	2.3874
9	0.3313	3.0180	0.0590	16.9451	2.4786
⅛	0.3437	2.9095	0.0612	16.3359	2.5710
¼	0.3562	2.8072	0.0634	15.7614	2.6647
⅜	0.3689	2.7106	0.0657	15.2189	2.7597
½	0.3818	2.6192	0.0680	14.7059	2.8560
⅝	0.3948	2.5327	0.0703	14.2203	2.9535
¾	0.4080	2.4507	0.0727	13.7599	3.0523
⅞	0.4214	2.3729	0.0751	13.3231	3.1524
10	0.4350	2.2990	0.0775	12.9080	3.2538
⅛	0.4487	2.2287	0.0799	12.5133	3.3564
¼	0.4626	2.1618	0.0824	12.1375	3.4603
⅜	0.4766	2.0980	0.0849	11.7795	3.5655
½	0.4909	2.0372	0.0874	11.4379	3.6720
⅝	0.5053	1.9791	0.0900	11.1119	3.7797
¾	0.5198	1.9236	0.0926	10.8004	3.8887
⅞	0.5346	1.8706	0.0952	10.5026	3.9990

4-½″ OD PIPE		VOLUME AND FILL BETWEEN PIPE AND HOLE			
HOLE DIAM (IN)	CU FT PER LIN FT	LIN FT PER CU FT	BARRELS PER LIN FT	LIN FT PER BARREL	GALLONS PER LIN FT
11	0.5495	1.8198	0.0979	10.2175	4.1106
⅛	0.5646	1.7712	0.1006	9.9445	4.2234
¼	0.5798	1.7246	0.1033	9.6829	4.3375
⅜	0.5953	1.6799	0.1060	9.4320	4.4529
½	0.6109	1.6370	0.1088	9.1912	4.5696
⅝	0.6266	1.5958	0.1116	8.9599	4.6875
¾	0.6426	1.5563	0.1144	8.7377	4.8067
⅞	0.6587	1.5182	0.1173	8.5241	4.9272
12	0.6750	1.4816	0.1202	8.3185	5.0490
⅛	0.6914	1.4463	0.1231	8.1206	5.1720
¼	0.7080	1.4124	0.1261	7.9300	5.2963
⅜	0.7248	1.3797	0.1291	7.7463	5.4219
½	0.7418	1.3481	0.1321	7.5692	5.5488
⅝	0.7589	1.3177	0.1352	7.3984	5.6769
(	0.7762	1.2883	0.1382	7.2335	5.8063
⅞	0.7937	1.2600	0.1414	7.0743	5.9370
13	0.8113	1.2326	0.1445	6.9204	6.0690
⅛	0.8291	1.2061	0.1477	6.7718	6.2022
¼	0.8471	1.1805	0.1509	6.6280	6.3367
⅜	0.8653	1.1557	0.1541	6.4890	6.4725
½	0.8836	1.1318	0.1574	6.3544	6.6096
⅝	0.9021	1.1086	0.1607	6.2241	6.7479
¾	0.9207	1.0861	0.1640	6.0980	6.8875
⅞	0.9396	1.0643	0.1673	5.9757	7.0284
14	0.9586	1.0432	0.1707	5.8573	7.1706
⅛	0.9777	1.0228	0.1741	5.7424	7.3140
¼	0.9971	1.0029	0.1776	5.6310	7.4587
⅜	1.0166	0.9837	0.1811	5.5229	7.6047
½	1.0363	0.9650	0.1846	5.4180	7.7520
⅝	1.0561	0.9468	0.1881	5.3161	7.9005
¾	1.0762	0.9292	0.1917	5.2172	8.0503
⅞	1.0964	0.9121	0.1953	5.1211	8.2014
15	1.1167	0.8955	0.1989	5.0277	8.3538
⅛	1.1373	0.8793	0.2026	4.9369	8.5074
¼	1.1580	0.8636	0.2062	4.8486	8.6623
⅜	1.1789	0.8483	0.2100	4.7627	8.8185
½	1.1999	0.8334	0.2137	4.6792	8.9760
⅝	1.2211	0.8189	0.2175	4.5978	9.1347
¾	1.2425	0.8048	0.2213	4.5187	9.2947
⅞	1.2641	0.7911	0.2251	4.4416	9.4560

4-¾" OD
CASING

VOLUME AND FILL BETWEEN PIPE AND HOLE

HOLE DIAM (IN)	CU FT PER LIN FT	LIN FT PER CU FT	BARRELS PER LIN FT	LIN FT PER BARREL	GALLONS PER LIN FT
6	0.0733	13.6444	0.0131	76.6076	0.5482
⅛	0.0816	12.2614	0.0145	68.8427	0.6101
¼	0.0900	11.1119	0.0160	62.3887	0.6732
⅜	0.0986	10.1419	0.0176	56.9425	0.7376
½	0.1074	9.3128	0.0191	52.2877	0.8032
⅝	0.1163	8.5965	0.0207	48.2656	0.8702
¾	0.1254	7.9716	0.0223	44.7571	0.9384
⅞	0.1347	7.4220	0.0240	41.6714	1.0079
7	0.1442	6.9351	0.0257	38.9377	1.0786
⅛	0.1538	6.5009	0.0274	36.5000	1.1507
¼	0.1636	6.1115	0.0291	34.3138	1.2240
⅜	0.1736	5.7605	0.0309	32.3429	1.2986
½	0.1837	5.4426	0.0327	30.5577	1.3744
⅝	0.1940	5.1533	0.0346	28.9339	1.4516
¾	0.2045	4.8892	0.0364	27.4510	1.5300
⅞	0.2152	4.6472	0.0383	26.0921	1.6097
8	0.2260	4.4247	0.0403	24.8426	1.6906
⅛	0.2370	4.2194	0.0422	23.6902	1.7729
¼	0.2482	4.0296	0.0442	22.6245	1.8564
⅜	0.2595	3.8536	0.0462	21.6363	1.9412
½	0.2710	3.6900	0.0483	20.7178	2.0272
⅝	0.2827	3.5376	0.0503	19.8621	2.1146
¾	0.2945	3.3953	0.0525	19.0632	2.2032
⅞	0.3065	3.2622	0.0546	18.3160	2.2931
9	0.3187	3.1375	0.0568	17.6156	2.3842
⅛	0.3311	3.0204	0.0590	16.9582	2.4767
¼	0.3436	2.9103	0.0612	16.3399	2.5704
⅜	0.3563	2.8065	0.0635	15.7576	2.6654
½	0.3692	2.7087	0.0658	15.2083	2.7616
⅝	0.3822	2.6163	0.0681	14.6895	2.8592
¾	0.3954	2.5289	0.0704	14.1988	2.9580
⅞	0.4088	2.4461	0.0728	13.7341	3.0581
10	0.4224	2.3677	0.0752	13.2935	3.1594
⅛	0.4361	2.2932	0.0777	12.8752	3.2621
¼	0.4500	2.2224	0.0801	12.4777	3.3660
⅜	0.4640	2.1550	0.0826	12.0996	3.4712
½	0.4783	2.0909	0.0852	11.7396	3.5776
⅝	0.4927	2.0298	0.0877	11.3964	3.6854
¾	0.5072	1.9715	0.0903	11.0690	3.7944
⅞	0.5220	1.9158	0.0930	10.7563	3.9047
11	0.5369	1.8626	0.0956	10.4575	4.0162
⅛	0.5520	1.8117	0.0983	10.1718	4.1291
¼	0.5672	1.7629	0.1010	9.8982	4.2432
⅜	0.5827	1.7163	0.1038	9.6362	4.3586
½	0.5983	1.6715	0.1066	9.3850	4.4752
⅝	0.6140	1.6286	0.1094	9.1440	4.5932
¾	0.6300	1.5874	0.1122	8.9127	4.7124
⅞	0.6461	1.5478	0.1151	8.6905	4.8329

VOLUME AND FILL BETWEEN PIPE AND HOLE

4-¾" OD CASING

HOLE DIAM (IN)	CU FT PER LIN FT	LIN FT PER CU FT	BARRELS PER LIN FT	LIN FT PER BARREL	GALLONS PER LIN FT
12	0.6623	1.5098	0.1180	8.4769	4.9546
⅛	0.6788	1.4732	0.1209	8.2715	5.0777
¼	0.6954	1.4380	0.1239	8.0738	5.2020
⅜	0.7122	1.4041	0.1268	7.8835	5.3276
½	0.7292	1.3715	0.1299	7.7002	5.4544
⅝	0.7463	1.3400	0.1329	7.5234	5.5826
¾	0.7636	1.3096	0.1360	7.3530	5.7120
⅞	0.7811	1.2803	0.1391	7.1885	5.8427
13	0.7987	1.2520	0.1423	7.0297	5.9746
⅛	0.8165	1.2247	0.1454	6.8764	6.1079
¼	0.8345	1.1983	0.1486	6.7282	6.2424
⅜	0.8526	1.1728	0.1519	6.5850	6.3782
½	0.8710	1.1482	0.1551	6.4464	6.5152
⅝	0.8895	1.1243	0.1584	6.3124	6.6536
¾	0.9081	1.1012	0.1617	6.1827	6.7932
⅞	0.9270	1.0788	0.1651	6.0570	6.9341
14	0.9460	1.0571	0.1685	5.9354	7.0762
⅛	0.9651	1.0361	0.1719	5.8174	7.2197
¼	0.9845	1.0158	0.1753	5.7031	7.3644
⅜	1.0040	0.9960	0.1788	5.5923	7.5104
½	1.0237	0.9769	0.1823	5.4847	7.6576
⅝	1.0435	0.9583	0.1859	5.3804	7.8062
¾	1.0636	0.9402	0.1894	5.2790	7.9560
⅞	1.0838	0.9227	0.1930	5.1807	8.1071
15	1.1041	0.9057	0.1967	5.0851	8.2594
⅛	1.1247	0.8892	0.2003	4.9922	8.4131
¼	1.1454	0.8731	0.2040	4.9020	8.5680
⅜	1.1663	0.8574	0.2077	4.8142	8.7242
½	1.1873	0.8422	0.2115	4.7289	8.8816
⅝	1.2085	0.8275	0.2152	4.6458	9.0404
¾	1.2299	0.8131	0.2191	4.5650	9.2004
⅞	1.2515	0.7991	0.2229	4.4864	9.3617
16	1.2732	0.7854	0.2268	4.4098	9.5242
¼	1.3172	0.7592	0.2346	4.2626	9.8532
½	1.3618	0.7343	0.2426	4.1228	10.1872
¾	1.4072	0.7106	0.2506	3.9900	10.5264

5" OD CASING	VOLUME AND FILL BETWEEN PIPE AND HOLE

HOLE DIAM (IN)	CU FT PER LIN FT	LIN FT PER CU FT	BARRELS PER LIN FT	LIN FT PER BARREL	GALLONS PER LIN FT
6	0.0600	16.6679	0.0107	93.5831	0.4488
1/8	0.0683	14.6494	0.0122	82.2503	0.5106
1/4	0.0767	13.0380	0.0137	73.2028	0.5737
3/8	0.0853	11.7225	0.0152	65.8167	0.6381
1/2	0.0941	10.6288	0.0168	59.6762	0.7038
5/8	0.1030	9.7057	0.0184	54.4934	0.7707
3/4	0.1122	8.9165	0.0200	50.0627	0.8389
7/8	0.1214	8.2345	0.0216	46.2333	0.9084
7	0.1309	7.6394	0.0233	42.8923	0.9792
1/8	0.1405	7.1159	0.0250	39.9530	1.0512
1/4	0.1503	6.6520	0.0268	37.3484	1.1245
3/8	0.1603	6.2383	0.0286	35.0253	1.1991
1/2	0.1704	5.8671	0.0304	32.9413	1.2750
5/8	0.1808	5.5324	0.0322	31.0620	1.3521
3/4	0.1912	5.2291	0.0341	29.3594	1.4305
7/8	0.2019	4.9532	0.0360	27.8103	1.5102
8	0.2127	4.7012	0.0379	26.3952	1.5912
1/8	0.2237	4.4702	0.0398	25.0981	1.6734
1/4	0.2349	4.2577	0.0418	23.9051	1.7569
3/8	0.2462	4.0617	0.0439	22.8046	1.8417
1/2	0.2577	3.8803	0.0459	21.7865	1.9278
5/8	0.2694	3.7122	0.0480	20.8423	2.0151
3/4	0.2812	3.5558	0.0501	19.9644	2.1037
7/8	0.2932	3.4101	0.0522	19.1463	2.1936
9	0.3054	3.2740	0.0544	18.3824	2.2848
1/8	0.3178	3.1467	0.0566	17.6676	2.3772
1/4	0.3303	3.0274	0.0588	16.9976	2.4709
3/8	0.3430	2.9153	0.0611	16.3683	2.5659
1/2	0.3559	2.8099	0.0634	15.7765	2.6622
5/8	0.3689	2.7106	0.0657	15.2189	2.7597
3/4	0.3821	2.6169	0.0681	14.6928	2.8585
7/8	0.3955	2.5284	0.0704	14.1958	2.9586
10	0.4091	2.4446	0.0729	13.7255	3.0600
1/8	0.4228	2.3653	0.0753	13.2801	3.1626
1/4	0.4367	2.2900	0.0778	12.8576	3.2665
3/8	0.4507	2.2186	0.0803	12.4565	3.3717
1/2	0.4650	2.1507	0.0828	12.0752	3.4782
5/8	0.4794	2.0861	0.0854	11.7124	3.5859
3/4	0.4939	2.0245	0.0880	11.3669	3.6949
7/8	0.5087	1.9659	0.0906	11.0374	3.8052
11	0.5236	1.9099	0.0933	10.7231	3.9168
1/8	0.5387	1.8564	0.0959	10.4228	4.0296
1/4	0.5539	1.8053	0.0987	10.1358	4.1437
3/8	0.5694	1.7564	0.1014	9.8612	4.2591
1/2	0.5850	1.7095	0.1042	9.5983	4.3758
5/8	0.6007	1.6647	0.1070	9.3464	4.4937
3/4	0.6167	1.6216	0.1098	9.1048	4.6129
7/8	0.6328	1.5804	0.1127	8.8731	4.7334

| 5" OD CASING | VOLUME AND FILL BETWEEN PIPE AND HOLE | | | | |

HOLE DIAM (IN)	CU FT PER LIN FT	LIN FT PER CU FT	BARRELS PER LIN FT	LIN FT PER BARREL	GALLONS PER LIN FT
12	0.6490	1.5407	0.1156	8.6505	4.8552
⅛	0.6655	1.5026	0.1185	8.4367	4.9782
¼	0.6821	1.4660	0.1215	8.2312	5.1025
⅜	0.6989	1.4308	0.1245	8.0335	5.2281
½	0.7159	1.3969	0.1275	7.8432	5.3550
⅝	0.7330	1.3643	0.1306	7.6599	5.4831
¾	0.7503	1.3328	0.1336	7.4832	5.6125
⅞	0.7678	1.3025	0.1367	7.3130	5.7432
13	0.7854	1.2732	0.1399	7.1487	5.8752
⅛	0.8032	1.2450	0.1431	6.9902	6.0084
¼	0.8212	1.2177	0.1463	6.8371	6.1429
⅜	0.8393	1.1914	0.1495	6.6893	6.2787
½	0.8577	1.1660	0.1528	6.5464	6.4158
⅝	0.8762	1.1413	0.1561	6.4082	6.5541
¾	0.8948	1.1175	0.1594	6.2745	6.6937
⅞	0.9137	1.0945	0.1627	6.1452	6.8346
14	0.9327	1.0722	0.1661	6.0200	6.9768
⅛	0.9518	1.0506	0.1695	5.8987	7.1202
¼	0.9712	1.0297	0.1730	5.7812	7.2649
⅜	0.9907	1.0094	0.1765	5.6673	7.4109
½	1.0104	0.9897	0.1800	5.5569	7.5582
⅝	1.0302	0.9706	0.1835	5.4498	7.7067
¾	1.0503	0.9521	0.1871	5.3459	7.8565
⅞	1.0705	0.9342	0.1907	5.2450	8.0076
15	1.0908	0.9167	0.1943	5.1471	8.1600
⅛	1.1114	0.8998	0.1979	5.0520	8.3136
¼	1.1321	0.8833	0.2016	4.9595	8.4685
⅜	1.1530	0.8673	0.2054	4.8697	8.6247
½	1.1740	0.8518	0.2091	4.7824	8.7822
⅝	1.1952	0.8367	0.2129	4.6975	8.9409
¾	1.2166	0.8220	0.2167	4.6149	9.1009
⅞	1.2382	0.8076	0.2205	4.5346	9.2622
16	1.2599	0.7937	0.2244	4.4563	9.4248
¼	1.3039	0.7669	0.2322	4.3060	9.7537
½	1.3485	0.7415	0.2402	4.1635	10.0878
¾	1.3939	0.7174	0.2483	4.0280	10.4269

VOLUME AND FILL BETWEEN PIPE AND HOLE

5-½''' OD CASING

HOLE DIAM (IN)	CU FT PER LIN FT	LIN FT PER CU FT	BARRELS PER LIN FT	LIN FT PER BARREL	GALLONS PER LIN FT
6	0.0314	31.8863	0.0056	179.0286	0.2346
⅛	0.0396	25.2348	0.0071	141.6828	0.2964
¼	0.0481	20.8053	0.0086	116.8130	0.3595
⅜	0.0567	17.6454	0.0101	99.0714	0.4239
½	0.0654	15.2789	0.0117	85.7845	0.4896
⅝	0.0744	13.4412	0.0133	75.4668	0.5565
¾	0.0835	11.9736	0.0149	67.2270	0.6247
⅞	0.0928	10.7752	0.0165	60.4982	0.6942
7	0.1023	9.7785	0.0182	54.9021	0.7650
⅛	0.1119	8.9369	0.0199	50.1771	0.8370
¼	0.1217	8.2172	0.0217	46.1362	0.9103
⅜	0.1317	7.5949	0.0235	42.6424	0.9849
½	0.1418	7.0518	0.0253	39.5929	1.0608
⅝	0.1521	6.5738	0.0271	36.9090	1.1379
¾	0.1626	6.1500	0.0290	34.5296	1.2163
⅞	0.1733	5.7719	0.0309	32.4065	1.2960
8	0.1841	5.4325	0.0328	30.5012	1.3770
⅛	0.1951	5.1263	0.0347	28.7822	1.4592
¼	0.2062	4.8488	0.0367	27.2242	1.5427
⅜	0.2176	4.5962	0.0388	25.8059	1.6275
½	0.2291	4.3654	0.0408	24.5099	1.7136
⅝	0.2407	4.1537	0.0429	23.3212	1.8009
¾	0.2526	3.9589	0.0450	22.2276	1.8895
⅞	0.2646	3.7791	0.0471	21.2182	1.9794
9	0.2768	3.6127	0.0493	20.2840	2.0706
⅛	0.2892	3.4583	0.0515	19.4172	2.1630
¼	0.3017	3.3147	0.0537	18.6109	2.2567
⅜	0.3144	3.1809	0.0560	17.8592	2.3517
½	0.3272	3.0558	0.0583	17.1569	2.4480
⅝	0.3403	2.9387	0.0606	16.4995	2.5455
¾	0.3535	2.8289	0.0630	15.8830	2.6443
⅞	0.3669	2.7257	0.0653	15.3037	2.7444
10	0.3804	2.6286	0.0678	14.7586	2.8458
⅛	0.3941	2.5371	0.0702	14.2449	2.9484
¼	0.4080	2.4507	0.0727	13.7599	3.0523
⅜	0.4221	2.3691	0.0752	13.3015	3.1575
½	0.4363	2.2918	0.0777	12.8677	3.2640
⅝	0.4507	2.2186	0.0803	12.4565	3.3717
¾	0.4653	2.1491	0.0829	12.0664	3.4807
⅞	0.4801	2.0831	0.0855	11.6958	3.5910
11	0.4950	2.0203	0.0882	11.3434	3.7026
⅛	0.5100	1.9606	0.09..8	11.0079	3.8154
¼	0.5253	1.9037	0.0936	10.6883	3.9295
⅜	0.5407	1.8494	0.0963	10.3834	4.0449
½	0.5563	1.7975	0.0991	10.0923	4.1616
⅝	0.5721	1.7480	0.1019	9.8142	4.2795
¾	0.5880	1.7006	0.1047	9.5482	4.3987
⅞	0.6041	1.6553	0.1076	9.2936	4.5192

5-½" OD CASING	**VOLUME AND FILL BETWEEN PIPE AND HOLE**				
HOLE DIAM (IN)	**CU FT PER LIN FT**	**LIN FT PER CU FT**	**BARRELS PER LIN FT**	**LIN FT PER BARREL**	**GALLONS PER LIN FT**
12	0.6204	1.6118	0.1105	9.0498	4.6410
⅛	0.6369	1.5702	0.1134	8.8161	4.7640
¼	0.6535	1.5303	0.1164	8.5919	4.8883
⅜	0.6703	1.4919	0.1194	8.3767	5.0139
½	0.6872	1.4551	0.1224	8.1700	5.1408
⅝	0.7044	1.4197	0.1255	7.9713	5.2689
¾	0.7217	1.3857	0.1285	7.7802	5.3983
⅞	0.7391	1.3530	0.1316	7.5963	5.5290
13	0.7568	1.3214	0.1348	7.4192	5.6610
⅛	0.7746	1.2910	0.1380	7.2486	5.7942
¼	0.7926	1.2617	0.1412	7.0841	5.9287
⅜	0.8107	1.2335	0.1444	6.9255	6.0645
½	0.8290	1.2062	0.1477	6.7725	6.2016
⅝	0.8475	1.1799	0.1510	6.6247	6.3399
¾	0.8662	1.1545	0.1543	6.4819	6.4795
⅞	0.8850	1.1299	0.1576	6.3440	6.6204
14	0.9040	1.1062	0.1610	6.2106	6.7626
⅛	0.9232	1.0832	0.1644	6.0816	6.9060
¼	0.9425	1.0610	0.1679	5.9568	7.0507
⅜	0.9621	1.0394	0.1714	5.8360	7.1967
½	0.9817	1.0186	0.1749	5.7190	7.3440
⅝	1.0016	0.9984	0.1784	5.6056	7.4925
¾	1.0216	0.9788	0.1820	5.4957	7.6423
⅞	1.0418	0.9599	0.1856	5.3892	7.7934
15	1.0622	0.9414	0.1892	5.2858	7.9458
⅛	1.0827	0.9236	0.1928	5.1856	8.0994
¼	1.1034	0.9063	0.1965	5.0882	8.2543
⅜	1.1243	0.8894	0.2003	4.9937	8.4105
½	1.1454	0.8731	0.2040	4.9020	8.5680
⅝	1.1666	0.8572	0.2078	4.8128	8.7267
¾	1.1880	0.8418	0.2116	4.7261	8.8867
⅞	1.2095	0.8268	0.2154	4.6419	9.0480
16	1.2313	0.8122	0.2193	4.5600	9.2106
¼	1.2752	0.7842	0.2271	4.4027	9.5395
½	1.3199	0.7576	0.2351	4.2538	9.8736
¾	1.3652	0.7325	0.2432	4.1125	10.2127

5-¾" OD
CASING

VOLUME AND FILL BETWEEN
PIPE AND HOLE

HOLE DIAM (IN)	CU FT PER LIN FT	LIN FT PER CU FT	BARRELS PER LIN FT	LIN FT PER BARREL	GALLONS PER LIN FT
7	0.0869	11.5041	0.0155	64.5907	0.6502
⅛	0.0966	10.3567	0.0172	58.1487	0.7223
¼	0.1064	9.4024	0.0189	52.7905	0.7956
⅜	0.1163	8.5965	0.0207	48.2656	0.8702
½	0.1265	7.9071	0.0225	44.3952	0.9460
⅝	0.1368	7.3110	0.0244	41.0483	1.0232
¾	0.1473	6.7906	0.0262	38.1265	1.1016
⅞	0.1579	6.3325	0.0281	35.5545	1.1813
8	0.1687	5.9264	0.0301	33.2740	1.2622
⅛	0.1797	5.5639	0.0320	31.2387	1.3445
¼	0.1909	5.2385	0.0340	29.4118	1.4280
⅜	0.2022	4.9449	0.0360	27.7634	1.5128
½	0.2137	4.6787	0.0381	26.2689	1.5988
⅝	0.2254	4.4364	0.0401	24.9083	1.6862
¾	0.2373	4.2149	0.0423	23.6647	1.7748
⅞	0.2493	4.0117	0.0444	22.5239	1.8647
9	0.2615	3.8247	0.0466	21.4741	1.9558
⅛	0.2738	3.6521	0.0488	20.5050	2.0483
¼	0.2863	3.4923	0.0510	19.6079	2.1420
⅜	0.2990	3.3440	0.0533	18.7753	2.2370
½	0.3119	3.2061	0.0556	18.0007	2.3332
⅝	0.3249	3.0774	0.0579	17.2784	2.4308
¾	0.3382	2.9572	0.0602	16.6035	2.5296
⅞	0.3515	2.8446	0.0626	15.9715	2.6297
10	0.3651	2.7391	0.0650	15.3787	2.7310
⅛	0.3788	2.6399	0.0675	14.8217	2.8337
¼	0.3927	2.5465	0.0699	14.2974	2.9376
⅜	0.4068	2.4584	0.0724	13.8032	3.0428
½	0.4210	2.3753	0.0750	13.3365	3.1492
⅝	0.4354	2.2968	0.0775	12.8954	3.2570
¾	0.4500	2.2224	0.0801	12.4777	3.3660
⅞	0.4647	2.1519	0.0828	12.0819	3.4763
11	0.4796	2.0850	0.0854	11.7062	3.5878
⅛	0.4947	2.0214	0.0881	11.3493	3.7007
¼	0.5100	1.9609	0.0908	11.0098	3.8148
⅜	0.5254	1.9034	0.0936	10.6865	3.9302
½	0.5410	1.8485	0.0964	10.3785	4.0468
⅝	0.5567	1.7961	0.0992	10.0846	4.1648
¾	0.5727	1.7462	0.1020	9.8039	4.2840
⅞	0.5888	1.6984	0.1049	9.5358	4.4045
12	0.6051	1.6527	0.1078	9.2792	4.5262
⅛	0.6215	1.6090	0.1107	9.0337	4.6493
¼	0.6381	1.5671	0.1137	8.7984	4.7736
⅜	0.6549	1.5269	0.1166	8.5729	4.8992
½	0.6719	1.4884	0.1197	8.3565	5.0260
⅝	0.6890	1.4514	0.1227	8.1487	5.1542
¾	0.7063	1.4158	0.1258	7.9491	5.2836
⅞	0.7238	1.3816	0.1289	7.7573	5.4143

5-¾" OD CASING	**VOLUME AND FILL BETWEEN PIPE AND HOLE**				
HOLE DIAM (IN)	CU FT PER LIN FT	LIN FT PER CU FT	BARRELS PER LIN FT	LIN FT PER BARREL	GALLONS PER LIN FT
13	0.7414	1.3488	0.1321	7.5727	5.5462
⅛	0.7592	1.3171	0.1352	7.3951	5.6795
¼	0.7772	1.2866	0.1384	7.2240	5.8140
⅜	0.7954	1.2573	0.1417	7.0591	5.9498
½	0.8137	1.2290	0.1449	6.9001	6.0868
⅝	0.8322	1.2017	0.1482	6.7468	6.2252
¾	0.8508	1.1753	0.1515	6.5988	6.3648
⅞	0.8697	1.1498	0.1549	6.4559	6.5057
14	0.8887	1.1253	0.1583	6.3178	6.6478
⅛	0.9079	1.1015	0.1617	6.1844	6.7913
¼	0.9272	1.0785	0.1651	6.0554	6.9360
⅜	0.9467	1.0563	0.1686	5.9306	7.0820
½	0.9664	1.0348	0.1721	5.8097	7.2292
⅝	0.9863	1.0139	0.1757	5.6928	7.3778
¾	1.0063	0.9937	0.1792	5.5795	7.5276
⅞	1.0265	0.9742	0.1828	5.4697	7.6787
15	1.0469	0.9552	0.1865	5.3633	7.8310
⅛	1.0674	0.9369	0.1901	5.2601	7.9847
¼	1.0881	0.9190	0.1938	5.1600	8.1396
⅜	1.1090	0.9017	0.1975	5.0628	8.2958
½	1.1300	0.8849	0.2013	4.9685	8.4532
⅝	1.1513	0.8686	0.2050	4.8769	8.6120
¾	1.1726	0.8528	0.2089	4.7880	8.7720
⅞	1.1942	0.8374	0.2127	4.7015	8.9333
16	1.2159	0.8224	0.2166	4.6175	9.0958
¼	1.2599	0.7937	0.2244	4.4563	9.4248
½	1.3046	0.7665	0.2324	4.3038	9.7588
¾	1.3499	0.7408	0.2404	4.1592	10.0980
17	1.3959	0.7164	0.2486	4.0221	10.4422
¼	1.4426	0.6932	0.2569	3.8919	10.7916
½	1.4900	0.6711	0.2654	3.7682	11.1460
¾	1.5381	0.6502	0.2739	3.6504	11.5056

6" OD CASING		VOLUME AND FILL BETWEEN PIPE AND HOLE			

HOLE DIAM (IN)	CU FT PER LIN FT	LIN FT PER CU FT	BARRELS PER LIN FT	LIN FT PER BARREL	GALLONS PER LIN FT
7	0.0709	14.1036	0.0126	79.1857	0.5304
1/8	0.0805	12.4171	0.0143	69.7169	0.6024
1/4	0.0903	11.0700	0.0161	62.1533	0.6757
3/8	0.1003	9.9696	0.0179	55.9749	0.7503
1/2	0.1104	9.0514	0.0197	50.8353	0.8262
5/8	0.1208	8.2810	0.0215	46.4944	0.9033
3/4	0.1312	7.6196	0.0234	42.7808	0.9817
7/8	0.1419	7.0476	0.0253	39.5691	1.0614
8	0.1527	6.5481	0.0272	36.7648	1.1424
1/8	0.1637	6.1084	0.0292	34.2959	1.2246
1/4	0.1749	5.7184	0.0311	32.1065	1.3081
3/8	0.1862	5.3703	0.0332	30.1522	1.3929
1/2	0.1977	5.0578	0.0352	28.3976	1.4790
5/8	0.2094	4.7758	0.0373	26.8142	1.5663
3/4	0.2212	4.5201	0.0394	25.3785	1.6549
7/8	0.2333	4.2872	0.0415	24.0711	1.7448
9	0.2454	4.0744	0.0437	22.8759	1.8360
1/8	0.2578	3.8791	0.0459	21.7793	1.9284
1/4	0.2703	3.6993	0.0481	20.7700	2.0221
3/8	0.2830	3.5333	0.0504	19.8382	2.1171
1/2	0.2959	3.3797	0.0527	18.9754	2.2134
5/8	0.3089	3.2370	0.0550	18.1745	2.3109
3/4	0.3221	3.1043	0.0574	17.4292	2.4097
7/8	0.3355	2.9805	0.0598	16.7342	2.5098
10	0.3491	2.8648	0.0622	16.0846	2.6112
1/8	0.3628	2.7564	0.0646	15.4763	2.7138
1/4	0.3767	2.6548	0.0671	14.9055	2.8177
3/8	0.3907	2.5593	0.0696	14.3691	2.9229
1/2	0.4050	2.4693	0.0721	13.8642	3.0294
5/8	0.4194	2.3845	0.0747	13.3880	3.1371
3/4	0.4339	2.3044	0.0773	12.9384	3.2461
7/8	0.4487	2.2287	0.0799	12.5133	3.3564
11	0.4636	2.1570	0.0826	12.1108	3.4680
1/8	0.4787	2.0890	0.0853	11.7291	3.5808
1/4	0.4939	2.0245	0.0880	11.3669	3.6949
3/8	0.5094	1.9632	0.0907	11.0227	3.8103
1/2	0.5250	1.9049	0.0935	10.6952	3.9270
5/8	0.5407	1.8494	0.0963	10.3834	4.0449
3/4	0.5567	1.7964	0.0991	10.0861	4.1641
7/8	0.5728	1.7459	0.1020	9.8025	4.2846
12	0.5890	1.6977	0.1049	9.5316	4.4064
1/8	0.6055	1.6515	0.1078	9.2727	4.5294
1/4	0.6221	1.6074	0.1108	9.0250	4.6537
3/8	0.6389	1.5652	0.1138	8.7878	4.7793
1/2	0.6559	1.5247	0.1168	8.5606	4.9062
5/8	0.6730	1.4859	0.1199	8.3427	5.0343
3/4	0.6903	1.4487	0.1229	8.1336	5.1637
7/8	0.7078	1.4129	0.1261	7.9329	5.2944

VOLUME AND FILL BETWEEN PIPE AND HOLE

6" OD CASING

HOLE DIAM (IN)	CU FT PER LIN FT	LIN FT PER CU FT	BARRELS PER LIN FT	LIN FT PER BARREL	GALLONS PER LIN FT
13	0.7254	1.3785	0.1292	7.7400	5.4264
⅛	0.7432	1.3455	0.1324	7.5545	5.5596
¼	0.7612	1.3137	0.1356	7.3760	5.6941
⅜	0.7793	1.2831	0.1388	7.2042	5.8299
½	0.7;77	1.2537	0.1421	7.0387	5.9670
⅝	0.8162	1.2252	0.1454	6.8792	6.1053
¾	0.8348	1.1979	0.1487	6.7254	6.2449
⅞	0.8537	1.1714	0.1520	6.5771	6.3858
14	0.8727	1.1459	0.1554	6.4338	6.5280
⅛	0.8918	1.1213	0.1588	6.2955	6.6714
¼	0.9112	1.0975	0.1623	6.1619	6.8161
⅜	0.9307	1.0745	0.1658	6.0326	6.9621
½	0.9504	1.0522	0.1693	5.9077	7.1094
⅝	0.9702	1.0307	0.1728	5.7868	7.2579
¾	0.9903	1.0098	0.1764	5.6698	7.4077
⅞	1.0105	0.9896	0.1800	5.5564	7.5588
15	1.0308	0.9701	0.1836	5.4466	7.7112
⅛	1.0514	0.9511	0.1873	5.3402	7.8648
¼	1.0721	0.9328	0.1909	5.2371	8.0197
⅜	1.0930	0.9149	0.1947	5.1370	8.1759
½	1.1140	0.8977	0.1984	5.0400	8.3334
⅝	1.1352	0.8809	0.2022	4.9458	8.4921
¾	1.1566	0.8646	0.2060	4.8543	8.6521
⅞	1.1782	0.8488	0.2098	4.7655	8.8134
16	1.1999	0.8334	0.2137	4.6792	8.9760
¼	1.2439	0.8039	0.2215	4.5137	9.3049
½	1.2885	0.7761	0.2295	4.3573	9.6390
¾	1.3339	0.7497	0.2376	4.2092	9.9781
17	1.3799	0.7247	0.2458	4.0688	10.3224
¼	1.4266	0.7010	0.2541	3.9356	10.6717
½	1.4740	0.6784	0.2625	3.8091	11.0262
¾	1.5220	0.6570	0.2711	3.6888	11.3857

6-⅝″ OD
CASING

VOLUME AND FILL BETWEEN PIPE AND HOLE

HOLE DIAM (IN)	CU FT PER LIN FT	LIN FT PER CU FT	BARRELS PER LIN FT	LIN FT PER BARREL	GALLONS PER LIN FT
8	0.1097	9.1175	0.0195	51.1908	0.8205
⅛	0.1207	8.2868	0.0215	46.5272	0.9027
¼	0.1318	7.5851	0.0235	42.5873	0.9862
⅜	0.1432	6.9846	0.0255	39.2158	1.0710
½	0.1547	6.4651	0.0275	36.2989	1.1571
⅝	0.1664	6.0114	0.0296	33.7513	1.2444
¾	0.1782	5.6118	0.0317	31.5077	1.3330
⅞	0.1902	5.2572	0.0339	29.5173	1.4229
9	0.2024	4.9407	0.0360	27.7400	1.5141
⅛	0.2148	4.6564	0.0382	26.1439	1.6065
¼	0.2273	4.3998	0.0405	24.7029	1.7002
⅜	0.2400	4.1670	0.0427	23.3958	1.7952
½	0.2529	3.9549	0.0450	22.2051	1.8915
⅝	0.2659	3.7610	0.0474	21.1162	1.9890
¾	0.2791	3.5830	0.0497	20.1168	2.0878
⅞	0.2925	3.4190	0.0521	19.1965	2.1879
10	0.3060	3.2677	0.0545	18.3466	2.2893
⅛	0.3197	3.1274	0.0569	17.5593	2.3919
¼	0.3336	2.9972	0.0594	16.8282	2.4958
⅜	0.3477	2.8760	0.0619	16.1477	2.6010
½	0.3619	2.7629	0.0645	15.5127	2.7075
⅝	0.3763	2.6572	0.0670	14.9190	2.8152
¾	0.3909	2.5581	0.0696	14.3629	2.9242
⅞	0.4057	2.4652	0.0722	13.8409	3.0345
11	0.4206	2.3777	0.0749	13.3501	3.1461
⅛	0.4357	2.2954	0.0776	12.8878	3.2589
¼	0.4509	2.2178	0.0803	12.4518	3.3730
⅜	0.4663	2.1444	0.0831	12.0399	3.4884
½	0.4819	2.0750	0.0858	11.6503	3.6051
⅝	0.4977	2.0093	0.0886	11.2813	3.7230
¾	0.5136	1.9469	0.0915	10.9312	3.8422
⅞	0.5297	1.8877	0.0943	10.5989	3.9627
12	0.5460	1.8315	0.0972	10.2829	4.0845
⅛	0.5625	1.7779	0.1002	9.9822	4.2075
¼	0.5791	1.7269	0.1031	9.6957	4.3318
⅜	0.5959	1.6782	0.1061	9.4226	4.4574
½	0.6128	1.6318	0.1091	9.1618	4.5843
⅝	0.6300	1.5874	0.1122	8.9127	4.7124
¾	0.6473	1.5450	0.1153	8.6745	4.8418
⅞	0.6647	1.5044	0.1184	8.4465	4.9725
13	0.6824	1.4655	0.1215	8.2281	5.1045
⅛	0.7002	1.4282	0.1247	8.0188	5.2377
¼	0.7182	1.3924	0.1279	7.8180	5.3722
⅜	0.7363	1.3581	0.1311	7.6253	5.5080
½	0.7546	1.3251	0.1344	7.4401	5.6450
⅝	0.7731	1.2934	0.1377	7.2622	5.7834
¾	0.7918	1.2630	0.1410	7.0910	5.9230
⅞	0.8106	1.2336	0.1444	6.9263	6.0639

6-⅝" OD CASING		VOLUME AND FILL BETWEEN PIPE AND HOLE			
HOLE DIAM (IN)	CU FT PER LIN FT	LIN FT PER CU FT	BARRELS PER LIN FT	LIN FT PER BARREL	GALLONS PER LIN FT
14	0.8296	1.2054	0.1478	6.7676	6.2060
⅛	0.8488	1.1781	0.1512	6.6147	6.3495
¼	0.8681	1.1519	0.1546	6.4673	6.4942
⅜	0.8877	1.1266	0.1581	6.3251	6.6402
½	0.9073	1.1021	0.1616	6.1879	6.7874
⅝	0.9272	1.0785	0.1651	6.0554	6.9360
¾	0.9472	1.0557	0.1687	5.9274	7.0858
⅞	0.9674	1.0337	0.1723	5.8036	7.2369
15	0.9878	1.0124	0.1759	5.6839	7.3892
⅛	1.0083	0.9917	0.1796	5.5682	7.5429
¼	1.0290	0.9718	0.1833	5.4561	7.6978
⅜	1.0499	0.9524	0.1870	5.3476	7.8540
½	1.0710	0.9337	0.1907	5.2425	8.0114
⅝	1.0922	0.9156	0.1945	5.1406	8.1702
¾	1.1136	0.8980	0.1983	5.0419	8.3302
⅞	1.1351	0.8809	0.2022	4.9461	8.4915
16	1.1569	0.8644	0.2060	4.8532	8.6540
¼	1.2009	0.8327	0.2139	4.6755	8.9830
½	1.2455	0.8029	0.2218	4.5079	9.3170
¾	1.2908	0.7747	0.2299	4.3495	9.6562
17	1.3369	0.7480	0.2381	4.1998	10.0004
¼	1.3836	0.7228	0.2464	4.0581	10.3498
½	1.4309	0.6988	0.2549	3.9237	10.7042
¾	1.4790	0.6761	0.2634	3.7962	11.0638
18	1.5278	0.6546	0.2721	3.6750	11.4284
¼	1.5772	0.6340	0.2809	3.5599	11.7982
½	1.6273	0.6145	0.2898	3.4502	12.1730
¾	1.6781	0.5959	0.2989	3.3458	12.5530

VOLUME AND FILL BETWEEN PIPE AND HOLE

7" OD
CASING

HOLE DIAM (IN)	CU FT PER LIN FT	LIN FT PER CU FT	BARRELS PER LIN FT	LIN FT PER BARREL	GALLONS PER LIN FT
8	0.0818	12.2231	0.0146	68.6276	0.6120
1/8	0.0928	10.7752	0.0165	60.4982	0.6942
1/4	0.1040	9.6182	0.0185	54.0021	0.7777
3/8	0.1153	8.6727	0.0205	48.6936	0.8625
1/2	0.1268	7.8859	0.0226	44.2759	0.9486
5/8	0.1385	7.2210	0.0247	40.5431	1.0359
3/4	0.1503	6.6520	0.0268	37.3484	1.1245
7/8	0.1623	6.1597	0.0289	34.5840	1.2144
9	0.1745	5.7296	0.0311	32.1692	1.3056
1/8	0.1869	5.3507	0.0333	30.0422	1.3980
1/4	0.1994	5.0146	0.0355	28.1549	1.4917
3/8	0.2121	4.7144	0.0378	26.4695	1.5867
1/2	0.2250	4.4448	0.0401	24.9555	1.6830
5/8	0.2380	4.2013	0.0424	23.5884	1.7805
3/4	0.2512	3.9804	0.0447	22.3482	1.8793
7/8	0.2646	3.7791	0.0471	21.2182	1.9794
10	0.2782	3.5950	0.0495	20.1846	2.0808
1/8	0.2919	3.4260	0.0520	19.2358	2.1834
1/4	0.3058	3.2704	0.0545	18.3619	2.2873
3/8	0.3198	3.1266	0.0570	17.5546	2.3925
1/2	0.3341	2.9934	0.0595	16.8068	2.4990
5/8	0.3485	2.8697	0.0621	16.1121	2.6067
3/4	0.3630	2.7545	0.0647	15.4654	2.7157
7/8	0.3778	2.6470	0.0673	14.8618	2.8260
11	0.3927	2.5465	0.0699	14.2974	2.9376
1/8	0.4078	2.4523	0.0726	13.7685	3.0504
1/4	0.4230	2.3639	0.0753	13.2721	3.1645
3/8	0.4385	2.2807	0.0781	12.8052	3.2799
1/2	0.4541	2.2024	0.0809	12.3653	3.3966
5/8	0.4698	2.1285	0.0837	11.9504	3.5145
3/4	0.4858	2.0586	0.0865	11.5583	3.6337
7/8	0.5019	1.9926	0.0894	11.1874	3.7542
12	0.5181	1.9300	0.0923	10.8359	3.8760
1/8	0.5346	1.8706	0.0952	10.5026	3.9990
1/4	0.5512	1.8142	0.0982	10.1859	4.1233
3/8	0.5680	1.7606	0.1012	9.8848	4.2489
1/2	0.5850	1.7095	0.1042	9.5983	4.3758
5/8	0.6021	1.6609	0.1072	9.3252	4.5039
3/4	0.6194	1.6145	0.1103	9.0647	4.6333
7/8	0.6369	1.5702	0.1134	8.8161	4.7640
13	0.6545	1.5279	0.1166	8.5785	4.8960
1/8	0.6723	1.4874	0.1197	8.3512	5.0292
1/4	0.6903	1.4487	0.1229	8.1336	5.1637
3/8	0.7084	1.4115	0.1262	7.9252	5.2995
1/2	0.7268	1.3760	0.1294	7.7254	5.4366
5/8	0.7453	1.3418	0.1327	7.5337	5.5749
3/4	0.7639	1.3090	0.1361	7.3497	5.7145
7/8	0.7828	1.2775	0.1394	7.1728	5.8554

7" OD CASING	VOLUME AND FILL BETWEEN PIPE AND HOLE				
HOLE DIAM (IN)	CU FT PER LIN FT	LIN FT PER CU FT	BARRELS PER LIN FT	LIN FT PER BARREL	GALLONS PER LIN FT
14	0.8018	1.2473	0.1428	7.0028	5.9976
⅛	0.8209	1.2181	0.1462	6.8393	6.1410
¼	0.8403	1.1901	0.1497	6.6818	6.2857
⅜	0.8598	1.1631	0.1531	6.5301	6.4317
½	0.8795	1.1370	0.1566	6.3840	6.5790
⅝	0.8993	1.1119	0.1602	6.2430	6.7275
¾	0.9194	1.0877	0.1637	6.1070	6.8773
⅞	0.9396	1.0643	0.1673	5.9757	7.0284
15	0.9599	1.0417	0.1710	5.8489	7.1808
⅛	0.9805	1.0199	0.1746	5.7264	7.3344
¼	1.0012	0.9988	0.1783	5.6080	7.4893
⅜	1.0221	0.9784	0.1820	5.4934	7.6455
½	1.0431	0.9587	0.1858	5.3826	7.8030
⅝	1.0643	0.9396	0.1896	5.2752	7.9617
¾	1.0857	0.9210	0.1934	5.1713	8.1217
⅞	1.1073	0.9031	0.1972	5.0706	8.2830
16	1.1290	0.8857	0.2011	4.9730	8.4456
¼	1.1730	0.8525	0.2089	4.7866	8.7745
½	1.2176	0.8213	0.2169	4.6110	9.1086
¾	1.2630	0.7918	0.2249	4.4455	9.4477
17	1.3090	0.7639	0.2331	4.2892	9.7920
¼	1.3557	0.7376	0.2415	4.1415	10.1413
½	1.4031	0.7127	0.2499	4.0016	10.4958
¾	1.4511	0.6891	0.2585	3.8691	10.8553
18	1.4999	0.6667	0.2671	3.7433	11.2200
¼	1.5493	0.6454	0.2759	3.6239	11.5897
½	1.5994	0.6252	0.2849	3.5104	11.9646
¾	1.6502	0.6060	0.2939	3.4023	12.3445

7-⅝″ OD
CASING

VOLUME AND FILL BETWEEN PIPE AND HOLE

HOLE DIAM (IN)	CU FT PER LIN FT	LIN FT PER CU FT	BARRELS PER LIN FT	LIN FT PER BARREL	GALLONS PER LIN FT
9	0.1247	8.0206	0.0222	45.0325	0.9327
⅛	0.1370	7.2974	0.0244	40.9717	1.0251
¼	0.1496	6.6861	0.0266	37.5399	1.1188
⅜	0.1623	6.1629	0.0289	34.6022	1.2138
½	0.1751	5.7101	0.0312	32.0596	1.3101
⅝	0.1882	5.3144	0.0335	29.8381	1.4076
¾	0.2014	4.9658	0.0359	27.8809	1.5064
⅞	0.2148	4.6564	0.0382	26.1439	1.6065
10	0.2283	4.3801	0.0407	24.5922	1.7079
⅛	0.2420	4.1318	0.0431	23.1981	1.8105
¼	0.2559	3.9075	0.0456	21.9389	1.9144
⅜	0.2700	3.7040	0.0481	20.7962	2.0196
½	0.2842	3.5185	0.0506	19.7549	2.1261
⅝	0.2986	3.3488	0.0532	18.8021	2.2338
¾	0.3132	3.1930	0.0558	17.9272	2.3428
⅞	0.3279	3.0494	0.0584	17.1212	2.4531
11	0.3428	2.9168	0.0611	16.3765	2.5647
⅛	0.3579	2.7939	0.0637	15.6863	2.6775
¼	0.3732	2.6796	0.0665	15.0451	2.7916
⅜	0.3886	2.5733	0.0692	14.4479	2.9070
½	0.4042	2.4740	0.0720	13.8905	3.0237
⅝	0.4200	2.3811	0.0748	13.3690	3.1416
¾	0.4359	2.2941	0.0776	12.8803	3.2608
⅞	0.4520	2.2123	0.0805	12.4213	3.3813
12	0.4683	2.1354	0.0834	11.9895	3.5031
⅛	0.4847	2.0630	0.0863	11.5827	3.6261
¼	0.5014	1.9946	0.0893	11.1988	3.7504
⅜	0.5181	1.9300	0.0923	10.8359	3.8760
½	0.5351	1.8688	0.0953	10.4925	4.0029
⅝	0.5522	1.8108	0.0984	10.1671	4.1310
¾	0.5695	1.7558	0.1014	9.8582	4.2604
⅞	0.5870	1.7036	0.1045	9.5648	4.3911
13	0.6046	1.6539	0.1077	9.2858	4.5231
⅛	0.6225	1.6065	0.1109	9.0201	4.6563
¼	0.6404	1.5614	0.1141	8.7668	4.7908
⅜	0.6586	1.5184	0.1173	8.5252	4.9266
½	0.6769	1.4773	0.1206	8.2944	5.0637
⅝	0.6954	1.4380	0.1239	8.0738	5.2020
¾	0.7141	1.4004	0.1272	7.8628	5.3416
⅞	0.7329	1.3644	0.1305	7.6608	5.4825
14	0.7519	1.3300	0.1339	7.4671	5.6246
⅛	0.7711	1.2969	0.1373	7.2814	5.7681
¼	0.7904	1.2651	0.1408	7.1032	5.9128
⅜	0.8099	1.2347	0.1443	6.9321	6.0588
½	0.8296	1.2054	0.1478	6.7676	6.2060
⅝	0.8495	1.1772	0.1513	6.6094	6.3546
¾	0.8695	1.1501	0.1549	6.4572	6.5044
⅞	0.8897	1.1240	0.1585	6.3106	6.6555

7-⅝" OD CASING		**VOLUME AND FILL BETWEEN PIPE AND HOLE**			
HOLE DIAM (IN)	**CU FT PER LIN FT**	**LIN FT PER CU FT**	**BARRELS PER LIN FT**	**LIN FT PER BARREL**	**GALLONS PER LIN FT**
15	0.9101	1.0988	0.1621	6.1694	6.8078
⅛	0.9306	1.0746	0.1657	6.0332	6.9615
¼	0.9513	1.0512	0.1694	5.9019	7.1164
⅜	0.9722	1.0286	0.1732	5.7751	7.2726
½	0.9933	1.0068	0.1769	5.6527	7.4300
⅝	1.0145	0.9857	0.1807	5.5345	7.5888
¾	1.0359	0.9654	0.1845	5.4202	7.7488
⅞	1.0574	0.9457	0.1883	5.3097	7.9101
16	1.0792	0.9267	0.1922	5.2028	8.0726
¼	1.1231	0.8904	0.2000	4.9991	8.4016
½	1.1678	0.8563	0.2080	4.8079	8.7356
¾	1.2131	0.8243	0.2161	4.6282	9.0748
17	1.2591	0.7942	0.2243	4.4591	9.4190
¼	1.3058	0.7658	0.2326	4.2996	9.7684
½	1.3532	0.7390	0.2410	4.1490	10.1228
¾	1.4013	0.7136	0.2496	4.0067	10.4824
18	1.4500	0.6896	0.2583	3.8720	10.8470
¼	1.4995	0.6669	0.2671	3.7444	11.2168
½	1.5496	0.6453	0.2760	3.6233	11.5916
¾	1.6004	0.6249	0.2850	3.5083	11.9716
19	1.6518	0.6054	0.2942	3.3990	12.3566
¼	1.7040	0.5869	0.3035	3.2949	12.7468
½	1.7568	0.5692	0.3129	3.1959	13.1420
¾	1.8104	0.5524	0.3224	3.1014	13.5424

7-¾" OD
CASING

VOLUME AND FILL BETWEEN PIPE AND HOLE

HOLE DIAM (IN)	CU FT PER LIN FT	LIN FT PER CU FT	BARRELS PER LIN FT	LIN FT PER BARREL	GALLONS PER LIN FT
9	0.1142	8.7568	0.0203	49.1660	0.8542
⅛	0.1266	7.9018	0.0225	44.3653	0.9467
¼	0.1391	7.1901	0.0248	40.3692	1.0404
⅜	0.1518	6.5885	0.0270	36.9919	1.1354
½	0.1646	6.0736	0.0293	34.1007	1.2316
⅝	0.1777	5.6279	0.0316	31.5983	1.3292
¾	0.1909	5.2385	0.0340	29.4118	1.4280
⅞	0.2043	4.8954	0.0364	27.4854	1.5281
10	0.2178	4.5908	0.0388	25.7756	1.6294
⅛	0.2315	4.3188	0.0412	24.2483	1.7321
¼	0.2454	4.0744	0.0437	22.8759	1.8360
⅜	0.2595	3.8536	0.0462	21.6363	1.9412
½	0.2737	3.6532	0.0488	20.5114	2.0476
⅝	0.2881	3.4706	0.0513	19.4861	2.1554
¾	0.3027	3.3035	0.0539	18.5480	2.2644
⅞	0.3174	3.1501	0.0565	17.6866	2.3747
11	0.3324	3.0088	0.0592	16.8930	2.4862
⅛	0.3474	2.8781	0.0619	16.1596	2.5991
¼	0.3627	2.7571	0.0646	15.4799	2.7132
⅜	0.3781	2.6446	0.0673	14.8484	2.8286
½	0.3937	2.5399	0.0701	14.2603	2.9452
⅝	0.4095	2.4421	0.0729	13.7112	3.0632
¾	0.4254	2.3506	0.0758	13.1976	3.1824
⅞	0.4415	2.2648	0.0786	12.7162	3.3029
12	0.4578	2.1843	0.0815	12.2641	3.4246
⅛	0.4743	2.1086	0.0845	11.8387	3.5477
¼	0.4909	2.0372	0.0874	11.4379	3.6720
⅜	0.5077	1.9698	0.0904	11.0597	3.7976
½	0.5246	1.9061	0.0934	10.7022	3.9244
⅝	0.5418	1.8459	0.0965	10.3638	4.0526
¾	0.5591	1.7887	0.0996	10.0431	4.1820
⅞	0.5765	1.7345	0.1027	9.7387	4.3127
13	0.5942	1.6830	0.1058	9.4496	4.4446
⅛	0.6120	1.6341	0.1090	9.1746	4.5779
¼	0.6300	1.5874	0.1122	8.9127	4.7124
⅜	0.6481	1.5430	0.1154	8.6631	4.8482
½	0.6664	1.5005	0.1187	8.4249	4.9852
⅝	0.6849	1.4600	0.1220	8.1974	5.1236
¾	0.7036	1.4213	0.1253	7.9800	5.2632
⅞	0.7224	1.3842	0.1287	7.7719	5.4041
14	0.7414	1.3488	0.1321	7.5727	5.5462
⅛	0.7606	1.3148	0.1355	7.3818	5.6897
¼	0.7799	1.2821	0.1389	7.1987	5.8344
⅜	0.7995	1.2508	0.1424	7.0230	5.9804
½	0.8191	1.2208	0.1459	6.8542	6.1276
⅝	0.8390	1.1919	0.1494	6.6920	6.2762
¾	0.8590	1.1641	0.1530	6.5360	6.4260
⅞	0.8792	1.1374	0.1566	6.3858	6.5771

7-¾" OD CASING					

VOLUME AND FILL BETWEEN PIPE AND HOLE

HOLE DIAM (IN)	CU FT PER LIN FT	LIN FT PER CU FT	BARRELS PER LIN FT	LIN FT PER BARREL	GALLONS PER LIN FT
15	0.8996	1.1116	0.1602	6.2412	6.7294
⅛	0.9201	1.0868	0.1639	6.1019	6.8831
¼	0.9408	1.0629	0.1676	5.9676	7.0380
⅜	0.9617	1.0398	0.1713	5.8381	7.1942
½	0.9828	1.0175	0.1750	5.7130	7.3516
⅝	1.0040	0.9960	0.1788	5.5923	7.5104
¾	1.0254	0.9752	0.1826	5.4756	7.6704
⅞	1.0469	0.9552	0.1865	5.3628	7.8317
16	1.0687	0.9357	0.1903	5.2538	7.9942
¼	1.1126	0.8988	0.1982	5.0461	8.3232
½	1.1573	0.8641	0.2061	4.8514	8.6572
¾	1.2026	0.8315	0.2142	4.6685	8.9964
17	1.2487	0.8009	0.2224	4.4965	9.3406
¼	1.2954	0.7720	0.2307	4.3344	9.6900
½	1.3427	0.7447	0.2392	4.1814	10.0444
¾	1.3908	0.7190	0.2477	4.0369	10.4040
18	1.4396	0.6947	0.2564	3.9002	10.7686
¼	1.4890	0.6716	0.2652	3.7707	11.1384
½	1.5391	0.6497	0.2741	3.6480	11.5132
¾	1.5899	0.6290	0.2832	3.5314	11.8932
19	1.6414	0.6093	0.2923	3.4207	12.2782
¼	1.6935	0.5905	0.3016	3.3153	12.6684
½	1.7464	0.5726	0.3110	3.2150	13.0636
¾	1.7999	0.5556	0.3206	3.1194	13.4640

8″ OD CASING	**VOLUME AND FILL BETWEEN PIPE AND HOLE**			

HOLE DIAM (IN)	CU FT PER LIN FT	LIN FT PER CU FT	BARRELS PER LIN FT	LIN FT PER BARREL	GALLONS PER LIN FT
9	0.0927	10.7851	0.0165	60.5538	0.6936
⅛	0.1051	9.5168	0.0187	53.4327	0.7860
¼	0.1176	8.5030	0.0209	47.7409	0.8797
⅜	0.1303	7.6744	0.0232	43.0886	0.9747
½	0.1432	6.9846	0.0255	39.2158	1.0710
⅝	0.1562	6.4016	0.0278	35.9424	1.1685
¾	0.1694	5.9025	0.0302	33.1401	1.2673
⅞	0.1828	5.4705	0.0326	30.7145	1.3674
10	0.1963	5.0930	0.0350	28.5948	1.4688
⅛	0.2101	4.7603	0.0374	26.7272	1.5714
¼	0.2240	4.4651	0.0399	25.0694	1.6753
⅜	0.2380	4.2013	0.0424	23.5884	1.7805
½	0.2523	3.9642	0.0449	22.2576	1.8870
⅝	0.2667	3.7501	0.0475	21.0555	1.9947
¾	0.2812	3.5558	0.0501	19.9644	2.1037
⅞	0.2960	3.3787	0.0527	18.9699	2.2140
11	0.3109	3.2166	0.0554	18.0599	2.3256
⅛	0.3260	3.0678	0.0581	17.2242	2.4384
¼	0.3412	2.9306	0.0608	16.4542	2.5525
⅜	0.3567	2.8039	0.0635	15.7425	2.6679
½	0.3722	2.6864	0.0663	15.0830	2.7846
⅝	0.3880	2.5772	0.0691	14.4701	2.9025
¾	0.4039	2.4756	0.0719	13.8993	3.0217
⅞	0.4201	2.3806	0.0748	13.3663	3.1422
12	0.4363	2.2918	0.0777	12.8677	3.2640
⅛	0.4528	2.2086	0.0806	12.4002	3.3870
¼	0.4694	2.1304	0.0836	11.9612	3.5113
⅜	0.4862	2.0568	0.0866	11.5482	3.6369
½	0.5031	1.9875	0.0896	11.1590	3.7638
⅝	0.5203	1.9221	0.0927	10.7916	3.8919
¾	0.5376	1.8602	0.0957	10.4443	4.0213
⅞	0.5550	1.8017	0.0989	10.1155	4.1520
13	0.5727	1.7462	0.1020	9.8039	4.2840
⅛	0.5905	1.6935	0.1052	9.5082	4.4172
¼	0.6085	1.6434	0.1084	9.2272	4.5517
⅜	0.6266	1.5958	0.1116	8.9599	4.6875
½	0.6450	1.5505	0.1149	8.7054	4.8246
⅝	0.6634	1.5073	0.1182	8.4627	4.9629
¾	0.6821	1.4660	0.1215	8.2312	5.1025
⅞	0.7009	1.4266	0.1248	8.0100	5.2434
14	0.7199	1.3890	0.1282	7.7986	5.3856
⅛	0.7391	1.3530	0.1316	7.5963	5.5290
¼	0.7585	1.3184	0.1351	7.4025	5.6737
⅜	0.7780	1.2854	0.1386	7.2168	5.8197
½	0.7977	1.2537	0.1421	7.0387	5.9670
⅝	0.8175	1.2232	0.1456	6.8678	6.1155
¾	0.8376	1.1940	0.1492	6.7036	6.2653
⅞	0.8578	1.1658	0.1528	6.5457	6.4164

8" OD CASING	**VOLUME AND FILL BETWEEN PIPE AND HOLE**				
HOLE DIAM (IN)	**CU FT PER LIN FT**	**LIN FT PER CU FT**	**BARRELS PER LIN FT**	**LIN FT PER BARREL**	**GALLONS PER LIN FT**
15	0.8781	1.1388	0.1564	6.3939	6.5688
⅛	0.8987	1.1128	0.1601	6.2477	6.7224
¼	0.9194	1.0877	0.1637	6.1070	6.8773
⅜	0.9402	1.0636	0.1675	5.9714	7.0335
½	0.9613	1.0403	0.1712	5.8406	7.1910
⅝	0.9825	1.0178	0.1750	5.7145	7.3497
¾	1.0039	0.9961	0.1788	5.5927	7.5097
⅞	1.0255	0.9752	0.1826	5.4752	7.6710
16	1.0472	0.9549	0.1865	5.3615	7.8336
¼	1.0912	0.9164	0.1943	5.1455	8.1625
½	1.1358	0.8804	0.2023	4.9432	8.4966
¾	1.1812	0.8466	0.2104	4.7534	8.8357
17	1.2272	0.8149	0.2186	4.5752	9.1800
¼	1.2739	0.7850	0.2269	4.4074	9.5293
½	1.3213	0.7568	0.2353	4.2494	9.8838
¾	1.3693	0.7303	0.2439	4.1002	10.2433
18	1.4181	0.7052	0.2526	3.9593	10.6080
¼	1.4675	0.6814	0.2614	3.8259	10.9777
½	1.5176	0.6589	0.2703	3.6996	11.3526
¾	1.5684	0.6376	0.2793	3.5798	11.7325
19	1.6199	0.6173	0.2885	3.4660	12.1176
¼	1.6720	0.5981	0.2978	3.3579	12.5077
½	1.7249	0.5798	0.3072	3.2551	12.9030
¾	1.7784	0.5623	0.3167	3.1571	13.3033

8-⅛" OD
CASING

VOLUME AND FILL BETWEEN PIPE AND HOLE

HOLE DIAM (IN)	CU FT PER LIN FT	LIN FT PER CU FT	BARRELS PER LIN FT	LIN FT PER BARREL	GALLONS PER LIN FT
10	0.1854	5.3950	0.0330	30.2908	1.3866
⅛	0.1991	5.0232	0.0355	28.2031	1.4892
¼	0.2130	4.6955	0.0379	26.3635	1.5931
⅜	0.2270	4.4047	0.0404	24.7307	1.6983
½	0.2413	4.1449	0.0430	23.2718	1.8048
⅝	0.2557	3.9114	0.0455	21.9608	1.9125
¾	0.2702	3.7005	0.0481	20.7766	2.0215
⅞	0.2850	3.5090	0.0508	19.7017	2.1318
11	0.2999	3.3345	0.0534	18.7219	2.2434
⅛	0.3150	3.1748	0.0561	17.8254	2.3562
¼	0.3302	3.0282	0.0588	17.0019	2.4703
⅜	0.3457	2.8930	0.0616	16.2432	2.5857
½	0.3613	2.7681	0.0643	15.5420	2.7024
⅝	0.3770	2.6524	0.0671	14.8921	2.8203
¾	0.3930	2.5448	0.0700	14.2881	2.9395
⅞	0.4091	2.4446	0.0729	13.7255	3.0600
12	0.4253	2.3511	0.0758	13.2003	3.1818
⅛	0.4418	2.2635	0.0787	12.7088	3.3048
¼	0.4584	2.1815	0.0816	12.2481	3.4291
⅜	0.4752	2.1044	0.0846	11.8154	3.5547
½	0.4922	2.0319	0.0877	11.4082	3.6816
⅝	0.5093	1.9636	0.0907	11.0245	3.8097
¾	0.5266	1.8990	0.0938	10.6623	3.9391
⅞	0.5441	1.8381	0.0969	10.3199	4.0698
13	0.5617	1.7803	0.1000	9.9958	4.2018
⅛	0.5795	1.7256	0.1032	9.6886	4.3350
¼	0.5975	1.6737	0.1064	9.3970	4.4695
⅜	0.6156	1.6243	0.1096	9.1199	4.6053
½	0.6340	1.5774	0.1129	8.8564	4.7424
⅝	0.6525	1.5327	0.1162	8.6053	4.8807
¾	0.6711	1.4901	0.1195	8.3660	5.0203
⅞	0.6900	1.4494	0.1229	8.1377	5.1612
14	0.7090	1.4105	0.1263	7.9195	5.3034
⅛	0.7281	1.3734	0.1297	7.7110	5.4468
¼	0.7475	1.3378	0.1331	7.5114	5.5915
⅜	0.7670	1.3038	0.1366	7.3203	5.7375
½	0.7867	1.2712	0.1401	7.1371	5.8847
⅝	0.8065	1.2399	0.1436	6.9614	6.0333
¾	0.8266	1.2098	0.1472	6.7927	6.1831
⅞	0.8468	1.1810	0.1508	6.6307	6.3342
15	0.8671	1.1532	0.1544	6.4749	6.4865
⅛	0.8877	1.1266	0.1581	6.3251	6.6402
¼	0.9084	1.1009	0.1618	6.1809	6.7951
⅜	0.9293	1.0761	0.1655	6.0420	6.9513
½	0.9503	1.0523	0.1693	5.9082	7.1087
⅝	0.9715	1.0293	0.1730	5.7792	7.2675
¾	0.9929	1.0071	0.1768	5.6547	7.4275
⅞	1.0145	0.9857	0.1807	5.5345	7.5888

8-⅛" OD
CASING

VOLUME AND FILL BETWEEN
PIPE AND HOLE

HOLE DIAM (IN)	CU FT PER LIN FT	LIN FT PER CU FT	BARRELS PER LIN FT	LIN FT PER BARREL	GALLONS PER LIN FT
16	1.0362	0.9651	0.1846	5.4184	7.7513
¼	1.0802	0.9258	0.1924	5.1978	8.0803
½	1.1248	0.8890	0.2003	4.9915	8.4143
¾	1.1702	0.8546	0.2084	4.7981	8.7535
17	1.2162	0.8222	0.2166	4.6165	9.0977
¼	1.2629	0.7918	0.2249	4.4458	9.4471
½	1.3103	0.7632	0.2334	4.2850	9.8015
¾	1.3583	0.7362	0.2419	4.1334	10.1611
18	1.4071	0.7107	0.2506	3.9902	10.5257
¼	1.4565	0.6866	0.2594	3.8548	10.8955
½	1.5066	0.6637	0.2683	3.7266	11.2703
¾	1.5574	0.6421	0.2774	3.6051	11.6503
19	1.6089	0.6215	0.2866	3.4897	12.0353
¼	1.6610	0.6020	0.2958	3.3802	12.4255
½	1.7139	0.5835	0.3053	3.2759	12.8207
¾	1.7674	0.5658	0.3148	3.1767	13.2211
20	1.8216	0.5490	0.3244	3.0822	13.6265
¼	1.8765	0.5329	0.3342	2.9921	14.0371
½	1.9320	0.5176	0.3441	2.9060	14.4527
¾	1.9883	0.5029	0.3541	2.8238	14.8735

8-⅝" OD
CASING

VOLUME AND FILL BETWEEN PIPE AND HOLE

HOLE DIAM (IN)	CU FT PER LIN FT	LIN FT PER CU FT	BARRELS PER LIN FT	LIN FT PER BARREL	GALLONS PER LIN FT
10	0.1397	7.1594	0.0249	40.1968	1.0449
⅛	0.1534	6.5190	0.0273	36.6014	1.1475
¼	0.1673	5.9777	0.0298	33.5622	1.2514
⅜	0.1814	5.5142	0.0323	30.9598	1.3566
½	0.1956	5.1129	0.0348	28.7070	1.4631
⅝	0.2100	4.7622	0.0374	26.7380	1.5708
¾	0.2246	4.4532	0.0400	25.0028	1.6798
⅞	0.2393	4.1788	0.0426	23.4624	1.7901
11	0.2542	3.9337	0.0453	22.0860	1.9017
⅛	0.2693	2.7133	0.0480	20.8489	2.0145
¼	0.2846	3.5143	0.0507	19.7312	2.1286
⅜	0.3000	3.3336	0.0534	18.7166	2.2440
½	0.3156	3.1688	0.0562	17.7917	2.3607
⅝	0.3313	3.0180	0.0590	16.9451	2.4786
¾	0.3473	2.8796	0.0619	16.1675	2.5978
⅞	0.3634	2.7519	0.0647	15.4509	2.7183
12	0.3797	2.6339	0.0676	14.7884	2.8401
⅛	0.3961	2.5246	0.0705	14.1744	2.9631
¼	0.4127	2.4229	0.0735	13.6037	3.0874
⅜	0.4295	2.3282	0.0765	13.0719	3.2130
½	0.4465	2.2398	0.0795	12.5754	3.3399
⅝	0.4636	2.1570	0.0826	12.1108	3.4680
¾	0.4809	2.0794	0.0857	11.6751	3.5974
⅞	0.4984	2.0065	0.0888	11.2658	3.7281
13	0.5160	1.9379	0.0919	10.8807	3.8601
⅛	0.5338	1.8733	0.0951	10.5176	3.9933
¼	0.5518	1.8122	0.0983	10.1749	4.1278
⅜	0.5700	1.7545	0.1015	9.8509	4.2636
½	0.5883	1.6999	0.1048	9.5440	4.4007
⅝	0.6068	1.6481	0.1081	9.2532	4.5390
¾	0.6254	1.5989	0.1114	8.9770	4.6786
⅞	0.6443	1.5521	0.1147	8.7146	4.8195
14	0.6633	1.5077	0.1181	8.4649	4.9617
⅛	0.6825	1.4653	0.1215	8.2271	5.1051
¼	0.7018	1.4249	0.1250	8.0003	5.2498
⅜	0.7213	1.3864	0.1285	7.7839	5.3958
½	0.7410	1.3495	0.1320	7.5771	5.5430
⅝	0.7609	1.3143	0.1355	7.3793	5.6916
¾	0.7809	1.2806	0.1391	7.1901	5.8414
⅞	0.8011	1.2483	0.1427	7.0088	5.9925
15	0.8214	1.2174	0.1463	6.8350	6.1448
⅛	0.8420	1.1877	0.1500	6.6683	6.2985
¼	0.8627	1.1592	0.1537	6.5082	6.4534
⅜	0.8836	1.1318	0.1574	6.3544	6.6096
½	0.9046	1.1054	0.1611	6.2065	6.7670
⅝	0.9258	1.0801	0.1649	6.0643	6.9258
¾	0.9472	1.0557	0.1687	5.9274	7.0858
⅞	0.9688	1.0322	0.1725	5.7954	7.2471

8-⅝″ OD CASING		**VOLUME AND FILL BETWEEN PIPE AND HOLE**			
HOLE DIAM (IN)	**CU FT PER LIN FT**	**LIN FT PER CU FT**	**BARRELS PER LIN FT**	**LIN FT PER BARREL**	**GALLONS PER LIN FT**
16	0.9905	1.0096	0.1764	5.6683	7.4096
¼	1.0345	0.9667	0.1843	5.4273	7.7386
½	1.0792	0.9267	0.1922	5.2028	8.0726
¾	1.1245	0.8893	0.2003	4.9930	8.4118
17	1.1705	0.8543	0.2085	4.7967	8.7560
¼	1.2172	0.8215	0.2168	4.6127	9.1054
½	1.2646	0.7908	0.2252	4.4398	9.4598
¾	1.3127	0.7618	0.2338	4.2773	9.8194
18	1.3614	0.7345	0.2425	4.1241	10.1840
¼	1.4108	0.7088	0.2513	3.9796	10.5538
½	1.4609	0.6845	0.2602	3.8431	10.9286
¾	1.5117	0.6615	0.2693	3.7140	11.3086
19	1.5632	0.6397	0.2784	3.5917	11.6936
¼	1.6154	0.6191	0.2877	3.4757	12.0838
½	1.6682	0.5994	0.2971	3.3656	12.4790
¾	1.7217	0.5808	0.3067	3.2610	12.8794
20	1.7759	0.5631	0.3163	3.1615	13.2848
¼	1.8308	0.5462	0.3261	2.0667	13.6954
½	1.8864	0.5301	0.3360	2.9764	14.1110
¾	1.9426	0.5148	0.3460	2.8902	14.5318

	9″ OD CASING		**VOLUME AND FILL BETWEEN PIPE AND HOLE**		

HOLE DIAM (IN)	CU FT PER LIN FT	LIN FT PER CU FT	BARRELS PER LIN FT	LIN FT PER BARREL	GALLONS PER LIN FT
10	0.1036	9.6498	0.0185	54.1797	0.7752
1/8	0.1173	8.5216	0.0209	47.8450	0.8778
1/4	0.1312	7.6196	0.0234	42.7808	0.9817
3/8	0.1453	6.8822	0.0259	38.6408	1.0869
1/2	0.1595	6.2683	0.0284	35.1936	1.1934
5/8	0.1739	5.7492	0.0310	32.2795	1.3011
3/4	0.1885	5.3048	0.0336	29.7841	1.4101
7/8	0.2033	4.9200	0.0362	27.6237	1.5204
11	0.2182	4.5837	0.0389	25.7354	1.6320
1/8	0.2333	4.2872	0.0415	24.0711	1.7448
1/4	0.2485	4.0241	0.0443	22.5935	1.8589
3/8	0.2639	3.7889	0.0470	21.2730	1.9743
1/2	0.2795	3.5775	0.0498	20.0861	2.0910
5/8	0.2953	3.3865	0.0526	19.0137	2.2089
3/4	0.3112	3.2131	0.0554	18.0401	2.3281
7/8	0.3273	3.0550	0.0583	17.1524	2.4486
12	0.3436	2.9103	0.0612	16.3399	2.5704
1/8	0.3601	2.7773	0.0641	15.5935	2.6934
1/4	0.3767	2.6548	0.0671	14.9055	2.8177
3/8	0.3935	2.5415	0.0701	14.2695	2.9433
1/2	0.4104	2.4365	0.0731	13.6799	3.0702
5/8	0.4276	2.3389	0.0762	13.1319	3.1983
3/4	0.4449	2.2479	0.0792	12.6212	3.3277
7/8	0.4623	2.1630	0.0823	12.1442	3.4584
13	0.4800	2.0835	0.0855	11.6979	3.5904
1/8	0.4978	2.0089	0.0887	11.2793	3.7236
1/4	0.5158	1.9389	0.0919	10.8861	3.8581
3/8	0.5339	1.8730	0.0951	10.5160	3.9939
1/2	0.5522	1.8108	0.0984	10.1671	4.1310
5/8	0.5707	1.7522	0.1017	9.8376	4.2693
3/4	0.5894	1.6967	0.1050	9.5261	4.4089
7/8	0.6082	1.6441	0.1083	9.2311	4.5498
14	0.6272	1.5943	0.1117	8.9514	4.6920
1/8	0.6464	1.5470	0.1151	8.6859	4.8354
1/4	0.6657	1.5021	0.1186	8.4335	4.9801
3/8	0.6853	1.4593	0.1221	8.1933	5.1261
1/2	0.7049	1.4185	0.1256	7.9645	5.2734
5/8	0.7248	1.3797	0.1291	7.7463	5.4219
3/4	0.7448	1.3426	0.1327	7.5380	5.5717
7/8	0.7650	1.3071	0.1363	7.3390	5.7228
15	0.7854	1.2732	0.1399	7.1487	5.8752
1/8	0.8059	1.2408	0.1435	6.9665	6.0288
1/4	0.8266	1.2097	0.1472	6.7920	6.1837
3/8	0.8475	1.1799	0.1510	6.6247	6.3399
1/2	0.8686	1.1513	0.1547	6.4641	6.4974
5/8	0.8898	1.1239	0.1585	6.3100	6.6561
3/4	0.9112	1.0975	0.1623	6.1619	6.8161
7/8	0.9327	1.0721	0.1661	6.0194	6.9774

9" OD CASING	**VOLUME AND FILL BETWEEN PIPE AND HOLE**				

HOLE DIAM (IN)	CU FT PER LIN FT	LIN FT PER CU FT	BARRELS PER LIN FT	LIN FT PER BARREL	GALLONS PER LIN FT
16	0.9545	1.0477	0.1700	5.8824	7.1400
¼	0.9985	1.0016	0.1778	5.6233	7.4689
½	1.0431	0.9587	0.1858	5.3826	7.8030
¾	1.0884	0.9187	0.1939	5.1584	8.1421
17	1.1345	0.8815	0.2021	4.9491	8.4864
¼	1.1812	0.8466	0.2104	4.7534	8.8357
½	1.2285	0.8140	0.2188	4.5701	9.1902
¾	1.2766	0.7833	0.2274	4.3980	9.5497
18	1.3254	0.7545	0.2361	4.2363	9.9144
¼	1.3748	0.7274	0.2449	4.0840	10.2841
½	1.4249	0.7018	0.2538	3.9403	10.6590
¾	1.4757	0.6776	0.2628	3.8047	11.0389
19	1.5272	0.6548	0.2720	3.6765	11.4240
¼	1.5793	0.6332	0.2813	3.5551	11.8141
½	1.6322	0.6127	0.2907	3.4400	12.2094
¾	1.6857	0.5932	0.3002	3.3308	12.6097
20	1.7399	0.5748	0.3099	3.2270	13.0152
¼	1.7948	0.5572	0.3197	3.1283	13.4257
½	1.8503	0.5404	0.3296	3.0344	13.8414
¾	1.9066	0.5245	0.3396	2.9449	14.2621

VOLUME AND FILL BETWEEN PIPE AND HOLE

9-5/8" OD CASING

HOLE DIAM (IN)	CU FT PER LIN FT	LIN FT PER CU FT	BARRELS PER LIN FT	LIN FT PER BARREL	GALLONS PER LIN FT
11	0.1547	6.4651	0.0275	36.2989	1.1571
1/8	0.1698	5.8907	0.0302	33.0735	1.2699
1/4	0.1850	5.4050	0.0330	30.3466	1.3840
3/8	0.2004	4.9890	0.0357	28.0113	1.4994
1/2	0.2160	4.6289	0.0385	25.9892	1.6161
5/8	0.2318	4.3140	0.0413	24.2215	1.7340
3/4	0.2477	4.0365	0.0441	22.6634	1.8532
7/8	0.2638	3.7901	0.0470	21.2799	1.9737
12	0.2801	3.5699	0.0499	20.0434	2.0955
1/8	0.2966	3.3719	0.0528	18.9318	2.2185
1/4	0.3132	3.1930	0.0558	17.9272	2.3428
3/8	0.3300	3.0305	0.0588	17.0151	2.4684
1/2	0.3469	2.8824	0.0618	16.1834	2.5953
5/8	0.3641	2.7468	0.0648	15.4219	2.7234
3/4	0.3814	2.6222	0.0679	14.7223	2.8528
7/8	0.3988	2.5073	0.0710	14.0775	2.9835
13	0.4165	2.4011	0.0742	13.4812	3.1155
1/8	0.4343	2.3026	0.0773	12.9283	3.2487
1/4	0.4523	2.2111	0.0806	12.4143	3.3832
3/8	0.4704	2.1258	0.0838	11.9352	3.5190
1/2	0.4887	2.0461	0.0870	11.4878	3.6561
5/8	0.5072	1.9715	0.0903	11.0690	3.7944
3/4	0.5259	1.9015	0.0937	10.6761	3.9340
7/8	0.5447	1.8358	0.0970	10.3070	4.0749
14	0.5637	1.7739	0.1004	9.9596	4.2171
1/8	0.5829	1.7155	0.1038	9.6319	4.3605
1/4	0.6023	1.6604	0.1073	9.3226	4.5052
3/8	0.6218	1.6083	0.1107	9.0299	4.6512
1/2	0.6415	1.5589	0.1142	8.7528	4.7985
5/8	0.6613	1.5121	0.1178	8.4900	4.9470
3/4	0.6813	1.4677	0.1214	8.2405	5.0968
7/8	0.7015	1.4254	0.1249	8.0032	5.2479
15	0.7219	1.3852	0.1286	7.7774	5.4002
1/8	0.7424	1.3469	0.1322	7.5623	5.5539
1/4	0.7632	1.3103	0.1359	7.3571	5.7088
3/8	0.7840	1.2755	0.1396	7.1611	5.8650
1/2	0.8051	1.2421	0.1434	6.9739	6.0224
5/8	0.8263	1.2102	0.1472	6.7948	6.1812
3/4	0.8477	1.1797	0.1510	6.6234	6.3412
7/8	0.8693	1.1504	0.1548	6.4591	6.5025
16	0.8910	1.1224	0.1587	6.3015	6.6650
1/4	0.9350	1.0696	0.1665	6.0052	6.9940
1/2	0.9796	1.0208	0.1745	5.7314	7.3280
3/4	1.0250	0.9757	0.1826	5.4779	7.6672
17	1.0710	0.9337	0.1907	5.2425	8.0114
1/4	1.1177	0.8947	0.1991	5.0234	8.3608
1/2	1.1651	0.8583	0.2075	4.8191	8.7152
3/4	1.2131	0.8243	0.2161	4.6282	9.0748

9-⅝″ OD CASING		VOLUME AND FILL BETWEEN PIPE AND HOLE			
HOLE DIAM (IN)	CU FT PER LIN FT	LIN FT PER CU FT	BARRELS PER LIN FT	LIN FT PER BARREL	GALLONS PER LIN FT
18	1.2619	0.7925	0.2247	4.4494	9.4394
¼	1.3113	0.7626	0.2336	4.2817	9.8092
½	1.3614	0.7345	0.2425	4.1241	10.1840
¾	1.4122	0.7081	0.2515	3.9758	10.5640
19	1.4637	0.6832	0.2607	3.8360	10.9490
¼	1.5158	0.6597	0.2700	3.7040	11.3392
½	1.5687	0.6375	0.2794	3.5792	11.7344
¾	1.6222	0.6165	0.2889	3.4611	12.1348

10″ OD CASING	**VOLUME AND FILL BETWEEN PIPE AND HOLE**				

HOLE DIAM (IN)	CU FT PER LIN FT	LIN FT PER CU FT	BARRELS PER LIN FT	LIN FT PER BARREL	GALLONS PER LIN FT
11	0.1145	8.7308	0.0204	49.0197	0.8568
⅛	0.1296	7.7148	0.0231	43.3153	0.9696
¼	0.1449	6.9025	0.0258	38.7544	1.0837
⅜	0.1603	6.2383	0.0286	35.0253	1.1991
½	0.1759	5.6852	0.0313	31.9198	1.3158
⅝	0.1917	5.2175	0.0341	29.2941	1.4337
¾	0.2076	4.8170	0.0370	27.0454	1.5529
⅞	0.2237	4.4702	0.0398	25.0981	1.6734
12	0.2400	4.1670	0.0427	23.3958	1.7952
⅛	0.2564	3.8997	0.0457	21.8951	1.9182
¼	0.2730	3.6624	0.0486	20.5626	2.0425
⅜	0.2898	3.4502	0.0516	19.3715	2.1681
½	0.3068	3.2595	0.0546	18.3007	2.2950
⅝	0.3239	3.0871	0.0577	17.3329	2.4231
¾	0.3412	2.9306	0.0608	16.4542	2.5525
⅞	0.3587	2.7879	0.0639	15.6528	2.6832
13	0.3763	2.6572	0.0670	14.9190	2.8152
⅛	0.3941	2.5371	0.0702	14.2449	2.9484
¼	0.4121	2.4264	0.0734	13.6233	3.0829
⅜	0.4303	2.3241	0.0766	13.0486	3.2187
½	0.4486	2.2291	0.0799	12.5157	3.3558
⅝	0.4671	2.1409	0.0832	12.0202	3.4941
¾	0.4858	2.0586	0.0865	11.5583	3.6337
⅞	0.5046	1.9818	0.0899	11.1269	3.7746
14	0.5236	1.9099	0.0933	10.7231	3.9168
⅛	0.5428	1.8424	0.0967	10.3442	4.0602
¼	0.5621	1.7790	0.1001	9.9883	4.2049
⅜	0.5816	1.7193	0.1036	9.6531	4.3509
½	0.6013	1.6630	0.1071	9.3371	4.4982
⅝	0.6212	1.6098	0.1106	9.0386	4.6467
¾	0.6412	1.5596	0.1142	8.7563	4.7965
⅞	0.6614	1.5119	0.1178	8.4889	4.9476
15	0.6818	1.4668	0.1214	8.2353	5.1000
⅛	0.7023	1.4239	0.1251	7.9945	5.2536
¼	0.7230	1.3831	0.1288	7.7655	5.4085
⅜	0.7439	1.3443	0.1325	7.5475	5.5647
½	0.7649	1.3073	0.1362	7.3399	5.7222
⅝	0.7862	1.2720	0.1400	7.1417	5.8809
¾	0.8076	1.2383	0.1438	6.9526	6.0409
⅞	0.8291	1.2061	0.1477	6.7718	6.2022
16	0.8508	1.1753	0.1515	6.5988	6.3648
¼	0.8948	1.1175	0.1594	6.2745	6.6937
½	0.9395	1.0644	0.1673	5.9763	7.0278
¾	0.9848	1.0154	0.1754	5.7012	7.3669
17	1.0308	0.9701	0.1836	5.4466	7.7112
¼	1.0775	0.9280	0.1919	5.2106	8.0605
½	1.1249	0.8890	0.2004	4.9911	8.4150
¾	1.1730	0.8525	0.2089	4.7866	8.7745

10″ OD CASING	**VOLUME AND FILL BETWEEN PIPE AND HOLE**				
HOLE DIAM (IN)	**CU FT PER LIN FT**	**LIN FT PER CU FT**	**BARRELS PER LIN FT**	**LIN FT PER BARREL**	**GALLONS PER LIN FT**
18	1.2217	0.8185	0.2176	4.5956	9.1392
¼	1.2712	0.7867	0.2264	4.4169	9.5089
½	1.3213	0.7568	0.2353	4.2494	9.8838
¾	1.3721	0.7288	0.2444	4.0921	10.2637
19	1.4235	0.7025	0.2535	3.9441	10.6488
¼	1.4757	0.6776	0.2628	3.8047	11.0389
½	1.5285	0.6542	0.2722	3.6732	11.4342
¾	1.5820	0.6321	0.2818	3.5489	11.8345

10-¾″ OD
CASING

VOLUME AND FILL BETWEEN PIPE AND HOLE

HOLE DIAM (IN)	CU FT PER LIN FT	LIN FT PER CU FT	BARRELS PER LIN FT	LIN FT PER BARREL	GALLONS PER LIN FT
12	0.1551	6.4473	0.0276	36.1992	1.1602
⅛	0.1716	5.8292	0.0306	32.7285	1.2833
¼	0.1882	5.3144	0.0335	29.8381	1.4076
⅜	0.2050	4.8791	0.0365	27.3940	1.5332
½	0.2219	4.5062	0.0395	25.3005	1.6600
⅝	0.2390	4.1833	0.0426	23.4875	1.7882
¾	0.2563	3.9010	0.0457	21.9024	1.9176
⅞	0.2738	3.6521	0.0488	20.5050	2.0483
13	0.2915	3.4310	0.0519	19.2639	2.1802
⅛	0.3093	3.2334	0.0551	18.1545	2.3135
¼	0.3272	3.0558	0.0583	17.1569	2.4480
⅜	0.3454	2.8952	0.0615	16.2552	2.5838
½	0.3637	2.7493	0.0648	15.4364	2.7208
⅝	0.3822	2.6163	0.0681	14.6895	2.8592
¾	0.4009	2.4945	0.0714	14.0056	2.9988
⅞	0.4197	2.3826	0.0748	13.3772	3.1397
14	0.4387	2.2794	0.0781	12.7977	3.2818
⅛	0.4579	2.1839	0.0816	12.2618	3.4253
¼	0.4772	2.0954	0.0850	11.7647	3.5700
⅜	0.4968	2.0131	0.0885	11.3025	3.7160
½	0.5164	1.9363	0.0920	10.8717	3.8632
⅝	0.5363	1.8646	0.0955	10.4692	4.0118
¾	0.5563	1.7975	0.0991	10.0923	4.1616
⅞	0.5765	1.7345	0.1027	9.7387	4.3127
15	0.5969	1.6754	0.1063	9.4064	4.4650
⅛	0.6174	1.6196	0.1100	9.0935	4.6187
¼	0.6381	1.5671	0.1137	8.7984	4.7736
⅜	0.6590	1.5174	0.1174	8.5197	4.9298
½	0.6801	1.4704	0.1211	8.2560	5.0872
⅝	0.7013	1.4260	0.1249	8.0061	5.2460
¾	0.7227	1.3837	0.1287	7.7692	5.4060
⅞	0.7442	1.3437	0.1326	7.5441	5.5673
16	0.7660	1.3055	0.1364	7.3301	5.7298
¼	0.8099	1.2347	0.1443	6.9321	6.0588
½	0.8546	1.1701	0.1522	6.5699	6.3928
¾	0.8999	1.1112	0.1603	6.2389	6.7320
17	0.9460	1.0571	0.1685	5.9354	7.0762
¼	0.9927	1.0074	0.1768	5.6561	7.4256
½	1.0400	0.9615	0.1852	5.3984	7.7800
¾	1.0881	0.9190	0.1938	5.1600	8.1396
18	1.1369	0.8796	0.2025	4.9387	8.5042
¼	1.1863	0.8430	0.2113	4.7329	8.8740
½	1.2364	0.8088	0.2202	4.5411	9.2488
¾	1.2872	0.7769	0.2293	4.3619	9.6288
19	1.3387	0.7470	0.2384	4.1942	10.0138
¼	1.3908	0.7190	0.2477	4.0369	10.4040
½	1.4436	0.6927	0.2571	3.8892	10.7992
¾	1.4972	0.6679	0.2667	3.7501	11.1996

| 11-¾" OD CASING | VOLUME AND FILL BETWEEN PIPE AND HOLE | | | | |

HOLE DIAM (IN)	CU FT PER LIN FT	LIN FT PER CU FT	BARRELS PER LIN FT	LIN FT PER BARREL	GALLONS PER LIN FT
13	0.1687	5.9264	0.0301	33.2740	1.2622
⅛	0.1865	5.3605	0.0332	30.0971	1.3955
¼	0.2045	4.8892	0.0364	27.4510	1.5300
⅜	0.2227	4.4907	0.0397	25.2134	1.6658
½	0.2410	4.1493	0.0429	23.2965	1.8028
⅝	0.2595	3.8536	0.0462	21.6363	1.9412
¾	0.2782	3.5950	0.0495	20.1846	2.0808
⅞	0.2970	3.3671	0.0529	18.9046	2.2217
14	0.3160	3.1646	0.0563	17.7677	2.3638
⅛	0.3352	2.9835	0.0597	16.7512	2.5073
¼	0.3545	2.8207	0.0631	15.8371	2.6520
⅜	0.3740	2.6735	0.0666	15.0108	2.7980
½	0.3937	2.5399	0.0701	14.2603	2.9452
⅝	0.4136	2.4179	0.0737	13.5756	3.0938
¾	0.4336	2.3062	0.0772	12.9486	3.2436
⅞	0.4538	2.2036	0.0808	12.3723	3.3947
15	0.4742	2.1089	0.0845	11.8409	3.5470
⅛	0.4947	2.0214	0.0881	11.3493	3.7007
¼	0.5154	1.9402	0.0918	10.8933	3.8556
⅜	0.5363	1.8646	0.0955	10.4692	4.0118
½	0.5573	1.7942	0.0993	10.0738	4.1692
⅝	0.5786	1.7284	0.1030	9.7043	4.3280
¾	0.6000	1.6668	0.1069	9.3583	4.4880
⅞	0.6215	1.6090	0.1107	9.0337	4.6493
16	0.6432	1.5546	0.1146	8.7285	4.8118
¼	0.6872	1.4551	0.1224	8.1700	5.1408
½	0.7319	1.3663	0.1304	7.6715	5.4748
¾	0.7772	1.2866	0.1384	7.2240	5.8140
17	0.8232	1.2147	0.1466	6.8201	6.1582
¼	0.8699	1.1495	0.1549	6.4540	6.5076
½	0.9173	1.0901	0.1634	6.1206	6.8620
¾	0.9654	1.0359	0.1719	5.8159	7.2216
18	1.0141	0.9861	0.1806	5.5363	7.5862
¼	1.0636	0.9402	0.1894	5.2790	7.9560
½	1.1137	0.8979	0.1984	5.0415	8.3308
¾	1.1645	0.8588	0.2074	4.8216	8.7108
19	1.2159	0.8224	0.2166	4.6175	9.0958
¼	1.2681	0.7886	0.2259	4.4276	9.4860
½	1.3209	0.7570	0.2353	4.2505	9.8812
¾	1.3744	0.7276	0.2448	4.0850	10.2816
20	1.4286	0.7000	0.2545	3.9300	10.6870
¼	1.4835	0.6741	0.2642	3.7846	11.0976
½	1.5391	0.6497	0.2741	3.6480	11.5132
¾	1.5953	0.6268	0.2841	3.5194	11.9340

Well Control Problems and Solutions

VOLUME AND FILL BETWEEN PIPE AND HOLE

12-¾" OD CASING

HOLE DIAM (IN)	CU FT PER LIN FT	LIN FT PER CU FT	BARRELS PER LIN FT	LIN FT PER BARREL	GALLONS PER LIN FT
14	0.1824	5.4833	0.0325	30.7862	1.3642
⅛	0.2015	4.9616	0.0359	27.8573	1.5077
¼	0.2209	4.5271	0.0393	25.4176	1.6524
⅜	0.2404	4.1596	0.0428	23.3543	1.7984
½	0.2601	3.8447	0.0463	21.5867	1.9456
⅝	0.2800	3.5720	0.0499	20.0556	2.0942
¾	0.3000	3.3336	0.0534	18.7166	2.2440
⅞	0.3202	3.1233	0.0570	17.5359	2.3951
15	0.3405	2.9365	0.0607	16.4871	2.5474
⅛	0.3611	2.7695	0.0643	15.5493	2.7011
¼	0.3818	2.6192	0.0680	14.7059	2.8560
⅜	0.4027	2.4834	0.0717	13.9434	3.0122
½	0.4237	2.3601	0.0755	13.2507	3.1696
⅝	0.4449	2.2475	0.0792	12.6188	3.3284
¾	0.4663	2.1444	0.0831	12.0399	3.4884
⅞	0.4879	2.0496	0.0869	11.5079	3.6497
16	0.5096	1.9622	0.0908	11.0171	3.8122
¼	0.5536	1.8064	0.0986	10.1420	4.1412
½	0.5983	1.6715	0.1066	9.3850	4.4752
¾	0.6436	1.5538	0.1146	8.7238	4.8144
17	0.6896	1.4501	0.1228	8.1417	5.1586
¼	0.7363	1.3581	0.1311	7.6253	5.5080
½	0.7837	1.2760	0.1396	7.1643	5.8624
¾	0.8318	1.2023	0.1481	6.7503	6.2220
18	0.8805	1.1357	0.1568	6.3765	6.5866
¼	0.9299	1.0753	0.1656	6.0376	6.9564
½	0.9800	1.0204	0.1746	5.7289	7.3312
¾	1.0308	0.9701	0.1836	5.4466	7.7112
19	1.0823	0.9240	0.1928	5.1876	8.0962
¼	1.1345	0.8815	0.2021	4.9491	8.4864
½	1.1873	0.8422	0.2115	4.7289	8.8816
¾	1.2408	0.8059	0.2210	4.5249	9.2820
20	1.2950	0.7722	0.2307	4.3355	9.6874
¼	1.3499	0.7408	0.2404	4.1592	10.0980
½	1.4055	0.7115	0.2503	3.9948	10.5136
¾	1.4617	0.6841	0.2603	3.8411	10.9344
21	1.5186	0.6585	0.2705	3.6971	11.3602
¼	1.5763	0.6344	0.2807	3.5620	11.7912
½	1.6345	0.6118	0.2911	3.4350	12.2272
¾	1.6935	0.5905	0.3016	3.3153	12.6684
22	1.7532	0.5704	0.3123	3.2025	13.1146
¼	1.8135	0.5514	0.3230	3.0960	13.5660
½	1.8745	0.5335	0.3339	2.9952	14.0224
¾	1.9362	0.5165	0.3449	2.8998	14.4840

13" OD CASING	**VOLUME AND FILL BETWEEN PIPE AND HOLE**				
HOLE DIAM (IN)	**CU FT PER LIN FT**	**LIN FT PER CU FT**	**BARRELS PER LIN FT**	**LIN FT PER BARREL**	**GALLONS PER LIN FT**
14	0.1473	6.7906	0.0262	38.1265	1.1016
⅛	0.1664	6.0083	0.0296	33.7340	1.2450
¼	0.1858	5.3826	0.0331	30.2213	1.3897
⅜	0.2053	4.8710	0.0366	27.3485	1.5357
½	0.2250	4.4448	0.0401	24.9555	1.6830
⅝	0.2448	4.0843	0.0436	22.9316	1.8315
¾	0.2649	3.7755	0.0472	21.1977	1.9813
⅞	0.2851	3.5080	0.0508	19.6958	2.1324
15	0.3054	3.2740	0.0544	18.3824	2.2848
⅛	0.3260	3.0678	0.0581	17.2242	2.4384
¼	0.3467	2.8845	0.0617	16.1953	2.5933
⅜	0.3676	2.7207	0.0655	15.2753	2.7495
½	0.3886	2.5733	0.0692	14.4479	2.9070
⅝	0.4098	2.4400	0.0730	13.6998	3.0657
¾	0.4312	2.3190	0.0768	13.0203	3.2257
⅞	0.4528	2.2086	0.0806	12.4002	3.3870
16	0.4745	2.1074	0.0845	11.8323	3.5496
¼	0.5185	1.9287	0.0923	10.8288	3.8785
½	0.5631	1.7758	0.1003	9.9701	4.2126
¾	0.6085	1.6434	0.1084	9.2272	4.5517
17	0.6545	1.5279	0.1166	8.5785	4.8960
¼	0.7012	1.4261	0.1249	8.0071	5.2453
½	0.7486	1.3359	0.1333	7.5003	5.5998
¾	0.7966	1.2553	0.1419	7.0478	5.9593
18	0.8454	1.1829	0.1506	6.6414	6.3240
¼	0.8948	1.1175	0.1594	6.2745	6.6937
½	0.9449	1.0583	0.1683	5.9418	7.0686
¾	0.9957	1.0043	0.1773	5.6387	7.4485
19	1.0472	0.9549	0.1865	5.3615	7.8336
¼	1.0994	0.9096	0.1958	5.1072	8.2237
½	1.1522	0.8679	0.2052	4.8730	8.6190
¾	1.2057	0.8294	0.2147	4.6567	9.0193
20	1.2599	0.7937	0.2244	4.4563	9.4248
¼	1.3148	0.7606	0.2342	4.2703	9.8353
½	1.3704	0.7297	0.2441	4.0972	10.2510
¾	1.4266	0.7010	0.2541	3.9356	10.6717
21	1.4835	0.6741	0.2642	3.7846	11.0976
¼	1.5411	0.6489	0.2745	3.6431	11.5285
½	1.5994	0.6252	0.2849	3.5104	11.9646
¾	1.6584	0.6030	0.2954	3.3855	12.4057
22	1.7181	0.5821	0.3060	3.2680	12.8520
¼	1.7784	0.5623	0.3167	3.1571	13.3033
½	1.8394	0.5437	0.3276	3.0524	13.7598
¾	1.9011	0.5260	0.3386	2.9533	14.2213

VOLUME AND FILL BETWEEN PIPE AND HOLE

13-⅜″ OD CASING

HOLE DIAM (IN)	CU FT PER LIN FT	LIN FT PER CU FT	BARRELS PER LIN FT	LIN FT PER BARREL	GALLONS PER LIN FT
15	0.2515	3.9763	0.0448	22.3255	1.8813
⅛	0.2720	3.6761	0.0484	20.6399	2.0349
¼	0.2927	3.4161	0.0521	19.1798	2.1898
⅜	0.3136	3.1886	0.0559	17.9029	2.3460
½	0.3347	2.9881	0.0596	16.7768	2.5035
⅝	0.3559	2.8099	0.0634	15.7765	2.6622
¾	0.3773	2.6506	0.0672	14.8820	2.8222
⅞	0.3988	2.5073	0.0710	14.0775	2.9835
16	0.4206	2.3777	0.0749	13.3501	3.1461
¼	0.4645	2.1527	0.0827	12.0863	3.4750
½	0.5092	1.9639	0.0907	11.0264	3.8091
¾	0.5545	1.8033	0.0988	10.1249	4.1482
17	0.6006	1.6651	0.1070	9.3490	4.4925
¼	0.6473	1.5450	0.1153	8.6745	4.8418
½	0.6946	1.4396	0.1237	8.0828	5.1963
¾	0.7427	1.3464	0.1323	7.5597	5.5558
18	0.7914	1.2635	0.1410	7.0941	5.9204
¼	0.8409	1.1892	0.1498	6.6771	6.2902
½	0.8910	1.1224	0.1587	6.3015	6.6650
¾	0.9418	1.0618	0.1677	5.9617	7.0450
19	0.9933	1.0068	0.1769	5.6527	7.4300
¼	1.0454	0.9566	0.1862	5.3707	7.8202
½	1.0982	0.9105	0.1956	5.1123	8.2154
¾	1.1518	0.8682	0.2051	4.8748	8.6158
20	1.2060	0.8292	0.2148	4.6557	9.0212
¼	1.2608	0.7931	0.2246	4.4530	9.4318
½	1.3164	0.7596	0.2345	4.2651	9.8474
¾	1.3727	0.7285	0.2445	4.0903	10.2682
21	1.4296	0.6995	0.2546	3.9274	10.6940
¼	1.4872	0.6724	0.2649	3.7753	11.1250
½	1.5455	0.6470	0.2753	3.6329	11.5610
¾	1.6045	0.6233	0.2858	3.4994	12.0022
22	1.6641	0.6009	0.2964	3.3739	12.4484
¼	1.7245	0.5799	0.3071	3.2559	12.8998
½	1.7855	0.5601	0.3180	3.1446	13.3562
¾	1.8472	0.5414	0.3290	3.0396	13.8178
23	1.9096	0.5237	0.3401	2.9403	14.2844
¼	1.9726	0.5069	0.3513	2.8463	14.7562
½	2.0364	0.4911	0.3627	2.7572	15.2330
¾	2.1008	0.4760	0.3742	2.6726	15.7150

VOLUME AND FILL BETWEEN PIPE AND HOLE

16" OD CASING

HOLE DIAM (IN)	CU FT PER LIN FT	LIN FT PER CU FT	BARRELS PER LIN FT	LIN FT PER BARREL	GALLONS PER LIN FT
18	0.3709	2.6963	0.0661	15.1384	2.7744
¼	0.4203	2.3792	0.0749	13.3582	3.1441
½	0.4704	2.1258	0.0838	11.9352	3.5190
¾	0.5212	1.9186	0.0928	10.7722	3.8989
19	0.5727	1.7462	0.1020	9.8039	4.2840
¼	0.6248	1.6004	0.1113	8.9856	4.6741
½	0.6777	1.4756	0.1207	8.2850	5.0694
¾	0.7312	1.3676	0.1302	7.6786	5.4697
20	0.7854	1.2732	0.1399	7.1487	5.8752
¼	0.8403	1.1901	0.1497	6.6818	6.2857
½	0.8958	1.1163	0.1596	6.2674	6.7014
¾	0.9521	1.0503	0.1696	5.8971	7.1221
21	1.0090	0.9911	0.1797	5.5644	7.5480
¼	1.0666	0.9375	0.1900	5.2639	7.9789
½	1.1249	0.8890	0.2004	4.9911	8.4150
¾	1.1839	0.8447	0.2109	4.7425	8.8561
22	1.2435	0.8042	0.2215	4.5150	9.3024
¼	1.3039	0.7669	0.2322	4.3060	9.7537
½	1.3649	0.7327	0.2431	4.1135	10.2102
¾	1.4266	0.7010	0.2541	3.9356	10.6717
23	1.4890	0.6716	0.2652	3.7707	11.1384
¼	1.5520	0.6443	0.2764	3.6175	11.6101
½	1.6158	0.6189	0.2878	3.4748	12.0870
¾	1.6802	0.5952	0.2993	3.3416	12.5689
24	1.7453	0.5730	0.3109	3.2169	13.0560
¼	1.8111	0.5521	0.3226	3.1001	13.5481
½	1.8776	0.5326	0.3344	2.9903	14.0454
¾	1.9447	0.5142	0.3464	2.8871	14.5477
25	2.0126	0.4969	0.3585	2.7897	15.0552
¼	2.0811	0.4805	0.3707	2.6979	15.5677
½	2.1503	0.4651	0.3830	2.6111	16.0854
¾	2.2202	0.4504	0.3954	2.5289	16.6081
26	2.2907	0.4365	0.4080	2.4510	17.1360
¼	2.3620	0.4234	0.4207	2.3771	17.6689
½	2.4339	0.4109	0.4335	2.3068	18.2070
¾	2.5065	0.3990	0.4464	2.2400	18.7501
27	2.5798	0.3876	0.4595	2.1764	19.2984
¼	2.6538	0.3768	0.4727	2.1157	19.8517
½	2.7284	0.3665	0.4860	2.0578	20.4102
¾	2.8038	0.3567	0.4994	2.0025	20.9737
28	2.8798	0.3472	0.5129	1.9496	21.5423
¼	2.9565	0.3382	0.5266	1.8991	22.1161
½	3.0339	0.3296	0.5404	1.8506	22.6949
¾	3.1119	0.3213	0.5543	1.8042	23.2789

VOLUME AND FILL BETWEEN PIPE AND HOLE

20" OD
CASING

HOLE DIAM (IN)	CU FT PER LIN FT	LIN FT PER CU FT	BARRELS PER LIN FT	LIN FT PER BARREL	GALLONS PER LIN FT
22	0.4581	2.1827	0.0816	12.2549	3.4272
¼	0.5185	1.9287	0.0923	10.8288	3.8785
½	0.5795	1.7256	0.1032	9.6886	4.3350
¾	0.6412	1.5596	0.1142	8.7563	4.7965
23	0.7036	1.4213	0.1253	7.9800	5.2632
¼	0.7666	1.3044	0.1365	7.3235	5.7349
½	0.8304	1.2042	0.1479	6.7613	6.2118
¾	0.8948	1.1175	0.1594	6.2745	6.6937
24	0.9599	1.0417	0.1710	5.8489	7.1808
¼	1.0257	0.9749	0.1827	5.4738	7.6729
½	1.0922	0.9156	0.1945	5.1406	8.1702
¾	1.1593	0.8626	0.2065	4.8429	8.6725
25	1.2272	0.8149	0.2186	4.5752	9.1800
¼	1.2957	0.7718	0.2308	4.3332	9.6925
½	1.3649	0.7327	0.2431	4.1135	10.2102
¾	1.4348	0.6970	0.2555	3.9132	10.7329
26	1.5053	0.6643	0.2681	3.7298	11.2608
¼	1.5766	0.6343	0.2808	3.5612	11.7937
½	1.6485	0.6066	0.2936	3.4058	12.3318
¾	1.7211	0.5810	0.3065	3.2622	12.8749
27	1.7944	0.5573	0.3196	3.1289	13.4232
¼	1.8684	0.5352	0.3328	3.0050	13.9765
½	1.9430	0.5147	0.3461	2.8896	14.5350
¾	2.0184	0.4954	0.3595	2.7817	15.0985
28	2.0944	0.4775	0.3730	2.6808	15.6672
¼	2.1711	0.4606	0.3867	2.5861	16.2409
½	2.2485	0.4447	0.4005	2.4971	16.8198
¾	2.3265	0.4298	0.4144	2.4133	17.4037
29	2.4053	0.4158	0.4284	2.3343	17.9928
¼	2.4847	0.4025	0.4425	2.2597	18.5869
½	2.5648	0.3899	0.4568	2.1891	19.1862
¾	2.6456	0.3780	0.4712	2.1222	19.7905
30	2.7271	0.3667	0.4857	2.0588	20.4000
¼	2.8092	0.3560	0.5003	1.9986	21.0145
½	2.8921	0.3458	0.5151	1.9414	21.6341
¾	2.9756	0.3361	0.5300	1.8869	22.2589
31	3.0598	0.3268	0.5450	1.8350	22.8887
¼	3.1447	0.3180	0.5601	1.7854	23.5237
½	3.2302	0.3096	0.5753	1.7381	24.1637
¾	3.3165	0.3015	0.5907	1.6929	24.8089
32	3.4034	0.2938	0.6062	1.6497	25.4591
¼	3.4910	0.2865	0.6218	1.6083	26.1145
½	3.5793	0.2794	0.6375	1.5686	26.7749
¾	3.6683	0.2726	0.6533	1.5306	27.4405

21-½" OD
CASING

VOLUME AND FILL BETWEEN
PIPE AND HOLE

HOLE DIAM (IN)	CU FT PER LIN FT	LIN FT PER CU FT	BARRELS PER LIN FT	LIN FT PER BARREL	GALLONS PER LIN FT
23	0.3641	2.7468	0.0648	15.4219	2.7234
¼	0.4271	2.3412	0.0761	13.1450	3.1951
½	0.4909	2.0372	0.0874	11.4379	3.6720
¾	0.5553	1.8008	0.0989	10.1109	4.1539
24	0.6204	1.6118	0.1105	9.0498	4.6410
¼	0.6862	1.4573	0.1222	8.1821	5.1331
½	0.7527	1.3286	0.1341	7.4595	5.6304
¾	0.8198	1.2198	0.1460	6.8485	6.1327
25	0.8877	1.1266	0.1581	6.3251	6.6402
¼	0.9562	1.0458	0.1703	5.8719	7.1527
½	1.0254	0.9752	0.1826	5.4756	7.6704
¾	1.0953	0.9130	0.1951	5.1262	8.1931
26	1.1658	0.8578	0.2076	4.8160	8.7210
¼	1.2371	0.8084	0.2203	4.5386	9.2539
½	1.3090	0.7639	0.2331	4.2892	9.7920
¾	1.3816	0.7238	0.2461	4.0638	10.3351
27	1.4549	0.6873	0.2591	3.8591	10.8834
¼	1.5289	0.6541	0.2723	3.6724	11.4367
½	1.6035	0.6236	0.2856	3.5014	11.9952
¾	1.6789	0.5956	0.2990	3.3443	12.5587
28	1.7549	0.5698	0.3126	3.1994	13.1274
¼	1.8316	0.5460	0.3262	3.0654	13.7011
½	1.9090	0.5238	0.3400	2.9412	14.2800
¾	1.9870	0.5033	0.3539	2.8256	14.8639
29	2.0658	0.4841	0.3679	2.7179	15.4530
¼	2.1452	0.4662	0.3821	2.6173	16.0471
½	2.2253	0.4494	0.3963	2.5231	16.6464
¾	2.3061	0.4336	0.4107	2.4347	17.2507
30	2.3876	0.4188	0.4252	2.3516	17.8602
¼	2.4697	0.4049	0.4399	2.2734	18.4747
½	2.5525	0.3918	0.4546	2.1996	19.0944
¾	2.6361	0.3794	0.4695	2.1299	19.7191
31	2.7203	0.3676	0.4845	2.0640	20.3490
¼	2.8051	0.3565	0.4996	2.0015	20.9839
½	2.8907	0.3459	0.5149	1.9423	21.6239
¾	2.9769	0.3359	0.5302	1.8860	22.2691
32	3.0639	0.3264	0.5457	1.8325	22.9193
¼	3.1515	0.3173	0.5613	1.7816	23.5747
½	3.2398	0.3087	0.5770	1.7330	24.2351
¾	3.3287	0.3004	0.5929	1.6867	24.9007
33	3.4184	0.2925	0.6088	1.6425	25.5713
¼	3.5087	0.2850	0.6249	1.6002	26.2471
½	3.5997	0.2778	0.6411	1.5597	26.9279
¾	3.6914	0.2709	0.6575	1.5210	27.6139

VOLUME AND FILL BETWEEN PIPE AND HOLE

24-½" OD
CASING

HOLE DIAM (IN)	CU FT PER LIN FT	LIN FT PER CU FT	BARRELS PER LIN FT	LIN FT PER BARREL	GALLONS PER LIN FT
26	0.4132	2.4204	0.0736	13.5896	3.0906
¼	0.4844	2.0644	0.0863	11.5909	3.6235
½	0.5563	1.7975	0.0991	10.0923	4.1616
¾	0.6289	1.5900	0.1120	8.9272	4.7047
27	0.7022	1.4241	0.1251	7.9954	5.2530
¼	0.7762	1.2883	0.1382	7.2335	5.8063
½	0.8508	1.1753	0.1515	6.5988	6.3648
¾	0.9262	1.0797	0.1650	6.0621	6.9283
28	1.0022	0.9978	0.1785	5.6023	7.4970
¼	1.0789	0.9269	0.1922	5.2040	8.0707
½	1.1563	0.8648	0.2059	4.8557	8.6496
¾	1.2343	0.8101	0.2198	4.5486	9.2335
29	1.3131	0.7616	0.2339	4.2759	9.8226
¼	1.3925	0.7181	0.2480	4.0320	10.4167
½	1.4726	0.6791	0.2623	3.8126	11.0160
¾	1.5534	0.6437	0.2767	3.6144	11.6203
30	1.6349	0.6117	0.2912	3.4342	12.2298
¼	1.7170	0.5824	0.3058	3.2699	12.8443
½	1.7999	0.5556	0.3206	3.1194	13.4640
¾	1.8834	0.5310	0.3354	2.9811	14.0887
31	1.9676	0.5082	0.3504	2.8535	14.7186
¼	2.0525	0.4872	0.3656	2.7355	15.3535
½	2.1380	0.4677	0.3808	2.6261	15.9936
¾	2.2243	0.4496	0.3962	2.5242	16.6387
32	2.3112	0.4327	0.4116	2.4293	17.2890
¼	2.3988	0.4169	0.4272	2.3406	17.9443
½	2.4871	0.4021	0.4430	2.2575	18.6048
¾	2.5761	0.3882	0.4588	2.1795	19.2703
33	2.6657	0.3751	0.4748	2.1062	19.9410
¼	2.7561	0.3628	0.4909	2.0372	20.6167
½	2.8471	0.3512	0.5071	1.9721	21.2976
¾	2.9388	0.3403	0.5234	1.9105	21.9835
34	3.0311	0.3299	0.5399	1.8523	22.6745
¼	3.1242	0.3201	0.5564	1.7971	23.3707
½	3.2180	0.3108	0.5731	1.7448	24.0719
¾	3.3124	0.3019	0.5900	1.6950	24.7783
35	3.4075	0.2935	0.6069	1.6477	25.4897
¼	3.5033	0.2854	0.6240	1.6027	26.2063
½	3.5997	0.2778	0.6411	1.5597	26.9279
¾	3.6969	0.2705	0.6584	1.5187	27.6547
36	3.7947	0.2635	0.6759	1.4796	28.3865
¼	3.8932	0.2569	0.6934	1.4421	29.1235
½	3.9924	0.2505	0.7111	1.4063	29.8655
¾	4.0923	0.2444	0.7289	1.3720	30.6127

VOLUME AND FILL BETWEEN PIPE AND HOLE

30" OD CASING

HOLE DIAM (IN)	CU FT PER LIN FT	LIN FT PER CU FT	BARRELS PER LIN FT	LIN FT PER BARREL	GALLONS PER LIN FT
32	0.6763	1.4786	0.1205	8.3017	5.0592
1/4	0.7639	1.3090	0.1361	7.3497	5.7145
1/2	0.8522	1.1734	0.1518	6.5883	6.3750
3/4	0.9412	1.0625	0.1676	5.9655	7.0405
33	1.0308	0.9701	0.1836	5.4466	7.7112
1/4	1.1212	0.8919	0.1997	5.0078	8.3869
1/2	1.2122	0.8250	0.2159	4.6318	9.0678
3/4	1.3039	0.7669	0.2322	4.3060	9.7537
34	1.3963	0.7162	0.2487	4.0211	10.4448
1/4	1.4893	0.6714	0.2653	3.7699	11.1409
1/2	1.5831	0.6317	0.2820	3.5466	11.8422
3/4	1.6775	0.5961	0.2988	3.3470	12.5485
35	1.7726	0.5641	0.3157	3.1674	13.2600
1/4	1.8684	0.5352	0.3328	3.0050	13.9765
1/2	1.9649	0.5089	0.3500	2.8575	14.6982
3/4	2.0620	0.4850	0.3673	2.7229	15.4249
36	2.1598	0.4630	0.3847	2.5995	16.1568
1/4	2.2584	0.4428	0.4022	2.4861	16.8937
1/2	2.3576	0.4242	0.4199	2.3815	17.6358
3/4	2.4574	0.4069	0.4377	2.2847	18.3829
37	2.5580	0.3909	0.4556	2.1949	19.1352
1/4	2.6592	0.3760	0.4736	2.1113	19.8925
1/2	2.7612	0.3622	0.4918	2.0334	20.6550
3/4	2.8638	0.3492	0.5101	1.9606	21.4225
38	2.9671	0.3370	0.5285	1.8923	22.1951
1/4	3.0710	0.3256	0.5470	1.8282	22.9729
1/2	3.1757	0.3149	0.5656	1.7680	23.7557
3/4	3.2810	0.3048	0.5844	1.7112	24.5437
39	3.3870	0.2952	0.6033	1.6577	25.3367
1/4	3.4937	0.2862	0.6223	1.6070	26.1349
1/2	3.6011	0.2777	0.6414	1.5591	26.9381
3/4	3.7092	0.2696	0.6606	1.5137	27.7465
40	3.8179	0.2619	0.6800	1.4706	28.5599
1/4	3.9273	0.2546	0.6995	1.4296	29.3785
1/2	4.0374	0.2477	0.7191	1.3906	30.2021
3/4	4.1482	0.2411	0.7388	1.3535	31.0309
41	4.2597	0.2348	0.7587	1.3181	31.8647
1/4	4.3718	0.2287	0.7787	1.2843	32.7037
1/2	4.4847	0.2230	0.7988	1.2519	33.5477
3/4	4.5982	0.2175	0.8190	1.2210	34.3969
42	4.7124	0.2122	0.8393	1.1915	35.2511
1/4	4.8273	0.2072	0.8598	1.1631	36.1105
1/2	4.9428	0.2023	0.8804	1.1359	36.9749
3/4	5.0591	0.1977	0.9011	1.1098	37.8445

Volume and fill between pipe and pipe

0.750" OD INSIDE TUBING			**VOLUME AND FILL BETWEEN TUBING AND TUBING**				
OUTSIDE TUBING							
SIZE OD INCHES	WEIGHT PER LIN FT	ID	CU FT PER LIN FT	LIN FT PER CU FT	BARRELS PER LIN FT	LIN FT PER BARREL	GALLONS PER LIN FT
	3.10	2.125	0.0216	46.3801	.0038	260.4052	0.1613
	3.32	2.107	0.0211	47.2914	.0038	265.5217	0.1582
	4.00	2.041	0.0197	50.8846	.0035	285.6959	0.1470
	4.60	1.995	0.0186	53.6489	.0033	301.2163	0.1394
	4.70	1.995	0.0186	53.6489	.0033	301.2163	0.1394
2⅜	5.00	1.947	0.0176	56.7934	.0031	318.8710	0.1317
	5.30	1.939	0.0174	57.3456	.0031	321.9715	0.1304
	5.80	1.867	0.0159	62.7214	.0028	352.1545	0.1193
	5.95	1.867	0.0159	62.7214	.0028	352.1545	0.1193
	6.20	1.853	0.0157	63.8591	.0028	358.5423	0.1171
	7.70	1.703	0.0128	78.4300	.0023	440.3517	0.0954
	4.36	2.579	0.0332	30.1124	.0059	169.0685	0.2484
	4.64	2.563	0.0328	30.5248	.0058	171.3842	0.2451
	5.90	2.469	0.0302	33.1341	.0054	186.0344	0.2258
	6.40	2.441	0.0294	33.9783	.0052	190.7742	0.2202
	6.50	2.441	0.0294	33.9783	.0052	190.7742	0.2202
	7.90	2.323	0.0264	37.9299	.0047	212.9604	0.1972
2⅞	8.60	2.259	0.0248	40.3795	.0044	226.7142	0.1853
	8.70	2.259	0.0248	40.3795	.0044	226.7142	0.1853
	9.50	2.195	0.0232	43.0843	.0041	241.9006	0.1736
	10.40	2.151	0.0222	45.1114	.0039	253.2820	0.1658
	10.70	2.091	0.0208	48.1252	.0037	270.2030	0.1554
	11.00	2.065	0.0202	49.5300	.0036	278.0904	0.1510
	11.65	1.995	0.0186	53.6489	.0033	301.2163	0.1394
	5.63	3.188	0.0524	19.0969	.0093	107.2212	0.3917
	7.70	3.068	0.0483	20.7168	.0086	116.3164	0.3611
	8.50	3.018	0.0466	21.4545	.0083	120.4582	0.3487
	8.90	3.018	0.0466	21.4545	.0083	120.4582	0.3487
	9.20	2.992	0.0458	21.8541	.0081	122.7017	0.3423
	9.30	2.992	0.0458	21.8541	.0081	122.7017	0.3423
	10.20	2.992	0.0458	21.8541	.0081	122.7017	0.3423
	10.30	2.922	0.0435	22.9885	.0077	129.0707	0.3254
3½	11.20	2.900	0.0428	23.3637	.0076	131.1773	0.3202
	12.70	2.750	0.0382	26.1924	.0068	147.0592	0.2856
	12.80	2.764	0.0386	24.9067	.0069	145.4551	0.2887
	12.95	2.750	0.0382	26.1924	.0068	147.0592	0.2856
	13.30	2.764	0.0386	25.9067	.0069	145.4551	0.2887
	15.50	2.602	0.0339	29.5344	.0060	165.8231	0.2533
	15.80	2.548	0.0323	30.9195	.0058	173.6000	0.2419
	16.70	2.480	0.0305	32.8113	.0054	184.2220	0.2280
	17.05	2.440	0.0294	34.0091	.0052	190.9470	0.2200

NOTE: No allowance has been made for couplings or upsets.
*Plain end weights are indicated by asterisk.

	1.000" OD INSIDE TUBING		VOLUME AND FILL BETWEEN TUBING AND TUBING				
	OUTSIDE TUBING						
SIZE OD INCHES	WEIGHT PER LIN FT	ID	CU FT PER LIN FT	LIN FT PER CU FT	BARRELS PER LIN FT	LIN FT PER BARREL	GALLONS PER LIN FT
2⅜	3.10	2.125	0.0192	52.1519	.0034	292.8111	0.1434
	3.32	2.107	0.0188	53.3069	.0033	299.2962	0.1403
	4.00	2.041	0.0173	57.9169	.0031	325.1794	0.1292
	4.60	1.995	0.0163	61.5252	.0029	345.4381	0.1216
	4.70	1.995	0.0163	61.5252	.0029	345.4381	0.1216
	5.00	1.947	0.0152	65.6965	.0027	368.8587	0.1139
	5.30	1.939	0.0151	66.4366	.0027	373.0139	0.1126
	5.80	1.867	0.0136	73.7608	.0024	414.1363	0.1014
	5.95	1.867	0.0136	73.7608	.0024	414.1363	0.1014
	6.20	1.853	0.0133	75.3393	.0024	422.9990	0.0993
	7.70	1.703	0.0104	96.4875	.0018	541.7373	0.0775
2⅞	4.36	2.579	0.0308	32.4436	.0055	182.1572	0.2306
	4.64	2.563	0.0304	32.9229	.0054	184.8482	0.2272
	5.90	2.469	0.0278	35.9788	.0050	202.0059	0.2079
	6.40	2.441	0.0270	36.9763	.0048	207.6068	0.2023
	6.50	2.441	0.0270	36.9763	.0048	207.6068	0.2023
	7.90	2.323	0.0240	41.7045	.0043	234.1531	0.1794
	8.60	2.259	0.0224	44.6851	.0040	250.8881	0.1674
	8.70	2.259	0.0224	44.6851	.0040	250.8881	0.1674
	9.50	2.195	0.0208	48.0213	.0037	269.6195	0.1558
	10.40	2.151	0.0198	50.5532	.0035	283.8353	0.1480
	10.70	2.091	0.0184	54.3687	.0033	305.2575	0.1376
	11.00	2.065	0.0178	56.1685	.0032	315.3625	0.1332
	11.65	1.995	0.0163	61.5252	.0029	345.4381	0.1216
3½	5.63	3.188	0.0500	20.0087	.0089	112.3404	0.3739
	7.70	3.068	0.0459	21.7942	.0082	122.3654	0.3432
	8.50	3.018	0.0442	22.6121	.0079	126.9577	0.3308
	8.90	3.018	0.0442	22.6121	.0079	126.9577	0.3308
	9.20	2.992	0.0434	23.0565	.0077	129.4524	0.3244
	9.30	2.992	0.0434	23.0565	.0077	129.4524	0.3244
	10.20	2.992	0.0434	23.0565	.0077	129.4524	0.3244
	10.30	2.922	0.0411	24.3227	.0073	136.5618	0.3076
	11.20	2.900	0.0404	24.7431	.0072	138.9223	0.3023
	12.70	2.750	0.0358	27.9385	.0064	156.8631	0.2677
	12.80	2.764	0.0362	27.6137	.0064	155.0394	0.2709
	12.95	2.750	0.0358	27.9385	.0064	156.8631	0.2677
	13.30	2.764	0.0362	27.6137	.0064	155.0394	0.2709
	15.50	2.602	0.0315	31.7736	.0056	178.3955	0.2354
	15.80	2.548	0.0300	33.3824	.0053	187.4285	0.2241
	16.70	2.480	0.0281	35.5985	.0050	199.8707	0.2101
	17.05	2.440	0.0270	37.0128	.0048	207.8113	0.2021

1.050″ OD
INSIDE TUBING

VOLUME AND FILL BETWEEN TUBING AND TUBING

OUTSIDE TUBING

SIZE OD INCHES	WEIGHT PER LIN FT	ID	CU FT PER LIN FT	LIN FT PER CU FT	BARRELS PER LIN FT	LIN FT PER BARREL	GALLONS PER LIN FT
2⅜	3.10	2.125	0.0186	53.7181	.0033	301.6046	0.1393
	3.32	2.107	0.0182	54.9444	.0032	308.4896	0.1361
	4.00	2.041	0.0167	59.8549	.0030	336.0605	0.1250
	4.60	1.995	0.0157	63.7167	.0028	357.7429	0.1174
	4.70	1.995	0.0157	63.7167	.0028	357.7429	0.1174
	5.00	1.947	0.0147	68.2014	.0026	382.9226	0.1097
	5.30	1.939	0.0145	68.9993	.0026	387.4025	0.1084
	5.80	1.867	0.0130	76.9333	.0023	431.9482	0.0972
	5.95	1.867	0.0130	76.9333	.0023	431.9482	0.0972
	6.20	1.853	0.0127	78.6520	.0023	441.5985	0.0951
	7.70	1.703	0.0098	01.9890	.0017	572.6256	0.0733
2⅞	4.36	2.579	0.0303	33.0429	.0054	185.5221	0.2264
	4.64	2.563	0.0298	33.5402	.0053	188.3143	0.2230
	5.90	2.469	0.0272	36.7173	.0049	206.1524	0.2037
	6.40	2.441	0.0265	37.7568	.0047	211.9889	0.1981
	6.50	2.441	0.0265	37.7568	.0047	211.9889	0.1981
	7.90	2.323	0.0234	42.7000	.0042	239.7427	0.1752
	8.60	2.259	0.0218	45.8300	.0039	257.3162	0.1632
	8.70	2.259	0.0218	45.8300	.0039	257.3162	0.1632
	9.50	2.195	0.0203	49.3461	.0036	277.0575	0.1516
	10.40	2.151	0.0192	52.0235	.0034	292.0903	0.1438
	10.70	2.091	0.0178	56.0730	.0032	314.8266	0.1334
	11.00	2.065	0.0172	57.9894	.0031	325.5862	0.1290
	11.65	1.995	0.0157	63.7167	.0028	357.7429	0.1174
3½	5.63	3.188	0.0494	20.2350	.0088	113.6113	0.3697
	7.70	3.068	0.0453	22.0630	.0081	123.8747	0.3391
	8.50	3.018	0.0437	22.9016	.0078	128.5832	0.3266
	8.90	3.018	0.0437	22.9016	.0078	128.5832	0.3266
	9.20	2.992	0.0428	23.3575	.0076	131.1428	0.3203
	9.30	2.992	0.0428	23.3575	.0076	131.1428	0.3203
	10.20	2.992	0.0428	23.3575	.0076	131.1428	0.3203
	10.30	2.922	0.0406	24.6580	.0072	138.4443	0.3034
	11.20	2.900	0.0399	25.0902	.0071	140.8709	0.2981
	12.70	2.750	0.0352	28.3818	.0063	159.3520	0.2636
	12.80	2.764	0.0357	28.0467	.0064	157.4703	0.2667
	12.95	2.750	0.0352	28.3818	.0063	159.3520	0.2636
	13.30	2.764	0.0357	28.0467	.0064	157.4703	0.2667
	15.50	2.602	0.0309	32.3482	.0055	181.6217	0.2312
	15.80	2.548	0.0294	34.0173	.0052	190.9929	0.2199
	16.70	2.480	0.0275	36.3213	.0049	203.9292	0.2060
	17.05	2.440	0.0265	37.7948	.0047	212.2022	0.1979

1.315" OD INSIDE TUBING			**VOLUME AND FILL BETWEEN TUBING AND TUBING**				
OUTSIDE TUBING							
SIZE OD INCHES	WEIGHT PER LIN FT	ID	CU FT PER LIN FT	LIN FT PER CU FT	BARRELS PER LIN FT	LIN FT PER BARREL	GALLONS PER LIN FT
	3.10	2.125	0.0152	65.8005	.0027	369.4424	0.1137
	3.32	2.107	0.0148	67.6499	.0026	379.8262	0.1106
	4.00	2.041	0.0133	75.2513	.0024	422.5047	0.0994
	4.60	1.995	0.0123	81.4584	.0022	457.3548	0.0918
	4.70	1.995	0.0123	81.4584	.0022	457.3548	0.0918
2⅜	5.00	1.947	0.0112	88.9348	.0020	499.3317	0.0841
	5.30	1.939	0.0111	90.2964	.0020	506.9767	0.0828
	5.80	1.867	0.0096	04.3839	.0017	586.0719	0.0717
	5.95	1.867	0.0096	04.3839	.0017	586.0719	0.0717
	6.20	1.853	0.0093	07.5735	.0017	603.9802	0.0695
	7.70	1.703	0.0064	56.5747	.0011	879.1018	0.0478
	4.36	2.579	0.0268	37.2503	.0048	209.1448	0.2008
	4.64	2.563	0.0264	37.8835	.0047	212.7001	0.1975
	5.90	2.469	0.0238	41.9871	.0042	235.7400	0.1782
	6.40	2.441	0.0231	43.3519	.0041	243.4031	0.1726
	6.50	2.441	0.0231	43.3519	.0041	243.4031	0.1726
	7.90	2.323	0.0200	49.9976	.0036	280.7158	0.1496
2⅞	8.60	2.259	0.0184	54.3433	.0033	305.1150	0.1377
	8.70	2.259	0.0184	54.3433	.0033	305.1150	0.1377
	9.50	2.195	0.0168	59.3585	.0030	333.2732	0.1260
	10.40	2.151	0.0158	63.2758	.0028	355.2674	0.1182
	10.70	2.091	0.0144	69.3691	.0026	389.4788	0.1078
	11.00	2.065	0.0138	72.3260	.0025	406.0805	0.1034
	11.65	1.995	0.0123	81.4584	.0022	457.3548	0.0918
	5.63	3.188	0.0460	21.7387	.0082	122.0536	0.3441
	7.70	3.068	0.0419	23.8627	.0075	133.9790	0.3135
	8.50	3.018	0.0402	24.8467	.0072	139.5040	0.3011
	8.90	3.018	0.0402	24.8467	.0072	139.5040	0.3011
	9.20	2.992	0.0394	25.3843	.0070	142.5221	0.2947
	9.30	2.992	0.0394	25.3843	.0070	142.5221	0.2947
	10.20	2.992	0.0394	25.3843	.0070	142.5221	0.2947
	10.30	2.922	0.0371	26.9276	.0066	151.1875	0.2778
3½	11.20	2.900	0.0364	27.4439	.0065	154.0860	0.2726
	12.70	2.750	0.0318	31.4311	.0057	176.4728	0.2380
	12.80	2.764	0.0322	31.0206	.0057	174.1679	0.2411
	12.95	2.750	0.0318	31.4311	.0057	176.4728	0.2380
	13.30	2.764	0.0322	31.0206	.0057	174.1679	0.2411
	15.50	2.602	0.0275	36.3698	.0049	204.2011	0.2057
	15.80	2.548	0.0260	38.4933	.0046	216.1237	0.1943
	16.70	2.480	0.0241	41.4701	.0043	232.8372	0.1804
	17.05	2.440	0.0230	43.4020	.0041	243.6844	0.1724

1.050" OD TUBING			**VOLUME AND FILL BETWEEN TUBING AND CASING**				
CASING							
SIZE OD INCHES	WEIGHT PER LIN FT	ID	CU FT PER LIN FT	LIN FT PER CU FT	BARRELS PER LIN FT	LIN FT PER BARREL	GALLONS PER LIN FT
	9.26	3.550	0.0627	15.9432	.0112	89.5143	0.4692
	9.50	3.550	0.0627	15.9432	.0112	89.5143	0.4692
4	11.00	3.480	0.0600	16.6559	.0107	93.5159	0.4491
	11.60	3.430	0.0582	17.1956	.0104	96.5462	0.4350
	12.60	3.364	0.0557	17.9505	.0099	100.7847	0.4167
	9.50	4.090	0.0852	11.7337	.0152	65.8800	0.6375
	10.50	4.052	0.0835	11.9708	.0149	67.2108	0.6249
	10.98	4.030	0.0826	12.1114	.0147	68.0002	0.6176
	11.00	4.026	0.0824	12.1372	.0147	68.1453	0.6163
	11.60	4.000	0.0813	12.3072	.0145	69.0998	0.6078
	11.75	3.990	0.0808	12.3736	.0144	69.4724	0.6046
	12.60	3.958	0.0794	12.5897	.0141	70.6857	0.5942
4½	12.75	3.960	0.0795	12.5760	.0142	70.6089	0.5948
	13.50	3.920	0.0778	12.8539	.0139	72.1692	0.5820
	15.10	3.826	0.0738	13.5453	.0131	76.0514	0.5523
	16.60	3.826	0.0738	13.5453	.0131	76.0514	0.5523
	18.80	3.640	0.0663	15.0938	.0118	84.7457	0.4956
	21.60	3.500	0.0608	16.4473	.0108	92.3448	0.4548
	24.60	3.380	0.0563	17.7629	.0100	99.7311	0.4211
	26.50	3.240	0.0512	19.5151	.0091	109.5693	0.3833
	16.00	4.082	0.0849	11.7830	.0151	66.1568	0.6349
	16.50	4.070	0.0843	11.8576	.0150	66.5753	0.6309
4¾	18.00	4.000	0.0813	12.3072	.0145	69.0998	0.6078
	20.00	3.910	0.0774	12.9248	.0138	72.5675	0.5788
	21.00	3.850	0.0748	13.3634	.0133	75.0302	0.5598
	11.50	4.560	0.1074	9.3111	.0191	52.2781	0.8034
	12.85	4.500	0.1044	9.5755	.0186	53.7623	0.7812
	13.00	4.494	0.1041	9.6025	.0185	53.9143	0.7790
	14.00	4.450	0.1020	9.8046	.0182	55.0489	0.7630
	15.00	4.408	0.1000	10.0036	.0178	56.1663	0.7478
5	18.00	4.276	0.0937	10.6710	.0167	59.9135	0.7010
	20.30	4.184	0.0895	11.1774	.0159	62.7563	0.6693
	21.00	4.154	0.0881	11.3505	.0157	63.7281	0.6590
	23.20	4.044	0.0832	12.0216	.0148	67.4962	0.6223
	24.20	4.000	0.0813	12.3072	.0145	69.0998	0.6078
5¼	16.00	4.650	0.1119	8.9350	.0199	50.1664	0.8372
	13.00	5.044	0.1328	7.5329	.0236	42.2941	0.9930
	14.00	5.012	0.1310	7.6338	.0233	42.8607	0.9799
	15.00	4.974	0.1289	7.7564	.0230	43.5488	0.9644
	15.50	4.950	0.1276	7.8353	.0227	43.9921	0.9547
	17.00	4.892	0.1245	8.0312	.0222	45.0921	0.9314
5½	20.00	4.778	0.1185	8.4387	.0211	47.3800	0.8865
	23.00	4.670	0.1129	8.8546	.0201	49.7148	0.8448
	25.00	4.580	0.1084	9.2255	.0193	51.7973	0.8109
	26.00	4.548	0.1068	9.3631	.0190	52.5699	0.7989
	32.30	4.276	0.0937	10.6710	.0167	59.9135	0.7010
	36.40	4.090	0.0852	11.7337	.0152	65.8800	0.6375

	1.050″ OD TUBING		VOLUME AND FILL BETWEEN TUBING AND CASING				
	CASING						
SIZE OD INCHES	WEIGHT PER LIN FT	ID	CU FT PER LIN FT	LIN FT PER CU FT	BARRELS PER LIN FT	LIN FT PER BARREL	GALLONS PER LIN FT
	14.00	5.290	0.1466	6.8205	.0261	38.2944	1.0968
	17.00	5.190	0.1409	7.0972	.0251	39.8479	1.0540
	19.50	5.090	0.1353	7.3913	.0241	41.4993	1.0121
5¾	20.00	5.090	0.1353	7.3913	.0241	41.4993	1.0121
	22.50	4.990	0.1298	7.7044	.0231	43.2571	0.9709
	23.00	4.990	0.1298	7.7044	.0231	43.2571	0.9709
	25.20	4.890	0.1244	8.0381	.0222	45.1307	0.9306
	15.00	5.524	0.1604	6.2337	.0286	34.9997	1.2000
	16.00	5.500	0.1590	6.2903	.0283	35.3174	1.1892
	18.00	5.424	0.1544	6.4747	.0275	36.3529	1.1553
6	20.00	5.352	0.1502	6.6571	.0268	37.3770	1.1237
	23.00	5.240	0.1437	6.9568	.0256	39.0594	1.0753
	26.00	5.140	0.1381	7.2420	.0246	40.6608	1.0329

1.315″ OD TUBING

VOLUME AND FILL BETWEEN TUBING AND CASING

SIZE OD INCHES	CASING WEIGHT PER LIN FT	ID	CU FT PER LIN FT	LIN FT PER CU FT	BARRELS PER LIN FT	LIN FT PER BARREL	GALLONS PER LIN FT
	9.26	3.550	0.0593	16.8621	.0106	94.6738	0.4436
	9.50	3.550	0.0593	16.8621	.0106	94.6738	0.4436
4	11.00	3.480	0.0566	17.6614	.0101	99.1616	0.4236
	11.60	3.430	0.0547	18.2695	.0097	102.5755	0.4095
	12.60	3.364	0.0523	19.1240	.0093	107.3730	0.3912
	9.50	4.090	0.0818	12.2240	.0146	68.6328	0.6120
	10.50	4.052	0.0801	12.4815	.0143	70.0783	0.5993
	10.98	4.030	0.0791	12.6344	.0141	70.9370	0.5921
	11.00	4.026	0.0790	12.6625	.0141	71.0948	0.5908
	11.60	4.000	0.0778	12.8477	.0139	72.1344	0.5822
	11.75	3.990	0.0774	12.9200	.0138	72.5406	0.5790
	12.60	3.958	0.0760	13.1558	.0135	73.8644	0.5686
4½	12.75	3.960	0.0761	13.1409	.0136	73.7806	0.5693
	13.50	3.920	0.0744	13.4446	.0132	75.4859	0.5564
	15.10	3.826	0.0704	14.2029	.0125	79.7436	0.5267
	16.60	3.826	0.0704	14.2029	.0125	79.7436	0.5267
	18.80	3.640	0.0628	15.9150	.0112	89.3560	0.4700
	21.60	3.500	0.0574	17.4271	.0102	97.8458	0.4292
	24.60	3.380	0.0529	18.9111	.0094	106.1780	0.3956
	26.50	3.240	0.0478	20.9100	.0085	117.4008	0.3577
	16.00	4.082	0.0814	12.2775	.0145	68.9332	0.6093
	16.50	4.070	0.0809	12.3585	.0144	69.3878	0.6053
4¾	18.00	4.000	0.0778	12.8477	.0139	72.1344	0.5822
	20.00	3.910	0.0740	13.5222	.0132	75.9218	0.5532
	21.00	3.850	0.0714	14.0031	.0127	78.6216	0.5342
	11.50	4.560	0.1040	9.6172	.0185	53.9967	0.7778
	12.85	4.500	0.1010	9.8995	.0180	55.5816	0.7556
	13.00	4.494	0.1007	9.9284	.0179	55.7440	0.7534
	14.00	4.450	0.0986	10.1446	.0176	56.9578	0.7374
	15.00	4.408	0.0965	10.3578	.0172	58.1549	0.7222
5	18.00	4.276	0.0903	11.0750	.0161	62.1816	0.6754
	20.30	4.184	0.0860	11.6214	.0153	65.2493	0.6437
	21.00	4.154	0.0847	11.8086	.0151	66.3005	0.6335
	23.20	4.044	0.0798	12.5368	.0142	70.3887	0.5967
	24.20	4.000	0.0778	12.8477	.0139	72.1344	0.5822
5¼	16.00	4.650	0.1085	9.2165	.0193	51.7468	0.8116
	13.00	5.044	0.1293	7.7320	.0230	43.4119	0.9675
	14.00	5.012	0.1276	7.8384	.0227	44.0091	0.9543
	15.00	4.974	0.1255	7.9676	.0224	44.7349	0.9389
	15.50	4.950	0.1242	8.0509	.0221	45.2027	0.9291
	17.00	4.892	0.1211	8.2579	.0216	46.3649	0.9059
5½	20.00	4.778	0.1151	8.6894	.0205	48.7873	0.8609
	23.00	4.670	0.1095	9.1309	.0195	51.2665	0.8192
	25.00	4.580	0.1050	9.5259	.0187	53.4839	0.7853
	26.00	4.548	0.1034	9.6727	.0184	54.3081	0.7734
	32.30	4.276	0.0903	11.0750	.0161	62.1816	0.6754
	36.40	4.090	0.0818	12.2240	.0146	68.6328	0.6120

VOLUME AND FILL BETWEEN TUBING AND CASING

1.315" OD TUBING

CASING							
SIZE OD INCHES	WEIGHT PER LIN FT	ID	CU FT PER LIN FT	LIN FT PER CU FT	BARRELS PER LIN FT	LIN FT PER BARREL	GALLONS PER LIN FT
5¾	14.00	5.290	0.1432	6.9833	.0255	39.2085	1.0712
	17.00	5.190	0.1375	7.2737	.0245	40.8386	1.0284
	19.50	5.090	0.1319	7.5829	.0235	42.5749	0.9865
	20.00	5.090	0.1319	7.5829	.0235	42.5749	0.9865
	22.50	4.990	0.1264	7.9128	.0225	44.4271	0.9454
	23.00	4.990	0.1264	7.9128	.0225	44.4271	0.9454
	25.20	4.890	0.1210	8.2652	.0215	46.4058	0.9051
6	15.00	5.524	0.1570	6.3694	.0280	35.7617	1.1744
	16.00	5.500	0.1556	6.4285	.0277	36.0935	1.1636
	18.00	5.424	0.1510	6.6213	.0269	37.1756	1.1298
	20.00	5.352	0.1468	6.8121	.0261	38.2473	1.0981
	23.00	5.240	0.1403	7.1262	.0250	40.0109	1.0497
	26.00	5.140	0.1347	7.4258	.0240	41.6929	1.0074

1.660" OD TUBING			**VOLUME AND FILL BETWEEN TUBING AND CASING**				
CASING							
SIZE OD INCHES	WEIGHT PER LIN FT	ID	CU FT PER LIN FT	LIN FT PER CU FT	BARRELS PER LIN FT	LIN FT PER BARREL	GALLONS PER LIN FT
	9.26	3.550	0.0537	18.6197	.0096	104.5420	0.4018
	9.50	3.550	0.0537	18.6197	.0096	104.5420	0.4018
4	11.00	3.480	0.0510	19.5992	.0091	110.0413	0.3817
	11.60	3.430	0.0491	20.3508	.0088	114.2613	0.3676
	12.60	3.364	0.0467	21.4167	.0083	120.2461	0.3493
	9.50	4.090	0.0762	13.1220	.0136	73.6743	0.5701
	10.50	4.052	0.0745	13.4191	.0133	75.3426	0.5575
	10.98	4.030	0.0736	13.5960	.0131	76.3360	0.5502
	11.00	4.026	0.0734	13.6286	.0131	76.5189	0.5489
	11.60	4.000	0.0722	13.8433	.0129	77.7245	0.5404
	11.75	3.990	0.0718	13.9273	.0128	78.1962	0.5371
	12.60	3.958	0.0704	14.2017	.0125	79.7367	0.5267
4½	12.75	3.960	0.0705	14.1843	.0126	79.6390	0.5274
	13.50	3.920	0.0688	14.5388	.0123	81.6296	0.5145
	15.10	3.826	0.0648	15.4297	.0115	86.6315	0.4848
	16.60	3.826	0.0648	15.4297	.0115	86.6315	0.4848
	18.80	3.640	0.0572	17.4716	.0102	98.0955	0.4282
	21.60	3.500	0.0518	19.3110	.0092	108.4233	0.3874
	24.60	3.380	0.0473	21.1502	.0084	118.7493	0.3537
	26.50	3.240	0.0422	23.6821	.0075	132.9649	0.3159
	16.00	4.082	0.0759	13.1836	.0135	74.0206	0.5674
	16.50	4.070	0.0753	13.2770	.0134	74.5450	0.5634
4¾	18.00	4.000	0.0722	13.8433	.0129	77.7245	0.5404
	20.00	3.910	0.0684	14.6297	.0122	82.1396	0.5113
	21.00	3.850	0.0658	15.1942	.0117	85.3089	0.4923
	11.50	4.560	0.0984	10.1645	.0175	57.0692	0.7359
	12.85	4.500	0.0954	10.4803	.0170	58.8425	0.7138
	13.00	4.494	0.0951	10.5127	.0169	59.0246	0.7116
	14.00	4.450	0.0930	10.7554	.0166	60.3872	0.6955
	15.00	4.408	0.0909	10.9954	.0162	61.7345	0.6803
5	18.00	4.276	0.0847	11.8070	.0151	66.2916	0.6336
	20.30	4.184	0.0805	12.4301	.0143	69.7896	0.6018
	21.00	4.154	0.0791	12.6445	.0141	70.9935	0.5916
	23.20	4.044	0.0742	13.4830	.0132	75.7015	0.5548
	24.20	4.000	0.0722	13.8433	.0129	77.7245	0.5404
5¼	16.00	4.650	0.1029	9.7179	.0183	54.5619	0.7698
	13.00	5.044	0.1237	8.0818	.0220	45.3760	0.9256
	14.00	5.012	0.1220	8.1981	.0217	46.0288	0.9125
	15.00	4.974	0.1199	8.3396	.0214	46.8233	0.8970
	15.50	4.950	0.1186	8.4309	.0211	47.3361	0.8873
	17.00	4.892	0.1155	8.6582	.0206	48.6122	0.8640
5½	20.00	4.778	0.1095	9.1337	.0195	51.2818	0.8190
	23.00	4.670	0.1039	9.6228	.0185	54.0281	0.7774
	25.00	4.580	0.0994	10.0625	.0177	56.4967	0.7434
	26.00	4.548	0.0978	10.2264	.0174	57.4171	0.7315
	32.30	4.276	0.0847	11.8070	.0151	66.2916	0.6336
	36.40	4.090	0.0762	13.1220	.0136	73.6743	0.5701

VOLUME AND FILL BETWEEN TUBING AND CASING

1.660" OD TUBING

CASING

SIZE OD INCHES	WEIGHT PER LIN FT	ID	CU FT PER LIN FT	LIN FT PER CU FT	BARRELS PER LIN FT	LIN FT PER BARREL	GALLONS PER LIN FT
5¾	14.00	5.290	0.1376	7.2674	.0245	40.8036	1.0293
	17.00	5.190	0.1319	7.5824	.0235	42.5721	0.9866
	19.50	5.090	0.1263	7.9191	.0225	44.4623	0.9446
	20.00	5.090	0.1263	7.9191	.0225	44.4623	0.9446
	22.50	4.990	0.1208	8.2795	.0215	46.4862	0.9035
	23.00	4.990	0.1208	8.2795	.0215	46.4862	0.9035
	25.20	4.890	0.1154	8.6662	.0206	48.6571	0.8632
6	15.00	5.524	0.1514	6.6049	.0270	37.0840	1.1326
	16.00	5.500	0.1500	6.6685	.0267	37.4409	1.1218
	18.00	5.424	0.1454	6.8761	.0259	38.6066	1.0879
	20.00	5.352	0.1412	7.0822	.0251	39.7637	1.0562
	23.00	5.240	0.1347	7.4223	.0240	41.6733	1.0078
	26.00	5.140	0.1291	7.7479	.0230	43.5013	0.9655
6⅝	13.00	6.260	0.1987	5.0326	.0354	28.2558	1.4864
	17.00	6.135	0.1903	5.2561	.0339	29.5108	1.4232
	20.00	6.049	0.1845	5.4189	.0329	30.4247	1.3805
	22.00	5.980	0.1800	5.5551	.0321	31.1898	1.3466
	24.00	5.921	0.1762	5.6759	.0314	31.8678	1.3179
	25.00	5.880	0.1735	5.7622	.0309	32.3524	1.2982
	26.00	5.855	0.1719	5.8158	.0306	32.6535	1.2862
	26.80	5.837	0.1708	5.8549	.0304	32.8729	1.2776
	28.00	5.791	0.1679	5.9567	.0299	33.4442	1.2558
	29.00	5.761	0.1660	6.0245	.0296	33.8250	1.2417
	31.80	5.675	0.1606	6.2257	.0286	34.9546	1.2016
	32.00	5.675	0.1606	6.2257	.0286	34.9546	1.2016
	34.00	5.595	0.1557	6.4223	.0277	36.0585	1.1648
7	17.00	6.538	0.2181	4.5848	.0388	25.7419	1.6316
	20.00	6.456	0.2123	4.7103	.0378	26.4465	1.5881
	22.00	6.398	0.2082	4.8023	.0371	26.9630	1.5577
	23.00	6.366	0.2060	4.8542	.0367	27.2546	1.5410
	24.00	6.336	0.2039	4.9037	.0363	27.5323	1.5255
	26.00	6.276	0.1998	5.0050	.0$56	28.1011	1.4946
	28.00	6.214	0.1956	5.1131	.0348	28.7079	1.4630
	29.00	6.184	0.1935	5.1667	.0345	29.0088	1.4478
	29.80	6.168	0.1925	5.1956	.0343	29.1713	1.4398
	30.00	6.154	0.1915	5.2211	.0341	29.3146	1.4327
	32.00	6.094	0.1875	5.3327	.0334	29.9412	1.4028
	35.00	6.004	0.1816	5.5072	.0323	30.9204	1.3583
	38.00	5.920	0.1761	5.6780	.0314	31.8795	1.3175
	40.20	5.836	0.1707	5.8571	.0304	32.8852	1.2772
	41.00	5.820	0.1697	5.8922	.0302	33.0823	1.2696
	43.00	5.736	0.1644	6.0819	.0293	34.1475	1.2300
	44.00	5.720	0.1634	6.1191	.0291	34.3564	1.2225
	49.50	5.540	0.1524	6.5631	.0271	36.8490	1.1398

VOLUME AND FILL BETWEEN TUBING AND CASING

1.660" OD TUBING

CASING

SIZE OD INCHES	WEIGHT PER LIN FT	ID	CU FT PER LIN FT	LIN FT PER CU FT	BARRELS PER LIN FT	LIN FT PER BARREL	GALLONS PER LIN FT
	20.00	7.125	0.2619	3.8189	.0466	21.4417	1.9588
	24.00	7.025	0.2541	3.9349	.0453	22.0928	1.9011
	26.40	6.969	0.2499	4.0022	.0445	22.4707	1.8691
	29.70	6.875	0.2428	4.1192	.0432	23.1277	1.8160
	33.70	6.765	0.2346	4.2629	.0418	23.9345	1.7548
7⅝	34.00	6.760	0.2342	4.2696	.0417	23.9722	1.7520
	35.50	6.710	0.2305	4.3377	.0411	24.3542	1.7246
	38.00	6.655	0.2265	4.4144	.0403	24.7852	1.6946
	39.00	6.625	0.2244	4.4572	.0400	25.0252	1.6783
	45.30	6.435	0.2108	4.7433	.0375	26.6318	1.5771
7¾	46.10	6.560	0.2197	4.5520	.0391	25.5577	1.6433
8	26.00	7.386	0.2825	3.5397	.0503	19.8739	2.1133
	28.00	7.485	0.2905	3.4419	.0517	19.3246	2.1734
	32.00	7.385	0.2824	3.5407	.0503	19.8795	2.1127
	35.50	7.285	0.2744	3.6439	.0489	20.4591	2.0529
8⅛	36.00	7.285	0.2744	3.6439	.0489	20.4591	2.0529
	39.50	7.185	0.2665	3.7518	.0475	21.0649	1.9938
	40.00	7.185	0.2665	3.7518	.0475	21.0649	1.9938
	42.00	7.125	0.2619	3.8189	.0466	21.4417	1.9588

1.900″ OD TUBING	VOLUME AND FILL BETWEEN TUBING AND CASING						
CASING							

SIZE OD INCHES	WEIGHT PER LIN FT	ID	CU FT PER LIN FT	LIN FT PER CU FT	BARRELS PER LIN FT	LIN FT PER BARREL	GALLONS PER LIN FT
	9.26	3.550	0.0490	20.3888	.0087	114.4747	0.3669
	9.50	3.550	0.0490	20.3888	.0087	114.4747	0.3669
4	11.00	3.480	0.0464	21.5692	.0083	121.1019	0.3468
	11.60	3.430	0.0445	22.4830	.0079	126.2326	0.3327
	12.60	3.364	0.0420	23.7912	.0075	133.5775	0.3144
	9.50	4.090	0.0715	13.9766	.0127	78.4728	0.5352
	10.50	4.052	0.0699	14.3142	.0124	80.3683	0.5226
	10.98	4.030	0.0689	14.5157	.0123	81.4997	0.5153
	11.00	4.026	0.0687	14.5528	.0122	81.7081	0.5140
	11.60	4.000	0.0676	14.7979	.0120	83.0843	0.5055
	11.75	3.990	0.0671	14.8940	.0120	83.6235	0.5023
	12.60	3.958	0.0658	15.2082	.0117	85.3877	0.4919
4½	12.75	3.960	0.0658	15.1883	.0117	85.2757	0.4925
	13.50	3.920	0.0641	15.5955	.0114	87.5620	0.4797
	15.10	3.826	0.0601	16.6251	.0107	93.3432	0.4500
	16.60	3.826	0.0601	16.6251	.0107	93.3432	0.4500
	18.80	3.640	0.0526	19.0201	.0094	106.7901	0.3933
	21.60	3.500	0.0471	21.2207	.0084	119.1452	0.3525
	24.60	3.380	0.0426	23.4626	.0076	131.7330	0.3188
	26.50	3.240	0.0376	26.6198	.0067	149.4591	0.2810
	16.00	4.082	0.0712	14.0466	.0127	78.8658	0.5325
	16.50	4.070	0.0707	14.1527	.0126	79.4614	0.5286
4¾	18.00	4.000	0.0676	14.7979	.0120	83.0843	0.5055
	20.00	3.910	0.0637	15.7000	.0113	88.1491	0.4765
	21.00	3.850	0.0612	16.3520	.0109	91.8095	0.4575
	11.50	4.560	0.0937	10.6699	.0167	59.9068	0.7011
	12.85	4.500	0.0908	11.0184	.0162	61.8638	0.6789
	13.00	4.494	0.0905	11.0543	.0161	62.0651	0.6767
	14.00	4.450	0.0883	11.3229	.0157	63.5735	0.6607
	15.00	4.408	0.0863	11.5892	.0154	65.0685	0.6455
5	18.00	4.276	0.0800	12.4945	.0143	70.1514	0.5987
	20.30	4.184	0.0758	13.1943	.0135	74.0807	0.5669
	21.00	4.154	0.0744	13.4362	.0133	75.4386	0.5567
	23.20	4.044	0.0695	14.3870	.0124	80.7768	0.5200
	24.20	4.000	0.0676	14.7979	.0120	83.0843	0.5055
5¼	16.00	4.650	0.0982	10.1788	.0175	57.1500	0.7349
	13.00	5.044	0.1191	8.3981	.0212	47.1518	0.8907
	14.00	5.012	0.1173	8.5237	.0209	47.8571	0.8776
	15.00	4.974	0.1152	8.6768	.0205	48.7166	0.8621
	15.50	4.950	0.1140	8.7757	.0203	49.2719	0.8524
	17.00	4.892	0.1108	9.0222	.0197	50.6560	0.8291
5½	20.00	4.778	0.1048	9.5397	.0187	53.5615	0.7841
	23.00	4.670	0.0993	10.0746	.0177	56.5646	0.7425
	25.00	4.580	0.0947	10.5575	.0169	59.2762	0.7085
	26.00	4.548	0.0931	10.7382	.0166	60.2903	0.6966
	32.30	4.276	0.0800	12.4945	.0143	70.1514	0.5987
	36.40	4.090	0.0715	13.9766	.0127	78.4728	0.5352

1.900" OD TUBING

VOLUME AND FILL BETWEEN TUBING AND CASING

SIZE OD INCHES	CASING WEIGHT PER LIN FT	ID	CU FT PER LIN FT	LIN FT PER CU FT	BARRELS PER LIN FT	LIN FT PER BARREL	GALLONS PER LIN FT
5¾	14.00	5.290	0.1329	7.5222	.0237	42.2339	0.9945
	17.00	5.190	0.1272	7.8601	.0227	44.1314	0.9517
	19.50	5.090	0.1216	8.2225	.0217	46.1660	0.9098
	20.00	5.090	0.1216	8.2225	.0217	46.1660	0.9098
	22.50	4.990	0.1161	8.6118	.0207	48.3518	0.8686
	23.00	4.990	0.1161	8.6118	.0207	48.3518	0.8686
	25.20	4.890	0.1107	9.0309	.0197	50.7048	0.8283
6	15.00	5.524	0.1467	6.8147	.0261	38.2617	1.0977
	16.00	5.500	0.1453	6.8824	.0259	38.6417	1.0869
	18.00	5.424	0.1408	7.1038	.0251	39.8847	1.0530
	20.00	5.352	0.1365	7.3239	.0243	41.1208	1.0214
	23.00	5.240	0.1301	7.6883	.0232	43.1664	0.9730
	26.00	5.140	0.1244	8.0381	0.222	45.1307	0.9306
6⅝	13.00	6.260	0.1940	5.1534	.0346	28.9343	1.4516
	17.00	6.135	0.1856	5.3881	.0331	30.2518	1.3883
	20.00	6.049	0.1799	5.5593	.0320	31.2129	1.3456
	22.00	5.980	0.1754	5.7028	.0312	32.0187	1.3117
	24.00	5.921	0.1715	5.8301	.0305	32.7336	1.2831
	25.00	5.880	0.1689	5.9212	.0301	33.2451	1.2633
	26.00	5.855	0.1673	5.9778	.0298	33.5631	1.2514
	26.80	5.837	0.1661	6.0191	.0296	33.7950	1.2428
	28.00	5.791	0.1632	6.1267	.0291	34.3990	1.2210
	29.00	5.761	0.1613	6.1985	.0287	34.8021	1.2068
	31.80	5.675	0.1560	6.4117	.0278	35.9990	1.1667
	32.00	5.675	0.1560	6.4117	.0278	35.9990	1.1667
	34.00	5.595	0.1510	6.6204	.0269	37.1710	1.1299
7	17.00	6.538	0.2135	4.6849	.0380	26.3039	1.5967
	20.00	6.456	0.2076	4.8160	.0370	27.0401	1.5532
	22.00	6.398	0.2036	4.9122	.0363	27.5802	1.5228
	23.00	6.366	0.2013	4.9666	.0359	27.8853	1.5062
	24.00	6.336	0.1993	5.0184	.0355	28.1762	1.4906
	26.00	6.276	0.1951	5.1245	.0348	28.7721	1.4597
	28.00	6.214	0.1909	5.2379	.0340	29.4086	1.4282
	29.00	6.184	0.1889	5.2942	.0336	29.7245	1.4130
	29.80	6.168	0.1878	5.3245	.0335	29.8951	1.4049
	30.00	6.154	0.1869	5.3514	.0333	30.0456	1.3979
	32.00	6.094	0.1829	5.4686	.0326	30.7042	1.3679
	35.00	6.004	0.1769	5.6522	.0315	31.7348	1.3235
	38.00	5.920	0.1715	5.8323	.0305	32.7459	1.2826
	40.20	5.836	0.1661	6.0214	.0296	33.8079	1.2423
	41.00	5.820	0.1651	6.0586	.0294	34.0163	1.2347
	43.00	5.736	0.1598	6.2593	.0285	35.1435	1.1951
	44.00	5.720	0.1588	6.2987	.0283	35.3648	1.1876
	49.50	5.540	0.1477	6.7701	.0263	38.0116	1.1049

1.900" OD TUBING	**VOLUME AND FILL BETWEEN TUBING AND CASING**					
CASING						

SIZE OD INCHES	WEIGHT PER LIN FT	ID	CU FT PER LIN FT	LIN FT PER CU FT	BARRELS PER LIN FT	LIN FT PER BARREL	GALLONS PER LIN FT
	20.00	7.125	0.2572	3.8881	.0458	21.8301	1.9239
	24.00	7.025	0.2495	4.0084	.0444	22.5055	1.8662
	26.40	6.969	0.2452	4.0783	.0437	22.8978	1.8342
	29.70	6.875	0.2381	4.1998	.0424	23.5803	1.7811
	33.70	6.765	0.2299	4.3493	.0410	24.4196	1.7199
7⅝	34.00	6.760	0.2296	4.3563	.0409	24.4588	1.7172
	35.50	6.710	0.2259	4.4272	.0402	24.8566	1.6897
	38.00	6.655	0.2219	4.5072	.0395	25.3058	1.6597
	39.00	6.625	0.2197	4.5517	.0391	25.5561	1.6434
	45.30	6.435	0.2062	4.8505	.0367	27.2337	1.5422
7¾	46.10	6.560	0.2150	4.6507	.0383	26.1116	1.6085
8	26.00	7.386	0.2779	3.5991	.0495	20.2072	2.0785
	28.00	7.485	0.2859	3.4980	.0509	19.6396	2.1385
	32.00	7.385	0.2778	3.6001	.0495	20.2130	2.0779
	35.50	7.285	0.2698	3.7069	.0480	20.8125	2.0180
8⅛	36.00	7.285	0.2698	3.7069	.0480	20.8125	2.0180
	39.50	7.185	0.2619	3.8186	.0466	21.4398	1.9590
	40.00	7.185	0.2619	3.8186	.0466	21.4398	1.9590
	42.00	7.125	0.2572	3.8881	.0458	21.8301	1.9239

VOLUME AND FILL BETWEEN TUBING AND CASING

2-⅜″ OD TUBING

CASING

SIZE OD INCHES	WEIGHT PER LIN FT	ID	CU FT PER LIN FT	LIN FT PER CU FT	BARRELS PER LIN FT	LIN FT PER BARREL	GALLONS PER LIN FT
	9.50	4.090	0.0605	16.5364	.0108	92.8448	0.4524
	10.50	4.052	0.0588	17.0111	.0105	95.5100	0.4397
	10.98	4.030	0.0578	17.2964	.0103	97.1120	0.4325
	11.00	4.026	0.0576	17.3491	.0103	97.4081	0.4312
	11.60	4.000	0.0565	17.6986	.0101	99.3703	0.4227
	11.75	3.990	0.0561	17.8362	.0100	100.1427	0.4194
	12.60	3.958	0.0547	18.2887	.0097	102.6833	0.4090
4½	12.75	3.960	0.0548	18.2598	.0098	102.5213	0.4097
	13.50	3.920	0.0530	18.8516	.0094	105.8439	0.3968
	15.10	3.826	0.0491	20.3772	.0087	114.4092a	0.3671
	16.60	3.826	0.0491	20.3772	.0087	114.4092	0.3671
	18.80	3.640	0.0415	24.0961	.0074	135.2895	0.3104
	21.60	3.500	0.0360	27.7404	.0064	155.7506	0.2697
	24.60	3.380	0.0315	31.7001	.0056	177.9831	0.2360
	26.50	3.240	0.0265	37.7491	.0047	211.9455	0.1982
	16.00	4.082	0.0601	16.6344	.0107	93.3955	0.4497
	16.50	4.070	0.0596	16.7834	.0106	94.2318	0.4457
4¾	18.00	4.000	0.0565	17.6986	.0101	99.3703	0.4227
	20.00	3.910	0.0526	19.0046	.0094	106.7030	0.3936
	21.00	3.850	0.0501	19.9683	.0089	112.1137	0.3746
	11.50	4.560	0.0826	12.0997	.0147	67.9348	0.6182
	12.85	4.500	0.0797	12.5499	.0142	70.4626	0.5961
	13.00	4.494	0.0794	12.5964	.0141	70.7238	0.5939
	14.00	4.450	0.0772	12.9465	.0138	72.6891	0.5778
	15.00	4.408	0.0752	13.2958	.0134	74.6502	0.5626
5	18.00	4.276	0.0690	14.5012	.0123	81.4181	0.5159
	20.30	4.184	0.0647	15.4524	.0115	86.7589	0.4841
	21.00	4.154	0.0634	15.7852	.0113	88.6273	0.4739
	23.20	4.044	0.0584	17.1139	.0104	96.0874	0.4371
	24.20	4.000	0.0565	17.6986	.0101	99.3703	0.4227
5¼	16.00	4.650	0.0872	11.4722	.0155	64.4114	0.6521
	13.00	5.044	0.1080	9.2593	.0192	51.9872	0.8079
	14.00	5.012	0.1062	9.4123	.0189	52.8460	0.7948
	15.00	4.974	0.1042	9.5993	.0186	53.8959	0.7793
	15.50	4.950	0.1029	9.7205	.0183	54.5764	0.7696
	17.00	4.892	0.0998	10.0238	.0178	56.2797	0.7463
5½	20.00	4.778	0.0937	10.6667	.0167	59.8891	0.7013
	23.00	4.670	0.0882	11.3399	.0157	63.6688	0.6597
	25.00	4.580	0.0836	11.9555	.0149	67.1250	0.6257
	26.00	4.548	0.0821	12.1876	.0146	68.4284	0.6138
	32.30	4.276	0.0690	14.5012	.0123	81.4181	0.5159
	36.40	4.090	0.0605	16.5364	.0108	92.8448	0.4524
	14.00	5.290	0.1219	8.2058	.0217	46.0722	0.9116
	17.00	5.190	0.1161	8.6096	.0207	48.3396	0.8689
	19.50	5.090	0.1105	9.0463	.0197	50.7914	0.8269
5¾	20.00	5.090	0.1105	9.0463	.0197	50.7914	0.8269
	22.50	4.990	0.1050	9.5198	.0187	53.4498	0.7858
	23.00	4.990	0.1050	9.5198	.0187	53.4498	0.7858
	25.20	4.890	0.0997	10.0346	.0177	56.3400	0.7455

| 2-3/8" OD TUBING | | | **VOLUME AND FILL BETWEEN TUBING AND CASING** | | | | |

SIZE OD INCHES	WEIGHT PER LIN FT	ID	CU FT PER LIN FT	LIN FT PER CU FT	BARRELS PER LIN FT	LIN FT PER BARREL	GALLONS PER LIN FT
	15.00	5.524	0.1357	7.3710	.0242	41.3852	1.0149
	16.00	5.500	0.1342	7.4503	.0239	41.8302	1.0041
	18.00	5.424	0.1297	7.7104	.0231	43.2906	0.9702
6	20.00	5.352	0.1255	7.9705	.0223	44.7508	0.9385
	23.00	5.240	0.1190	8.4038	.0212	47.1841	0.8901
	26.00	5.140	0.1133	8.8237	.0202	49.5411	0.8478
	13.00	6.260	0.1830	5.4654	.0326	30.6858	1.3687
	17.00	6.135	0.1745	5.7300	.0311	32.1716	1.3055
	20.00	6.049	0.1688	5.9240	.0301	33.2608	1.2627
	22.00	5.980	0.1643	6.0872	.0293	34.1774	1.2289
	24.00	5.921	0.1604	6.2325	.0286	34.9931	1.2002
	25.00	5.880	0.1578	6.3368	.0281	35.5783	1.1805
6⅝	26.00	5.855	0.1562	6.4017	.0278	35.9427	1.1685
	26.80	5.837	0.1551	6.4491	.0276	36.2088	1.1599
	28.00	5.791	0.1521	6.5727	.0271	36.9031	1.1381
	29.00	5.761	0.1503	6.6554	.0268	37.3673	1.1240
	31.80	5.675	0.1449	6.9018	.0258	38.7508	1.0838
	32.00	5.675	0.1449	6.9018	.0258	38.7508	1.0838
	34.00	5.595	0.1400	7.1443	.0249	40.1121	1.0471
	17.00	6.538	0.2024	4.9413	.0360	27.7434	1.5139
	20.00	6.456	0.1966	5.0874	.0350	28.5636	1.4704
	22.00	6.398	0.1925	5.1949	.0343	29.1670	1.4400
	23.00	6.366	0.1903	5.2557	.0339	29.5085	1.4233
	24.00	6.336	0.1882	5.3137	.0335	29.8344	1.4078
	26.00	6.276	0.1841	5.4329	.0328	30.5034	1.3769
	28.00	6.214	0.1798	5.5605	.0320	31.2198	1.3453
	29.00	6.184	0.1778	5.6239	.0317	31.5759	1.3301
	29.80	6.168	0.1767	5.6582	.0315	31.7685	1.3221
7	30.00	6.154	0.1758	5.6885	.0313	31.9385	1.3150
	32.00	6.094	0.1718	5.8212	.0306	32.6837	1.2850
	35.00	6.004	0.1658	6.0297	.0295	33.8541	1.2406
	38.00	5.920	0.1604	6.2351	.0286	35.0072	1.1998
	40.20	5.836	0.1550	6.4517	.0276	36.2237	1.1595
	41.00	5.820	0.1540	6.4943	.0274	36.4630	1.1519
	43.00	5.736	0.1487	6.7256	.0265	37.7613	1.1122
	44.00	5.720	0.1477	6.7711	.0263	38.0169	1.1048
	49.50	5.540	0.1366	7.3189	.0243	41.0928	1.0221
	20.00	7.125	0.2461	4.0631	.0438	22.8125	1.8411
	24.00	7.025	0.2384	4.1946	.0425	23.5510	1.7834
	26.40	6.969	0.2341	4.2712	.0417	23.9809	1.7514
	29.70	6.875	0.2270	4.4047	.0404	24.7307	1.6983
	33.70	6.765	0.2188	4.5694	.0390	25.6554	1.6371
7⅝	34.00	6.760	0.2185	4.5771	.0389	25.6987	1.6343
	35.50	6.710	0.2148	4.6554	.0383	26.1382	1.6068
	38.00	6.655	0.2108	4.7440	.0375	26.6354	1.5769
	39.00	6.625	0.2086	4.7934	.0372	26.9128	1.5606
	45.30	6.435	0.1951	5.1259	.0347	28.7798	1.4594
7¾	46.10	6.560	0.2039	4.9032	.0363	27.5296	1.5256
8	26.00	7.386	0.2668	3.7485	.0475	21.0461	1.9956

VOLUME AND FILL BETWEEN TUBING AND CASING

2-³⁄₈″ OD TUBING

CASING

SIZE OD INCHES	WEIGHT PER LIN FT	ID	CU FT PER LIN FT	LIN FT PER CU FT	BARRELS PER LIN FT	LIN FT PER BARREL	GALLONS PER LIN FT
	28.00	7.485	0.2748	3.6389	.0489	20.4311	2.0557
	32.00	7.385	0.2667	3.7496	.0475	21.0524	1.9950
	35.50	7.285	0.2587	3.8656	.0461	21.7036	1.9352
8⅛	36.00	7.285	0.2587	3.8656	.0461	21.7036	1.9352
	39.50	7.185	0.2508	3.9872	.0447	22.3865	1.8761
	40.00	7.185	0.2508	3.9872	.0447	22.3865	1.8761
	42.00	7.125	0.2461	4.0631	.0438	22.8125	1.8411
	24.00	8.097	0.3268	3.0598	.0582	17.1796	2.4448
	28.00	8.017	0.3198	3.1271	.0570	17.5573	2.3922
	32.00	7.921	0.3114	3.2109	.0555	18.0278	2.3297
	36.00	7.825	0.3032	3.2982	.0540	18.5180	2.2681
	38.00	7.775	0.2989	3.3451	.0532	18.7815	2.2362
8⅝	40.00	7.725	0.2947	3.3931	.0525	19.0509	2.2046
	43.00	7.651	0.2885	3.4661	.0514	19.4607	2.1582
	44.00	7.625	0.2863	3.4923	.0510	19.6079	2.1420
	49.00	7.511	0.2769	3.6110	.0493	20.2742	2.0716
	52.00	7.435	0.2707	3.6936	.0482	20.7382	2.0253
	34.00	8.290	0.3441	2.9064	.0613	16.3183	2.5738
	38.00	8.196	0.3356	2.9796	.0598	16.7292	2.5106
	40.00	8.150	0.3315	3.0165	.0590	16.9362	2.4799
	41.20	8.150	0.3315	3.0165	.0590	16.9362	2.4799
9	45.00	8.032	0.3211	3.1143	.0572	17.4855	2.4020
	46.10	8.032	0.3211	3.1143	.0572	17.4855	2.4020
	54.00	7.810	0.3019	3.3122	.0538	18.5964	2.2585
	55.20	7.812	0.3021	3.3103	.0538	18.5859	2.2598
	29.30	9.063	0.4172	2.3968	.0743	13.4569	3.1211
	32.30	9.001	0.4111	2.4324	.0732	13.6568	3.0754
	36.00	8.921	0.4033	2.4795	.0718	13.9216	3.0169
	38.00	8.885	0.3998	2.5012	.0712	14.0434	2.9907
	40.00	8.835	0.3950	2.5318	.0703	14.2152	2.9546
	42.00	8.799	0.3915	2.5542	.0697	14.3409	2.9287
	43.50	8.755	0.3873	2.5820	.0690	14.4969	2.8972
9⅝	44.30	8.750	0.3868	2.5852	.0689	14.5148	2.8936
	47.00	8.681	0.3803	2.6298	.0677	14.7652	2.8445
	47.20	8.680	0.3802	2.6304	.0677	14.7688	2.8438
	53.50	8.535	0.3665	2.7281	.0653	15.3174	2.7420
	57.40	8.450	0.3587	2.7880	.0639	15.6537	2.6831
	58.40	8.435	0.3573	2.7988	.0636	15.7142	2.6727
	61.10	8.375	0.3518	2.8426	.0627	15.9599	2.6316
	33.00	9.384	0.4495	2.2246	.0801	12.4900	3.3627
10	60.00	8.780	0.3897	2.5662	.0694	14.4079	2.9151

VOLUME AND FILL BETWEEN TUBING AND CASING

2-⅜″ OD TUBING

CASING

SIZE OD INCHES	WEIGHT PER LIN FT	ID	CU FT PER LIN FT	LIN FT PER CU FT	BARRELS PER LIN FT	LIN FT PER BARREL	GALLONS PER LIN FT
	32.75	10.192	0.5358	1.8664	.0954	10.4790	4.0080
	35.75	10.140	0.5300	1.8867	.0944	10.5930	3.9649
	40.50	10.050	0.5201	1.9226	.0926	10.7948	3.8908
	45.50	9.950	0.5092	1.9638	.0907	11.0261	3.8092
	46.20	9.950	0.5092	1.9638	.0907	11.0261	3.8092
	48.00	9.902	0.5040	1.9841	.0898	11.1398	3.7703
	49.50	9.850	0.4984	2.0064	.0888	11.2650	3.7284
10¾	51.00	9.850	0.4984	2.0064	.0888	11.2650	3.7284
	54.00	9.784	0.4913	2.0352	.0875	11.4270	3.6755
	55.50	9.760	0.4888	2.0459	.0871	11.4868	3.6564
	60.70	9.660	0.4782	2.0912	.0852	11.7413	3.5771
	65.70	9.560	0.4677	2.1381	.0833	12.0044	3.4987
	71.10	9.450	0.4563	2.1915	.0813	12.3045	3.4134
	76.00	9.350	0.4461	2.2419	.0794	12.5873	3.3367
	81.00	9.250	0.4359	2.2941	.0776	12.8803	3.2608

514

2-⅞″ OD
TUBING

VOLUME AND FILL BETWEEN TUBING AND CASING

CASING

SIZE OD INCHES	WEIGHT PER LIN FT	ID	CU FT PER LIN FT	LIN FT PER CU FT	BARRELS PER LIN FT	LIN FT PER BARREL	GALLONS PER LIN FT
	9.50	4.090	0.0462	21.6658	.0082	121.6446	0.3453
	10.50	4.052	0.0445	22.4880	.0079	126.2608	0.3326
	10.98	4.030	0.0435	22.9894	.0077	129.0757	0.3254
	11.00	4.026	0.0433	23.0826	.0077	129.5993	0.3241
	11.60	4.000	0.0422	23.7054	.0075	133.0960	0.3156
	11.75	3.990	0.0417	23.9529	.0074	134.4853	0.3123
	12.60	3.958	0.0404	24.7761	.0072	139.1074	0.3019
4½	12.75	3.960	0.0404	24.7232	.0072	138.8104	0.3026
	13.50	3.920	0.0387	25.8206	.0069	144.9721	0.2897
	15.10	3.826	0.0348	28.7708	.0062	161.5363	0.2600
	16.60	3.826	0.0348	28.7708	.0062	161.5363	0.2600
	18.80	3.640	0.0272	36.7872	.0048	206.5448	0.2033
	21.60	3.500	0.0217	46.0164	.0039	258.3628	0.1626
	24.60	3.380	0.0172	58.0435	.0031	325.8903	0.1289
	26.50	3.240	0.0122	82.1454	.0022	461.2122	0.0911
	16.00	4.082	0.0458	21.8345	.0082	122.5916	0.3426
	16.50	4.070	0.0453	22.0919	.0081	124.0366	0.3386
4¾	18.00	4.000	0.0422	23.7054	.0075	133.0960	0.3156
	20.00	3.910	0.0383	26.1085	.0068	146.5885	0.2865
	21.00	3.850	0.0358	27.9625	.0064	156.9977	0.2675
	11.50	4.560	0.0683	14.6350	.0122	82.1692	0.5111
	12.85	4.500	0.0654	15.2988	.0116	85.8964	0.4890
	13.00	4.494	0.0651	15.3680	.0116	86.2849	0.4868
	14.00	4.450	0.0629	15.8922	.0112	89.2282	0.4707
	15.00	4.408	0.0609	16.4218	.0108	92.2014	0.4555
5	18.00	4.276	0.0546	18.3007	.0097	102.7508	0.4088
	20.30	4.184	0.0504	19.8422	.0090	111.4057	0.3770
	21.00	4.154	0.0490	20.3943	.0087	114.5054	0.3668
	23.20	4.044	0.0441	22.6681	.0079	127.2718	0.3300
	24.20	4.000	0.0422	23.7054	.0075	133.0960	0.3156
5¼	16.00	4.650	0.0729	13.7268	.0130	77.0700	0.5450
	13.00	5.044	0.0937	10.6744	.0167	59.9322	0.7008
	14.00	5.012	0.0919	10.8782	.0164	61.0764	0.6877
	15.00	4.974	0.0899	11.1287	.0160	62.4832	0.6722
	15.50	4.950	0.0886	11.2920	.0158	63.3998	0.6625
	17.00	4.892	0.0854	11.7034	.0152	65.7099	0.6392
5½	20.00	4.778	0.0794	12.5893	.0141	70.6838	0.5942
	23.00	4.670	0.0739	13.5378	.0132	76.0092	0.5526
	25.00	4.580	0.0693	14.4245	.0123	80.9875	0.5186
	26.00	4.548	0.0677	14.7638	.0121	82.8924	0.5067
	32.30	4.276	0.0546	18.3007	.0097	102.7508	0.4088
	36.40	4.090	0.0462	21.6658	.0082	121.6446	0.3453
	14.00	5.290	0.1075	9.2982	.0192	52.2056	0.8045
	17.00	5.190	0.1018	9.8201	.0181	55.1359	0.7618
	19.50	5.090	0.0962	10.3923	.0171	58.3486	0.7198
5¾	20.00	5.090	0.0962	10.3923	.0171	58.3486	0.7198
	22.50	4.990	0.0907	11.0221	.0162	61.8844	0.6787
	23.00	4.990	0.0907	11.0221	.0162	61.8844	0.6787
	25.20	4.890	0.0853	11.7181	.0152	65.7921	0.6384

VOLUME AND FILL BETWEEN TUBING AND CASING

2-⅞" OD TUBING

CASING

SIZE OD INCHES	WEIGHT PER LIN FT	ID	CU FT PER LIN FT	LIN FT PER CU FT	BARRELS PER LIN FT	LIN FT PER BARREL	GALLONS PER LIN FT
6	15.00	5.524	0.1213	8.2407	.0216	46.2680	0.9078
	16.00	5.500	0.1199	8.3399	.0214	46.8248	0.8970
	18.00	5.424	0.1154	8.6672	.0205	48.6625	0.8631
	20.00	5.352	0.1111	8.9972	.0198	50.5153	0.8314
	23.00	5.240	0.1047	9.5533	.0186	53.6377	0.7830
	26.00	5.140	0.0990	10.0995	.0176	56.7046	0.7407
6⅝	13.00	6.260	0.1687	5.9293	.0300	33.2097	1.2616
	17.00	6.135	0.1602	6.2421	.0285	35.0467	1.1984
	20.00	6.049	0.1545	6.4730	.0275	36.3432	1.1556
	22.00	5.980	0.1500	6.6684	.0267	37.4404	1.1218
	24.00	5.921	0.1461	6.8432	.0260	38.4216	1.0931
	25.00	5.880	0.1435	6.9690	.0256	39.1282	1.0734
	26.00	5.855	0.1419	7.0476	.0253	39.5694	1.0614
	26.80	5.837	0.1407	7.1051	.0251	39.8921	1.0528
	28.00	5.791	0.1378	7.2555	.0245	40.7365	1.0310
	29.00	5.761	0.1359	7.3564	.0242	41.3030	1.0169
	31.80	5.675	0.1306	7.6586	.0233	42.9998	0.9767
	32.00	5.675	0.1306	7.6586	.0233	42.9998	0.9767
	34.00	5.595	0.1257	7.9583	.0224	44.6825	0.9400
7	17.00	6.538	0.1881	5.3175	.0335	29.8556	1.4068
	20.00	6.456	0.1822	5.4871	.0325	30.8076	1.3633
	22.00	6.398	0.1782	5.6123	.0317	31.5106	1.3329
	23.00	6.366	0.1760	5.6833	.0313	31.9096	1.3162
	24.00	6.336	0.1739	5.7513	.0310	32.2910	1.3007
	26.00	6.276	0.1697	5.8911	.0302	33.0761	1.2698
	28.00	6.214	0.1655	6.0414	.0295	33.9201	1.2382
	29.00	6.184	0.1635	6.1164	.0291	34.3410	1.2230
	29.80	6.168	0.1624	6.1570	.0289	34.5689	1.2150
	30.00	6.154	0.1615	6.1929	.0288	34.7704	1.2079
	32.00	6.094	0.1575	6.3505	.0280	35.6554	1.1779
	35.00	6.004	0.1515	6.5994	.0270	37.0528	1.1335
	38.00	5.920	0.1461	6.8462	.0260	38.4385	1.0927
	40.20	5.836	0.1407	7.1083	.0251	39.9102	1.0524
	41.00	5.820	0.1397	7.1601	.0249	40.2009	1.0448
	43.00	5.736	0.1344	7.4422	.0239	41.7848	1.0051
	44.00	5.720	0.1334	7.4980	.0238	42.0981	0.9977
	49.50	5.540	0.1223	8.1756	.0218	45.9028	0.9150
7⅝	20.00	7.125	0.2318	4.3140	.0413	24.2215	1.7340
	24.00	7.025	0.2241	4.4626	.0399	25.0557	1.6763
	26.40	6.969	0.2198	4.5494	.0391	25.5429	1.6443
	29.70	6.875	0.2127	4.7012	.0379	26.3952	1.5912
	33.70	6.765	0.2045	4.8893	.0364	27.4513	1.5300
	34.00	6.760	0.2042	4.8981	.0364	27.5009	1.5272
	35.50	6.710	0.2005	4.9879	.0357	28.0048	1.4997
	38.00	6.655	0.1965	5.0896	.0350	28.5763	1.4698
	39.00	6.625	0.1943	5.1466	.0346	28.8958	1.4535
	45.30	6.435	0.1808	5.5319	.0322	31.0592	1.3523
7¾	46.10	6.560	0.1896	5.2734	.0338	29.6081	1.4185
8	26.00	7.386	0.2525	3.9610	.0450	22.2396	1.8885

Well Control Problems and Solutions

2-⅞" OD TUBING			VOLUME AND FILL BETWEEN TUBING AND CASING				
CASING							
SIZE OD INCHES	WEIGHT PER LIN FT	ID	CU FT PER LIN FT	LIN FT PER CU FT	BARRELS PER LIN FT	LIN FT PER BARREL	GALLONS PER LIN FT
8⅛	28.00	7.485	0.2605	3.8389	.0464	21.5541	1.9486
	32.00	7.385	0.2524	3.9623	.0450	22.2467	1.8879
	35.50	7.285	0.2444	4.0920	.0435	22.9751	1.8281
	36.00	7.285	0.2444	4.0920	.0435	22.9751	1.8281
	39.50	7.185	0.2365	4.2286	.0421	23.7419	1.7690
	40.00	7.185	0.2365	4.2286	.0421	23.7419	1.7690
	42.00	7.125	0.2318	4.3140	.0413	24.2215	1.7340
8⅝	24.00	8.097	0.3125	3.2000	.0557	17.9667	2.3377
	28.00	8.017	0.3055	3.2737	.0544	18.3802	2.2851
	32.00	7.921	0.2971	3.3656	.0529	18.8964	2.2226
	36.00	7.825	0.2889	3.4617	.0515	19.4357	2.1610
	38.00	7.775	0.2846	3.5134	.0507	19.7262	2.1291
	40.00	7.725	0.2804	3.5664	.0499	20.0236	2.0975
	43.00	7.651	0.2742	3.6471	.0488	20.4768	2.0511
	44.00	7.625	0.2720	3.6761	.0484	20.6399	2.0349
	49.00	7.511	0.2626	3.8079	.0468	21.3795	1.9645
	52.00	7.435	0.2564	3.8999	.0457	21.8961	1.9182
9	34.00	8.290	0.3297	3.0326	.0587	17.0268	2.4667
	38.00	8.196	0.3213	3.1124	.0572	17.4747	2.4035
	40.00	8.150	0.3172	3.1526	.0565	17.7006	2.3728
	41.20	8.150	0.3172	3.1526	.0565	17.7006	2.3728
	45.00	8.032	0.3068	3.2596	.0546	18.3015	2.2949
	46.10	8.032	0.3068	3.2596	.0546	18.3015	2.2949
	54.00	7.810	0.2876	3.4770	.0512	19.5222	2.1514
	55.20	7.812	0.2878	3.4750	.0513	19.5106	2.1527
9⅝	29.30	9.063	0.4029	2.4819	.0718	13.9350	3.0140
	32.30	9.001	0.3968	2.5201	.0707	14.1496	2.9683
	36.00	8.921	0.3890	2.5708	.0693	14.4340	2.9098
	38.00	8.885	0.3855	2.5941	.0687	14.5649	2.8836
	40.00	8.835	0.3807	2.6271	.0678	14.7498	2.8475
	42.00	8.799	0.3772	2.6512	.0672	14.8852	2.8216
	43.50	8.755	0.3730	2.6811	.0664	15.0533	2.7901
	44.30	8.750	0.3725	2.6846	.0663	15.0726	2.7865
	47.00	8.681	0.3659	2.7327	.0652	15.3428	2.7374
	47.20	8.680	0.3658	2.7334	.0652	15.3468	2.7367
	53.50	8.535	0.3522	2.8390	.0627	15.9400	2.6349
	57.40	8.450	0.3444	2.9040	.0613	16.3045	2.5760
	58.40	8.435	0.3430	2.9156	.0611	16.3702	2.5656
	61.10	8.375	0.3375	2.9632	.0601	16.6370	2.5245
10	33.00	9.384	0.4352	2.2978	.0775	12.9009	3.2556
	60.00	8.780	0.3754	2.6640	.0669	14.9575	2.8080

2-⅞" OD TUBING			VOLUME AND FILL BETWEEN TUBING AND CASING				
CASING							
SIZE OD INCHES	WEIGHT PER LIN FT	ID	CU FT PER LIN FT	LIN FT PER CU FT	BARRELS PER LIN FT	LIN FT PER BARREL	GALLONS PER LIN FT
	32.75	10.192	0.5215	1.9176	.0929	10.7667	3.9009
	35.75	10.140	0.5157	1.9391	.0919	10.8871	3.8578
	40.50	10.050	0.5058	1.9771	.0901	11.1004	3.7837
	45.50	9.950	0.4949	2.0206	.0881	11.3450	3.7021
	46.20	9.950	0.4949	2.0206	.0881	11.3450	3.7021
	48.00	9.902	0.4897	2.0421	.0872	11.4655	3.6632
	49.50	9.850	0.4841	2.0657	.0862	11.5981	3.6213
10¾	51.00	9.850	0.4841	2.0657	.0862	11.5981	3.6213
	54.00	9.784	0.4770	2.0963	.0850	11.7700	3.5684
	55.50	9.760	0.4745	2.1076	.0845	11.8334	3.5493
	60.70	9.660	0.4639	2.1558	.0826	12.1036	3.4700
	65.70	9.560	0.4534	2.2056	.0808	12.3835	3.3916
	71.10	9.450	0.4420	2.2625	.0787	12.7030	3.3063
	76.00	9.350	0.4317	2.3162	.0769	13.0047	3.2296
	81.00	9.250	0.4216	2.3720	.0751	13.3177	3.1537

VOLUME AND FILL BETWEEN TUBING AND CASING

3-½" OD TUBING

SIZE OD INCHES	WEIGHT PER LIN FT	ID	CU FT PER LIN FT	LIN FT PER CU FT	BARRELS PER LIN FT	LIN FT PER BARREL	GALLONS PER LIN FT
					CASING		
5½	13.00	5.044	0.0720	13.8984	.0128	78.0336	0.5382
	14.00	5.012	0.0702	14.2459	.0125	79.9847	0.5251
	15.00	4.974	0.0681	14.6787	.0121	82.4146	0.5096
	15.50	4.950	0.0668	14.9640	.0119	84.0167	0.4999
	17.00	4.892	0.0637	15.6952	.0113	88.1222	0.4766
	20.00	4.778	0.0577	17.3307	.0103	97.3047	0.4316
	23.00	4.670	0.0521	19.1807	.0093	107.6917	0.3900
	25.00	4.580	0.0476	21.0106	.0085	117.9655	0.3560
	26.00	4.548	0.0460	21.7382	.0082	122.0509	0.3441
	32.30	4.276	0.0329	30.3847	.0059	170.5973	0.2462
	36.40	4.090	0.0244	40.9429	.0044	229.8775	0.1827
5¾	14.00	5.290	0.0858	11.6528	.0153	65.4257	0.6419
	17.00	5.190	0.0801	12.4844	.0143	70.0945	0.5992
	19.50	5.090	0.0745	13.4240	.0133	75.3702	0.5572
	20.00	5.090	0.0745	13.4240	.0133	75.3702	0.5572
	22.50	4.990	0.0690	14.4937	.0123	81.3760	0.5161
	23.00	4.990	0.0690	14.4937	.0123	81.3760	0.5161
	25.20	4.890	0.0636	15.7216	.0113	88.2701	0.4758
6	15.00	5.524	0.0996	10.0384	.0177	56.3612	0.7452
	16.00	5.500	0.0982	10.1859	.0175	57.1897	0.7344
	18.00	5.424	0.0936	10.6784	.0167	59.9550	0.7005
	20.00	5.352	0.0894	11.1838	.0159	62.7925	0.6689
	23.00	5.240	0.0829	12.0562	.0148	67.6908	0.6205
	26.00	5.140	0.0773	12.9394	.0138	72.6495	0.5781
6⅝	13.00	6.260	0.1469	6.8063	.0262	38.2148	1.0991
	17.00	6.135	0.1385	7.2217	.0247	40.5469	1.0358
	20.00	6.049	0.1328	7.5326	.0236	42.2924	0.9931
	22.00	5.980	0.1282	7.7985	.0228	43.7855	0.9592
	24.00	5.921	0.1244	8.0386	.0222	45.1334	0.9306
	25.00	5.880	0.1218	8.2128	.0217	46.1116	0.9108
	26.00	5.855	0.1202	8.3222	.0214	46.7257	0.8989
	26.80	5.837	0.1190	8.4025	.0212	47.1763	0.8903
	28.00	5.791	0.1161	8.6136	.0207	48.3618	0.8685
	29.00	5.761	0.1142	8.7562	.0203	49.1622	0.8543
	31.80	5.675	0.1088	9.1877	.0194	51.5852	0.8142
	32.00	5.675	0.1088	9.1877	.0194	51.5852	0.8142
	34.00	5.595	0.1039	9.6225	.0185	54.0261	0.7774

| 3-½" OD TUBING | | | **VOLUME AND FILL BETWEEN TUBING AND CASING** | | | | |
| CASING | | | | | | | |

SIZE OD INCHES	WEIGHT PER LIN FT	ID	CU FT PER LIN FT	LIN FT PER CU FT	BARRELS PER LIN FT	LIN FT PER BARREL	GALLONS PER LIN FT
7	17.00	6.538	0.1663	6.0123	.0296	33.7563	1.2442
	20.00	6.456	0.1605	6.2299	.0286	34.9785	1.2007
	22.00	6.398	0.1564	6.3919	.0279	35.8876	1.1703
	23.00	6.366	0.1542	6.4842	.0275	36.4060	1.1537
	24.00	6.336	0.1521	6.5728	.0271	36.9033	1.1381
	26.00	6.276	0.1480	6.7560	.0264	37.9323	1.1072
	28.00	6.214	0.1438	6.9545	.0256	39.0465	1.0756
	29.00	6.184	0.1418	7.0540	.0252	39.6053	1.0605
	29.80	6.168	.01407	7.1080	.0251	39.9087	1.0524
	30.00	6.154	0.1397	7.1559	.0249	40.1774	1.0454
	32.00	6.094	0.1357	7.3672	.0242	41.3638	1.0154
	35.00	6.004	0.1298	7.7043	.0231	43.2563	0.9710
	38.00	5.920	0.1243	8.0428	.0221	45.1569	0.9301
	40.20	5.836	0.1189	8.4070	.0212	47.2016	0.8898
	41.00	5.820	0.1179	8.4795	.0210	47.6087	0.8822
	43.00	5.736	0.1126	8.8780	.0201	49.8465	0.8426
	44.00	5.720	0.1116	8.9575	.0199	50.2928	0.8351
	49.50	5.540	0.1006	9.9420	.0179	55.8202	0.7524
7 ⅝	20.00	7.125	0.2101	4.7603	.0374	26.7272	1.5714
	24.00	7.025	0.2024	4.9419	.0360	27.7465	1.5137
	26.40	6.969	0.1981	5.0485	.0353	28.3453	1.4817
	29.70	6.875	0.1910	5.2361	.0340	29.3987	1.4286
	33.70	6.765	0.1828	5.4705	.0326	30.7148	1.3674
	34.00	6.760	0.1824	5.4816	.0325	30.7769	1.3647
	35.50	6.710	0.1788	5.5942	.0318	31.4094	1.3372
	38.00	6.655	0.1747	5.7226	.0311	32.1300	1.3072
	39.00	6.625	0.1726	5.7947	.0307	32.5346	1.2909
	45.30	6.435	0.1590	6.2878	.0283	35.3032	1.1897
7¾	46.10	6.560	0.1679	5.9560	.0299	33.4403	1.2560
8	26.00	7.386	0.2307	4.3341	.0411	24.3343	1.7260
8⅛	28.00	7.485	0.2388	4.1884	.0425	23.5159	1.7860
	32.00	7.385	0.2306	4.3356	.0411	24.3428	1.7254
	35.50	7.285	0.2226	4.4915	.0397	25.2176	1.6655
	36.00	7.285	0.2226	4.4915	.0397	25.2176	1.6655
	39.50	7.185	0.2148	4.6565	.0382	26.1444	1.6065
	40.00	7.185	0.2148	4.6565	.0382	26.1444	1.6065
	42.00	7.125	0.2101	4.7603	.0374	26.7272	1.5714
8⅝	24.00	8.097	0.2908	3.4392	.0518	19.3095	2.1751
	28.00	8.017	0.2837	3.5244	.0505	19.7879	2.1225
	32.00	7.921	0.2754	3.6312	.0490	20.3876	2.0601
	36.00	7.825	0.2671	3.7432	.0476	21.0168	1.9984
	38.00	7.775	0.2629	3.8038	.0468	21.3569	1.9666
	40.00	7.725	0.2587	3.8660	.0461	21.7059	1.9350
	43.00	7.651	0.2525	3.9610	.0450	22.2394	1.8885
	44.00	7.625	0.2503	3.9953	.0446	22.4319	1.8723
	49.00	7.511	0.2409	4.1514	.0429	23.3083	1.8019
	52.00	7.435	0.2347	4.2610	.0418	23.9236	1.7556

VOLUME AND FILL BETWEEN TUBING AND CASING

3-½" OD TUBING

SIZE OD INCHES	CASING WEIGHT PER LIN FT	ID	CU FT PER LIN FT	LIN FT PER CU FT	BARRELS PER LIN FT	LIN FT PER BARREL	GALLONS PER LIN FT
9	34.00	8.290	0.3080	3.2466	.0549	18.2281	2.3041
	38.00	8.196	0.2996	3.3382	.0534	18.7424	2.2409
	40.00	8.150	0.2955	3.3845	.0526	19.0025	2.2102
	41.20	8.150	0.2955	3.3845	.0526	19.0025	2.2102
	45.00	8.032	0.2851	3.5081	.0508	19.6968	2.1323
	46.10	8.032	0.2851	3.5081	.0508	19.6968	2.1323
	54.00	7.810	0.2659	3.7613	.0474	21.1179	1.9888
	55.20	7.812	0.2660	3.7588	.0474	21.1044	1.9901
9⅝	29.30	9.063	0.3812	2.6234	.0679	14.7295	2.8514
	32.30	9.001	0.3751	2.6662	.0668	14.9694	2.8057
	36.00	8.921	0.3673	2.7229	.0654	15.2881	2.7472
	38.00	8.885	0.3638	2.7491	.0648	15.4351	2.7211
	40.00	8.835	0.3589	2.7861	.0639	15.6429	2.6849
	42.00	8.799	0.3555	2.8133	.0633	15.7952	2.6590
	43.50	8.755	0.3512	2.8470	.0626	15.9847	2.6275
	44.30	8.750	0.3508	2.8509	.0625	16.0064	2.6239
	47.00	8.681	0.3442	2.9052	.0613	16.3115	2.5749
	47.20	8.680	0.3441	2.9060	.0613	16.3160	2.5742
	53.50	8.535	0.3305	3.0257	.0589	16.9881	2.4723
	57.40	8.450	0.3226	3.0996	.0575	17.4027	2.4134
	58.40	8.435	0.3212	3.1129	.0572	17.4776	2.4031
	61.10	8.375	0.3157	3.1671	.0562	17.7821	2.3619
10	33.00	9.384	0.4135	2.4185	.0736	13.5790	3.0930
	60.00	8.780	0.3536	2.8277	.0630	15.8766	2.6454
10 ¾	32.75	10.192	0.4997	2.0010	.0890	11.2349	3.7384
	35.75	10.140	0.4940	2.0244	.0880	11.3660	3.6952
	40.50	10.050	0.4841	2.0658	.0862	11.5987	3.6211
	45.50	9.950	0.4732	2.1134	.0843	11.8661	3.5395
	46.20	9.950	0.4732	2.1134	.0843	11.8661	3.5395
	48.00	9.902	0.4680	2.1369	.0833	11.9979	3.5006
	49.50	9.850	0.4624	2.1628	.0824	12.1433	3.4587
	51.00	9.850	0.4624	2.1628	.0824	12.1433	3.4587
	54.00	9.784	0.4553	2.1964	.0811	12.3318	3.4058
	55.50	9.760	0.4527	2.2088	.0806	12.4014	3.3867
	60.70	9.660	0.4421	2.2617	.0787	12.6985	3.3075
	65.70	9.560	0.4317	2.3166	.0769	13.0069	3.2291
	71.10	9.450	0.4203	2.3795	.0749	13.3599	3.1437
	76.00	9.350	0.4100	2.4390	.0730	13.6940	3.0670
	81.00	9.250	0.3999	2.5009	.0712	14.0415	2.9911
11¾	38.00	11.150	0.6113	1.6360	.1089	9.1853	4.5725
	42.00	11.084	0.6033	1.6577	.1074	9.3071	4.5127
	47.00	11.000	0.5931	1.6859	.1056	9.4659	4.4370
	50.00	10.950	0.5872	1.7031	.1046	9.5624	4.3922
	54.00	10.880	0.5788	1.7277	.1031	9.7001	4.3299
	60.00	10.772	0.5661	1.7666	.1008	9.9186	4.2345
	61.00	10.770	0.5658	1.7673	.1008	9.9227	4.2327
	65.00	10.682	0.5555	1.8001	.0989	10.1067	4.1557
12	40.00	11.384	0.6400	1.5625	.1140	8.7725	4.7877

3-½″ OD TUBING							
CASING		**VOLUME AND FILL BETWEEN TUBING AND CASING**					
SIZE OD INCHES	WEIGHT PER LIN FT	ID	CU FT PER LIN FT	LIN FT PER CU FT	BARRELS PER LIN FT	LIN FT PER BARREL	GALLONS PER LIN FT
	33.38*	12.250	0.7517	1.3304	.1339	7.4697	5.6227
	37.42*	12.188	0.7434	1.3452	.1324	7.5527	5.5609
	41.45*	12.126	0.7352	1.3602	.1309	7.6372	5.4994
12¾	43.77*	12.090	0.7304	1.3691	.1301	7.6869	5.4638
	45.58*	12.062	0.7267	1.3760	.1294	7.7259	5.4363
	49.56*	12.000	0.7186	1.3916	.1280	7.8134	5.3754
	53.00	11.970	0.7147	1.3993	.1273	7.8563	5.3460
	40.00	12.438	0.7770	1.2871	.1384	7.2263	5.8121
	45.00	12.360	0.7664	1.3048	.1365	7.3258	5.7332
13	50.00	12.282	0.7559	1.3229	.1346	7.4274	5.6548
	54.00	12.200	0.7450	1.3423	.1327	7.5365	5.5729
	48.00	12.715	0.8150	1.2270	.1452	6.8893	6.0964
	54.50	12.615	0.8012	1.2482	.1427	7.0081	5.9930
	61.00	12.515	0.7874	1.2699	.1402	7.1301	5.8905
	68.00	12.415	0.7738	1.2922	.1378	7.2554	5.7888
	72.00	12.347	0.7647	1.3078	.1362	7.3426	5.7201
13⅜	77.00	12.275	0.7550	1.3245	.1345	7.4366	5.6478
	83.50	12.175	0.7417	1.3483	.1321	7.5703	5.5480
	85.00	12.159	0.7395	1.3522	.1317	7.5920	5.5321
	92.00	12.031	0.7226	1.3838	.1287	7.7695	5.4058
	98.00	11.937	0.7104	1.4077	.1265	7.9039	5.3139
14	50.00	13.344	0.9044	1.1057	.1611	6.2083	6.7651
	55.00	15.375	1.2225	0.8180	.2177	4.5927	9.1449
	65.00	15.250	1.2016	0.8322	.2140	4.6725	8.9887
	70.00	15.198	1.1930	0.8382	.2125	4.7063	8.9241
16	75.00	15.125	1.1809	0.8468	.2103	4.7545	8.8338
	84.00	15.010	1.1620	0.8606	.2070	4.8318	8.6924
	109.00	14.688	1.1099	0.9010	.1977	5.0589	8.3023
	118.00	14.570	1.0910	0.9166	.1943	5.1462	8.1614
18	80.00	17.180	1.5430	0.6481	.2748	3.6388	11.5424

| 4″ OD TUBING | | | | | | | |

VOLUME AND FILL BETWEEN TUBING AND CASING

| CASING | | | | | | | |

SIZE OD INCHES	WEIGHT PER LIN FT	ID	CU FT PER LIN FT	LIN FT PER CU FT	BARRELS PER LIN FT	LIN FT PER BARREL	GALLONS PER LIN FT
5½	13.00	5.044	0.0515	19.4183	.0092	109.0257	0.3852
	14.00	5.012	0.0497	20.1035	.0089	112.8726	0.3721
	15.00	4.974	0.0477	20.9762	.0085	117.7728	0.3566
	15.50	4.950	0.0464	21.5638	.0083	121.0719	0.3469
	17.00	4.892	0.0433	23.1158	.0077	129.7854	0.3236
	20.00	4.778	0.0372	26.8471	.0066	150.7353	0.2786
	23.00	4.670	0.0317	31.5630	.0056	177.2133	0.2370
	25.00	4.580	0.0271	36.8432	.0048	206.8592	0.2030
	26.00	4.548	0.0255	39.1406	.0046	219.7582	0.1911
	32.30	4.276	0.0125	80.2681	.0022	450.6720	0.0932
5¾	14.00	5.290	0.0654	15.2991	.0116	85.8983	0.4890
	17.00	5.190	0.0596	16.7653	.0106	94.1299	0.4462
	19.50	5.090	0.0540	18.5047	.0096	103.8962	0.4042
	20.00	5.090	0.0540	18.5047	.0096	103.8962	0.4042
	22.50	4.990	0.0485	20.6005	.0086	115.6632	0.3631
	23.00	4.990	0.0485	20.6005	.0086	115.6632	0.3631
	25.20	4.890	0.0432	23.1729	.0077	130.1063	0.3228
6	15.00	5.524	0.0792	12.6319	.0141	70.9228	0.5922
	16.00	5.500	0.0777	12.8664	.0138	72.2396	0.5814
	18.00	5.424	0.0732	13.6624	.0130	76.7087	0.5475
	20.00	5.352	0.0690	14.5008	.0123	81.4158	0.5159
	23.00	5.240	0.0625	16.0022	.0111	89.8455	0.4675
	26.00	5.140	0.0568	17.5963	.0101	98.7959	0.4251
6⅝	13.00	6.260	0.1265	7.9071	.0225	44.3950	0.9461
	17.00	6.135	0.1180	8.4733	.0210	47.5739	0.8828
	20.00	6.049	0.1123	8.9045	.0200	49.9949	0.8401
	22.00	5.980	0.1078	9.2785	.0192	52.0948	0.8062
	24.00	5.921	0.1039	9.6203	.0185	54.0141	0.7776
	25.00	5.880	0.1013	9.8709	.0180	55.4211	0.7578
	26.00	5.855	0.0997	10.0293	.0178	56.3105	0.7459
	26.80	5.837	0.0986	10.1461	.0176	56.9663	0.7373
	28.00	5.791	0.0956	10.4556	.0170	58.7040	0.7155
	29.00	5.761	0.0938	10.6664	.0167	59.8875	0.7013
	31.80	5.675	0.0884	11.3138	.0157	63.5220	0.6612
	32.00	5.675	0.0884	11.3138	.0157	63.5220	0.6612
	34.00	5.595	0.0835	11.9803	.0149	67.2643	0.6244

VOLUME AND FILL BETWEEN TUBING AND CASING

4" OD TUBING

| CASING | | | | | | | |

SIZE OD INCHES	WEIGHT PER LIN FT	ID	CU FT PER LIN FT	LIN FT PER CU FT	BARRELS PER LIN FT	LIN FT PER BARREL	GALLONS PER LIN FT
7	17.00	6.538	0.1459	6.8552	.0260	38.4893	1.0912
	20.00	6.456	0.1401	7.1397	.0249	40.0863	1.0477
	22.00	6.398	0.1360	7.3532	.0242	41.2849	1.0173
	23.00	6.366	0.1338	7.4756	.0238	41.9724	1.0007
	24.00	6.336	0.1317	7.5936	.0235	42.6349	0.9851
	26.00	6.276	0.1276	7.8393	.0227	44.0143	0.9542
	28.00	6.214	0.1233	8.1077	.0220	45.5215	0.9226
	29.00	6.184	0.1213	8.2433	.0216	46.2827	0.9075
	29.80	6.168	0.1202	8.3172	.0214	46.6977	0.8994
	30.00	6.154	0.1193	8.3828	.0212	47.0660	0.8924
	32.00	6.094	0.1153	8.6743	.0205	48.7024	0.8624
	35.00	6.004	0.1093	9.1454	.0195	51.3474	0.8180
	38.00	5.920	0.1039	9.6263	.0185	54.0477	0.7771
	40.20	5.836	0.0985	10.1527	.0175	57.0032	0.7368
	41.00	5.820	0.0975	10.2586	.0174	57.5980	0.7292
	43.00	5.736	0.0922	10.8478	.0164	60.9060	0.6896
	44.00	5.720	0.0912	10.9667	.0162	61.5737	0.6821
	49.50	5.540	0.0801	12.4797	.0143	70.0682	0.5994
7⅝	20.00	7.125	0.1896	5.2738	.0338	29.6101	1.4184
	24.00	7.025	0.1819	5.4975	.0324	30.8664	1.3607
	26.40	6.969	0.1776	5.6298	.0316	31.6092	1.3287
	29.70	6.875	0.1705	5.8642	.0304	32.9248	1.2756
	33.70	6.765	0.1623	6.1598	.0289	34.5845	1.2144
	34.00	6.760	0.1620	6.1738	.0288	34.6632	1.2117
	35.50	6.710	0.1583	6.3170	.0282	35.4676	1.1842
	38.00	6.655	0.1543	6.4812	.0275	36.3892	1.1542
	39.00	6.625	0.1521	6.5738	.0271	39.9090	1.1379
	45.30	6.435	0.1386	7.2157	.0247	40.5134	1.0367
7¾	46.10	6.560	0.1474	6.7822	.0263	38.0791	1.1030
8	26.00	7.386	0.2103	4.7557	.0375	26.7013	1.5730
8⅛	28.00	7.485	0.2183	4.5808	.0389	25.7191	1.6330
	32.00	7.385	0.2102	4.7575	.0374	26.7115	1.5724
	35.50	7.285	0.2022	4.9458	.0360	27.7685	1.5125
	36.00	7.285	0.2022	4.9458	.0360	27.7685	1.5125
	39.50	7.185	0.1943	5.1467	.0346	28.8965	1.4535
	40.00	7.185	0.1943	5.1467	.0346	28.8965	1.4535
	42.00	7.125	0.1896	5.2738	.0338	29.6101	1.4184
8⅝	24.00	8.097	0.2703	3.6994	.0481	20.7705	2.0221
	28.00	8.017	0.2633	3.7982	.0469	21.3252	1.9695
	32.00	7.921	0.2549	3.9225	.0454	22.0232	1.9071
	36.00	7.825	0.2467	4.0536	.0439	22.7592	1.8454
	38.00	7.775	0.2424	4.1247	.0432	23.1586	1.8136
	40.00	7.725	0.2382	4.1979	.0424	23.5695	1.7820
	43.00	7.651	0.2320	4.3102	.0413	24.2000	1.7355
	44.00	7.625	0.2298	4.3508	.0409	24.4281	1.7193
	49.00	7.511	0.2204	4.5366	.0393	25.4710	1.6489
	52.00	7.435	0.2142	4.6678	.0382	26.2076	1.6026

VOLUME AND FILL BETWEEN TUBING AND CASING

4" OD TUBING

CASING							
SIZE OD INCHES	WEIGHT PER LIN FT	ID	CU FT PER LIN FT	LIN FT PER CU FT	BARRELS PER LIN FT	LIN FT PER BARREL	GALLONS PER LIN FT
	34.00	8.290	0.2876	3.4775	.0512	19.5245	2.1511
	38.00	8.196	0.2791	3.5828	.0497	20.1158	2.0879
	40.00	8.150	0.2750	3.6362	.0490	20.4158	2.0572
	41.20	8.150	0.2750	3.6362	.0490	20.4158	2.0572
9	45.00	8.032	0.2646	3.7793	.0471	21.2193	1.9793
	46.10	8.032	0.2646	3.7793	.0471	21.2193	1.9793
	54.00	7.810	0.2454	4.0747	.0437	22.8779	1.8358
	55.20	7.812	0.2456	4.0719	.0437	22.8620	1.8371
	29.30	9.063	0.3607	2.7722	.0642	15.5646	2.6984
	32.30	9.001	0.3546	2.8199	.0632	15.8328	2.6527
	36.00	8.921	0.3468	2.8835	.0618	16.1898	2.5942
	38.00	8.885	0.3433	2.9129	.0611	16.3546	2.5681
	40.00	8.835	0.3385	2.9545	.0603	16.5881	2.5319
	42.00	8.799	0.3350	2.9850	.0597	16.7596	2.5060
	43.50	8.755	0.3308	3.0230	.0589	16.9730	2.4745
9⅝	44.30	8.750	0.3303	3.0274	.0588	16.9976	2.4709
	47.00	8.681	0.3238	3.0887	.0577	17.3420	2.4219
	47.20	8.680	0.3237	3.0896	.0576	17.3470	2.4212
	53.50	8.535	0.3100	3.2253	.0552	18.1088	2.3193
	57.40	8.450	0.3022	3.3094	.0538	18.5806	2.2604
	58.40	8.435	0.3008	3.3246	.0536	18.6660	2.2501
	61.10	8.375	0.2953	3.3865	.0526	19.0137	2.2089
10	33.00	9.384	0.3930	2.5444	.0700	14.2856	2.9400
	60.00	8.780	0.3332	3.0013	.0593	16.8512	2.4924
	32.75	10.192	0.4793	2.0864	.0854	11.7143	3.5854
	35.75	10.140	0.4735	2.1118	.0843	11.8569	3.5422
	40.50	10.050	0.4636	2.1570	.0826	12.1104	3.4681
	45.50	9.950	0.4527	2.2089	.0806	12.4022	3.3865
	46.20	9.950	0.4527	2.2089	.0806	12.4022	3.3865
	48.00	9.902	0.4475	2.2346	.0797	12.5462	3.3476
	49.50	9.850	0.4419	2.2629	.0787	12.7053	3.3057
10¾	51.00	9.850	0.4419	2.2629	.0787	12.7053	3.3057
	54.00	9.784	0.4348	2.2997	.0774	12.9118	3.2528
	55.50	9.760	0.4323	2.3133	.0770	12.9882	3.2337
	60.70	9.660	0.4217	2.3714	.0751	13.3144	3.1545
	65.70	9.560	0.4112	2.4319	.0732	13.6539	3.0761
	71.10	9.450	0.3998	2.5012	.0712	14.0434	2.9907
	76.00	9.350	0.3895	2.5671	.0694	14.4130	2.9140
	81.00	9.250	0.3794	2.6357	.0676	14.7984	2.8381
	38.00	11.150	0.5908	1.6926	.1052	9.5032	4.4195
	42.00	11.084	0.5828	1.7158	.1038	9.6337	4.3597
	47.00	11.000	0.5727	1.7462	.1020	9.8039	4.2840
	50.00	10.950	0.5667	1.7646	.1009	9.9075	4.2392
11¾	54.00	10.880	0.5584	1.7909	.0994	10.0554	4.1769
	60.00	10.772	0.5456	1.8328	.0972	10.2904	4.0815
	61.00	10.770	0.5454	1.8336	.0971	10.2949	4.0797
	65.00	10.682	0.5351	1.8689	.0953	10.4930	4.0027
12	40.00	11.384	0.6196	1.6140	.1103	9.0621	4.6347

4" OD TUBING							

VOLUME AND FILL BETWEEN TUBING AND CASING

SIZE OD INCHES	WEIGHT PER LIN FT	ID	CU FT PER LIN FT	LIN FT PER CU FT	BARRELS PER LIN FT	LIN FT PER BARREL	GALLONS PER LIN FT
	33.38*	12.250	0.7312	1.3676	.1302	7.6786	5.4697
	37.42*	12.188	0.7229	1.3833	.1288	7.7664	5.4079
	41.45*	12.126	0.7147	1.3992	.1273	7.8557	5.3464
12¾	43.77*	1.090	0.7100	1.4085	.1264	7.9083	5.3108
	45.58*	12.062	0.7063	1.4159	.1258	7.9496	5.2833
	49.56*	12.000	0.6981	1.4324	.1243	8.0423	5.2224
	53.00	11.970	0.6942	1.4405	.1236	8.0877	5.1930
	40.00	12.438	0.7565	1.3219	.1347	7.4217	5.6591
	45.00	12.360	0.7460	1.3406	.1329	7.5266	5.5802
13	50.00	12.282	0.7355	1.3597	.1310	7.6339	5.5018
	54.00	12.200	0.7245	1.3802	.1290	7.7493	5.4199
	48.00	12.715	0.7945	1.2586	.1415	7.0667	5.9434
	54.50	12.615	0.7807	1.2809	.1390	7.1917	5.8400
	61.00	12.515	0.7670	1.3038	.1366	7.3203	5.7375
	68.00	12.415	0.7534	1.3273	.1342	7.4524	5.6358
13⅜	72.00	12.347	0.7442	1.3437	.1325	7.5443	5.5671
	77.00	12.275	0.7345	1.3614	.1308	7.6437	5.4948
	83.50	12.175	0.7212	1.3866	.1285	7.7850	5.3950
	85.00	12.159	0.7191	1.3907	.1281	7.8080	5.3791
	92.00	12.031	0.7022	1.4241	.1251	7.9958	5.2528
	98.00	11.937	0.6899	1.4495	.1229	8.1382	5.1609
14	50.00	13.344	0.8839	1.1313	.1574	6.3520	6.6121
	55.00	15.375	1.2020	0.8319	.2141	4.6709	8.9919
	65.00	15.250	1.1812	0.8466	.2104	4.7534	8.8357
	70.00	15.198	1.1725	0.8529	.2088	4.7884	8.7711
16	75.00	15.125	1.1605	0.8617	.2067	4.8383	8.6808
	84.00	15.010	1.1416	0.8760	.2033	4.9184	8.5394
	109.00	14.688	1.0894	0.9179	.1940	5.1538	8.1493
	118.00	14.570	1.0706	0.9341	.1907	5.2445	8.0084
18	80.00	17.180	1.5225	0.6568	.2712	3.6876	11.3894

VOLUME AND FILL BETWEEN TUBING AND CASING

4-½″ OD TUBING

CASING SIZE OD INCHES	WEIGHT PER LIN FT	ID	CU FT PER LIN FT	LIN FT PER CU FT	BARRELS PER LIN FT	LIN FT PER BARREL	GALLONS PER LIN FT
6⅝	13.00	6.260	0.1033	9.6816	.0184	54.3582	0.7727
	17.00	6.135	0.0948	10.5443	.0169	59.2018	0.7094
	20.00	6.049	0.0891	11.2204	.0159	62.9981	0.6667
	22.00	5.980	0.0846	11.8209	.0151	66.3693	0.6328
	24.00	5.921	0.0808	12.3814	.0144	69.5163	0.6042
	25.00	5.880	0.0781	12.7996	.0139	71.8644	0.5844
	26.00	5.855	0.0765	13.0672	.0136	73.3670	0.5725
	26.80	5.837	0.0754	13.2662	.0134	74.4842	0.5639
	28.00	5.791	0.0725	13.8003	.0129	77.4830	0.5421
	29.00	5.761	0.0706	14.1699	.0126	79.5583	0.5279
	31.80	5.675	0.0652	15.3356	.0116	86.1029	0.4878
	32.00	5.675	0.0652	15.3356	.0116	86.1029	0.4878
	34.00	5.595	0.0603	16.5864	.0107	93.1257	0.4510
7	17.00	6.538	0.1227	8.1504	.0219	45.7610	0.9178
	20.00	6.456	0.1169	8.5556	.0208	48.0363	0.8743
	22.00	6.398	0.1128	8.8640	.0201	49.7676	0.8439
	23.00	6.366	0.1106	9.0426	.0197	50.7702	0.8273
	24.00	6.336	0.1085	9.2158	.0193	51.7426	0.8117
	26.00	6.276	0.1044	9.5801	.0186	53.7885	0.7808
	28.00	6.214	0.1002	9.9841	.0178	56.0567	0.7492
	29.00	6.184	0.0981	10.1905	.0175	57.2156	0.7341
	29.80	6.168	0.0971	10.3037	.0173	57.8510	0.7260
	30.00	6.154	0.0961	10.4046	.0171	58.4174	0.7190
	32.00	6.094	0.0921	10.8574	.0164	60.9596	0.6890
	35.00	6.004	0.0862	11.6057	.0153	65.1610	0.6446
	38.00	5.920	0.0807	12.3913	.0144	69.5719	0.6037
	40.20	5.836	0.0753	13.2774	.0134	74.5472	0.5634
	41.00	5.820	0.0743	13.4592	.0132	75.5678	0.5558
	43.00	5.736	0.0690	14.4919	.0123	81.3657	0.5126
	44.00	5.720	0.0680	14.7049	.0121	82.5618	0.5087
	49.50	5.540	0.0570	17.5592	.0101	98.5878	0.4260
7⅝	20.00	7.125	0.1664	6.0083	.0296	33.7340	1.2450
	24.00	7.025	0.1587	6.3004	.0283	35.3743	1.1873
	26.40	6.969	0.1544	6.4748	.0275	36.3533	1.1553
	29.70	6.875	0.1473	6.7867	.0262	38.1044	1.1022
	33.70	6.765	0.1392	7.1858	.0248	40.3451	1.0410
	34.00	6.760	0.1388	7.2049	.0247	40.4523	1.0383
	35.50	6.710	0.1351	7.4007	.0241	41.5520	1.0108
	38.00	6.655	0.1311	7.6270	.0234	42.8226	0.9808
	39.00	6.625	0.1289	7.7556	.0230	43.5443	0.9645
	45.30	6.435	0.1154	8.6651	.0206	48.6508	0.8633
7¾	46.10	6.560	0.1243	8.0473	.0221	45.1822	0.9296
8	26.00	7.386	0.1871	5.3449	.0333	30.0095	1.3996
8⅛	28.00	7.485	0.1951	5.1250	.0348	28.7745	1.4596
	32.00	7.385	0.1870	5.3472	.0333	30.0224	1.3990
	35.50	7.285	0.1790	5.5862	.0319	31.3643	1.3391
	36.00	7.285	0.1790	5.5862	.0319	31.3643	1.3391
	39.50	7.185	0.1711	5.8439	.0305	32.8108	1.2801
	40.00	7.185	0.1711	5.8439	.0305	32.8108	1.2801
	42.00	7.125	0.1664	6.0083	.0296	33.7340	1.2450

VOLUME AND FILL BETWEEN TUBING AND CASING

4-½" OD TUBING

CASING

SIZE OD INCHES	WEIGHT PER LIN FT	ID	CU FT PER LIN FT	LIN FT PER CU FT	BARRELS PER LIN FT	LIN FT PER BARREL	GALLONS PER LIN FT
8⅝	24.00	8.097	0.2471	4.0464	.0440	22.7187	1.8487
	28.00	8.017	0.2401	4.1649	.0428	23.3839	1.7961
	32.00	7.921	0.2318	4.3148	.0413	24.2259	1.7337
	36.00	7.825	0.2235	4.4740	.0398	25.1195	1.6720
	38.00	7.775	0.2193	4.5608	.0391	25.6069	1.6402
	40.00	7.725	0.2150	4.6504	.0383	26.1103	1.6086
	43.00	7.651	0.2088	4.7886	.0372	26.8862	1.5621
	44.00	7.625	0.2067	4.8388	.0368	27.1680	1.5459
	49.00	7.511	0.1973	5.0697	.0351	28.4643	1.4755
	52.00	7.435	0.1911	5.2341	.0340	29.3873	1.4292
9	34.00	8.290	0.2644	3.7824	.0471	21.2364	1.9777
	38.00	8.196	0.2559	3.9073	.0456	21.9377	1.9145
	40.00	8.150	0.2518	3.9709	.0449	22.2950	1.8838
	41.20	8.150	0.2518	3.9709	.0449	22.2950	1.8838
	45.00	8.032	0.2414	4.1422	.0430	23.2568	1.8059
	46.10	8.032	0.2414	4.1422	.0430	23.2568	1.8059
	54.00	7.810	0.2222	4.4997	.0396	25.2641	1.6624
	55.20	7.812	0.2224	4.4963	.0396	25.2448	1.6637
9⅝	29.30	9.063	0.3375	2.9626	.0601	16.6335	2.5250
	32.30	9.001	0.3314	3.0172	.0590	16.9401	2.4793
	36.00	8.921	0.3236	3.0901	.0576	17.3494	2.4208
	38.00	8.885	0.3201	3.1238	.0570	17.5389	2.3947
	40.00	8.835	0.3153	3.1717	.0562	17.8077	2.3585
	42.00	8.799	0.3118	3.2069	.0555	18.0054	2.3326
	43.50	8.755	0.3076	3.2508	.0548	18.2520	2.3011
	44.30	8.750	0.3071	3.2559	.0547	18.2804	2.2975
	47.00	8.681	0.3006	3.3269	.0535	18.6793	2.2485
	47.20	8.680	0.3005	3.3280	.0535	18.6852	2.2478
	53.50	8.535	0.2869	3.4859	.0511	19.5720	2.1459
	57.40	8.450	0.2790	3.5843	.0497	20.1244	2.0870
	58.40	8.435	0.2776	3.6021	.0494	20.2246	2.0767
	61.10	8.375	0.2721	3.6750	.0485	20.6334	2.0355
10	33.00	9.384	0.3698	2.7038	.0659	15.1810	2.7666
	60.00	8.780	0.3100	3.2258	.0552	18.1112	2.3190
10¾	32.75	10.192	0.4561	2.1924	.0812	12.3096	3.410
	35.75	10.140	0.4503	2.2205	.0802	12.4672	3.3688
	40.50	10.050	0.4404	2.2705	.0784	12.7478	3.2947
	45.50	9.950	0.4295	2.3281	.0765	13.0715	3.2131
	46.20	9.950	0.4295	2.3281	.0765	13.0715	3.2131
	48.00	9.902	0.4243	2.3567	.0756	13.2616	3.1742
	49.50	9.850	0.4187	2.3882	.0746	13.4086	3.1323
	51.00	9.850	0.4187	2.3882	.0746	13.4086	3.1323
	54.00	9.784	0.4117	2.4292	.0733	13.6388	3.0794
	55.50	9.760	0.4091	2.4444	.0729	13.7241	3.0603
	60.70	9.660	0.3985	2.5093	.0710	14.0889	2.9811
	65.70	9.560	0.3880	2.5771	.0691	14.4695	2.9027
	71.10	9.450	0.3766	2.6552	.0671	14.9077	2.8173
	76.00	9.350	0.3664	2.7295	.0653	15.3249	2.7406
	81.00	9.250	0.3562	2.8072	.0634	15.7614	2.6647

4-¾" OD
INSIDE CASING

VOLUME AND FILL BETWEEN CASING AND CASING

OUTSIDE CASING

SIZE OD INCHES	WEIGHT PER LIN FT	ID	CU FT PER LIN FT	LIN FT PER CU FT	BARRELS PER LIN FT	LIN FT PER BARREL	GALLONS PER LIN FT
6⅝	13.00	6.260	0.0907	11.0283	.0162	61.9193	0.6783
	17.00	6.135	0.0822	12.1617	.0146	68.2829	0.6151
	20.00	6.049	0.0765	13.0701	.0136	73.3833	0.5723
	22.00	5.980	0.0720	13.8921	.0128	77.9983	0.5385
	24.00	5.921	0.0682	14.6727	.0121	82.3812	0.5098
	25.00	5.880	0.0655	15.2637	.0117	85.6995	0.4901
	26.00	5.855	0.0639	15.6459	.0114	87.8450	0.4781
	26.80	5.837	0.0628	15.9320	.0112	89.4515	0.4695
	28.00	5.791	0.0598	16.7086	.0107	93.8118	0.4477
	29.00	5.761	0.0580	17.2535	.0103	96.8713	0.4336
	31.80	5.675	0.0526	19.0132	.0094	106.7511	0.3934
	32.00	5.675	0.0526	19.0132	.0094	106.7511	0.3934
	34.00	5.595	0.0477	20.9742	.0085	117.7614	0.3567
7	17.00	6.538	0.1101	9.0842	.0196	51.0042	0.8235
	20.00	6.456	0.1043	9.5905	.0186	53.8469	0.7800
	22.00	6.398	0.1002	9.9797	.0178	56.0320	0.7496
	23.00	6.366	0.0980	10.2066	.0175	57.3060	0.7329
	24.00	6.336	0.0959	10.4278	.0171	58.5480	0.7174
	26.00	6.276	0.0918	10.8968	.0163	61.1811	0.6865
	28.00	6.214	0.0875	11.4225	.0156	64.1328	0.6549
	29.00	6.184	0.0855	11.6935	.0152	65.6541	0.6397
	29.80	6.168	0.0844	11.8428	.0150	66.4922	0.6317
	30.00	6.154	0.0835	11.9762	.0149	67.2415	0.6246
	32.00	6.094	0.0795	12.5801	.0142	70.6320	0.9546
	35.00	6.004	0.0736	13.5958	.0131	76.3348	0.5502
	38.00	5.920	0.0681	14.6866	.0121	82.4593	0.5093
	40.20	5.836	0.0627	15.9482	.0112	89.5423	0.4691
	41.00	5.820	0.0617	16.2112	.0110	91.0189	0.4614
	43.00	5.736	0.0564	17.7331	.0100	99.5642	0.4218
	44.00	5.720	0.0554	18.0532	.0099	101.3612	0.4144
	49.50	5.540	0.0443	22.5543	.0079	126.6332	0.3317
7⅝	20.00	7.125	0.1538	6.5009	.0274	36.5000	1.1507
	24.00	7.025	0.1461	6.8443	.0260	38.4280	1.0930
	26.40	6.969	0.1418	7.0506	.0253	39.5861	1.0610
	29.70	6.875	0.1347	7.4220	.0240	41.6714	1.0079
	33.70	6.765	0.1266	7.9019	.0225	44.3661	0.9467
	34.00	6.760	0.1262	7.9250	.0225	44.4958	0.9439
	35.50	6.710	0.1225	8.1627	.0218	45.8300	0.9164
	38.00	6.655	0.1185	8.4388	.0211	47.3805	0.8864
	39.00	6.625	0.1163	8.5965	.0207	48.2656	0.8702
	45.30	6.435	0.1028	9.7283	.0183	54.6203	0.7689
7¾	46.10	6.560	0.1117	8.9564	.0199	50.2862	0.8352
8	26.00	7.386	0.1745	5.7313	.0311	32.1788	1.3052
8⅛	28.00	7.485	0.1825	5.4791	.0325	30.7630	1.3653
	32.00	7.385	0.1744	5.7339	.0311	32.1936	1.3046
	35.50	7.285	0.1664	6.0096	.0296	33.7416	1.2448
	36.00	7.285	0.1664	6.0096	.0296	33.7416	1.2448
	39.50	7.185	0.1585	6.3089	.0282	35.4216	1.1857
	40.00	7.185	0.1585	6.3089	.0282	35.4216	1.1857
	42.00	7.125	0.1538	6.5009	.0274	36.5000	1.1507

VOLUME AND FILLBETWEEN CASING AND CASING

5" OD
INSIDE CASING

| OUTSIDE CASING | | | | | | | |

SIZE OD INCHES	WEIGHT PER LIN FT	ID	CU FT PER LIN FT	LIN FT PER CU FT	BARRELS PER LIN FT	LIN FT PER BARREL	GALLONS PER LIN FT
	17.00	6.538	0.0968	10.3320	.0172	58.0101	0.7240
	20.00	6.456	0.0910	10.9920	.0162	61.7157	0.6805
	22.00	6.398	0.0869	11.5063	.0155	64.6032	0.6501
	23.00	6.366	0.0847	11.8090	.0151	66.3028	0.6335
	24.00	6.336	0.0826	12.1062	.0147	67.9710	0.6179
	26.00	6.276	0.0785	12.7429	.0140	71.5458	0.5870
	28.00	6.214	0.0743	13.4677	.0132	75.6155	0.5554
	29.00	6.184	0.0722	13.8460	.0129	77.7394	0.5403
	29.80	6.168	0.0711	14.0058	.0127	78.9172	0.5322
7	30.00	6.154	0.0702	14.2441	.0125	79.9749	0.5252
	32.00	6.094	0.0662	15.1066	.0118	84.8173	0.4952
	35.00	6.004	0.0603	16.5954	.0107	93.1764	0.4508
	38.00	5.920	0.0548	18.2500	.0098	102.4660	0.4099
	40.20	5.836	0.0494	20.2394	.0088	113.6357	0.3696
	41.00	5.820	0.0484	20.6648	.0086	116.0243	0.3620
	43.00	5.736	0.0431	23.2034	.0077	130.2776	0.3224
	44.00	5.720	0.0421	23.7545	.0075	133.3714	0.3149
	49.50	5.540	0.0310	32.2135	.0055	180.8655	0.2322
	20.00	7.125	0.1405	7.1159	.0250	39.9530	1.0512
	24.00	7.025	0.1328	7.5294	.0237	42.2747	0.9935
	26.40	6.969	0.1285	7.7798	.0229	43.6804	0.9615
	29.70	6.875	0.1214	8.2345	.0216	46.2333	0.9084
	33.70	6.765	0.1133	8.8295	.0202	49.5739	0.8472
7⅝	34.00	6.760	0.1129	8.8583	.0201	49.7359	0.8445
	35.50	6.710	0.1092	9.1563	.0195	51.4088	0.8170
	38.00	6.655	0.1052	9.5052	.0187	53.3679	0.7870
	39.00	6.625	0.1030	9.7057	.0184	54.4934	0.7707
	45.30	6.435	0.0895	11.1734	.0159	62.7339	0.6695
7¾	46.10	6.560	0.0984	10.1669	.0175	57.0831	0.7358
8	26.00	7.386	0.1612	6.2040	.0287	34.8328	1.2058
	28.00	7.485	0.1692	5.9096	.0301	33.1799	1.2658
	32.00	7.385	0.1611	6.2071	.0287	34.8502	1.2052
	35.50	7.285	0.1531	6.5315	.0273	36.6715	1.1453
8⅛	36.00	7.285	0.1531	6.5315	.0273	36.6715	1.1453
	39.50	7.185	0.1452	6.8865	.0259	38.6646	1.0863
	40.00	7.185	0.1452	6.8865	.0259	38.6646	1.0863
	42.00	7.125	0.1405	7.1159	.0250	39.9530	1.0512
	24.00	8.097	0.2212	4.5202	.0394	25.3792	1.6549
	28.00	8.017	0.2142	4.6686	.0382	26.2122	1.6023
	32.00	7.921	0.2059	4.8579	.0367	27.2749	1.5399
	36.00	7.825	0.1976	5.0605	.0352	28.4128	1.4782
	38.00	7.775	0.1934	5.1719	.0344	29.0380	1.4464
8⅝	40.00	7.725	0.1891	5.2875	.0337	29.6870	1.4148
	43.00	7.651	0.1829	5.4669	.0326	30.6941	1.3683
	44.00	7.625	0.1808	5.5324	.0322	31.0620	1.3521
	49.00	7.511	0.1713	5.8362	.0305	32.7681	1.2817
	52.00	7.435	0.1651	6.0552	.0294	33.9974	1.2354

5″ OD INSIDE CASING			**VOLUME AND FILLBETWEEN CASING AND CASING**				
OUTSIDE CASING							
SIZE OD INCHES	WEIGHT PER LIN FT	ID	CU FT PER LIN FT	LIN FT PER CU FT	BARRELS PER LIN FT	LIN FT PER BARREL	GALLONS PER LIN FT
	34.00	8.290	0.2385	4.1933	.0425	23.5434	1.7839
	38.00	8.196	0.2300	4.3473	.0410	24.4085	1.7207
	40.00	8.150	0.2259	4.4263	.0402	24.8516	1.6900
	41.20	8.150	0.2259	4.4263	.0402	24.8516	1.6900
9	45.00	8.032	0.2155	4.6402	.0384	26.0525	1.6121
	46.10	8.032	0.2155	4.6402	.0384	26.0525	1.6121
	54.00	7.810	0.1963	5.0935	.0350	28.5979	1.4686
	55.20	7.812	0.1965	5.0891	.0350	28.5731	1.4699

5-½″ OD
INSIDE CASING

VOLUME AND FILL BETWEEN CASING AND CASING

OUTSIDE CASING

SIZE OD INCHES	WEIGHT PER LIN FT	ID	CU FT PER LIN FT	LIN FT PER CU FT	BARRELS PER LIN FT	LIN FT PER BARREL	GALLONS PER LIN FT
	17.00	6.538	0.0682	14.6731	.0121	82.3832	0.5098
	20.00	6.456	0.0623	16.0409	.0111	90.0630	0.4663
	22.00	6.398	0.0583	17.1602	.0104	96.3474	0.4359
	23.00	6.366	0.0560	17.8423	.0100	100.1770	0.4193
	24.00	6.336	0.0540	18.5294	.0096	104.0349	0.4037
	26.00	6.276	0.0498	20.0638	.0089	112.6498	0.3728
	28.00	6.214	0.0456	21.9214	.0081	123.0798	0.3412
	29.00	6.184	0.0436	22.9417	.0078	128.8079	0.3261
	29.80	6.168	0.0425	23.5234	.0076	132.0740	0.3180
7	30.00	6.154	0.0416	24.0558	.0074	135.0633	0.3110
	32.00	6.094	0.0376	26.6227	.0067	149.4756	0.2810
	35.00	6.004	0.0316	31.6223	.0056	177.5459	0.2366
	38.00	5.920	0.0262	38.2259	.0047	214.6222	0.1957
	40.20	5.836	0.0208	48.1364	.0037	270.2658	0.1554
	41.00	5.820	0.0198	50.6146	.0035	284.1802	0.1478
	43.00	5.736	0.0145	69.1431	.0026	388.2097	0.1082
	44.00	5.720	0.0135	74.2775	.0024	417.0369	0.1007
	20.00	7.125	0.1119	8.9369	.0199	50.1771	0.8370
	24.00	7.025	0.1042	9.5990	.0186	53.8943	0.7793
	26.40	6.969	0.0999	10.0097	.0178	56.2001	0.7473
	29.70	6.875	0.0928	10.7752	.0165	60.4982	0.6942
	33.70	6.765	0.0846	11.8172	.0151	66.3486	0.6330
7⅝	34.00	6.760	0.0843	11.8689	.0150	66.6391	0.6303
	35.50	6.710	0.0806	12.4100	.0144	69.6769	0.6028
	38.00	6.655	0.0766	13.0598	.0136	73.3252	0.5728
	39.00	6.625	0.0744	13.4412	.0133	75.4668	0.5565
	45.30	6.435	0.0609	16.4300	.0108	92.2478	0.4553
7¾	46.10	6.560	0.0697	14,3423	.0124	80.5262	0.5216
8	26.00	7.386	0.1326	7.5442	.0236	42.3575	0.9916
	28.00	7.485	0.1406	7.1133	.0250	39.9381	1.0516
	32.00	7.385	0.1325	7.5488	.0236	42.3833	0.9910
	35.50	7.285	0.1245	8.0340	.0222	45.1078	0.9311
8⅛	36.00	7.285	0.1245	8.0340	.0222	45.1078	0.9311
	39.50	7.185	0.1166	8.5779	.0208	48.1615	0.8721
	40.00	7.185	0.1166	8.5779	.0208	48.1615	0.8721
	42.00	7.125	0.1119	8.9369	.0199	50.1771	0.8370
	24.00	8.097	0.1926	5.1923	.0343	29.1525	1.4407
	28.00	8.017	0.1856	5.3890	.0331	30.2571	1.3881
	32.00	7.921	0.1772	5.6428	.0316	31.6818	1.3257
	36.00	7.825	0.1690	5.9181	.0301	33.2277	1.2640
	38.00	7.775	0.1647	6.0710	.0293	34.0859	1.2322
8⅝	40.00	7.725	0.1605	6.2308	.0286	34.9836	1.2006
	43.00	7.651	0.1543	6.4815	.0275	36.3907	1.1541
	44.00	7.625	0.1521	6.5738	.0271	36.9090	1.1379
	49.00	7.511	0.1427	7.0073	.0254	39.3430	1.0675
	52.00	7.435	0.1365	7.3253	.0243	41.1285	1.0212

5-½" OD
INSIDE CASING

VOLUME AND FILL BETWEEN CASING AND CASING

OUTSIDE CASING							
SIZE OD INCHES	WEIGHT PER LIN FT	ID	CU FT PER LIN FT	LIN FT PER CU FT	BARRELS PER LIN FT	LIN FT PER BARREL	GALLONS PER LIN FT
	34.00	8.290	0.2098	4.7655	.0374	26.7560	1.5697
	38.00	8.196	0.2014	4.9655	.0359	27.8790	1.5065
	40.00	8.150	0.1973	5.0687	.0351	28.4585	1.4758
	41.20	8.150	0.1973	5.0687	.0351	28.4585	1.4758
9	45.00	8.032	0.1869	5.3511	.0333	30.0445	1.3979
	46.10	8.032	0.1869	5.3511	.0333	30.0445	1.3979
	54.00	7.810	0.1677	5.9632	.0299	33.4811	1.2544
	55.20	7.812	0.1679	5.9572	.0299	33.4471	1.2557

5-¾" OD
INSIDE CASING

VOLUME AND FILL BETWEEN CASING AND CASING

OUTSIDE CASING

SIZE OD INCHES	WEIGHT PER LIN FT	ID	CU FT PER LIN FT	LIN FT PER CU FT	BARRELS PER LIN FT	LIN FT PER BARREL	GALLONS PER LIN FT
7⅝	20.00	7.125	0.0966	10.3567	.0172	58.1487	0.7223
	24.00	7.025	0.0888	11.2565	.0158	63.2003	0.6646
	26.40	6.969	0.0846	11.8254	.0151	66.3947	0.6326
	29.70	6.875	0.0775	12.9089	.0138	72.4780	0.5795
	33.70	6.765	0.0693	14.4336	.0123	81.0388	0.5183
	34.00	6.760	0.0689	14.5109	.0123	81.4726	0.5155
	35.50	6.710	0.0652	15.3279	.0116	86.0599	0.4880
	38.00	6.655	0.0612	16.3315	.0109	91.6948	0.4580
	39.00	6.625	0.0591	16.9324	.0105	95.0686	0.4418
	45.30	6.435	0.0455	21.9663	.0081	123.3315	0.3405
7¾	46.10	6.560	0.0544	18.3878	.0097	103.2398	0.4068
8	26.00	7.386	0.1172	8.5315	.0209	47.9009	0.8768
8⅛	28.00	7.485	0.1252	7.9845	.0223	44.8298	0.9369
	32.00	7.385	0.1171	8.5374	.0209	47.9338	0.8762
	35.50	7.285	0.1091	9.1633	.0194	51.4483	0.8164
	36.00	7.285	0.1091	9.1633	.0194	51.4483	0.8164
	39.50	7.185	0.1012	9.8777	.0180	55.4590	0.7573
	40.00	7.185	0.1012	9.8777	.0180	55.4590	0.7573
	42.00	7.125	0.0966	10.3567	.0172	58.1487	0.7223
8⅝	24.00	8.097	0.1773	5.6416	.0316	31.6753	1.3260
	28.00	8.017	0.1702	5.8746	.0303	32.9837	1.2734
	32.00	7.921	0.1619	6.1775	.0288	34.6841	1.2109
	36.00	7.825	0.1536	6.5090	.0274	36.5454	1.1493
	38.00	7.775	0.1494	6.6944	.0266	37.5862	1.1174
	40.00	7.725	0.1452	6.8893	.0259	38.6807	1.0858
	43.00	7.651	0.1389	7.1970	.0247	40.4083	1.0394
	44.00	7.625	0.1368	7.3110	.0244	41.0483	1.0232
	49.00	7.511	0.1274	7.8512	.0227	44.0813	0.9528
	52.00	7.435	0.1212	8.2526	.0216	46.3351	0.9064
9	34.00	8.290	0.1945	5.1413	.0346	28.8662	1.4550
	38.00	8.196	0.1861	5.3749	.0331	30.1776	1.3918
	40.00	8.150	0.1820	5.4960	.0324	30.8577	1.3611
	41.20	8.150	0.1820	5.4960	.0324	30.8577	1.3611
	45.00	8.032	0.1715	5.8297	.0306	32.7312	1.2832
	46.10	8.032	0.1715	5.8297	.0306	32.7312	1.2832
	54.00	7.810	0.1524	6.5637	.0271	36.8522	1.1397
	55.20	7.812	0.1525	6.5563	.0272	36.8110	1.1410
9⅝	29.30	9.063	0.2677	3.7360	.0477	20.9761	2.0023
	32.30	9.001	0.2616	3.8233	.0466	21.4660	1.9566
	36.00	8.921	0.2537	3.9411	.0452	22.1276	1.8981
	38.00	8.885	0.2502	3.9962	.0446	22.4367	1.8719
	40.00	8.835	0.2454	4.0748	.0437	22.8786	1.8358
	42.00	8.799	0.2419	4.1332	.0431	23.2060	1.8099
	43.50	8.755	0.2377	4.2064	.0423	23.6172	1.7784
	44.30	8.750	0.2373	4.2149	.0423	23.6647	1.7748
	47.00	8.681	0.2307	4.3347	.0411	24.3376	1.7257
	47.20	8.680	0.2306	4.3365	.0411	24.3476	1.7250
	53.50	8.535	0.2170	4.6086	.0386	25.8753	1.6232
	57.40	8.450	0.2091	4.7821	.0372	26.8496	1.5643
	58.40	8.435	0.2077	4.8139	.0370	27.0282	1.5539
	61.10	8.375	0.2022	4.9449	.0360	27.7634	1.5128

VOLUME AND FILL BETWEEN CASING AND CASING

6" OD
INSIDE CASING

OUTSIDE CASING

SIZE OD INCHES	WEIGHT PER LIN FT	ID	CU FT PER LIN FT	LIN FT PER CU FT	BARRELS PER LIN FT	LIN FT PER BARREL	GALLONS PER LIN FT
7⅝	20.00	7.125	0.0805	12.4171	.0143	69.7169	0.6024
	24.00	7.025	0.0728	13.7332	.0130	77.1061	0.5447
	26.40	6.969	0.0685	14.5896	.0122	81.9143	0.5127
	29.70	6.875	0.0614	16.2749	.0109	91.3766	0.4596
	33.70	6.765	0.0533	18.7755	.0095	105.4163	0.3984
	34.00	6.760	0.0529	18.9064	.0094	106.1514	0.3957
	35.50	6.710	0.0492	20.3174	.0088	114.0739	0.3682
	38.00	6.655	0.0452	22.1192	.0081	124.1900	0.3382
	39.00	6.625	0.0430	23.2360	.0077	130.4604	0.3219
	45.30	6.435	0.0295	33.8952	.0053	190.3071	0.2207
7¾	46.10	6.560	0.0384	26.0672	.0068	146.3567	0.2870
8	26.00	7.386	0.1012	9.8823	.0180	55.4851	0.7570
8⅛	28.00	7.485	0.1092	9.1558	.0195	51.4059	0.8170
	32.00	7.385	0.1011	9.8902	.0180	55.5293	0.7564
	35.50	7.285	0.0931	10.7401	.0166	60.3011	0.6965
	36.00	7.285	0.0931	10.7401	.0166	60.3011	0.6965
	39.50	7.185	0.0852	11.7348	.0152	65.8858	0.6375
	40.00	7.185	0.0852	11.7348	.0152	65.8858	0.6375
	42.00	7.125	0.0805	12.4171	.0143	69.7169	0.6024
8⅝	24.00	8.097	0.1612	6.2022	.0287	34.8229	1.2061
	28.00	8.017	0.1542	6.4850	.0275	36.4107	1.1535
	32.00	7.921	0.1459	6.8561	.0260	38.4939	1.0911
	36.00	7.825	0.1376	7.2668	.0245	40.8002	1.0294
	38.00	7.775	0.1334	7.4986	.0238	42.1018	0.9976
	40.00	7.725	0.1291	7.7441	.0230	43.4799	0.9660
	43.00	7.651	0.1229	8.1351	.0219	45.6750	0.9195
	44.00	7.625	0.1208	8.2810	.0215	46.4944	0.9033
	49.00	7.511	0.1113	8.9809	.0198	50.4241	0.8329
	52.00	7.435	0.1052	9.5101	.0187	53.3950	0.7866
9	34.00	8.290	0.1785	5.6028	.0318	31.4574	1.3351
	38.00	8.196	0.1700	5.8813	.0303	33.0211	1.2719
	40.00	8.150	0.1659	6.0267	.0296	33.8373	1.2412
	41.20	8.150	0.1659	6.0267	.0296	33.8373	1.2412
	45.00	8.032	0.1555	6.4303	.0277	36.1033	1.1633
	46.10	8.032	0.1555	6.4303	.0277	36.1033	1.1633
	54.00	7.810	0.1363	7.3350	.0243	41.1830	1.0198
	55.20	7.812	0.1365	7.3258	.0243	41.1316	1.0211
9⅝	29.30	9.063	0.2516	3.9739	.0448	22.3116	1.8824
	32.30	9.001	0.2455	4.0727	.0437	22.8667	1.8367
	36.00	8.921	0.2377	4.2067	.0423	23.6190	1.7782
	38.00	8.885	0.2342	4.2695	.0417	23.9715	1.7521
	40.00	8.835	0.2294	4.3595	.0409	24.4765	1.7159
	42.00	8.799	0.2259	4.4263	.0402	24.8516	1.6900
	43.50	8.755	0.2217	4.5104	.0395	25.3238	1.6585
	44.30	8.750	0.2212	4.5201	.0394	25.3785	1.6549
	47.00	8.681	0.2147	4.6582	.0382	26.1540	1.6059
	47.20	8.680	0.2146	4.6603	.0382	26.1655	1.6052
	53.50	8.535	0.2010	4.9760	.0358	27.9381	1.5033
	57.40	8.450	0.1931	5.1789	.0344	29.0774	1.4444
	58.40	8.435	0.1917	5.2162	.0341	29.2870	1.4341
	61.10	8.375	0.1862	5.3703	.0332	30.1522	1.3929

6-⅝" OD			**VOLUME AND FILL BETWEEN**				
INSIDE CASING			**CASING AND CASING**				
OUTSIDE CASING							
SIZE OD INCHES	**WEIGHT PER LIN FT**	**ID**	**CU FT PER LIN FT**	**LIN FT PER CU FT**	**BARRELS PER LIN FT**	**LIN FT PER BARREL**	**GALLONS PER LIN FT**
	24.00	8.097	0.1182	8.4605	.0211	47.5024	0.8842
	28.00	8.017	0.1112	8.9957	.0198	50.5069	0.8316
	32.00	7.921	0.1028	9.7258	.0183	54.6062	0.7691
	36.00	7.825	0.0946	10.5736	.0168	59.3664	0.7075
	38.00	7.775	0.0903	11.0716	.0161	62.1627	0.6756
8⅝	40.00	7.725	0.0861	11.6152	.0153	65.2147	0.6440
	43.00	7.651	0.0799	12.5175	.0142	70.2807	0.5976
	44.00	7.625	0.0777	12.8664	.0138	72.2396	0.5814
	49.00	7.511	0.0683	14.6390	.0122	82.1921	0.5110
	52.00	7.435	0.0621	16.0991	.0111	90.3899	0.4647
	34.00	8.290	0.1354	7.3830	.0241	41.4527	1.0132
	38.00	8.196	0.1270	7.8744	.0226	44.2116	0.9500
	40.00	8.150	0.1229	8.1372	.0219	45.6870	0.9193
	41.20	8.150	0.1229	8.1372	.0219	45.6870	0.9193
9	45.00	8.032	0.1125	8.8906	.0200	49.9173	0.8414
	46.10	8.032	0.1125	8.8906	.0200	49.9173	0.8414
	54.00	7.810	0.0933	10.7186	.0166	60.1804	0.6979
	55.20	7.812	0.0935	10.6990	.0166	60.0707	0.6992
	29.30	9.063	0.2086	4.7937	.0372	26.9147	1.5605
	32.30	9.001	0.2025	4.9383	.0361	27.7266	1.5148
	36.00	8.921	0.1947	5.1367	.0347	28.8403	1.4563
	38.00	8.885	0.1912	5.2306	.0341	29.3677	1.4301
	40.00	8.835	0.1863	5.3662	.0332	30.1293	1.3940
	42.00	8.799	0.1829	5.4678	.0326	30.6997	1.3681
	43.50	8.755	0.1787	5.5968	.0318	31.4235	1.3366
9⅝	44.30	8.750	0.1782	5.6118	.0317	31.5077	1.3330
	47.00	8.681	0.1716	5.8262	.0306	32.7119	1.2839
	47.20	8.680	0.1715	5.8294	.0306	32.7299	1.2832
	53.50	8.535	0.1579	6.3320	.0281	35.5515	1.1814
	57.40	8.450	0.1501	6.6643	.0267	37.4171	1.1225
	58.40	8.435	0.1487	6.7262	.0265	37.7647	1.1121
	61.10	8.375	0.1432	6.9846	.0255	39.2158	1.0710
	33.00	9.384	0.2409	4.1510	.0429	23.3063	1.8021
10	60.00	8.780	0.1811	5.5229	.0322	31.0085	1.3545
	32.75	10.192	0.3272	3.0565	.0583	17.1608	2.4474
	35.75	10.140	0.3214	3.1113	.0572	17.4687	2.4043
	40.50	10.050	0.3115	3.2103	.0555	18.0245	2.3302
	45.50	9.950	0.3006	3.3268	.0535	18.6786	2.2486
	46.20	9.950	0.3006	3.3268	.0535	18.6786	2.2486
	48.00	9.902	0.2954	3.3853	.0526	19.0073	2.2097
	49.50	9.850	0.2898	3.4508	.0516	19.3747	2.1678
10¾	51.00	9.850	0.2898	3.4508	.0516	19.3747	2.1678
	54.00	9.784	0.2827	3.5370	.0504	19.8590	2.1149
	55.50	9.760	0.2802	3.5693	.0499	20.0404	2.0958
	60.70	9.660	0.2696	3.7096	.0480	20.8278	2.0165
	65.70	9.560	0.2591	3.8597	.0461	21.6705	1.9381
	71.10	9.450	0.2477	4.0374	.0441	22.6684	1.8528
	76.00	9.350	0.2374	4.2118	.0423	23.6474	1.7761
	81.00	9.250	0.2273	4.3998	.0405	24.7029	1.7002

| 7″ OD |
| INSIDE CASING |

VOLUME AND FILL BETWEEN CASING AND CASING

| OUTSIDE CASING |

SIZE OD INCHES	WEIGHT PER LIN FT	ID	CU FT PER LIN FT	LIN FT PER CU FT	BARRELS PER LIN FT	LIN FT PER BARREL	GALLONS PER LIN FT
	29.30	9.063	0.1807	5.5328	.0322	31.0645	1.3520
	32.30	9.001	0.1746	5.7264	.0311	32.1511	1.3063
	36.00	8.921	0.1668	5.9948	.0297	33.6583	1.2478
	38.00	8.885	0.1633	6.1231	.0291	34.3789	1.2217
	40.00	8.835	0.1585	6.3098	.0282	35.4271	1.1855
	42.00	8.799	0.1550	6.4508	.0276	36.2184	1.1596
	43.50	8.755	0.1508	6.6310	.0269	37.2301	1.1281
9⅝	44.30	8.750	0.1503	6.6520	.0268	37.3484	1.1245
	47.00	8.681	0.1438	6.9555	.0256	39.0525	1.0755
	47.20	8.680	0.1437	6.9601	.0256	39.0782	1.0748
	53.50	8.535	0.1301	7.6887	.0232	43.1689	0.9729
	57.40	8.450	0.1222	8.1842	.0218	45.9509	0.9140
	58.40	8.435	0.1208	8.2778	.0215	46.4763	0.9037
	61.10	8.375	0.1153	8.6727	.0205	48.6936	0.8625
	33.00	9.384	0.2130	4.6940	.0379	26.3551	1.5936
10	60.00	8.780	0.1532	6.5275	.0273	36.6491	1.1460
	32.75	10.192	0.2993	3.3411	.0533	18.7586	2.2390
	35.75	10.140	0.2935	3.4067	.0523	19.1271	2.1958
	40.50	10.050	0.2836	3.5257	.0505	19.7955	2.1217
	45.50	9.950	0.2727	3.6667	.0486	20.5873	2.0401
	46.20	9.950	0.2727	3.6667	.0486	20.5873	2.0401
	48.00	9.902	0.2675	3.7380	.0476	20.9872	2.0012
	49.50	9.850	0.2619	3.8179	.0467	21.4361	1.9593
10¾	51.00	9.850	0.2619	3.8179	.0467	21.4361	1.9593
	54.00	9.784	0.2549	3.9238	.0454	22.0306	1.9064
	55.50	9.760	0.2523	3.9636	.0449	22.2539	1.8873
	60.70	9.660	0.2417	4.1373	.0430	23.2292	1.8081
	65.70	9.560	0.2312	4.3249	.0412	24.2823	1.7297
	71.10	9.450	0.2198	4.5493	.0392	25.5422	1.6443
	76.00	9.350	0.2096	4.7719	.0373	26.7920	1.5676
	81.00	9.250	0.1994	5.0146	.0355	28.1549	1.4917
	38.00	11.150	0.4108	2.4342	.0732	13.6668	3.0732
	42.00	11.084	0.4028	2.4825	.0717	13.9383	3.0133
	47.00	11.000	0.3927	2.5465	.0699	14.2974	2.9376
	50.00	10.950	0.3867	2.5859	.0689	14.5187	2.8928
11¾	54.00	10.880	0.3784	2.6429	.0674	14.8385	2.8305
	60.00	10.772	0.3656	2.7350	.0651	15.3561	2.7351
	61.00	10.770	0.3654	2.7368	.0651	15.3660	2.7333
	65.00	10.682	0.3551	2.8162	.0632	15.8116	2.6563
12	40.00	11.384	0.4396	2.2749	.0783	12.7726	3.2883
	33.38*	12.250	0.5512	1.8142	.0982	10.1859	4.1233
	37.42*	12.188	0.5429	1.8418	.0967	10.3410	4.0615
	41.45*	12.126	0.5347	1.8701	.0952	10.5000	4.0000
12¾	43.77*	12.090	0.5300	1.8869	.0944	10.5942	3.9644
	45.58*	12.062	0.5263	1.9001	.0937	10.6684	3.9369
	49.56*	12.000	0.5181	1.9300	.0923	10.8359	3.8760
	53.00	11.970	0.5142	1.9447	.0916	10.9186	3.8467
	40.00	12.438	0.5765	1.7345	.1027	9.7387	4.3127
	45.00	12.360	0.5660	1.7669	.1008	9.9202	4.2338
13	50.00	12.282	0.5555	1.8002	.0989	10.1074	4.1554
	54.00	12.200	0.5445	1.8364	.0970	10.3106	4.0735

7-⅝" OD INSIDE CASING		VOLUME AND FILL BETWEEN CASING AND CASING					
OUTSIDE CASING							
SIZE OD INCHES	WEIGHT PER LIN FT	ID	CU FT PER LIN FT	LIN FT PER CU FT	BARRELS PER LIN FT	LIN FT PER BARREL	GALLONS PER LIN FT
	32.75	10.192	0.2495	4.0088	.0444	22.5076	1.8660
	35.75	10.140	0.2437	4.1036	.0434	23.0402	1.8229
	40.50	10.050	0.2338	4.2776	.0416	24.0170	1.7488
	45.50	9.950	0.2229	4.4870	.0397	25.1925	1.6672
	46.20	9.950	0.2229	4.4870	.0397	25.1925	1.6672
	48.00	9.902	0.2177	4.5941	.0388	25.7940	1.6283
	49.50	9.850	0.2121	4.7155	.0378	26.4754	1.5864
10¾	51.00	9.850	0.2121	4.7155	.0378	26.4754	1.5864
	54.00	9.784	0.2050	4.8780	.0365	27.3882	1.5335
	55.50	9.760	0.2024	4.9397	.0361	27.7343	1.5144
	60.70	9.660	0.1918	5.2124	.0342	29.2655	1.4351
	65.70	9.560	0.1814	5.5137	.0323	30.9571	1.3567
	71.10	9.450	0.1700	5.8837	.0303	33.0344	1.2714
	76.00	9.350	0.1597	6.2614	.0284	35.1553	1.1947
	81.00	9.250	0.1496	6.6861	.0266	37.5399	1.1188
	38.00	11.150	0.3610	2.7703	.0643	15.5543	2.7002
	42.00	11.084	0.3530	2.8332	.0629	15.9070	2.6403
	47.00	11.000	0.3428	2.9168	.0611	16.3765	2.5647
	50.00	10.950	0.3369	2.9686	.0600	16.6675	2.5199
11¾	54.00	10.880	0.3285	3.0439	.0585	17.0903	2.4575
	60.00	10.772	0.3158	3.1669	.0562	17.7806	2.3621
	61.00	10.770	0.3155	3.1692	.0562	17.7938	2.3604
	65.00	10.682	0.3052	3.2761	.0544	18.3941	2.2833
12	40.00	11.384	0.3897	2.5659	.0694	14.4065	2.9154
	33.38*	12.250	0.5014	1.9946	.0893	11.1988	3.7504
	37.42*	12.188	0.4931	2.0280	.0878	11.3865	3.6886
	41.45*	12.126	0.4849	2.0624	.0864	11.5796	3.6271
12¾	43.77*	12.090	0.4801	2.0828	.0855	11.6942	3.5915
	45.58*	12.062	0.4764	2.0990	.0849	11.7848	3.5639
	49.56*	12.000	0.4683	2.1354	.0834	11.9895	3.5031
	53.00	11.970	0.4644	2.1535	.0827	12.0908	3.4737
	40.00	12.438	0.5267	1.8987	.0938	10.6605	3.9398
	45.00	12.360	0.5161	1.9375	.0919	10.8784	3.8609
13	50.00	12.282	0.5056	1.9777	.0901	11.1040	3.7824
	54.00	12.200	0.4947	2.0215	.0881	11.3497	3.7005
	48.00	12.715	0.5647	1.7709	.1006	9.9431	4.2240
	54.50	12.615	0.5509	1.8154	.0981	10.1925	4.1207
	61.00	12.515	0.5372	1.8617	.0957	10.4525	4.0182
	68.00	12.415	0.5236	1.9100	.0932	10.7240	3.9164
	72.00	12.347	0.5144	1.9441	.0916	10.9155	3.8477
13⅜	77.00	12.275	0.5047	1.9814	.0899	11.1246	3.7754
	83.50	12.175	0.4914	2.0351	.0875	11.4265	3.6757
	85.00	12.159	0.4892	2.0440	.0871	11.4761	3.6598
	92.00	12.031	0.4724	2.1171	.0841	11.8864	3.5334
	98.00	11.937	0.4601	2.1736	.0819	12.2039	3.4415
14	50.00	13.344	0.6541	1.5289	.1165	8.5841	4.8928

7-⅝″ OD
INSIDE CASING

VOLUME AND FILL BETWEEN CASING AND CASING

OUTSIDE CASING							
SIZE OD INCHES	WEIGHT PER LIN FT	ID	CU FT PER LIN FT	LIN FT PER CU FT	BARRELS PER LIN FT	LIN FT PER BARREL	GALLONS PER LIN FT
16	55.00	15.375	0.9722	1.0286	.1732	5.7751	7.2726
	65.00	15.250	0.9513	1.0512	.1694	5.9019	7.1164
	70.00	15.198	0.9427	1.0608	.1679	5.9559	7.0518
	75.00	15.125	0.9306	1.0746	.1657	6.0332	6.9615
	84.00	15.010	0.9117	1.0968	.1624	6.1583	6.8201
	109.00	14.688	0.8596	1.1634	.1531	6.5320	6.4299
	118.00	14.570	0.8407	1.1894	.1497	6.6783	6.2891
18	80.00	17.180	1.2927	0.7736	.2302	4.3433	9.6701
18⅝	78.00	17.855	1.4217	0.7034	.2532	3.9493	10.6349
	87.50	17.755	1.4023	0.7131	.2498	4.0040	10.4896
	96.50	17.655	1.3829	0.7231	.2463	4.0599	10.3452

7-¾" OD
INSIDE CASING

| OUTSIDE CASING |

VOLUME AND FILL BETWEEN CASING AND CASING

SIZE OD INCHES	WEIGHT PER LIN FT	ID	CU FT PER LIN FT	LIN FT PER CU FT	BARRELS PER LIN FT	LIN FT PER BARREL	GALLONS PER LIN FT
	38.00	11.150	0.3505	2.8532	.0624	16.0195	2.6218
	42.00	11.084	0.3425	2.9199	.0610	16.3939	2.5619
	47.00	11.000	0.3324	3.0088	.0592	16.8930	2.4862
	50.00	10.950	0.3264	3.0639	.0581	17.2028	2.4415
11¾	54.00	10.880	0.3180	3.1442	.0566	17.6536	2.3791
	60.00	10.772	0.3053	3.2756	.0544	18.3911	2.2837
	61.00	10.770	0.3051	3.2781	.0543	18.4053	2.2820
	65.00	10.682	0.2948	3.3926	.0525	19.0482	2.2049
12	40.00	11.384	0.3792	2.6368	.0675	14.8047	2.8369
	33.38*	12.250	0.4909	2.0372	.0874	11.4379	3.6720
	37.42*	12.188	0.4826	2.0721	.0860	11.6338	3.6102
	41.45*	12.126	0.4744	2.1080	.0845	11.8354	3.5487
12¾	43.77*	12.090	0.4696	2.1293	.0836	11.9553	3.5131
	45.58*	12.062	0.4659	2.1462	.0830	12.0499	3.4855
	49.56*	12.000	0.4578	2.1843	.0815	12.2641	3.4246
	53.00	11.970	0.4539	2.2032	.0808	12.3700	3.3953
	40.00	12.438	0.5162	1.9373	.0919	10.8770	3.8614
	45.00	12.360	0.5056	1.9777	.0901	11.1039	3.7824
13	50.00	12.282	0.4952	2.0196	.0882	11.3390	3.7040
	54.00	12.200	0.4842	2.0652	.0862	11.5954	3.6221
	48.00	12.715	0.5542	1.8044	.0987	10.1312	4.1456
	54.50	12.615	0.5404	1.8506	.0962	10.3902	4.0423
	61.00	12.515	0.5267	1.8987	.0938	10.6606	3.9397
	68.00	12.415	0.5131	1.9490	.0914	10.9431	3.8380
	72.00	12.347	0.5039	1.9846	.0897	11.1425	3.7693
13⅜	77.00	12.275	0.4942	2.0234	.0880	11.3605	3.6970
	83.50	12.175	0.4809	2.0795	.0856	11.6756	3.5973
	85.00	12.159	0.4788	2.0887	.0853	11.7274	3.5814
	92.00	12.031	0.4619	2.1651	.0823	12.1562	3.4550
	98.00	11.937	0.4496	2.2243	.0801	12.4884	3.3631
14	50.00	13.344	0.6436	1.5538	.1146	8.7239	4.8144
	55.00	15.375	0.9617	1.0398	.1713	5.8381	7.1942
	65.00	15.250	0.9408	1.0629	.1676	5.9676	7.0380
	70.00	15.198	0.9322	1.0727	.1660	6.0229	6.9734
16	75.00	15.125	0.9201	1.0868	.1639	6.1019	6.8831
	84.00	15.010	0.9012	1.1096	.1605	6.2299	6.7417
	109.00	14.688	0.8491	1.1778	.1512	6.6126	6.3515
	118.00	14.570	0.8302	1.2045	.1479	6.7626	6.2107
18	80.00	17.180	1.2822	0.7799	.2284	4.3788	9.5916
	78.00	17.855	1.4112	0.7086	.2513	3.9786	10.5565
18⅝	87.50	17.755	1.3918	0.7185	.2479	4.0341	10.4112
	96.50	17.655	1.3725	0.7286	.2444	4.0909	10.2667
	90.00	19.190	1.6809	0.5949	.2994	3.3402	12.5743
	94.00	19.124	1.6671	0.5998	.2969	3.3678	12.4711
20	106.50	19.000	1.6414	0.6093	.2923	3.4207	12.2782
	133.00	18.730	1.5858	0.6306	.2824	3.5405	11.8626
	169.00	18.376	1.5142	0.6604	.2697	3.7081	11.3267

VOLUME AND FILL BETWEEN CASING AND CASING

8" OD
INSIDE CASING

OUTSIDE CASING

SIZE OD INCHES	WEIGHT PER LIN FT	ID	CU FT PER LIN FT	LIN FT PER CU FT	BARRELS PER LIN FT	LIN FT PER BARREL	GALLONS PER LIN FT
	38.00	11.150	0.3290	3.0394	.0586	17.0652	2.4612
	42.00	11.084	0.3210	3.1152	.0572	17.4907	2.4013
	47.00	11.000	0.3109	3.2166	.0554	18.0599	2.3256
	50.00	10.950	0.3049	3.2798	.0543	18.4145	2.2808
11¾	54.00	10.880	0.2966	3.3719	.0528	18.9320	2.2185
	60.00	10.772	0.2838	3.5235	.0505	19.7827	2.1231
	61.00	10.770	0.2836	3.5264	.0505	19.7991	2.1213
	65.00	10.682	0.2733	3.6592	.0487	20.5451	2.0443
12	40.00	11.384	0.3578	2.7951	.0637	15.6934	2.6763
	33.38*	12.250	0.4694	2.1304	.0836	11.9612	3.5113
	37.42*	12.188	0.4611	2.1686	.0821	12.1756	3.4495
	41.45*	12.126	0.4529	2.2079	.0807	12.3966	3.3880
12¾	43.77*	12.090	0.4482	2.2314	.0798	12.5281	3.3525
	45.58*	12.062	0.4445	2.2499	.0792	12.6321	3.3249
	49.56*	12.000	0.4363	2.2918	.0777	12.8677	3.2640
	53.00	11.970	0.4324	2.3126	.0770	12.9844	3.2347
	40.00	12.438	0.4947	2.0214	.0881	11.3492	3.7007
	45.00	12.360	0.4842	2.0654	.0862	11.5965	3.6218
13	50.00	12.282	0.4737	2.1111	.0844	11.8531	3.5434
	54.00	12.200	0.4627	2.1611	.0824	12.1336	3.4615
	48.00	12.715	0.5327	1.8772	.0949	10.5396	3.9850
	54.50	12.615	0.5189	1.9272	.0924	10.8202	3.8816
	61.00	12.515	0.5052	1.9794	.0900	11.1138	3.7791
	68.00	12.415	0.4916	2.0342	.0876	11.4212	3.6774
	72.00	12.347	0.4824	2.0729	.0859	11.6386	3.6087
13⅜	77.00	12.275	0.4727	2.1153	.0842	11.8766	3.5364
	83.50	12.175	0.4594	2.1767	.0818	12.2214	3.4366
	85.00	12.159	0.4573	2.1868	.0814	12.2781	3.4207
	92.00	12.031	0.4404	2.2707	.0784	12.7490	3.2944
	98.00	11.937	0.4281	2.3359	.0762	13.1149	3.2025
14	50.00	13.344	0.6221	1.6074	.1108	9.0250	4.6537
	55.00	15.375	0.9402	1.0636	.1675	5.9714	7.0335
	65.00	15.250	0.9194	1.0877	.1637	6.1070	6.8773
	70.00	15.198	0.9107	1.0980	.1622	6.1649	6.8127
16	75.00	15.125	0.8987	1.1128	.1601	6.2477	6.7224
	84.00	15.010	0.8798	1.1367	.1567	6.3820	6.5810
	109.00	14.688	0.8276	1.2083	.1474	6.7842	6.1909
	118.00	14.570	0.8088	1.2364	.1440	6.9421	6.0500
18	80.00	17.180	1.2607	0.7932	.2245	4.4534	9.4310
	78.00	17.855	1.3897	0.7196	.2475	4.0401	10.3959
18⅝	87.50	17.755	1.3703	0.7298	.2441	4.0973	10.2506
	96.50	17.655	1.3510	0.7402	.2406	4.1559	10.1061
	90.00	19.190	1.6595	0.6026	.2956	3.3834	12.4136
	94.00	19.124	1.6457	0.6077	.2931	3.4117	12.3104
20	106.50	19.000	1.6199	0.6173	.2885	3.4660	12.1176
	133.00	18.730	1.5643	0.6393	.2786	3.5891	11.7019
	169.00	18.376	1.4927	0.6699	.2659	3.7614	11.1660

8″ OD
INSIDE CASING

VOLUME AND FILL BETWEEN CASING AND CASING

OUTSIDE CASING

SIZE OD INCHES	WEIGHT PER LIN FT	ID	CU FT PER LIN FT	LIN FT PER CU FT	BARRELS PER LIN FT	LIN FT PER BARREL	GALLONS PER LIN FT
21½	92.50	20.710	1.9902	0.5025	.3545	2.8211	14.8881
	103.00	20.610	1.9677	0.5082	.3505	2.8534	14.7195
	114.00	20.510	1.9453	0.5141	.3465	2.8863	14.5517
24½	88.00	23.850	2.7534	0.3632	.4904	2.0392	20.5967
	100.50	23.750	2.7274	0.3666	.4858	2.0586	20.4025
	113.00	23.650	2.7016	0.3702	.4812	2.0783	20.2091

VOLUME AND FILL BETWEEN CASING AND CASING

8-⅛″ OD
INSIDE CASING

OUTSIDE CASING

SIZE OD INCHES	WEIGHT PER LIN FT	ID	CU FT PER LIN FT	LIN FT PER CU FT	BARRELS PER LIN FT	LIN FT PER BARREL	GALLONS PER LIN FT
	38.00	11.150	0.3180	3.1445	.0566	17.6551	2.3789
	42.00	11.084	0.3100	3.2257	.0552	18.1109	2.3190
	47.00	11.000	0.2999	3.3345	.0534	18.7219	2.2434
	50.00	10.950	0.2939	3.4024	.0523	19.1032	2.1986
11¾	54.00	10.880	0.2856	3.5017	.0509	19.6608	2.1362
	60.00	10.772	0.2728	3.6654	.0486	20.5799	2.0408
	61.00	10.770	0.2726	3.6686	.0485	20.5976	2.0391
	65.00	10.682	0.2623	3.8126	.0467	21.4062	1.9620
12	40.00	11.384	0.3468	2.8837	.0618	16.1909	2.5941
	33.38*	12.250	0.4584	2.1815	.0816	12.2481	3.4291
	37.42*	12.188	0.4501	2.2215	.0802	12.4730	3.3673
	41.45*	12.126	0.4419	2.2629	.0787	12.7050	3.3058
12¾	43.77*	12.090	0.4372	2.2875	.0779	12.8432	3.2702
	45.58*	12.062	0.4335	2.3069	.0772	12.9525	3.2426
	49.56*	12.000	0.4253	2.3511	.0758	13.2003	3.1818
	53.00	11.970	0.4214	2.3729	.0751	13.3231	3.1524
	40.00	12.438	0.4837	2.0673	.0862	11.6071	3.6185
	45.00	12.360	0.4732	2.1134	.0843	11.8659	3.5396
13	50.00	12.282	0.4627	2.1613	.0824	12.1348	3.4611
	54.00	12.200	0.4517	2.2137	.0805	12.4289	3.3792
	48.00	12.715	0.5217	1.9167	.0929	10.7617	3.9027
	54.50	12.615	0.5079	1.9689	.0905	11.0544	3.7994
	61.00	12.515	0.4942	2.0235	.0880	11.3610	3.6969
	68.00	12.415	0.4806	2.0807	.0856	11.6824	3.5951
	72.00	12.347	0.4714	2.1213	.0840	11.9100	3.5264
13⅜	77.00	12.275	0.4617	2.1657	.0822	12.1594	3.4541
	83.50	12.175	0.4484	2.2301	.0799	12.5210	3.3544
	85.00	12.159	0.4463	2.2407	.0795	12.5806	3.3385
	92.00	12.031	0.4294	2.3288	.0765	13.0754	3.2121
	98.00	11.937	0.4171	2.3974	.0743	13.4606	3.1202
14	50.00	13.344	0.6111	1.6363	.1088	9.1874	4.5715
	55.00	15.375	0.9293	1.0761	.1655	6.0420	6.9513
	65.00	15.250	0.9084	1.1009	.1618	6.1809	6.7951
	70.00	15.198	0.8997	1.1114	.1602	6.2403	6.7305
16	75.00	15.125	0.8877	1.1266	.1581	6.3251	6.6402
	84.00	15.010	0.8688	1.1511	.1547	6.4627	6.4988
	109.00	14.688	0.8166	1.2246	.1454	6.8755	6.1086
	118.00	14.570	0.7978	1.2535	.1421	7.0378	5.9678
18	80.00	17.180	1.2497	0.8002	.2226	4.4926	9.3488
	78.00	17.855	1.3787	0.7253	.2456	4.0723	10.3136
18⅝	87.50	17.755	1.3593	0.7357	.2421	4.1305	10.1683
	96.50	17.655	1.3400	0.7463	.2387	4.1900	10.0239
	90.00	19.190	1.6485	0.6066	.2936	3.4059	12.3314
	94.00	19.124	1.6347	0.6117	.2911	3.4347	12.2282
20	106.50	19.000	1.6089	0.6215	.2866	3.4897	12.0353
	133.00	18.730	1.5533	0.6438	.2767	3.6146	11.6197
	169.00	18.376	1.4817	0.6749	.2639	3.7893	11.0838

<table>
<tr><td colspan="2">8-⅛″ OD
INSIDE CASING</td><td colspan="6" align="center">VOLUME AND FILL BETWEEN
CASING AND CASING</td></tr>
</table>

OUTSIDE CASING							
SIZE OD INCHES	WEIGHT PER LIN FT	ID	CU FT PER LIN FT	LIN FT PER CU FT	BARRELS PER LIN FT	LIN FT PER BARREL	GALLONS PER LIN FT
	92.50	20.710	1.9792	0.5052	.3525	2.8367	14.8058
21½	103.00	20.610	1.9567	0.5111	.3485	2.8694	14.6372
	114.00	20.510	1.9343	0.5170	.3445	2.9027	14.4695
	88.00	23.850	2.7424	0.3646	.4884	2.0473	20.5145
24½	100.50	23.750	2.7164	0.3681	.4838	2.0669	20.3203
	113.00	23.650	2.6906	0.3717	.4792	2.0868	20.1269

VOLUME AND FILL BETWEEN CASING AND CASING

8-⅝" OD
INSIDE CASING

OUTSIDE CASING

SIZE OD INCHES	WEIGHT PER LIN FT	ID	CU FT PER LIN FT	LIN FT PER CU FT	BARRELS PER LIN FT	LIN FT PER BARREL	GALLONS PER LIN FT
11¾	38.00	11.150	0.2723	3.6719	.0485	20.6164	2.0372
	42.00	11.084	0.2643	3.7831	.0471	21.2406	1.9773
	47.00	11.000	0.2542	3.9337	.0453	22.0860	1.9017
	50.00	10.950	0.2482	4.0285	.0442	22.6186	1.8569
	54.00	10.880	0.2399	4.1685	.0427	23.4044	1.7945
	60.00	10.772	0.2271	4.4026	.0405	24.7186	1.6991
	61.00	10.770	0.2269	4.4071	.0404	24.7442	1.6974
	65.00	10.682	0.2166	4.6166	.0386	25.9204	1.6203
12	40.00	11.384	0.3011	3.3212	.0536	18.6472	2.2524
12¾	33.38*	12.250	0.4127	2.4229	.0735	13.6037	3.0874
	37.42*	12.188	0.4045	2.4724	.0720	13.8816	3.0256
	41.45*	12.126	0.3962	2.5237	.0706	14.1696	2.9641
	43.77*	12.090	0.3915	2.5544	.0697	14.3417	2.9285
	45.58*	12.062	0.3878	2.5787	.0691	14.4782	2.9009
	49.56*	12.000	0.3797	2.6339	.0676	14.7884	2.8401
	53.00	11.970	0.3757	2.6614	.0669	14.9428	2.8107
13	40.00	12.438	0.4380	2.2829	.0780	12.8175	3.2768
	45.00	12.360	0.4275	2.3392	.0761	13.1338	3.1979
	50.00	12.282	0.4170	2.3980	.0743	13.4640	3.1194
	54.00	12.200	0.4061	2.4627	.0723	13.8270	3.0375
13⅜	48.00	12.715	0.4760	2.1007	.0848	11.7943	3.5610
	54.50	12.615	0.4622	2.1634	.0823	12.1468	3.4577
	61.00	12.515	0.4485	2.2296	.0799	12.5180	3.3552
	68.00	12.415	0.4349	2.2993	.0775	12.9094	3.2534
	72.00	12.347	0.4257	2.3489	.0758	13.1878	3.1848
	77.00	12.275	0.4161	2.4034	.0741	13.4943	3.1124
	83.50	12.175	0.4027	2.4830	.0717	13.9411	3.0127
	85.00	12.159	0.4006	2.4962	.0714	14.0150	2.9968
	92.00	12.031	0.3837	2.6060	.0683	14.6319	2.8705
	98.00	11.937	0.3714	2.6923	.0662	15.1159	2.7785
14	50.00	13.344	0.5654	1.7685	.1007	9.9296	4.2298
16	55.00	15.375	0.8836	1.1318	.1574	6.3544	6.6096
	65.00	15.250	0.8627	1.1592	.1537	6.5082	6.4534
	70.00	15.198	0.8541	1.1709	.1521	6.5740	6.3888
	75.00	15.125	0.8420	1.1877	.1500	6.6683	6.2985
	84.00	15.010	0.8231	1.2149	.1466	6.8214	6.1571
	109.00	14.688	0.7709	1.2971	.1373	7.2829	5.7669
	118.00	14.570	0.7521	1.3296	.1340	7.4652	5.6261
18	80.00	17.180	1.2041	0.8305	.2145	496630	9.0071
18⅝	78.00	17.855	1.3331	0.7502	.2374	4.2118	9.9719
	87.50	17.755	1.3136	0.7612	.2340	4.2741	9.8266
	96.50	17.655	1.2943	0.7726	.2305	4.3379	9.6822
20	90.00	19.190	1.6028	0.6239	.2855	3.5030	11.9897
	94.00	19.124	1.5890	0.6293	.2830	3.5334	11.8865
	106.50	19.000	1.5632	0.6397	.2784	3.5917	11.6936
	133.00	18.730	1.5076	0.6633	.2685	3.7241	11.2780
	169.00	18.376	1.4360	0.6964	.2558	3.9099	10.7421

| 8-⅝" OD INSIDE CASING | **VOLUME AND FILL BETWEEN** | | | | | |
| OUTSIDE CASING | **CASING AND CASING** | | | | | |

SIZE OD INCHES	WEIGHT PER LIN FT	ID	CU FT PER LIN FT	LIN FT PER CU FT	BARRELS PER LIN FT	LIN FT PER BARREL	GALLONS PER LIN FT
21½	92.50	20.710	1.9336	0.5172	.3444	2.9037	14.4641
	103.00	20.610	1.9110	0.5233	.3404	2.9380	14.2955
	114.00	20.510	1.8886	0.5295	.3364	2.9729	14.1278
24½	88.00	23.850	2.6967	0.3708	.4803	2.0820	20.1728
	100.50	23.750	2.6707	0.3744	.4757	2.1023	19.9786
	113.00	23.650	2.6449	0.3781	.4711	2.1228	19.7852

VOLUME AND FILL BETWEEN CASING AND CASING

9" OD
INSIDE CASING

OUTSIDE CASING

SIZE OD INCHES	WEIGHT PER LIN FT	ID	CU FT PER LIN FT	LIN FT PER CU FT	BARRELS PER LIN FT	LIN FT PER BARREL	GALLONS PER LIN FT
11¾	38.00	11.150	0.2363	4.2321	.0421	23.7617	1.7676
	42.00	11.084	0.2283	4.3805	.0407	24.5947	1.7077
	47.00	11.000	0.2182	4.5837	.0389	25.7354	1.6320
	50.00	10.950	0.2122	4.7130	.0378	26.4614	1.5872
	54.00	10.880	0.2038	4.9057	.0363	27.5433	1.5249
	60.00	10.772	0.1911	5.2331	.0340	29.3816	1.4295
	61.00	10.770	0.1909	5.2395	.0340	29.4178	1.4277
	65.00	10.682	0.1806	5.5383	.0322	31.0953	1.3507
12	40.00	11.384	0.2650	3.7729	.0472	21.1833	1.9827
12¾	33.38*	12.250	0.3767	2.6548	.0671	14.9055	2.8177
	37.42*	12.188	0.3684	2.7143	.0656	15.2399	2.7559
	41.45*	12.126	0.3602	2.7763	.0642	15.5878	2.6944
	43.77*	12.090	0.3554	2.8134	.0633	15.7963	2.6589
	45.58*	12.062	0.3517	2.8429	.0626	15.9619	2.6313
	49.56*	12.000	0.3436	2.9103	.0612	16.3399	2.5704
	53.00	11.970	0.3397	2.9439	.0605	16.5286	2.5411
13	40.00	12.438	0.4020	2.4876	.0716	13.9669	3.0071
	45.00	12.360	0.3914	2.5547	.0697	14.3433	2.9282
	50.00	12.282	0.3810	2.6250	.0679	14.7380	2.8498
	54.00	12.200	0.3700	2.7026	.0659	15.1741	2.7679
13⅜	48.00	12.715	0.4400	2.2728	.0784	12.7606	3.2914
	54.50	12.615	0.4262	2.3464	.0759	13.1743	3.1880
	61.00	12.515	0.4125	2.4244	.0735	13.6120	3.0855
	68.00	12.415	0.3989	2.5071	.0710	14.0761	2.9838
	72.00	12.347	0.3897	2.5661	.0694	14.4078	2.9151
	77.00	12.275	0.3800	2.6314	.0677	14.7744	2.8428
	83.50	12.175	0.3667	2.7271	.0653	15.3117	2.7430
	85.00	12.159	0.3646	2.7430	.0649	15.4009	2.7271
	92.00	12.031	0.3477	2.8763	.0619	16.1489	2.6008
	98.00	11.937	0.3354	2.9816	.0597	16.7406	2.5089
14	50.00	13.344	0.5294	1.8890	.0943	10.6057	3.9601
16	55.00	15.375	0.8475	1.1799	.1510	6.6247	6.3399
	65.00	15.250	0.8266	1.2097	.1472	6.7920	6.1837
	70.00	15.198	0.8180	1.2225	.1457	6.8637	6.1191
	75.00	15.125	0.8059	1.2408	.1435	6.9665	6.0288
	84.00	15.010	0.7870	1.2706	.1402	7.1338	5.8874
	109.00	14.688	0.7349	1.3608	.1309	7.6402	5.4973
	118.00	14.570	0.7160	1.3966	.1275	7.8411	5.3564
18	80.00	17.180	1.1680	0.8561	.2080	4.8069	8.7374
18⅝	78.00	17.855	1.2970	0.7710	.2310	4.3289	9.7023
	87.50	17.755	1.2776	0.7827	.2275	4.3947	9.5570
	96.50	17.655	1.2583	0.7947	.2241	4.4622	9.4125
20	90.00	19.190	1.5667	0.6383	.2790	3.5836	11.7200
	94.00	19.124	1.5529	0.6439	.2766	3.6154	11.6168
	106.50	19.000	1.5272	0.6548	.2720	3.6765	11.4240
	133.00	18.730	1.4716	0.6795	.2621	3.8153	11.0083
	169.00	18.376	1.4000	0.7143	.2493	4.0105	10.4724

VOLUME AND FILL BETWEEN CASING AND CASING

9" OD
INSIDE CASING

OUTSIDE CASING

SIZE OD INCHES	WEIGHT PER LIN FT	ID	CU FT PER LIN FT	LIN FT PER CU FT	BARRELS PER LIN FT	LIN FT PER BARREL	GALLONS PER LIN FT
21½	92.50	20.710	1.8975	0.5270	.3380	2.9589	14.1945
	103.00	20.610	1.8750	0.5333	.3339	2.9945	14.0259
	114.00	20.510	1.8526	0.5398	.3300	3.0307	13.8581
24½	88.00	23.850	2.6607	0.3758	.4739	2.1102	19.9031
	100.50	23.750	2.6347	0.3796	.4693	2.1310	19.7089
	113.00	23.650	2.6088	0.3833	.4647	2.1521	19.5155

VOLUME AND FILL BETWEEN CASING AND CASING

9-⅝" OD
INSIDE CASING

OUTSIDE CASING

SIZE OD INCHES	WEIGHT PER LIN FT	ID	CU FT PER LIN FT	LIN FT PER CU FT	BARRELS PER LIN FT	LIN FT PER BARREL	GALLONS PER LIN FT
11¾	38.00	11.150	0.1728	5.7871	.0308	32.4922	1.2926
	42.00	11.084	0.1648	6.0682	.0294	34.0703	1.2327
	47.00	11.000	0.1547	6.4651	.0275	36.2989	1.1571
	50.00	10.950	0.1487	6.7254	.0265	37.7602	1.1123
	54.00	10.880	0.1404	7.1247	.0250	40.0025	1.0499
	60.00	10.772	0.1276	7.8369	.0227	44.0008	0.9545
	61.00	10.770	0.1274	7.8513	.0227	44.0820	0.9528
	65.00	10.682	0.1171	8.5418	.0209	47.9589	0.8757
12	40.00	11.384	0.2016	4.9614	.0359	27.8560	1.5078
12¾	33.38*	12.250	0.3132	3.1930	.0558	17.9272	2.3428
	37.42*	12.188	0.3049	3.2795	.0543	18.4131	2.2810
	41.45*	12.126	0.2967	3.3704	.0528	18.9233	2.2195
	43.77*	12.090	0.2919	3.4253	.0520	19.2315	2.1839
	45.58*	12.062	0.2883	3.4691	.0513	19.4776	2.1563
	49.56*	12.000	0.2801	3.5699	.0499	20.0434	2.0955
	53.00	11.970	0.2762	3.6206	.0492	20.3280	2.0661
13	40.00	12.438	0.3385	2.9542	.0603	16.5865	2.5322
	45.00	12.360	0.3280	3.0492	.0584	17.1201	2.4533
	50.00	12.282	0.3175	3.1499	.0565	17.6854	2.3748
	54.00	12.200	0.3065	3.2624	.0546	18.3172	2.2929
13⅜	48.00	12.715	0.3765	2.6560	.0671	14.9124	2.8164
	54.50	12.615	0.3627	2.7572	.0646	15.4805	2.7131
	61.00	12.515	0.3490	2.8655	.0622	16.0885	2.6106
	68.00	12.415	0.3354	2.9817	.0597	16.7407	2.5089
	72.00	12.347	0.3262	3.0656	.0581	17.2120	2.4402
	77.00	12.275	0.3165	3.1592	.0564	17.7378	2.3678
	83.50	12.175	0.3032	3.2982	.0540	18.5180	2.2681
	85.00	12.159	0.3011	3.3215	.0536	18.6486	2.2522
	92.00	12.031	0.2842	3.5188	.0506	19.7568	2.1259
	98.00	11.937	0.2719	3.6779	.0484	20.6497	2.0339
14	50.00	13.344	0.4659	2.1464	.0830	12.0510	3.4852
16	55.00	15.375	0.7840	1.2755	.1396	7.1611	5.8650
	65.00	15.250	0.7632	1.3103	.1359	7.3571	5.7088
	70.00	15.198	0.7545	1.3253	.1344	7.4413	5.6442
	75.00	15.125	0.7424	1.3469	.1322	7.5623	5.5539
	84.00	15.010	0.7235	1.3821	.1289	7.7598	5.4125
	109.00	14.688	0.6714	1.4895	.1196	8.3626	5.0223
	118.00	14.570	0.6526	1.5324	.1162	8.6040	4.8815
18	80.00	17.180	1.1045	0.9054	.1967	5.0832	8.2625
18⅝	78.00	17.855	1.2335	0.8107	.2197	4.5517	9.2273
	87.50	17.755	1.2141	0.8237	.2162	4.6245	9.0820
	96.50	17.655	1.1948	0.8370	.2128	4.6993	8.9376
20	90.00	19.190	1.5032	0.6652	.2677	3.7350	11.2451
	94.00	19.124	1.4895	0.6714	.2653	3.7695	11.1419
	106.50	19.000	1.4637	0.6832	.2607	3.8360	10.9490
	133.00	18.730	1.4081	0.7102	.2508	3.9873	10.5334
	169.00	18.376	1.3365	0.7482	.2380	4.2011	9.9975

9-⅝″ OD
INSIDE CASING

VOLUME AND FILL BETWEEN CASING AND CASING

OUTSIDE CASING

SIZE OD INCHES	WEIGHT PER LIN FT	ID	CU FT PER LIN FT	LIN FT PER CU FT	BARRELS PER LIN FT	LIN FT PER BARREL	GALLONS PER LIN FT
21½	92.50	20.710	1.8340	0.5452	.3267	3.0613	13.7195
	103.00	20.610	1.8115	0.5520	.3226	3.0994	13.5509
	114.00	20.510	1.7891	0.5590	.3186	3.1383	13.3832
24½	88.00	23.850	2.5972	0.3850	.4626	2.1618	19.4282
	100.50	23.750	2.5712	0.3889	.4580	2.1836	19.2340
	113.00	23.650	2.5454	0.3929	.4533	2.2058	19.0406

VOLUME AND FILL BETWEEN CASING AND CASING

10" OD
INSIDE CASING

OUTSIDE CASING							
SIZE OD INCHES	WEIGHT PER LIN FT	ID	CU FT PER LIN FT	LIN FT PER CU FT	BARRELS PER LIN FT	LIN FT PER BARREL	GALLONS PER LIN FT
	48.00	12.715	0.3364	2.9730	.0599	16.6920	2.5162
	54.50	12.615	0.3225	3.1003	.0574	17.4069	2.4128
	61.00	12.515	0.3088	3.2379	.0550	18.1794	2.3103
	68.00	12.415	0.2952	3.3870	.0526	19.0167	2.2086
	72.00	12.347	0.2861	3.4957	.0509	19.6272	2.1399
13⅜	77.00	12.275	0.2764	3.6180	.0492	20.3138	2.0676
	83.50	12.175	0.2631	3.8015	.0469	21.3436	1.9678
	85.00	12.159	0.2609	3.8324	.0465	21.5173	1.9519
	92.00	12.031	0.2440	4.0976	.0435	23.0063	1.8256
	98.00	11.937	0.2318	4.3149	.0413	24.2261	1.7337
14	50.00	13.344	0.4258	2.3487	.0758	13.1871	3.1849
	55.00	15.375	0.7439	1.3443	.1325	7.5475	5.5647
	65.00	15.250	0.7230	1.3831	.1288	7.7655	5.4085
	70.00	15.198	0.7144	1.3998	.1272	7.8594	5.3439
16	75.00	15.125	0.7023	1.4239	.1251	7.9945	5.2536
	84.00	15.010	0.6834	1.4633	.1217	8.2156	5.1122
	109.00	14.688	0.6312	1.5842	.1124	8.8944	4.7221
	118.00	14.570	0.6124	1.6329	.1091	9.1679	4.5812
18	80.00	17.180	1.0644	0.9395	.1896	5.2749	7.9622
	78.00	17.855	1.1934	0.8380	.2125	4.7048	8.9271
18⅝	97.50	17.755	1.1740	0.8518	.2091	4.7826	8.7818
	96.50	17.655	1.1546	0.8661	.2057	4.8636	8.6373
	90.00	19.190	1.4631	0.6835	.2606	3.8374	10.9448
	94.00	19.124	1.4493	0.6900	.2581	3.8739	10.8417
20	106.50	19.000	1.4235	0.7025	.2535	3.9441	10.6488
	133.00	18.730	1.3680	0.7310	.2436	4.1043	10.2331
	169.00	18.376	1.2963	0.7714	.2309	4.3311	9.6972
	92.50	20.710	1.7939	0.5574	.3195	3.1298	13.4193
21½	103.00	20.610	1.7714	0.5645	.3155	3.1697	13.2507
	114.00	20.510	1.7489	0.5718	.3115	3.2103	13.0829
	88.00	23.850	2.5570	0.3911	.4554	2.1957	19.1279
24½	100.50	23.750	2.5311	0.3951	.4508	2.2183	18.9337
	113.00	23.650	2.5052	0.3992	.4462	2.2412	18.7403
	98.93*	29.376	4.1612	0.2403	.7411	1.3493	31.1283
	118.65*	29.250	4.1210	0.2427	.7340	1.3624	30.8269
	157.53*	29.000	4.0415	0.2474	.7198	1.3892	30.2327
	196.08*	28.750	3.9628	0.2523	.7058	1.4168	29.6437
30	234.29*	28.500	3.8847	0.2574	.6919	1.4453	29.0597
	309.72*	28.000	3.7306	0.2681	.6645	1.5050	27.9071
	346.93*	27.750	3.6546	0.2736	.6509	1.5363	27.3385
	383.81*	27.500	3.5793	0.2794	.6375	1.5686	26.7749
	456.57*	27.000	3.4307	0.2915	.6110	1.6366	25.6631

10-¾″ OD INSIDE CASING		**VOLUME AND FILL BETWEEN CASING AND CASING**					
OUTSIDE CASING							
SIZE OD INCHES	WEIGHT PER LIN FT	ID	CU FT PER LIN FT	LIN FT PER CU FT	BARRELS PER LIN FT	LIN FT PER BARREL	GALLONS PER LIN FT
	48.00	12.715	0.2515	3.9764	.0448	22.3258	1.8812
	54.50	12.615	0.2377	4.2075	.0423	23.6236	1.7779
	61.00	12.515	0.2240	4.4650	.0399	25.0693	1.6754
	68.00	12.415	0.2104	4.7536	.0375	26.6897	1.5736
	72.00	12.347	0.2012	4.9706	.0358	27.9081	1.5049
13⅜	77.00	12.275	0.1915	5.2216	.0341	29.3171	1.4326
	83.50	12.175	0.1782	5.6124	.0317	31.5113	1.3329
	85.00	12.159	0.1761	5.6801	.0314	31.8914	1.3170
	92.00	12.031	0.1592	6.2828	.0283	35.2751	1.1906
	98.00	11.937	0.1469	6.8084	.0262	38.2263	1.0987
14	50.00	13.344	0.3409	2.9336	.0607	16.4707	2.5500
	55.00	15.375	0.6590	1.5174	.1174	8.5197	4.9298
	65.00	15.250	0.6381	1.5671	.1137	8.7984	4.7736
	70.00	15.198	0.6295	1.5886	.1121	8.9191	4.7090
16	75.00	15.125	0.6174	1.6196	.1100	9.0935	4.6187
	84.00	15.010	0.5985	1.6708	.1066	9.3807	4.4773
	109.00	14.688	0.5464	1.8303	.0973	10.2762	4.0871
	118.00	14.570	0.5275	1.8956	.0940	10.6430	3.9463
18	80.00	17.180	0.9795	1.0209	.1745	5.7320	7.3273
	78.00	17.855	1.1085	0.9021	.1974	5.0651	8.2921
18⅝	87.50	17.755	1.0891	0.9182	.1940	5.1554	8.1468
	96.50	17.655	1.0698	0.9348	.1905	5.2485	8.0024
	90.00	19.190	1.3782	0.7256	.2455	4.0738	10.3099
	94.00	19.124	1.3644	0.7329	.2430	4.1149	10.2067
20	106.50	19.000	1.3387	0.7470	.2384	4.1942	10.0138
	133.00	18.730	1.2831	0.7794	.2285	4.3758	9.5982
	169.00	18.376	1.2114	0.8255	.2158	4.6346	9.0623
	92.50	20.710	1.7090	0.5851	.3044	3.2853	12.7843
21½	103.00	20.610	1.6865	0.5930	.3004	3.3292	12.6157
	114.00	20.510	1.6640	0.6009	.2964	3.3740	12.4480
	88.00	23.850	2.4721	0.4045	.4403	2.2711	18.4930
24½	100.50	23.750	2.4462	0.4088	.4357	2.2952	18.2988
	113.00	23.650	2.4203	0.4132	.4311	2.3198	18.1054
	98.93*	29.376	4.0764	0.2453	.7260	1.3774	30.4933
	118.65*	29.250	4.0361	0.2478	.7189	1.3911	30.1919
	157.53*	29.000	3.9566	0.2527	.7047	1.4190	29.5978
	196.08*	28.750	3.8779	0.2579	.6907	1.4478	29.0087
30	234.29*	28.500	3.7998	0.2632	.6768	1.4776	28.4248
	309.72*	28.000	3.6458	0.2743	.6493	1.5400	27.2722
	346.93*	27.750	3.5697	0.2801	.6358	1.5728	26.7035
	383.81*	27.500	3.4944	0.2862	.6224	1.6067	26.1400
	456.57*	27.000	3.3458	0.2989	.5959	1.6781	25.0282

VOLUME AND FILL BETWEEN CASING AND CASING

11-¾" OD
INSIDE CASING

OUTSIDE CASING

SIZE OD INCHES	WEIGHT PER LIN FT	ID	CU FT PER LIN FT	LIN FT PER CU FT	BARRELS PER LIN FT	LIN FT PER BARREL	GALLONS PER LIN FT
14	50.00	13.344	0.2182	4.5837	.0389	25.7355	1.6320
16	55.00	15.375	0.5363	1.8646	.0955	10.4692	4.0118
	65.00	15.250	0.5154	1.9402	.0918	10.8933	3.8556
	70.00	15.198	0.5068	1.9732	.0903	11.0789	3.7910
	75.00	15.125	0.4947	2.0214	.0881	11.3493	3.7007
	84.00	15.010	0.4758	2.1017	.0847	11.8001	3.5593
	109.00	14.688	0.4237	2.3604	.0755	13.2529	3.1691
	118.00	14.570	0.4048	2.4702	.0721	13.8693	3.0283
18	80.00	17.180	0.8568	1.1671	.1526	6.5530	6.4093
18⅝	78.00	17.855	0.9858	1.0144	.1756	5.6956	7.3741
	87.50	17.755	0.9664	1.0348	.1721	5.8101	7.2288
	96.50	17.655	0.9470	1.0559	.1687	5.9286	7.0844
20	90.00	19.190	1.2555	0.7965	.2236	4.4719	9.3919
	94.00	19.124	1.2417	0.8053	.2212	4.5216	9.2887
	106.50	19.000	1.2159	0.8224	.2166	4.6175	9.0958
	133.00	18.730	1.1604	0.8618	.2067	4.8386	8.6802
	169.00	18.376	1.0887	0.9185	.1939	5.1570	8.1443
21½	92.50	20.710	1.5863	0.6304	.2825	3.5394	11.8663
	103.00	20.610	1.5638	0.6395	.2785	3.5904	11.6977
	114.00	20.510	1.5413	0.6488	.2745	3.6427	11.5300
24½	88.00	23.850	2.3494	0.4256	.4185	2.3898	17.5750
	100.50	23.750	2.3235	0.4304	.4138	2.4165	17.3808
	113.00	23.650	2.2976	0.4352	.4092	2.4437	17.1874
30	98.93*	29.376	3.9536	0.2529	.7042	1.4201	29.5753
	118.65*	29.250	3.9134	0.2555	.6970	1.4347	29.2739
	157.53*	29.000	3.8339	0.2608	.6829	1.4644	28.6798
	196.08*	28.750	3.7552	0.2663	.6688	1.4952	28.0907
	234.29*	28.500	3.6771	0.2720	.6549	1.5269	27.5068
	309.72*	28.000	3.5230	0.2838	.6275	1.5937	26.3542
	346.93*	27.750	3.4470	0.2901	.6139	1.6288	25.7855
	383.81*	27.500	3.3717	0.2966	.6005	1.6652	25.2220
	456.57*	27.000	3.2231	0.3103	.5741	1.7420	24.1102

VOLUME AND FILL BETWEEN CASING AND CASING

12-¾" OD INSIDE CASING

OUTSIDE CASING

SIZE OD INCHES	WEIGHT PER LIN FT	ID	CU FT PER LIN FT	LIN FT PER CU FT	BARRELS PER LIN FT	LIN FT PER BARREL	GALLONS PER LIN FT
	55.00	15.375	0.4027	2.4834	.0717	13.9434	3.0122
	65.00	15.250	0.3818	2.6192	.0680	14.7059	2.8560
	70.00	15.198	0.3732	2.6798	.0665	15.0462	2.7914
16	75.00	15.125	0.3611	2.7695	.0643	15.5493	2.7011
	84.00	15.010	0.3422	2.9224	.0609	16.4082	2.5597
	109.00	14.688	0.2900	3.4480	.0517	19.3590	2.1695
	118.00	14.570	0.2712	3.6874	.0483	20.7032	2.0287
18	80.00	17.180	0.7232	1.3828	.1288	7.7639	5.4097
	78.00	17.855	0.8521	1.1735	.1518	6.5887	6.3745
18⅝	87.50	17.755	0.8327	1.2009	.1483	6.7424	6.2292
	96.50	17.655	0.8134	1.2294	.1449	6.9025	6.0848
	90.00	19.190	1.1219	0.8914	.1998	5.0046	8.3923
	94.00	19.124	1.1081	0.9025	.1974	5.0669	8.2891
20	106.50	19.000	1.0823	0.9240	.1928	5.1876	8.0962
	133.00	18.730	1.0267	0.9740	.1829	5.4683	7.6806
	169.00	18.376	0.9551	1.0470	.1701	5.8785	7.1447
	92.50	20.710	1.4527	0.6884	.2587	3.8650	10.8667
21½	103.00	20.610	1.4301	0.6992	.2547	3.9259	10.6981
	114.00	20.510	1.4077	0.7104	.2507	3.9885	10.5304
	88.00	23.850	2.2158	0.4513	.3947	2.5339	16.5754
24½	100.50	23.750	2.1898	0.4567	.3900	2.5639	16.3812
	113.00	23.650	2.1640	0.4621	.3854	2.5946	16.1878
	98.93*	29.376	3.8200	0.2618	.6804	1.4698	28.5757
	118.65*	29.250	3.7797	0.2646	.6732	1.4854	28.2743
	157.53*	29.000	3.7003	0.2702	.6591	1.5173	27.6802
	196.08*	28.750	3.6216	0.2761	.6450	1.5503	27.0911
30	234.29*	28.500	3.5435	0.2822	.6311	1.5845	26.5072
	309.72*	28.000	3.3894	0.2950	.6037	1.6565	25.3546
	346.93*	27.750	3.3134	0.3018	.5901	1.6945	24.7859
	383.81*	27.500	3.2381	0.3088	.5767	1.7339	24.2224
	456.57*	27.000	3.0894	0.3237	.5503	1.8173	23.1106

13-⅜″ OD
INSIDE CASING

VOLUME AND FILL BETWEEN CASING AND CASING

| OUTSIDE CASING | | | | | | | |
SIZE OD INCHES	WEIGHT PER LIN FT	ID	CU FT PER LIN FT	LIN FT PER CU FT	BARRELS PER LIN FT	LIN FT PER BARREL	GALLONS PER LIN FT
16	55.00	15.375	0.3136	3.1886	.0559	17.9029	2.3460
	65.00	15.250	0.2927	3.4161	.0521	19.1798	2.1898
	70.00	15.198	0.2841	3.5199	.0506	19.7628	2.1252
	75.00	15.125	0.2720	3.6761	.0484	20.6399	2.0349
	84.00	15.010	0.2531	3.9506	.0451	22.1811	1.8935
	109.00	14.688	0.2010	4.9759	.0358	27.9377	1.5033
	118.00	14.570	0.1821	5.4904	.0324	30.8261	1.3625
18	80.00	17.180	0.6341	1.5770	.1129	8.8543	4.7435
18⅝	78.00	17.855	0.7631	1.3105	.1359	7.3577	5.7083
	87.50	17.755	0.7437	1.3447	.1325	7.5498	5.5630
	96.50	17.655	0.7244	1.3805	.1290	7.7511	5.4186
20	90.00	19.190	1.0328	0.9682	.1840	5.4361	7.7261
	94.00	19.124	1.0190	0.9813	.1815	5.5097	7.6229
	106.50	19.000	0.9933	1.0068	.1769	5.6527	7.4300
	133.00	18.730	0.9377	1.0664	.1670	5.9877	7.0144
	169.00	18.376	0.8660	1.1547	.1542	6.4830	6.4785
21½	92.50	20.710	1.3636	0.7333	.2429	4.1174	10.2005
	103.00	20.610	1.3411	0.7457	.2389	4.1866	10.0319
	114.00	20.510	1.3186	0.7584	.2349	4.2578	9.8642
24½	88.00	23.850	2.1267	0.4702	.3788	2.6400	15.9092
	100.50	23.750	2.1008	0.4760	.3742	2.6726	15.7150
	113.00	23.650	2.0749	0.4819	.3696	2.7059	15.5216
30	98.93*	29.376	3.7310	0.2680	.6645	1.5049	27.9095
	118.65*	29.250	3.6907	0.2710	.6573	1.5213	27.6081
	157.53*	29.000	3.6112	0.2769	.6432	1.5547	27.0140
	196.08*	28.750	3.5325	0.2831	.6292	1.5894	26.4250
	234.29*	28.500	3.4544	0.2895	.6153	1.6253	25.8410
	309.72*	28.000	3.3004	0.3030	.5878	1.7012	24.6884
	346.93*	27.750	3.2243	0.3101	.5743	1.7413	24.1198
	383.81*	27.500	3.1490	0.3176	.5609	1.7830	23.5562
	456.57*	27.000	3.0004	0.3333	.5344	1.8713	22.4444

16" OD INSIDE CASING							
OUTSIDE CASING			**VOLUME AND FILL BETWEEN CASING AND CASING**				
SIZE OD INCHES	WEIGHT PER LIN FT	ID	CU FT PER LIN FT	LIN FT PER CU FT	BARRELS PER LIN FT	LIN FT PER BARREL	GALLONS PER LIN FT
	90.00	19.190	0.6123	1.6333	.1090	9.1702	4.5800
	94.00	19.124	0.5985	1.6709	.1066	9.3816	4.4769
20	106.50	19.000	0.5727	1.7462	.1020	9.8039	4.2840
	133.00	18.730	0.5171	1.9338	.0921	10.8573	3.8684
	169.00	18.376	0.4455	2.2448	.0793	12.6034	3.3324
	92.50	20.710	0.9430	1.0604	.1680	5.9537	7.0545
21½	103.00	20.610	0.9205	1.0864	.1639	6.0994	6.8859
	114.00	20.510	0.8981	1.1135	.1600	6.2518	6.7181
	88.00	23.850	1.7062	0.5861	.3039	3.2907	12.7631
24½	100.50	23.750	1.6802	0.5952	.2993	3.3416	12.5689
	113.00	23.650	1.6544	0.6045	.2947	3.3938	12.3755
	98.93*	29.376	3.3104	0.3021	.5896	1.6960	24.7635
	118.65*	29.250	3.2701	0.3058	.5824	1.7169	24.4621
	157.53*	29.000	3.1907	0.3134	.5683	1.7597	23.8679
	196.08*	28.750	3.1119	0.3213	.5543	1.8042	23.2789
30	234.29*	28.500	3.0339	0.3296	.5404	1.8506	22.6949
	309.72*	28.000	2.8798	0.3472	.5129	1.9496	21.5423
	346.93*	27.750	2.8038	0.3567	.4994	2.0025	20.9737
	383.81*	27.500	2.7284	0.3665	.4860	2.0578	20.4102
	456.57*	27.000	2.5798	0.3876	.4595	2.1764	19.2984

VOLUME AND FILL BETWEEN CASING AND CASING

20" OD
INSIDE CASING

OUTSIDE CASING

SIZE OD INCHES	WEIGHT PER LIN FT	ID	CU FT PER LIN FT	LIN FT PER CU FT	BARRELS PER LIN FT	LIN FT PER BARREL	GALLONS PER LIN FT
24½	88.00	23.850	0.9208	1.0860	.1640	6.0976	6.8879
	100.50	23.750	0.8948	1.1175	.1594	6.2745	6.6937
	113.00	23.650	0.8690	1.1508	.1548	6.4612	6.5003
30	98.93*	29.376	2.5250	0.3960	.4497	2.2236	18.8883
	118.65*	29.250	2.4847	0.4025	.4425	2.2597	18.5869
	157.53*	29.000	2.4053	0.4158	.4284	2.3343	17.9928
	196.08*	28.750	2.3265	0.4298	.4144	2.4133	17.4037
	234.29*	28.500	2.2485	0.4447	.4005	2.4971	16.8198
	309.72*	28.000	2.0944	0.4775	.3730	2.6808	15.6672
	346.93*	27.750	2.0184	0.4954	.3595	2.7817	15.0985
	383.81*	27.500	1.9430	0.5147	.3461	2.8896	14.5350
	456.57*	27.000	1.7944	0.5573	.3196	3.1289	13.4232

Pump Output Tables

Notes: 1. Volume shown are for one complete cycle or revolution.
2. To get output in volume/minute, multiply output/cycle by pump rpm.

DOUBLE ACTING DUPLEX PUMP

Note: For triplex double acting pump multiply output by 1.5.

Stroke, in.	Bore, in.	Rod D, in.	100% Efficiency		90% Efficiency	
			cu ft	bbl	cu ft	bbl
6.	4.00	1.5	0.1623	0.0289	0.1460	0.0260
8.	4.00	1.5	0.2163	0.0385	0.1947	0.0347
8.	4.50	1.5	0.2782	0.0495	0.2503	0.0446
8.	5.00	1.5	0.3472	0.0618	0.3125	0.0557
10.	4.00	1.5	0.2704	0.0482	0.2434	0.0433
10.	4.50	1.5	0.3477	0.0619	0.3129	0.0557
10.	5.00	2.0	0.4182	0.0745	0.3763	0.0670
12.	4.00	1.5	0.3245	0.0578	0.2921	0.0520
12.	4.50	1.5	0.4172	0.0743	0.3755	0.0669
12.	5.00	2.0	0.5018	0.0894	0.4516	0.0804
12.	5.50	2.0	0.6163	0.1098	0.5547	0.0988
14.	4.50	1.5	0.4868	0.0867	0.4381	0.0780
14.	5.00	2.0	0.5854	0.1043	0.5269	0.0938
14.	5.50	2.0	0.7190	0.1281	0.6471	0.1153
14.	6.00	2.0	0.8654	0.1541	0.7789	0.1387
14.	6.25	2.0	0.9433	0.1680	0.8490	0.1512
14.	6.50	2.0	1.0245	0.1825	0.9220	0.1642
14.	6.75	2.0	1.1088	0.1975	0.9979	0.1777
14.	7.00	2.0	1.1963	0.2131	1.0766	0.1918
14.	7.25	2.5	1.2583	0.2241	1.1325	0.2017
14.	7.50	2.5	1.3522	0.2408	1.2170	0.2167
14.	7.75	2.5	1.4492	0.2581	1.3043	0.2323
16.	5.00	2.5	0.6363	0.1133	0.5727	0.1020
16.	5.50	2.5	0.7890	0.1405	0.7101	0.1265
16.	6.00	2.5	0.9563	0.1703	0.8607	0.1533
16.	6.25	2.5	1.0454	0.1862	0.9408	0.1676
16.	6.50	2.5	1.1381	0.2027	1.0243	0.1824
16.	6.75	2.5	1.2345	0.2199	1.1110	0.1979
16.	7.00	2.5	1.3344	0.2377	1.2010	0.2139
16.	7.25	2.5	1.4381	0.2561	1.2943	0.2305
16.	7.50	2.5	1.5453	0.2752	1.3908	0.2477
16.	7.75	2.5	1.6562	0.2950	1.4906	0.2655
18.	5.00	2.5	0.7159	0.1275	0.6443	0.1147
18.	5.50	2.5	0.8877	0.1581	0.7989	0.1423
18.	6.00	2.5	1.0758	0.1916	0.9682	0.1725
18.	6.25	2.5	1.1761	0.2095	1.0584	0.1885
18.	6.50	2.5	1.2804	0.2280	1.1523	0.2052
18.	6.75	2.5	1.3888	0.2473	1.2499	0.2226
18.	7.00	2.5	1.5013	0.2674	1.3511	0.2406
18.	7.25	2.5	1.6178	0.2881	1.4561	0.2593
18.	7.50	2.5	1.7385	0.3096	1.5647	0.2787
18.	7.75	2.5	1.8633	0.3319	1.6769	0.2987

DOUBLE ACTING DUPLEX PUMP (Continued)

Stroke, in.	Bore, in.	Rod D, in.	100% Efficiency		90% Efficiency	
			cu ft	bbl	cu ft	bbl
20.	6.50	2.5	1.4226	0.2534	1.2804	0.2280
20.	6.75	2.5	1.5431	0.2748	1.3888	0.2473
20.	7.00	2.5	1.6681	0.2971	1.5013	0.2674
20.	7.25	2.5	1.7976	0.3202	1.6178	0.2881
20.	7.50	2.5	1.9317	0.3440	1.7385	0.3096
20.	7.75	2.5	2.0703	0.3687	1.8633	0.3319
20.	8.00	2.5	2.2135	0.3942	1.9921	0.3548

SINGLE ACTING TRIPLEX PUMP

Note: For single acting quintuplex pump, multiply output by 1.67.

Stroke, in.	Bore, in.	100% Efficiency		90% Efficiency	
		cu ft	bbl	cu ft	bbl
4.	3.00	0.0491	0.0087	0.0442	0.0078
4.	3.75	0.0576	0.0103	0.0518	0.0093
4.	3.50	0.0668	0.0119	0.0601	0.0107
4.	3.75	0.0767	0.0137	0.0690	0.0123
4.	4.00	0.0873	0.0155	0.0786	0.0140
4.	4.50	0.1104	0.0197	0.0994	0.0177
4.	5.00	0.1364	0.0243	0.1228	0.0219
4.	6.00	0.1963	0.0350	0.1767	0.0315
4.	8.00	0.3491	0.0622	0.3142	0.0560
6.	3.00	0.0737	0.0131	0.0663	0.0117
6.	3.25	0.0864	0.0155	0.0777	0.0140
6.	3.50	0.1002	0.0179	0.0902	0.0161
6.	3.75	0.1151	0.0206	0.1035	0.0185
6.	4.00	0.1310	0.0233	0.1179	0.0210
6.	4.50	0.1656	0.0296	0.1491	0.0266
6.	5.00	0.2046	0.0365	0.1842	0.0329
6.	6.00	0.2945	0.0525	0.2651	0.0473
6.	8.00	0.5237	0.0933	0.4713	0.0840
8.	3.00	0.0982	0.0174	0.0884	0.0156
8.	3.25	0.1152	0.0206	0.1036	0.0186
8.	3.50	0.1336	0.0238	0.1202	0.0214
8.	3.75	0.1534	0.0274	0.1380	0.0246
8.	4.00	0.1746	0.0310	0.1572	0.0280
8.	4.50	0.2208	0.0394	0.1988	0.0354
8.	5.00	0.2728	0.0486	0.2456	0.0438
8.	6.00	0.3926	0.0700	0.3534	0.0630
8.	8.00	0.6982	0.1244	0.6284	0.1120
10.	3.00	0.1228	0.0218	0.1105	0.0195
10.	3.25	0.1440	0.0258	0.1295	0.0233
10.	3.50	0.1670	0.0298	0.1503	0.0268
10.	3.75	0.1918	0.0343	0.1725	0.0308
10.	4.00	0.2183	0.0388	0.1965	0.0350
10.	4.50	0.2760	0.0493	0.2485	0.0443
10.	5.00	0.3410	0.0608	0.3070	0.0548
10.	6.00	0.4908	0.0875	0.4418	0.0788
10.	8.00	0.8728	0.1555	0.7855	0.1400

Dimensional Data and Minimum

OD (in.) (mm)	Weight w/Cplg (lb/ft)	Wall Thickness (in.) (mm)	ID (in.) (mm)	Drift Dia. (in.) (mm)	Round or Buttress (in.) (mm)	Other (in.) (mm)	Bored Pin ID (in.) (mm)	Grade	Collapse Resistance (psi) **
								H-40	2,770
	9.50	.205	4.090	3.965	5.000	5.000(7)	3.990(8)	J-55	3,310
		5,21	103,9	100,7	127,0	127,0	101,3	K-55	3,310
	10.50	.224	4.052	3.927	5.000	5.000(7)	3.952(8)	J-55	4,010
		5,69	102,9	99,75	127,0	127,0	100,4	K-55	4,010
						4.719(1)		J-55	4,960
						119,9	3.920(1)	K-55	4,960
	11.60	.250	4.000	3.875	5.000	5.000(7)	99,57	C-75	6,130
		6,35	101,6	98,43	127,0	127,0	3.925(8)	N-80	6,350
						4.862(9)	99,69	C-95	7,010
						123,5		P-110	7,560
						4.719(1)	3.878(1)	K-55	5,720
4-½	12.60	.271	3.958	3.833		119,9	98,50	C-75	7,200
114,3		6,88	100,5	97,36		5.000(7)	3.883(8)	N-80	7,500
						127,0	98,63	C-95	8,400
						4.719(1)			
						119,9		K-55	6,420
						4.875(5)	3.840(1)	C-75	8,170
	13.50	.290	3.920	3.795	5.000	123,8	97,54	N-80	8,540
	13.60	7,37	99,60	96,39	127,0	5.000(7)	3.845(8)	C-95	9,650
						127,0	97,66	P-110	10,670
						4.961(9)		V-150	12,880
						126,0			
						4.750(1)			
						120,7			
						4.594(3)		K-55	7,620
						116,7	3.746(1)	C-75	10,390
	15.10	.337	3.826	3.701	5.000	4.875(5)	95,15(3)	N-80	11,080
		8,56	97,20	94,01	127,0	123,8	3.751(8)	C-95	
						5.250(7)	95,27	P-110	14,320
						133,3		V-150	18,110
						4.961(9)			
						126,0			

Performance Properties of Casing

Internal Yield Pressure (psi) **				Body Yield Strength (1000 lb) **	Joint Yield Strength (1000 lb) ** — Threaded and Coupled Joint			
Plain End or Extreme Line	Round Thread Short	Round Thread Long	Buttress Thread		Round Thread Short	Round Thread Long	Other*	Other*
3,190	3,190			111	77			
4,380	4,380			152	101			118(8)
4,380	4,380			152	112			118(8)
4,790	4,790			165	132			129(8)
4,790	4,790			165	146			129(8)
5,350	5,350	5,350	5,350	184	154	162		
5,350	5,350	5,350	5,350	184	170	180	174(8)	
7,290		7,290	7,290	250		212	174	210(1)
7,780		7,780	7,780	267		223	(8)	221(1)
9,240		9,240	9,240	317		234	184(8)	
10,690		10,690	10,690	367		279	193(8)	276(1)
							229(8)	
5,790				199			148(8)	
7,900				271			188(8)	235(1)
8,440				289			198(8)	248(1)
10,010				343			208(8)	
6,200				211			200(8)	
8,460		8,460	8,460	288		257	200(8)	258(1)
9,020		9,020	9,020	307		270	211(8)	271(1)
10,710		10,710	10,710	364		284	222(8)	
12,410		12,410	12,410	422		338	264(8)	339(1)
16,920				576				434(5)
7,210				242			230(8)	222(2)
9,830				331			230(8)	312(1)
10,480				353			242(8)	328(1)
				419			255(8)	
14,420		14,420	13,460	485		406	303(8)	411(1)
19,660				661			525(5)	400(3)

OD (in.) (mm)	Weight w/Cplg (lb/ft)	Wall Thickness (in.) (mm)	ID (in.) (mm)	Drift Dia. (in.) (mm)	Coupling or Joint OD		Bored Pin ID (in.) (mm)	Grade	Collapse Resistance (psi) **
					Round or Buttress (in.) (mm)	Other (in.) (mm)			
	16.6 16.8 17.1	.373 9,47	3.754 95,35	3.629 92,17		4.750(1) 120,6 4.00(2)(8) 114,3 4.625(3) 117,5 4.875(5) 123,8 5.250(7) 133,3	3.674(1)(3) 93,32(5) 3.679(8) 93,45	K-55 C-75 N-80 C-95 P-110 V-150	8,360 11,400 12,160 14,440 16,720 22,110
	16.9	.380 9,65	3.740 95,00	3.615 91,83		5.106(9) 129,7		C-95 P-110 V-150	14,690 17,010 22,890
	17.7	.402 10,20	3.696 93,90	3.571 90,72		5.106(9) 129,7		C-95 P-110 V-150	15,450 17,890 24,390
4-½ 114,3	18.8 20.0	.430 10,92	3.640 92,46	3.515 89,29		4.750(1) 120,6 4.500(2)(8) 114,3 4.625(3) 117,5 4.875(5) 123,8 5.250(7) 133,3 5.106(9) 129,7	3.560(1)(3) 90,42(5) 3.565(8) 90,55	K-55 C-75 N-80 C-95 P-110 V-150	9,510 12,960 13,830 16,420 19,010 25,930
	21.6	.500 12,7	3.500 88,9	3.375 85,72		5.000(6) 127,0 5.375(7) 136,5		K-55 C-75 N-80 P-110 V-150	10,860 14,810 15,800 21,730 29,630
	24.6	.560 14,22	3.380 85,85	3.255 82,68		5.000(6) 127,0		N-80 P-110 V-150	17,430 23,970 32,690

Plain End or Extreme Line	Internal Yield Pressure (psi) **			Body Yield Strength (1000 lb) **	Joint Yield Strength (1000 lb) **		
	Round Thread		Buttress Thread		Threaded and Coupled Joint		Other*
					Round Thread		
	Short	Long			Short	Long	
7,980				266			241(2)
10,880				363		299(8)	318(1)
11,600				387		314(8)	335(1)
13,780				459		330(8)	
15,960				532		393(8)	418(1)
21,760				725			594(5)
14,040				467			
16,260				541			
22,170				738			
14,810				492			
17,140				569			
23,380				776			
9,200				302		340(8)	291(2)
12,540				412			318(1)
13,380				440		357(8)	359(3)
15,890				522		375(8)	
18,390				605		447(8)	547(5)
25,080				825			700(5)
10,690				345			
14,580				471			
15,560				503			
21,390				691			
29,170				942			
17,420				554			
23,960				762			
32,670				1,040			

OD (in.) (mm)	Weight w/Cplg (lb/ft)	Wall Thickness (in.) (mm)	ID (in.) (mm)	Drift Dia. (in.) (mm)	Coupling or Joint OD		Bored Pin ID (in.) (mm)	Grade	Collapse Resistance (psi) **
					Round or Buttress (in.) (mm)	Other (in.) (mm)			
	26.5	.630 16,0	3.240 82,3	3.115 79,12		5.000(6) 127,0		N-80 P-110 V-150	19,260 26,490 36,120
	11.5	.220 5,59	4.560 115,8	4.435 112,6	5.563 141,3	5.500(7) 139,7	4.460(8) 113,3	J-55 K-55	3,060 3,060
5 127,0	13.0	.253 6,43	4.494 114,1	4.283 108,8	5.563 141,3	5.219(1) 132,6 5.500(7) 139,7	4.414(1) 112,1 4.419(8) 112,2	J-55 K-55 C-75 M-80	4,140 4,140 4,990 5,140
	15.0	.296 7,52	4.408 112,0	4.283 108,8	5.563 141,3	5.219(1) 132,6 5.500(7) 139,7 5.370(4) 136,4 5.375(5) 136,5 5.563(9) 141,3 5.360(10) 136,1	4.328(1) 109,9(4)(5) 4.333(8) 110,1	J-55 K-55 C-75 N-80 C-95 P-110 V-150	5,550 5,550 6,970 7,250 8,090 8,830 10,260
5 127,0	18.0	.36 9,19	4.276 108,6	4.151 105,4	5.563 141,3	5.250(1) 133,3 5.094(3) 129,4 5.420(4) 137,7 5.375(5) 136,5 5.875(7) 149,2 5.563(9) 141,3 5.360(10) 136,1	4.196(1)(3) 106,6(4)(5) 4.201(8) 106,7	K-55 C-75 N-80 C-95 P-110 V-150	7,390 10,000 10,490 12,010 13,450 16,860
	20.3	.408 10,36	4.184 106,3	4.059 103,1		5.250(1) 133,3 5.094(3) 129,4 5.420(4) 137,7 5.375(5) 136,5	4.104(1)(3) 104,2(4)(5)	K-55 C-75 N-80 P-110 V-150	8,240 11,240 11,990 16,490 21,470

	Internal Yield Pressure (psi) **				Joint Yield Strength (1000 lb) **			
					Threaded and Coupled Joint			
Plain End or Extreme Line	Round Thread		Buttress Thread	Body Yield Strength (1000 lb) **	Round Thread		Other*	
	Short	Long			Short	Long		
19,600				613				
26,950				842				
36,750				1,149				
4,240	4,240			182	133			
4,420	4,240			182	147			141(8)
4,870	4,870	4,870	4,870	208	169	182		
4,870	4,870	4,870	4,870	208	186	201		197(8)
6,640				283				239(1)
7,090				302				
5,700	5,700	5,700	5,700	241	207	223		
5,700	5,700	5,700	5,700	241	228	246		
7,700		7,700	7,700	328		295		
8,290		8,290	8,290	350		311	228(8)	297(1)
9,840		9,840	9,840	416		326	240(8)	312(1)
11,400				481			252(8)	390(1)
15,540				656			300(8)	500(5)
6,970				290				269(2)
9,500		9,500	9,290	396		376	326(8)	369(1)
10,140		10,140	9,910	422		396	343(8)	303(3)
12,040		12,040	11,770	501		416	360(8)	
13,940		13,940	13,620	580		495	428(8)	354(2)
19,000				791				644(5)
7850				324				308(2)
10,710				441			369(1)	
11,420				471			388(1)	325(2)
15,710				647			579(5)	406(2)
21,420				883			741(5	583(3)

OD (in.) (mm)	Weight w/Cplg (lb/ft)	Wall Thickness (in.) (mm)	ID (in.) (mm)	Drift Dia. (in.) (mm)	Coupling or Joint OD Round or Buttress (in.) (mm)	Other (in.) (mm)	Bored Pin ID (in.) (mm)	Grade	Collapse Resistance (psi) **
5 127,0	20.8	.422 10,72	4.156 105,6	4.031 102,4		5.094(3) 129,4 5.563(9) 141,3	4.076(3) 103,5	K-55 C-75 N-80 P-110 V-150	8,500 11,590 12,360 17,000 22,870
	23.2 23.6	.478 12,14	4.044 102,7	3.919 99,54		5.250(1) 133,3 5.094(3) 129,4 5.420(4) 137,7 5.500(5) 139,7 5.875(7) 149,2	3.964(1)(3) 100,7(4)(5)	K-55 C-75 N-80 C-95 P-110 V-150	9,510 12,970 13,830 16,430 19,020 25,940
	24.2	.500 12,7	4.000 101,6	3.875 98,42		5.125(3) 130,2	3.920(3) 99,57	K-55 N-80 P-110 V-150	9,900 14,400 19,800 27,000
5-½ 139,7	14.0	.244 6,20	5.012 127,7	4.887 124,1	6.050 153,7	5.750(1) 146,1 6.050(7) 153,7	4.932(1) 125,3 4.912(8) 124,8	H-40 J-55 K-55 C-75	2,630 3,120 3,120 3,560
	15.5	.275 6.98	4.950 125,7	4.825 122,6 4.653 (10) 118,2	153,7	5.750(1) 146,1 6.050(7) 153,7(9) 5.860(10) 148,8	4.870(1) 123,7 4.875(8) 123,8	J-55 K-55 C-75 N-80 P-110	4,040 4,040 4,860 4,990 5,620
	17.0	.304 7,72	4.892 124,3	4.767 121,1 4.653 (10) 118,2	6.050 153,7	5.781 146,8 5.900(4) 149,9 6.050(7)(9) 153,7 5.860(10) 148,8	4.812(1)(4) 122,2 4.817(8) 122,4	J-55 K-55 C-75 N-80 C-95 P-110 B-150	4,910 4,910 6,070 6,280 6,930 7,460 8,300

Plain End or Extreme Line	Round Thread — Short	Round Thread — Long	Buttress Thread	Body Yield Strength (1000 lb) **	Round Thread — Short	Round Thread — Long	Other*	
Internal Yield Pressure (psi) **					*Joint Yield Strength (1000 lb) ** — Threaded and Coupled Joint*			
8,120				334				308(2)
11,080				455				
11,820				486			383(3)	325(2)
16,250				668			479(3)	406(2)
22,160				910			612(3)	
9,200				373				352(2)
12,550				509			419(8)	369(1)
13,380				543			441(8)	442(3)
15,890				645			463(8)	
18,400				747			692(5)	463(2)
25,100				1,019			886(5)	707(3)
9,630								352(2)
14,000							470(3)	370(2)
19,250							587(3)	463(2)
26,250							751(3)	
3,110	3,110			161	130			
4,270	4,270			222	172			
4,270	4,270			222	189		172(8)	
5,820				302				252(1)
4,810	4,810	4,810	4,810	248	202	217		
4,810	4,810	4,810	4,810	248	222	239	236(8)	
6,560				339				298(1)
7,000				362				
9,620				497				
5,320	5.320	5,320	5,320	273	229	247		
5,320	5.320	5,320	5,320	273	252	272		
7,250		7,250	7,250	372		327	259(8)	340(1)
7,740		7,740	7,740	397		348	273(8)	358(1)
9,190		9,190	9,190	471		374	287(8)	
10,640		10,640	10,640	546		445	341(8)	448(1)
14,510				744				

OD (in.) (mm)	Weight w/Cplg (lb/ft)	Wall Thickness (in.) (mm)	ID (in.) (mm)	Drift Dia. (in.) (mm)	Coupling or Joint OD Round or Buttress (in.) (mm)	Other (in.) (mm)	Bored Pin ID (in.) (mm)	Grade	Collapse Resistance (psi) **
5-½ 139,7	20.0	.361 9,17	4.778 121,4	4.653 118,2	6.050 153,7	5.781(1) 146,8 5.625(3) 142,9 6.000(4) 152,4 5.938(5) 150,8 6.375(7) 161,9 6.050(9) 153,7 5.860(10) 148,8	4.698(1)(3) 119,3(4)(5) 4.703(8) 119,5	K-55 C-75 N-80 C-95 P-110 V-150	6,610 8,440 8,830 10,000 11,080 13,480
	23.0	.415 10,54	4.670 118,6	4.545 115,4	6.050 153,7	5.781(1) 146,8 5.625(3) 142,9 6.000(4) 152,4 5.938(5) 150,8 6.375(7) 161,9 6.050(9) 153,7 5.860(10) 148,8	4.590(1)(3) 116,6(4)(5) 4.595(8) 116,7	K-55 C-75 N-80 C-95 P-110 V-150	7,670 10,400 11,160 12,920 14,520 18,390
	26.0	.476 12,01	4.548 115,5	4.423 112,3		5.781(1) 146,8 5.625(3) 142,9 6.068(4) 154,1 6.000(5) 152,4 6.375(7) 161,9	4.468(1)(3) 113,5(4)(5) 4.473(8) 113,6	K-55 C-75 N-80 C-95 P-110 V-150	8,700 11,860 12,650 15,020 17,390 23,720
	28.4	.530 13,46	4.440 112,8	4.315 109,6		6.375(7) 161,9	4.365(8) 110,9	C-75 N-80 P-110	

	Internal Yield Pressure (psi) **			Body Yield Strength (1000 lb) **	Joint Yield Strength (1000 lb) **			
	Round Thread				Threaded and Coupled Joint			
Plain End or Extreme Line			Buttress Thread		Round Thread		Other*	
	Short	Long			Short	Long		
6,310				321				299(2)
8,610		8,610	8,430	437		403	360(8)	423(1)
9,190		9,190	8,990	466		428	379(8)	315(2)
10,910		10,910	10,680	554		460	398(8)	
12,640		12,640	12,360	641		548	474(8)	418(3)
17,220				874			712(5)	536(3)
7,270				365				299(2)
9,900		9,260	8,430	497		473	409(8)	432(1)
10,560		9,880	8,990	530		502	431(8)	315(2)
12,540		11,730	10,680	630		540	452(8)	
14,520		13,580	12,360	729		643	539(8)	519(3)
19,810				994			840(5)	664(3)
8,330				413			368(2)	
11,360				563	464(8)			432(1)
12,120				601	488(8)	503(3)	387(2)	455(1)
14,390				714	513(8)			
16,660				826	610(8)	629(3)	484(2)	569(1)
22,720				1,127		805(3)	981(5)	
				621				
				662				
				910				

OD (in.) (mm)	Weight w/Cplg (lb/ft)	Wall Thickness (in.) (mm)	ID (in.) (mm)	Drift Dia. (in.) (mm)	Coupling or Joint OD Round or Buttress (in.) (mm)	Other (in.) (mm)	Bored Pin ID (in.) (mm)	Grade	Collapse Resistance (psi) **
5-½	32.3	.612 15,54	4.276 108,6	4.151 105,4		6.187(6) 157,1		N-80 P-110 V-150	15,820 21,760 29,670
139,7	36.4	.705 17,91	4.090 103,9	3.965 100,7		6.187(6) 157,1		N-80 P-110 V-150	17,880 24,590 33,530
	18.0	.288 7,32	5.424 137,8	5.299 134,6	6.625 168,3			H-40 J-55 N-80	2,780 3,620 4,740
6	20.0	.324 8,23	5.352 135,9	5.227 132,8	6.625 168,3			N-80	5,690
152,4	23.0	.380 9,65	5.240 133,1	5.115 129,9	6.625 168,3			N-80 P-110	7,180 10,380
	26.0	.434 11,02	5.132 130,4	5.007 127,2	6.625 168,3			P-110	12,380
	20.0	.288 7,32	6.049 153,7	5.924 150,5	7.390 187,7	6.938(1) 176,2 7.390(7)(9) 187,7	5.970(1) 151,6 5.974(8) 151,7	H-40 J-55 K-55 N-80 C-95	2,520 2,970 2,970 3,480 3,830
6-⅝ 168.3	24.0	.352 8,94	5.921 150,4	5.796 147,2 5.730 (10) 145,5	7.390 187,7	6.938(1) 176,2 7.072(4) 179,6 7.390(7)(9) 187,7 7.000(10) 177,8	5.840(1)(4) 148,3 5.846(8) 148,5	J-55 K-55 C-75 N-80 C-95 P-110 V-150	4,560 4,560 5,570 5,760 6,290 6,710 7,350
	28.0	.417 10,59	5.791 147,1	5.666 143,9	7.390 187,7	6.969(1) 177,0 6.750(3) 171,5 7.072(4) 179,6 7.390(7)(9) 187,7 7.000(10) 177,8	5.710(1)(3) 145,0(4) 5.716(8) 145,2	K-55 C-75 N-80 C-95 P-110 V-150	6,170 7,830 8,170 9,200 10,140 12,130

Internal Yield Pressure (psi) **				Body Yield Strength (1000 lb) **	Joint Yield Strength (1000 lb) **			
Plain End or Extreme Line	Round Thread		Buttress Thread		Threaded and Coupled Joint			Other*
	Short	Long			Round Thread			
					Short	Long		
15,580				752				
21,420				1,034				952(6)
29,210				1,410				1,219(6)
17,950				850				
24,680				1,168				1,105(6)
33,650				1,593				1,414(6)
	3,360			206	179			
	4,620			283	239	279		
		6,720		412		323		
		7,560				366		
				461				
		8,870		737		432		
		12,190		536		565		
		13,920				646		
				833				
3,040	3,040			229	184			
4,180	4,180	4,180	4,180	315	245	266		
4,180	4,180	4,180	4,180	315	267	290		300(8)
6,090				459				392(1)
7,230				545				
5,110	5,110	5,110	5,110	382	314	340		
5,110	5,110	5,110	5,110	382	342	372		362(8)
6,970		6,970	6,970	520		453		486(1)
7,440		7,440	7,440	555		481		512(1)
8,830		8,830	8,830	659		546		
10,230		10,230	10,230	763		641		640(1)
13,960				1,041				
6,060				447				413(2)
8,260		8,260	8,260	610		552	502(8)	600(1)
8,810		8,810	8,810	651		586	529(8)	435(2)
10,460		10,460	10,460	773		665	555(8)	
12,120		12,120	12,120	895		781	661(8)	591(3)
16,510				1,220				756(3)

OD (in.) (mm)	Weight w/Cplg (lb/ft)	Wall Thick-ness (in.) (mm)	ID (in.) (mm)	Drift Dia. (in.) (mm)	Coupling or Joint OD		Bored Pin ID (in.) (mm)	Grade	Col-lapse Resis-tance (psi) **
					Round or Buttress (in.) (mm)	Other (in.) (mm)			
6-⅝ 168,3	32.0	.475 12,06	5.675 144,2	5.550 141,0	7.390 187,7	6.969(1) 177,0 6.781(3) 172,2 7.152(4) 181,7 7.500(7) 190,5 7.390(9) 187,7 7.000(10) 177,8	5.595(1)(3) 142,1(4) 5.600(8) 142,2	K-55 C-75 N-80 C-95 P-110 V-150	7,320 9,830 10,320 11,800 13,200 16,510
7 177,8	17.0	.231 5,87	6.538 166,1	6.413 162,9	7.656 194,5			H-40	1,450
	20.0	.272 6,91	6.456 164,0	6.331 160,8	7.656 194,5	7.312(1) 185,7 7.656(7) 194,5	6.376(1) 161,9 6.381(8) 162,1	H-40 J-55 K-55 C-75	1,980 2,270 2,270 2,660
	23.0	.317 8,05	6.366 161,7	6.241 158,5 6.151 (10) 156,2	7.656 194,5	7.312(1) 185,7 7.444(4) 189,1 7.656(7)(9) 194,5 7.390(10) 187,7	6.286(1)(4) 159,7 6.291(8) 159,8	J-55 K-55 C-75 N-80 C-95	3,270 3,270 3,770 3,830 4,150
	26.0	.362 9,19	6.276 159,4	6.151 156,2	7.656 194,5	7.312(1) 185,7 7.125(3) 181,0 7.444(4) 189,1 7.656(7)(9) 194,5 7.390(1) 187,7	6.196(1)(3) 157,4(4) 6.201(8) 157,5	J-55 K-55 C-75 N-80 C-95 P-110 V-150	4,320 4,320 5,250 5,410 5,870 6,210 6,890

Internal Yield Pressure (psi) **				Body Yield Strength (1000 lb) **	Joint Yield Strength (1000 lb) **			
Plain End or Extreme Line	Round Thread Short	Round Thread Long	Buttress Thread		Threaded and Coupled Joint — Round Thread Short	Threaded and Coupled Joint — Round Thread Long	Other*	Other*
6,900				504				
9,410		9,410	9,200	688		638	567(8)	439(2)
10,040		10,040	9,820	734		677	596(8)	608(1)
11,920		11,920	11,660	872		769	626(8)	462(2)
13,800		13,800	13,500	1,009		904	746(8)	721(3)
18,820				1,377				923(3)
2,310	2,310			196	122			
2,720	2,720			230	176			
3,740	3,740			316	234			
3,740	3,740			316	234			300(8)
5,100				431				364(1)
4,360	4,360	4,360	4,360	366	284	313		
4,360	4,360	4,360	4,360	366	309	341		
5,940		5,940	5,940	499		416	348(8)	450(1)
6,340		6,340	6,340	532		442	366(8)	473(1)
7,530		7,530	7,530	632		505	384(8)	
4,980	4,980	4,980	4,980	415	334	367		
4,980	4,980	4,980	4,980	415	364	401		
6,790		6,790	6,790	566		489	466(8)	384(2)
7,240		7,240	7,240	604		519	491(8)	535(1)
8,600		8,600	8,600	717		593	515(8)	404(2)
9,960		9,960	9,960	830		693	613(8)	703(1)
13,580				1,132				694(3)

OD (in.) (mm)	Weight w/Cplg (lb/ft)	Wall Thick-ness (in.) (mm)	ID (in.) (mm)	Drift Dia. (in.) (mm)	Coupling or Joint OD		Bored Pin ID (in.) (mm)	Grade	Col-lapse Resis-tance (psi) **
					Round or Buttress (in.) (mm)	Other (in.) (mm)			
7 177,8	29.0	.408 10,36	6.184 157,1	6.059 153,9	7.656 194.5	7.312(1) 185,7 7.125(3) 181,0 7.572(4) 192,3 7.562(5) 192,1 7.875(7) 200,0 7.656(9) 194,5 7.390(10) 187,7	6.104(1)(3) 155,0(4)(5) 6.109(8) 155,2	K-55 C-75 N-80 C-95 P-110 V-150	5,410 6,760 7,020 7,820 8,510 9,800
	32.0	.453 11,51	6.094 154,8	5.969 151,6	7.656 194,5	7.344(1) 186,5 7.156(3) 181,8 7.572(4) 192,3 7.562(5) 192,1 7.875(7) 200,0 7.656(9) 194,5 7.390(10) 187,7	6.014(1)(3) 152,8(4)(5) 6.019(8) 152,9	K-55 C-75 N-80 C-95 P-110 V-150	6,460 8,230 8,600 9,730 10,760 13,020
	35.0	.498 12,65	6.004 152,5	5.879 149,3	7.656 194,5	7.344(1) 186,5 7.187(3) 182,5 7.572(7) 192,3 7.562(5) 192,1 7.875(7) 200,0 7.656(9) 194,5 7.530(10) 191,3	5.924(1)(3) 150,5(4)(5) 5.929(8) 150,6	K-55 C-75 N-80 C-95 P-110 V-150	7,270 9,710 10,180 11,640 13,010 16,230

					Joint Yield Strength (1000 lb) **			
Internal Yield Pressure (psi) **					Threaded and Coupled Joint			
Plain End or Extreme Line	Round Thread		Buttress Thread	Body Yield Strength (1000 lb) **	Round Thread		Other*	
	Short	Long			Short	Long		
5,610				465				423(2)
7,650		7,650	7,650	634		562	522(8)	620(1)
8,160		8,160	8,160	676		597	549(8)	445(2)
9,690		9,690	9,690	803		683	577(8)	
11,220		11,220	11,220	929		797	686(8)	816(1)
15,300				1,267			1,044(5)	808(3)
6,230				512				461(2)
8,490		8,490	7,930	699		633	575(8)	664(1)
9,060		9,060	8,460	745		672	606(8)	485(2)
10,760		10,760	10,050	885		768	636(8)	
12,460		12,460	11,640	1,025		897	757(8)	874(1)
16,980				1,397			1,183(5)	918(3)
6,850				559				511(2)
9,340		8,680	7,930	763		703	628(8)	664(1)
9,960		9,240	8,460	814		746	661(8)	538(2)
11,830		10,970	10,050	966		853	694(8)	905(9)
13,700		12,700	11,640	1,119		996	826(8)	874(1)
18,660				1,526			1,320(5)	932(3)

OD (in.) (mm)	Weight w/Cplg (lb/ft)	Wall Thickness (in.) (mm)	ID (in.) (mm)	Drift Dia. (in.) (mm)	Coupling or Joint OD		Bored Pin ID (in.) (mm)	Grade	Collapse Resistance (psi) **
					Round or Buttress (in.) (mm)	Other (in.) (mm)			
7 177,8	38.0	.540 13,72	5.920 150,4	5.795 147,2	7.656 194,5	7.344(1) 186,5 7.187(3) 182,5 7.635(4) 193,9 7.562(5) 192,1 8.000(7) 203,2 7.656(9) 194,5 7.530(10) 191,3	5.840(1)(3) 148,3(4)(5) 5.845(8) 148,5	K-55 C-75 N-80 C-95 P-110 V-150	7,830 10,680 11,390 13,420 15,110 19,240
	41.0	.590 14,98	5.820 147,8	5.695 144,7		7.562(6) 192,1 8.000(7) 203,2 7.656(9) 194,5		C-95 P-110 V-150	14,670 16,990 23,160
	44.0	.640 16,25	5.720 145,3	5.595 142,1		7.750(6) 196,8 7.656(9) 194,5		C-95 P-110 V-150	15,780 18,280 24,920
	49.5	.730 18,54	5.540 140,7	5.415 137,5		7.750(6) 196,8		P-110 V-150	20,550 28,020
	24.0	.300 7,62	7.025 178,4	6.900 175,3	8.500 215,9			H-40	2,040
7-⅝ 193,7	26.4	.328 8,33	6.969 177,0	6.844 173,8 6.750(10) 171,5	8.500 215,9	7.938(1) 201,6 7.750(3) 196,8 8.125(4) 206,4 8.500(7) 215,9 8.504(9) 216,0 8.010(10) 203,5	6.889(1)(3) 175,0(4) 6.894(8) 175,1	J-55 K-55 C-75 N-80 C-95 P-110 V-150	2,890 2,890 3,280 3,400 3,710 3,930 4,080

	Internal Yield Pressure (psi) **				Joint Yield Strength (1000 lb) **			
					Threaded and Coupled Joint			
Plain End or Extreme Line	Round Thread		Buttress Thread	Body Yield Strength (1000 lb) **	Round Thread		Other*	
	Short	Long			Short	Long		
7,420				603				627(2)
10,120		8,660	7,930	822		767	677(8)	664(1)
10,800		9,240	8,460	877		814	712(8)	660(2)
12,820		10,970	10,050	1,041		931	740(8)	905(9)
14,850		12,700	11,640	1,205		1,087	890(8)	874(1)
20,240				1,644			1,446(5)	1,058(3)
14,010				1,129			905(9)	
16,220				1,306			1,078(9)	1,168(6)
22,210				1,782			1,379(9)	1,495(6)
15,200				1,215			905(9)	
17,600				1,407			1,078(9)	1,281(6)
24,000				1,918			1,379(9)	1,640(6)
20,080				1,582				1,480(6)
27,380				2,157				1,894(6)
2,750	2,750			276	212			
4,140	4,140	4,140	4,140	414	315	346		
4,410	4,140	4,140	4,140	414	342	377		357(2)
5,650		5,650	5,650	564		461	393(8)	515(1)
6,020		6,020	6,020	602		490	414(8)	404(3)
7,150		7,150	7,150	714		560	434(8)	
8,280				827				
11,290				1,128				

OD (in.) (mm)	Weight w/Cplg (lb/ft)	Wall Thickness (in.) (mm)	ID (in.) (mm)	Drift Dia. (in.) (mm)	Coupling or Joint OD		Bored Pin ID (in.) (mm)	Grade	Collapse Resistance (psi) **
					Round or Buttress (in.) (mm)	Other (in.) (mm)			
7-⅝ 193,7	29.7	.375 9,52	6.875 174,7	6.750 171,5	8.500 215,9	7.938(1) 201,6 7.750(3) 196,8 8.250(4) 209,6 8.500(7) 215,9 8.504(9) 216,0 8.010(10) 203,5	6,795(1)(3) 172,6(4) 6.800(8) 172,7	C-75 N-80 C-95 P-110 V-150	4,670 4,790 5,120 5,340 6,060
	33.7	.430 10,92	6.765 171,9	6.640 168,7	8.500 215,9	8.000(1) 203,2 7.750(3) 196,8 8.250(4) 209,6 8.198(5) 208,2 8.500(7) 215,9 8.504(9) 216,0 8.010(10) 203,5	6.685(1)(3) 169,8(4)(5) 6,690(8) 169,9	K-55 C-75 N-80 C-95 P-110 V-150	5,090 6,320 6,560 7,260 7,850 8,860
	39	.500 12,70	6.625 168,3	6.500 165,1	8.500 215,9	8.000(1) 203,2 7.812(3) 198,4 8.250(4) 209,6 8.198(5) 208,2 8.500(7) 215,9 8.504(9) 216,0 8.010(10) 203,5	6.545(1)(3) 166,2(4)(5) 6.550(8) 166,4	K-55 C-75 N-80 C-95 P-110 V-150	8,430 8,810 9,980 11,060 13,450

Internal Yield Pressure (psi) **				Body Yield Strength (1000 lb) **	Joint Yield Strength (1000 lb) **			
					Threaded and Coupled Joint		Other*	
Plain End or Extreme Line	Round Thread		Buttress Thread		Round Thread			
	Short	Long			Short	Long		
6,450 6,890 8,180 9,470 12,910		6,450 6,890 8,180 9,470	6,450 6,890 8,180 9,470	641 683 714 940 1,281		542 575 659 769	527(8) 555(8) 583(8) 694(8)	612(1) 461(2) 806(1) 792(3)
5,430 7,400 7,900 8,180 10,860 14,800		7,400 7,900 8,180 10,860	7,400 7,900 8,180 10,860	535 729 778 923 1,069 1,458		635 674 772 901	600(8) 632(8) 663(8) 790(8) 1,220(5)	510(2) 724(1) 576(3) 671(2) 922(3)
8,610 9,180 9,380 12,620 17,220		8,610 9,180 9,380 12,620	8,610 9,180 9,380 12,620	615 839 895 923 1,231 1,679			691(8) 727(8) 764(8) 1,137(5) 1,456(5)	567(2) 761(1) 636(3) 746(2) 1,017(3)

OD (in.) (mm)	Weight w/Cplg (lb/ft)	Wall Thickness (in.) (mm)	ID (in.) (mm)	Drift Dia. (in.) (mm)	Coupling or Joint OD		Bored Pin ID (in.) (mm)	Grade	Collapse Resistance (psi) **
					Round or Buttress (in.) (mm)	Other (in.) (mm)			
7-5/8 193,7	45.3	.595 15,11	6.435 163,5	6.310 160,3		8.000(1) 203,2 7.812(3) 198,4 8.312(4) 211,1(5) 8.504(9) 216.0	6.355(1)(3) 161,4(4)(5)	K-55 C-75 N-80 P-110 V-150	7,910 10,790 11,510 15,420 19,680
7-3/4 196,8	46.1	.595 15,11	6.560 166,6	6.500 165,1	8.500 215,9	7.938(3) 201,6	(11) (11)	K-55 N-80 S95 S-105 P-110 V-150	7,800 11,340 12,650 13,960 14,980 19,050
8-5/8 219.1	24.0	.264 6,71	8.097 205,7	7.972 202,5	9.625 244,5	9.625(7) 244,5	8.022(8) 203,8	J-55 K-55	1,370 1,370
	28.0	.304 7.72	8.017 203,7	7.892 200,5	9.625 244,5	9.625(7) 244,5(9)	7.942(8) 2201,7	H-40 K-55	1,640 1,880
	32.0	.352 8,94	7.921 201,2	7.796 198,0 .700(10) 195,6	9.625 244,5	8.938(1) 227,0 8.750(3) 222,2 9.135(4) 232,0 9.625(7)(9) 244,5 9.120(10) 231,6	7.811(1)(3) 198,4(4) 7.846(8) 199,3	H-40 J-55 K-55 C-75 N-80 P-110	2,210 2,530 2,530 2,950 3,050 3,430
	36.0	.400 10,16	7.825 198,8	7.700 195,6	9.625 244,5	8.938(1) 227,0 8.750(3) 222,2 9.135(4) 232,0 9.625(7)(9) 244,5 9.120(10) 231,6	7.745(1)(3) 196,7(4) 7.750(8) 196,9	J-55 K-55 C-75 N-80 C-95 P-110	3,450 3,450 4,020 4,100 4,360 4,700

	Internal Yield Pressure (psi) **			Body Yield Strength (1000 lb) **	Joint Yield Strength (1000 lb) **			
	Round Thread				Threaded and Coupled Joint			
Plain End or Extreme Line			Buttress Thread		Round Thread		Other*	
	Short	Long			Short	Long		
7,510				723				
10,240				986			694(2)	761(1)
10,920				1,051			730(2)	801(1)
15,020				1,446			1,266(5)	1,038(3)
20,480				1,971			1,620(5)	1,329(3)
7,390								668(2)
10,750							703(3)	703(2)
12,760		12,460	11,620	1,070		992		
12,760		12,460	11,620	1,271		1,065		
14,780							879(3)	879(2)
20,150							1,125(3)	
2,950	2,950			381	244			
2,950	2,950			381	263			362(8)
2,470	2,470			318	233			
3,390				437				415(8)
2,860	2,860			366	279			
3,930	3,930	3,930	3,930	503	372	417		
3,930	3,930	3,930	3,930	503	402	452	478(8)	443(2)
5,360				686				583(1)
5,710				732				501(3)
7,860				1,006				
4,460	4,460	4,460	4,460	568	434	486		
4,460	4,460	4,460	4,460	568	468	526		
6,090		6,090	6,090	775		648		523(2)
6,490		6,490	6,490	827		688	638(8)	696(1)
7,710		7,710	7,710	982		789	672(8)	587(3)
8,930				1,137			705(8)	

OD (in.) (mm)	Weight w/Cplg (lb/ft)	Wall Thickness (in.) (mm)	ID (in.) (mm)	Drift Dia. (in.) (mm)	Coupling or Joint OD — Round or Buttress (in.) (mm)	Other (in.) (mm)	Bored Pin ID (in.) (mm)	Grade	Collapse Resistance (psi) **
8-⅝ 219,1	40.0	.450 11,43	7.725 196,2	7.600 193,0	9.625 244,5	8.938(1) 227,0 8.750(3) 222,2 9.135(4) 232,0 9.625(7)(9) 244,5 9.120(10) 231,6	7.645(1)(3) 194,2(4) 7.650(8) 194,3	K-55 C-75 N-80 C-95 P-110 V-150	4,400 5,350 5,520 6,010 6,380 7,040
	44.0	.500 12,70	7.625 193,7	7.500 190,5	9.625 244,5	9.031(1) 229,4 8.750(3) 222,2 9.300(4) 236,2 9.625(7)(9) 244,5 9.120(10) 231,6	7.545(1)(3) 191,6(4) 7.550(8) 191,8	K-55 C-75 N-80 C-95 P-110 V-150	5,350 6,680 6,950 7,730 8,400 9,645
	49.0	.557 14,15	7.511 190,8	7.386 187,6	9.625 244,5	9.032(1) 229,4 8.750(3) 222,2 9.300(4) 236,2 9.625(7)(9) 244,5 9.120(10) 231,6	7.431(1)(3) 188,7(4) 7.436(8) 188,9	K-55 C-75 N-80 C-95 P-110 V-150	6,440 8,200 8,570 8,690 10,720 12,950
	52.0	.595	7.435	7.310		8.812(3) 223,8	7.355(3) 186,8	K-55 C-75 N-80 P-110 V-150	7,060 9,210 9,650 12,260 15,160
8-¾ ‡ 22,3	49.7	.557 14.15	7.636 194,0	7.500 190,5	9.625 244,5		(11) (11)	S-95 S-95	10,260 11,100
9-⅝ 244,5	32.3	.312 7,92	9.001 228,7	8.845 224,7	10.625 269,9			H-40	1,400

| | Internal Yield Pressure (psi) ** | | | | Joint Yield Strength (1000 lb) ** | | | |
| | Round Thread | | | Body Yield Strength (1000 lb) ** | Threaded and Coupled Joint — Round Thread | | Other* | |
Plain End or Extreme Line	Short	Long	Buttress Thread		Short	Long		
5,020				636				556(2)
6,850		6,850	6,850	867		742	714(8)	812(1)
7,300		7,300	7,300	925		788	751(8)	585(2)
8,670		8,670	8,670	1,098		904	789(8)	
10,040		10,040	10,040	1,271		1,055	931(8)	859(3)
13,700				1,734				1,100(3)
5,580				702				650(2)
7,610		7,610	7,610	957		834	788(8)	927(1)
8,120		8,120	8,120	1,021		887	830(8)	685(2)
9,640		9,640	9,640	1,212		1,017	871(8)	
11,160		11,160	11,160	1,404		1,186	1,037(8)	956(3)
15,210				1,915				1,224(3)
6,220				776				742(2
8,480		8,480	8,480	1,059		939	866(8)	950(1)
9,040		9,040	9,040	1,129		997	912(8)	781(2)
10,740		10,740	10,740	1,341		1,144	957(8)	1,113(10)
12,430		12,430	12,430	1,553		1,335	1,140(8)	1,061(3)
16,940				2,120			1,501(3)	1,358(3)
6,640				826				742(2)
9,050				1,126				
9,660				1,201				781(2)
13,280				1,651				1,173(3)
18,110				2,252				1,501(3)
10,580		10,580	10,580	1,362		1,017	1,232(12)	
10,580		10,580	10,580	1,362		1,095	1,273(12)	
2,270	2,270			365	254			

OD (in.) (mm)	Weight w/Cplg (lb/ft)	Wall Thickness (in.) (mm)	ID (in.) (mm)	Drift Dia. (in.) (mm)	Round or Buttress (in.) (mm)	Other (in.) (mm)	Bored Pin ID (in.) (mm)	Grade	Collapse Resistance (psi) **
9-5⁄8 244,5	36.0	.352 8.94	8.921 226,6	8.765 222,6	10.625 269,9	10.000(4) 254,0 9.750(3) 247,7 10.172(4) 258,4 10.625(9) 269,9	8.781(1)(3) 223,0 8.811(4) 223,8	H-40 J-55 K-55 C-75 N-80 P-110	1,740 2,020 2,020 2,320 2,370 2,470
	40.0	.395 10,03	8.835 224,4	8.679 220,4 8.599 (10) 218,4	10.625 269,9	10.000(1) 254,0 9.750(3) 247,7 10.172(4) 258,4 10.625(9) 269,9 10.100(10) 256,5	8.755(1)(3) 222,4 8.725(4) 221,6	J-55 K-55 C-75 N-80 C-95 P-110	2,570 2,570 2,980 3,090 3,330 3,480
	43.5	.435 11,05	8.755 222,4	8.599 218,4	10.625 269,9	10.000(1) 254,0 9.750(3) 247,7 10.172(4) 258,4 10.625(9) 269,9 10.100(10) 256,5	8.675(1)(3) 220,3 8.655(4)	K-55 C-75 N-80 C-95 P-110 V-150	3,250 3,750 3,810 4,130 4,430 4,750
	47.0	.472 11,99	8.681 220,5	8.525 216,5	10.625 269,9	10.000(1) 254.0 9.750(3) 247,7 10.270(4) 260,9 10.625(9) 269,9 10.100(10) 256,5	8.601(1) 218,5(3) 8.581(4) 218,0	K-55 C-75 N-80 C-95 P-110 V-150	3,880 4,630 4,750 5,080 5,310 6,020

Internal Yield Pressure (psi) **				Body Yield Strength (1000 lb) **	Joint Yield Strength (1000 lb) **			
Plain End or Extreme Line	Round Thread		Buttress Thread		Threaded and Coupled Joint			
					Round Thread		Other*	
	Short	Long			Short	Long		
2,560 3,520 3,520 4,800 5,120 7,040	2,560 3,520 3,520	3,520 3,520	3,520 3,520	410 564 564 769 820 1,128	294 394 423	453 489		495(2) 655(1) 561(3)
3,950 3,950 5,390 5,750 6,820 7,900	3,950 3,950	3,950 3,950 5,390 5,750 6,820	3,950 3,950 5,390 5,750 6,820	630 630 859 916 1,088 1,260	452 486	520 561 694 737 847	571(2) 601(2)	769(1) 645(3)
4,350 5,930 6,330 7,510 8,700 11,860		5,930 6,330 7,510 8,700	5,930 6,330 7,510 8,700	691 942 1,005 1,193 1,381 1,818		776 825 948 1,106	722(3) 902(3) 1,155(3)	641(2) 874(1) 675(2) 1,150(1)
4,720 6,440 6,870 8,150 9,440 12,870		6,440 6,870 8,150 9,440	6,440 6,870 8,150 9,440	746 1,018 1,086 1,289 1,493 2,036		852 905 1,040 1,213	794(3) 999(3) 1,271(3)	705(2) 970(1) 739(2) 1,276(1)

OD (in.) (mm)	Weight w/Cplg (lb/ft)	Wall Thickness (in.) (mm)	ID (in.) (mm)	Drift Dia. (in.) (mm)	Coupling or Joint OD		Bored Pin ID (in.) (mm)	Grade	Collapse Resistance (psi) **
					Round or Buttress (in.) (mm)	Other (in.) (mm)			
9-⅝ 244,5	53.5	.545 13,84	8.535 216,8	8.379 212,8	10.625 269,9	10.062(1) 255,6 9.750(3) 247,7 10.270(4) 250,9 10.188(5) 258,8 10.625(9) 269,9 10.100(10) 256,5	8.525(1) 216,5(5) 8.455(3) 214,8 8.435(4) 214,2	K-55 C-75 N-80 C-95 P-110 V-150	5,130 6,380 6,620 7,330 7,930 8,970
	58.4	.595 15,11	8.435 214,2	8.279 210,3 8.375 (11) 212,7	10.625 269,9	9.844(3) 250,0 10.188(5) 258,8 10.625(9) 269,9	8.355(3) 212,2(5)	K-55 C-75 N-80 S-95 P-110 V-150	5,990 7,570 7,890 9,950 9,750 11,570
	61.1	.625 15,87	8.375 212,7	8.219 208,8	10.625 269,9	10.625(9) 269,9	(11)	C-95 S-95 P-110 V-150	9,800 10,500 10,840 13,130
	71.8	.750 19,05	8.125 206,4	7.969 202,4	10.625 269,9	10.625(9) 269,9		P-110 V-150	15,810 19,640
9-¾ ‡ 247,7	59.2	.595 15,11	8.560 217,4	8.500 (11) 215,9	10.625 269,9		(11) (11)	S-95 S-105	9,750 10,470
9-⅞ ‡ 250,8	62.8	.625 15,88	8.625 219,1	8.500 (11) 215,9	10.625 269,9		(11) (11)	S-95 S-105	10,180 11,010
10-¾ 273,0	32.75	.279 7,09	10.192 258,8	10.036 254,9	11.750 298,5			H-40	880
	40.5	.350 8,89	10.050 255,3	9.894 251,3	11.750 298,5	11.188(1) 284,2 11.750(9) 298.5	9.910(1) 251,7	H-40 J-55 K-55 C-75	1,420 1,580 1,580 1,720

Plain End or Extreme Line	Internal Yield Pressure (psi) **			Body Yield Strength (1000 lb) **	Joint Yield Strength (1000 lb) **			
	Round Thread		Buttress Thread		Threaded and Coupled Joint			
					Round Thread		Other*	
	Short	Long			Short	Long		
5,450 7,430 7,930 9,410 10,900 14,860		7,430 7,930 9,410 10,900	7,430 7,930 9,410 10,900	855 1,166 1,244 1,477 1,710 2,332		999 1,062 1,220 1,422	925(3) 1,297(10) 1,544(10) 1,480(3)	831(2) 1,112(1) 875(2) 1,464(1) 1,950(5)
5,950 8,110 8,650 10,280 11,900 16,230		10,280	10,280	928 1,266 1,350 1,604 1,857 2,530		1,357	1,690(5) 2,163(5)	831(2) 875(2) 1,323(3) 1,693(3)
10,800 10,800 12,500 17,050		10,800	10,490	1,680 1,679 1,944 2,650		1,430		
15,000 20,450		18,060	16,560	2,300 3,137		2,672	2,692(12)	
10,150		10,150	10,150	1,626		1,204	1,469(12)	
10,150		10,150	10,150	1,626		1,297	1,524(12)	
10,520		10,520	10,490	1,725		1,123	1,385(12)	
10,520		10,520	10,490	1,725		1,210	1,437(12)	
1,820	1,820			367	205			
2,280 3,130 3,130 4,270	2,280 3,130 3,130		3,130 3,130	457 629 629 858	314 420 450			729(1)

OD (in.) (mm)	Weight w/Cplg (lb/ft)	Wall Thickness (in.) (mm)	ID (in.) (mm)	Drift Dia. (in.) (mm)	Coupling or Joint OD		Bored Pin ID (in.) (mm)	Grade	Collapse Resistance (psi) **
					Round or Buttress (in.) (mm)	Other (in.) (mm)			
10-¾ 273,0	45.5	.400 10,16	9.950 252,7	9.794 248,8	11.750 298,5	11.188(1) 284,2 10.875(3) 276,2 11.250(4) 285,7 11.750(9) 298,5 11.460(10) 291,1	9.870(1) 250,7(3) 9.850(4) 250,2	J-55 K-55 C-75 N-80 P-110	2,090 2,090 2,410 2,480 2,610
	51.0	.450 11,43	9.850 250,1	9.694 246,2	11.750 298,5	11.188(1) 284,2 10.875(3) 276,2 11.375(4) 288,9 11.750(9) 298,5 11.460(10) 291,1	9.770(1)(3) 248,2 9.750(4) 247,7	J-55 K-55 C-75 N-80 C-95 P-110	2,700 2,700 3,100 3,220 3,490 3,670
	55.5	.495 12,57	9.760 247,9	9.604 243,9	11.750 298,5	11.188(1) 284,2 10.875(3) 276,2 11.375(4) 288,9 11.750(9) 298,5 11.460(10) 291,1	9.680(1)(3) 245,9 9.660(4) 245.4	K-55 C-75 N-80 C-95 P-110 V-150	3,320 3,950 4,020 4,300 4,630 5,040
	60.7	.545 13,84	9.660 245,4	9.504 241,4	11.750 298,5	11.250(1) 285,7 10.906(3) 277,0 11.500(4) 292,1 11.312(5) 287,3 11.750(9) 298,5 11.460(10) 291,1	9.580(1)(3) 243,3(5) 9.560(4) 242,8	K-55 C-75 N-80 C-95 P-110 V-150	4,160 5,020 5,160 5,566 5,860 6,560

Internal Yield Pressure (psi) **				Body Yield Strength (1000 lb) **	Joint Yield Strength (1000 lb) **			
Plain End or Extreme Line	Round Thread		Buttress Thread		Threaded and Coupled Joint		Other*	
	Short	Long			Round Thread			
					Short	Long		
3,580	3,580		3,580	715	493			
3,580	3,580		3,580	715	528			647(2)
4,880				975				878(1)
5,210				1,041				732(3)
7,160				1,430				
4,030	4,030		4,030	801	565			
4,030	4,030		4,030	801	606			745(2)
5,490	5,490		5,490	1,092	756			1,026(1)
5,860	5,860		5,860	1,165	804		842(3)	784(2)
6,960	6,960		6,960	1,383	927			
8,060	8,060		8,060	1,602	1,080			1,350(1)
4,430				877				831(2)
6,040	6,040		6,040	1,196	843			1,158(1)
6,450	6,450		6,450	1,276	895		937(2)	875(2)
7,660	7,660		7,660	1,515	1,032			
8,860	8,860		8,860	1,754	1,203			1,523(1)
12,090				2,392				
4,880				961				883(2)
6,650				1,310				1,303(1)
7,100				1,390		986(3)		929(2)
8,436				1,648				
9,760	9,760		9,760	1,922	1,338		1,232(3)	1,714(1)
13,310				2,620			1,577(3)	2,194(5)

OD (in.) (mm)	Weight w/Cplg (lb/ft)	Wall Thickness (in.) (mm)	ID (in.) (mm)	Drift Dia. (in.) (mm)	Coupling or Joint OD		Bored Pin ID (in.) (mm)	Grade	Collapse Resistance (psi) **
					Round or Buttress (in.) (mm)	Other (in.) (mm)			
10-¾ 273,0	65,7	.595 15,11	9.560 242,8	9.404 238,9	11.750 298,5	11.312(5) 287,3	9.525(5) 241,9	K-55 C-75 N-80 C-95 P-110 V-150	4,920 6,080 6,300 6,950 7,490 8,330
	71.1	.650 16,51	9.450 240,0	9.294 236,1	11.750 298,5	11.750(9) 298,5	(11)	C-95 S-95 P-110 V-150	8,470 9,600 9,280 10,890
	76.0	.700 17,78	9.350 237,5	9.194 233,5		11.750(9) 298,5		C-95 P-110 V-150	9,850 10,900 13,220
	81.0	.750 19.05	9.250 234,9	9.094 231,0		11.750(9) 298,5		P-110 V-150	12,530 12,550
11-¾ 298,5	42.0	.333 8.46	11.084 281,5	10.928 277,6	12,750 323,9			H-40	1,070
	47.0	.375 9,52	11.000 279,4	10.844 275,4	12.750 323,9	12.188(1) 309,6 11.938(3) 303,2	10.860(1) 275,8 10.890(3) 276,6	J-55 K-55 C-75 N-80	1,510 1,510 1,620
	54.0	.435 11,05	10.880 276,3	10.724 272,4	12.750 323,9	12.188(1) 309,6 11.938(3) 303,2	10.800(1)(3) 274,3	J-55 K-55 C-75 N-80	2,070 2,070 2,380
	60.0	.489 12,42	10.772 273,6	10.616 269,6	12.750 323,9	12.188(1) 309,6 11.938(3) 303,2	10.692(1) 271,6(3)	J-55 K-55 C-75 N-80 C-95 P-110 V-150	2,660 2,660 3,070 3,180 3,440 3,610 3,680
	65.0	.534 13,56	10.682 271,3	10.526 267,4	12.750 323,9	12.188(1) 309,6 12.000(3) 304,8	10.650(1) 270,5 10.600(3) 269,2	K-55 C-75 N-80 S-95 P-110 V-150	3,290 3,810 3,870 5,740 4,490 4,850
	71.0	.582 14,78	10.586 268,9	10.430 264,9	12.750 323,9		(11)	S-95	7,280
11-⅞ ‡ 301,6	71.8	.582 14,78	10.711 272,1	10.625(11) 269,9	12.750 323,9		(11)	S-95	7,190

	Internal Yield Pressure (psi) **				Joint Yield Strength (1000 lb) **			
					Threaded and Coupled Joint			
Plain End or Extreme Line	Round Thread		Buttress Thread	Body Yield Strength (1000 lb) **	Round Thread		Other*	
	Short	Long			Short	Long		
5,330				1,134				883(2)
7,260				1,424				1,446(1)
7,750				1,519			1,137(3)	929(2)
9,200				1,803				
10,650	10,650		10,650	2,088	1,472		1,421(3)	1,903(1)
14,530				2,847			1,818(3)	2,436(5)
10,050				1,856				
10,050	9,710		9,480	1,959	1,403			
11,640	11,200		10,980	2,269	1,618			
15,070	15,070		14,970	3,094	2,174			
10,830				2,100				2,023(9)
12,530				2,431				2,408(9)
17,090				3,315				3,083(9)
13,430				2,592				2,408(9)
18,310				3,534				3,083(9)
1,980	1,980			478	307			
3,070	3,070		3,070	737	477			
3,070	3,070		3,070	737	509			630(2)
4,190				1,005				882(1)
				1,072				716(3)
3,560	3,560		3,560	850	568			
3,560	3,560		3,560	850	606			786(2)
4,860				1,160				1,078(1)
				1,237				887(3)
4,010	4,010		4,010	952	649			
4,010	4,010		4,010	952	693			786(2)
5,460	5,460		5,460	1,298	869			1,253(1)
5,830	5,830		5,830	1,384	924			827(2)
6,920	6,920		6,920	1,644	1,066		1,070(3)	
8,010				1,903			1,596(12)	
10,920				2,595			1,338(3)	1,648(1)
							1,712(3)	
4,370				1,035				786(2)
5,960				1,411				1,349(1)
6,360				1,505			1,222(3)	827(2)
7,560	7,560		7,560	1,788	1,189		1,781(12)	
8,750				2,070			1,528(3)	1,775(1)
11,930				2,822			1,955(3)	
8,230	8,230		8,230	1,940	1,306		1,933(12)	
8,150	8,150		8,150	1,962	1,153		1,735(12)	

OD (in.) (mm)	Weight w/Cplg (lb/ft)	Wall Thickness (in.) (mm)	ID (in.) (mm)	Drift Dia. (in.) (mm)	Coupling or Joint OD		Bored Pin ID (in.) (mm)	Grade	Collapse Resistance (psi) **
					Round or Buttress (in.) (mm)	Other (in.) (mm)			
13-⅜ 339,7	48.0	.330 8,38	12.715 322,9	12.559 319,0	14.375 365,1			H-40	770
	54.5	.380 9,65	12.615 320,4	12.459 316,5	14.375 365,1	14.375(9) 365,1		J-55 K-55 N-80 P-110	1,130 1,130 1,130 1,130
	61.0	.430 10,92	12.515 317,9	12.359 313,9	14.375 365,1	14.375(9) 365,1 13.562(3) 344,5	12.435(3) 315,8	J-55 K-55 C-75 N-80	1,540 1,540 1,660 1,670
	68.0	.480 12,19	12.415 315,3	12.259 311,4	14.375 365,1	13.812(1) 350,8 13.562(3) 344,5 14.375(9) 365,1	12.305(1) 312,(3)	J-55 K-55 C-75 N-80 C-95	1,950 1,950 2,220 2,270 2,330
	72.0	.514 13,06	12.347 313,6	12.191 309,7	14.375 365,1	13.812(1) 350,8 13.562(3) 344,5 14.375(9) 365,1	12.237(1) 310,8(3)	K-55 C-75 N-80 C-95 P-110	2,230 2,590 2,670 2,820 2,880
	77.0	.550 13,97	12.275 311,8	12.119 307,8	14.375 365,1	13.812(1) 350,8	12.165(1) 309,0	C-75 N-80	2,990 3,100
	80.7	.580 14,73	12.215 310,3	12.059 306,3	14.375 365,1		(11) (11)	S-80 S-95	4,800 4,990
	85.0	.608 15,44	12.159 308,8	12.003 304,9	14.375 365,1	13.812(1) 350,8	12.049(1) 306,0	C-75 N-80 P-110	3,810 3,870 4,490
	86.0	.625 15,87	12.125 308,0	11.969 304,0	14.375 365,1		(11)	S-95	6,240
	92.0	.672 17,07	12.031 305,6	11.875 301,6		14.000(1) 355,6	11.921(1) 302,8	C-75 N-80 P-110	4,910 5,050 5,700
	98.0	.719 18,26	11.937 303,2	11.781 299,2	14.375 365,1	14.000(1) 355,6	11.827(1) 300,4	C-75 N-80 P-110	5,720 5,910 6,930
13-½ ‡ 342,9	81.4	.580 14,73	12.340 313,4	12.250 (11) 311,2	14.375 365,1		(11)	S-95	4,860
13-⅝ ‡ 346,1	88.2	.625 15,88	12.375 314,3	12.250 (11) 311,2	14.375 365,1		(11)	S-95	5,930

	Internal Yield Pressure (psi) **					Joint Yield Strength (1000 lb) **		
		Round Thread			Body Yield Strength (1000 lb) **	Threaded and Coupled Joint — Round Thread		Other*
Plain End or Extreme Line	Short	Long	Buttress Thread		Short	Long		
1,730	1,730			541	322			
2,730	2,730		2,730	853	514			
2,730	2,730		2,730	853	547			
3,980				1,241				
5,470				1,629				
3,090	3,090		3,090	962	595			
3,090	3,090		3,090	962	633			
4,220				1,312				
4,500				1,400				968(3)
3,450	3,450		3,450	1,069	675			
3,450	3,450		3,450	1,069	718			1,008(2)
4,710				1,458				1,284(1)
5,020				1,556			1,164(3)	1,061(2)
5,970				1,847			1,770(9)	
3,700				1,142				1,008(2)
5,040	5,040		5,040	1,558	978			1,410(1)
5,380	5,380		5,380	1,661	1,040		1,296(3)	1,061(2)
6,390	6,390		6,390	1,973	1,204			
7,400	7,400		7,400	2,284	1,402			
5,400	5,400		5,400	1,662	1,054			1,542(1)
5,760	5,760		5,760	1,773	1,122			1,624(1)
4,170	4,170		4,170	1,282	1,118			
7,210	7,210		7,210	2,215	1,389			
5,970	5,970		5,970	1,829	1,177			1,754(1)
6,360	6,360		6,360	1,951	1,252			1,846(1)
8,750				2,682				2,308(1)
7,770	7,770		7,750	2,378	1,507		2,333(12)	
6,590				2,011				1,985(1)
7,030				2,145				2,089(1)
9,670				2,950				2,612(1)
7,060				2,144				2,153(1)
7,530				2,287				2.266(1)
10,350				3,145				2,833(1)
7,140	7,140		7,140	2,236	1,225		1,948(12)	
7,630	7,630		7,630	2,425	1,178		1,885(12)	

OD (in.) (mm)	Weight w/Cplg (lb/ft)	Wall Thick-ness (in.) (mm)	ID (in.) (mm)	Drift Dia. (in.) (mm)	Coupling or Joint OD			Bored Pin ID (in.) (mm)	Grade	Col-lapse Resis-tance (psi) **
					Round or Buttress (in.) (mm)	Other (in.) (mm)				
16 406,4	65.0	.375 9,52	15.250 387,4	15.062 382,6	17.000 431,8				H-40	670
	75.0	.438 11,3	15.124 384,1	14.936 379,4	17.000 431,8				J-55 K-55	1,020 1,020
	84.0	.495 12,57	15.010 381,3	14.822 376,5	17.000 431,8				J-55 K-55	1,410 1,410
	109.0	.656 16,67	14.688 373,1	14.500 368,3	17.000 431,8				K-55 C-75 N-80	2,560 2,980 3,080
18-⅝ 473,1	87.5	.435 11,05	17.755 451,0	17.567 446,2	19.625 498,5				H-40 J-55 K-55	630 630 630
20 508,0	94.0	.438 11,13	19.124 485,7	18.936 481,0	21.000 533,4				H-40 J-55 K-55	520 520 520
	106.5	.500 12,70	19.000 482,6	18.812 477,8	21.000 533,4				J-55 K-55	770 770
	133.0	.635 16,13	18.730 475,7	18.542 471,0	21.000 533,4				J-55 K-55	1,500 1,500

* Joint strength of buttress, extreme line, Vallourec VAM, Rucker Atlas-Bradford TC-4S and Hydril Super EU casing threads is generally comparable to or greater than the pipe body yield strength and is not given in this listing except for a few cases of higher grade steels where it is less. In these cases the joint strength is listed and identified.
Joint strength of other premium threads is generally less than pipe body yield strength and is listed and identified for some representative grades. This will provide an approximation of joint efficiency in other grades.

** Collapse resistance, internal yield pressure, and joint yield strengths are minimum values with no safety factor, reproduced from API Bulletin 5C2, Bulletin on Performance Properties of Casing and Tubing, and from published literature of manufacturers of premium threads and tubular goods.

Internal Yield Pressure (psi) **				Body Yield Strength (1000 lb) **	Joint Yield Strength (1000 lb) **			
	Round Thread				Threaded and Coupled Joint		Other*	
					Round Thread			
Plain End or Extreme Line	Short	Long	Buttress Thread		Short	Long		
1,640	1,640		1,640	736	439		614(12)	
2,630	2,630		2,630	1,178	710			
2,630	2,630		2,630	1,178	752			
2,980	2,980		2,980	1,326	817			
2,980	2,980		2,980	1,326	865			
3,950	3,950		3,950	1,739	1,181			
5,380	5,380			2,372	1,499			
5,740	5,740			2,530	1,594			
1,630	1,630			1,994	559			
2,250	2,250			1,367	754			
2,250	2,250			1,367	794			
1,530	1,530			1,077	581			
2,110	2,110	2,110		1,480	784	907	1,402(12)	
2,110	2,110	2,110		1,480	824	955		
2,410	2,410	2,410		1,685	913	1,057	1,596(12)	
2,410	2,410	2,410		1,685	960	1,113		
3,060	3,060	3,060		2,125	1,192	1,380	2,012(12)	
3,060	3,060	3,060		2,125	1,253	1,453		

‡ These casing sizes have standard API 8 Round or Buttress threads or, in some sizes, certain Hydril or Rucker Atlas-Bradford premium threads, of the next smaller OD and the threads are interchangeable. For example, 7-3/4 in. casing will have the same thread as standard 7-5/8 in. casing and will use standard casing collars.

Data for premium threads and tubular goods are identified as follows:
(1) Hydril TS Tripleseal
(2) Hydril FJ-P Flush Joint (joint OD and ID same as pipe)
(3) Hydril Super FJ-P
(4) Hydril Super EU
(5) Hydril CTS
(6) Hydril CTS-4
(7) Rucker Atlas-Bradford TC-4S
(8) Rucker Atlas-Bradford FL-3S and FL-4S
(9) Vallourec VAM
(10) Extreme Line
(11) Lone Star Steel Company
(12) Buttress Thread

Kill Mud Weight Increases

This table of kill mud weight increases gives the amount of mud weight increase necessary to balance the bottom hole formation pressure based on a known vertical bit depth and a true shut in drill pipe pressure. The formula for calculating these mud weight increases should be used in actual kick situations with this table being used to check calculated increase. As noted in previous chapters, it is advantageous to use no safety factors in addition to the calculated mud weight increases.

True vertical* depth	Shut-in Drill Pipe Pressure																						
	100	150	200	250	300	350	400	450	500	550	600	650	700	750	800	850	900	950	1,000	1,050	1,100	1,150	1,200
1,000	1.9	2.9	3.9	4.8	5.8	6.7	7.7																
1,500	1.3	1.9	2.6	3.2	3.9	4.5	5.1	5.8	6.4	7.1	7.7												
2,000	1.0	1.5	1.9	2.4	2.9	3.4	3.9	4.3	4.8	5.3	5.8	6.2	6.7	7.2	7.7								
2,500	.8	1.2	1.5	1.9	2.3	2.7	3.1	3.5	3.9	4.3	4.6	5.0	5.4	5.8	6.2	6.6	7.0	7.3	7.7				
3,000	.7	1.0	1.3	1.6	1.9	2.3	2.6	2.9	3.2	3.5	3.9	4.2	4.5	4.8	5.1	5.5	5.8	6.1	6.4	6.7	7.1	7.4	7.7
3,500	.6	.8	1.1	1.4	1.7	1.9	2.2	2.5	2.8	3.0	3.3	3.6	3.9	4.1	4.4	4.7	5.0	5.2	5.5	5.8	6.0	6.3	6.6
4,000	.5	.7	1.0	1.2	1.5	1.7	1.9	2.2	2.4	2.7	2.9	3.1	3.4	3.6	3.9	4.1	4.3	4.6	4.8	5.1	5.3	5.5	5.8
4,250	.5	.7	.9	1.1	1.4	1.6	1.8	2.0	2.3	2.5	2.7	3.0	3.2	3.4	3.6	3.9	4.1	4.3	4.5	4.8	5.0	5.2	5.4
4,500	.4	.7	.9	1.1	1.3	1.5	1.7	1.9	2.2	2.4	2.6	2.8	3.0	3.2	3.4	3.6	3.9	4.1	4.3	4.5	4.7	4.9	5.1
4,750	.4	.6	.8	1.0	1.2	1.4	1.6	1.8	2.0	2.2	2.4	2.6	2.8	3.0	3.2	3.5	3.7	3.9	4.1	4.3	8.7	4.7	4.9
5,000	.4	.6	.8	1.0	1.2	1.4	1.6	1.7	1.9	2.1	2.3	2.5	2.7	2.9	3.1	3.3	3.5	3.7	3.9	4.2	4.2	4.4	4.6
5,250	.4	.6	.8	.9	1.1	1.3	1.5	1.7	1.9	2.0	2.2	2.4	2.6	2.8	2.9	3.1	3.3	3.5	3.7	3.9	4.0	4.2	4.4

5,500	5,750	6,000	6,250	6,500	6,750	7,000	7,250	7,500	7,750	8,000	8,250	8,500	8,750	9,000	9,250	9,500	9,750
4.2	4.0	3.9	3.7	3.6	3.4	3.3	3.2	3.1	3.0	2.9	2.8	2.7	2.7	2.6	2.5	2.4	2.4
4.0	3.9	3.7	3.6	3.4	3.3	3.2	3.1	3.0	2.9	2.8	2.7	2.6	2.5	2.5	2.4	2.3	2.3
3.9	3.7	3.5	3.4	3.3	3.2	3.2	2.9	2.8	2.7	2.7	2.6	2.5	2.4	2.4	2.3	2.2	2.2
3.7	3.5	3.4	3.2	3.1	3.1	2.9	2.8	2.7	2.6	2.5	2.5	2.4	2.3	2.3	2.2	2.1	2.1
3.5	3.4	3.2	3.1	3.0	2.9	2.8	2.7	2.6	2.5	2.4	2.3	2.3	2.2	2.2	2.1	2.0	2.0
3.3	3.2	3.1	2.9	2.8	2.7	2.6	2.5	2.5	2.4	2.3	2.2	2.2	2.1	2.0	2.0	1.9	1.9
3.2	3.0	2.9	2.8	2.7	2.6	2.5	2.4	2.3	2.3	2.2	2.1	2.1	2.0	1.9	1.9	1.8	1.8
3.0	2.9	2.7	2.6	2.5	2.4	2.4	2.3	2.2	2.1	2.1	2.0	1.9	1.9	1.8	1.8	1.7	1.7
2.8	2.7	2.6	2.5	2.4	2.3	2.2	2.1	2.1	2.0	1.9	1.9	1.8	1.8	1.7	1.7	1.6	1.6
2.6	2.5	2.4	2.3	2.2	2.2	2.1	2.0	1.9	1.9	1.8	1.8	1.7	1.7	1.6	1.6	1.5	1.5
2.5	2.4	2.3	2.1	2.1	2.0	1.9	1.9	1.8	1.8	1.7	1.7	1.6	1.6	1.5	1.5	1.4	1.4
2.3	2.2	2.1	2.0	1.9	1.9	1.8	1.7	1.7	1.6	1.6	1.5	1.5	1.4	1.4	1.4	1.3	1.3
2.1	2.0	1.9	1.9	1.8	1.7	1.7	1.6	1.5	1.4	1.4	1.4	1.3	1.3	1.3	1.2	1.2	1.2
1.9	1.9	1.8	1.7	1.6	1.6	1.6	1.5	1.5	1.4	1.4	1.3	1.3	1.2	1.2	1.2	1.1	1.1
1.8	1.7	1.6	1.6	1.5	1.4	1.4	1.3	1.3	1.3	1.2	1.2	1.2	1.1	1.1	1.0	1.0	1.0
1.6	1.5	1.5	1.4	1.4	1.3	1.3	1.2	1.2	1.2	1.1	1.1	1.0	1.0	1.0	1.0	.9	.9
1.4	1.4	1.3	1.2	1.2	1.2	1.1	1.1	1.0	1.0	1.0	1.0	.9	.9	.9	.8	.8	.8
1.2	1.2	1.1	1.1	1.1	1.0	1.0	.9	.9	.9	.9	.8	.8	.8	.8	.7	.7	.7
1.1	1.0	1.0	.9	.9	.9	.8	.8	.8	.8	.7	.7	.7	.7	.7	.6	.6	.6
.9	.9	.8	.8	.8	.7	.7	.7	.7	.7	.6	.6	.6	.6	.6	.5	.5	.5
.7	.7	.7	.6	.6	.6	.6	.6	.5	.5	.5	.5	.5	.5	.4	.4	.4	.4
.5	.5	.5	.5	.5	.4	.4	.4	.4	.4	.4	.4	.4	.3	.3	.3	.3	.3
.4	.4	.3	.3	.3	.3	.3	.3	.3	.3	.3	.3	.3	.2	.2	.2	.2	.2

Kill Mud Weight Increases

True vertical depth	Shut-in Drill Pipe Pressure																						
	100	150	200	250	300	350	400	450	500	550	600	650	700	750	800	850	900	950	1,000	1,050	1,100	1,150	1,200
10,000	.2	.3	.4	.5	.6	.7	.8	.9	1.0	1.1	1.2	1.3	1.4	1.5	1.6	1.7	1.8	1.9	1.9	2.0	2.1	2.2	2.3
10,250	.2	.3	.4	.5	.6	.7	.8	.9	1.0	1.1	1.1	1.2	1.3	1.4	1.5	1.6	1.7	1.8	1.9	2.0	2.1	2.2	2.3
10,500	.2	.3	.4	.5	.6	.7	.8	.8	.9	1.0	1.1	1.2	1.3	1.4	1.5	1.6	1.7	1.8	1.9	1.9	2.0	2.1	2.2
10,750	.2	.3	.4	.5	.6	.6	.7	.8	.9	1.0	1.1	1.2	1.3	1.4	1.5	1.5	1.6	1.7	1.8	1.9	2.0	2.1	2.2
11,000	.2	.3	.4	.4	.5	.6	.7	.8	.9	1.0	1.1	1.2	1.2	1.3	1.4	1.5	1.6	1.7	1.8	1.9	1.9	2.0	2.2
11,250	.2	.3	.4	.4	.5	.6	.7	.8	.9	1.0	1.0	1.1	1.2	1.3	1.4	1.5	1.6	1.6	1.7	1.8	1.9	2.0	2.1
11,500	.2	.3	.4	.4	.5	.6	.7	.8	.9	1.0	1.0	1.1	1.2	1.3	1.4	1.4	1.5	1.6	1.7	1.8	1.9	1.9	2.0
11,750	.2	.3	.3	.4	.5	.6	.7	.8	.9	.9	1.0	1.1	1.2	1.2	1.3	1.4	1.5	1.6	1.7	1.7	1.8	1.9	2.0
12,000	.2	.3	.3	.4	.5	.6	.7	.7	.8	.9	1.0	1.1	1.1	1.2	1.3	1.4	1.5	1.5	1.6	1.7	1.8	1.9	1.9
12,250	.2	.3	.3	.4	.5	.6	.7	.7	.8	.9	1.0	1.0	1.1	1.2	1.3	1.4	1.4	1.5	1.6	1.7	1.7	1.8	1.9
12,500	.2	.3	.3	.4	.5	.6	.6	.7	.8	.9	.9	1.0	1.1	1.2	1.2	1.3	1.4	1.5	1.6	1.6	1.7	1.8	1.9
12,750	.2	.2	.3	.4	.5	.5	.6	.7	.8	.8	.9	1.0	1.1	1.2	1.2	1.3	1.4	1.5	1.5	1.6	1.7	1.8	1.8
13,000	.2	.2	.3	.4	.5	.5	.6	.7	.8	.8	.9	1.0	1.1	1.1	1.2	1.3	1.4	1.4	1.5	1.6	1.6	1.7	1.8
13,250	.2	.2	.3	.4	.5	.5	.6	.7	.8	.8	.9	1.0	1.0	1.1	1.2	1.3	1.3	1.4	1.5	1.6	1.6	1.7	1.8

*True vertical depth

13,500	1.7	1.7	1.6	1.6	1.4	1.4	1.3	1.2	1.2	1.2	1.0	.9	.9	.8	.7	.7	.6	.5	.4	.4	.2	.2	.2
13,750	1.7	1.6	1.6	1.5	1.4	1.3	1.3	1.2	1.1	1.1	1.0	.9	.9	.8	.7	.6	.6	.5	.4	.4	.3	.2	.2
14,000	1.7	1.6	1.5	1.5	1.4	1.3	1.3	1.2	1.1	1.1	1.0	.9	.8	.8	.7	.6	.6	.5	.4	.4	.3	.2	.2
14,250	1.6	1.6	1.5	1.4	1.4	1.3	1.2	1.2	1.1	1.0	1.0	.9	.8	.8	.7	.6	.6	.5	.4	.4	.3	.2	.2
14,500	1.6	1.5	1.5	1.4	1.3	1.3	1.2	1.1	1.1	1.0	1.0	.9	.8	.7	.7	.6	.6	.5	.4	.4	.3	.2	.2
14,750	1.6	1.6	1.5	1.4	1.4	1.3	1.2	1.1	1.1	1.0	.9	.9	.8	.7	.7	.6	.6	.5	.4	.4	.3	.2	.2
15,000	1.6	1.5	1.4	1.4	1.3	1.2	1.2	1.1	1.0	1.0	.9	.9	.8	.7	.7	.6	.5	.5	.4	.3	.3	.2	.2
15,250	1.5	1.5	1.4	1.3	1.3	1.2	1.2	1.1	1.0	1.0	.9	.9	.8	.7	.7	.6	.5	.5	.4	.3	.3	.2	.1
15,500	1.5	1.4	1.4	1.3	1.3	1.2	1.1	1.1	1.0	1.0	.9	.8	.8	.7	.7	.6	.5	.5	.4	.3	.3	.2	.1
15,750	1.5	1.4	1.4	1.3	1.2	1.2	1.1	1.0	1.0	.9	.9	.8	.7	.7	.6	.6	.5	.4	.4	.3	.3	.2	.1
16,000	1.5	1.4	1.3	1.3	1.2	1.2	1.1	1.0	1.0	.9	.9	.8	.7	.7	.6	.6	.5	.4	.4	.3	.3	.2	.1
16,250	1.4	1.4	1.3	1.3	1.2	1.1	1.1	1.0	.9	.9	.8	.8	.7	.7	.6	.5	.5	.4	.4	.3	.3	.2	.1
16,500	1.4	1.3	1.3	1.2	1.2	1.1	1.1	1.0	.9	.9	.8	.8	.7	.7	.6	.5	.5	.4	.4	.3	.3	.2	.1
16,750	1.4	1.3	1.3	1.2	1.2	1.1	1.1	1.0	.9	.9	.8	.8	.7	.7	.6	.5	.5	.4	.4	.3	.2	.2	.1
17,000	1.4	1.3	1.2	1.2	1.2	1.1	1.0	1.0	.9	.9	.8	.7	.7	.6	.6	.5	.5	.4	.4	.3	.2	.2	.1
17,250	1.4	1.3	1.2	1.2	1.1	1.1	1.0	1.0	.9	.8	.8	.7	.7	.6	.6	.5	.5	.4	.4	.3	.2	.2	.1
17,500	1.3	1.3	1.2	1.2	1.1	1.1	1.0	1.0	.9	.8	.8	.7	.7	.6	.6	.5	.5	.4	.3	.3	.2	.2	.1
17,750	1.3	.3	1.2	1.2	1.1	1.1	1.0	.9	.9	.8	.8	.7	.7	.6	.6	.5	.5	.4	.3	.3	.2	.2	.1

Kill Mud Weight Increases

True vertical* depth	Shut-in Drill Pipe Pressure																						
	100	150	200	250	300	350	400	450	500	550	600	650	700	750	800	850	900	950	1,000	1,050	1,100	1,150	1,200
18,000	.1	.2	.2	.3	.3	.4	.4	.5	.6	.6	.7	.7	.8	.8	.9	.9	1.0	1.0	1.1	1.1	1.2	1.2	1.3
18,250	.1	.2	.2	.3	.3	.4	.4	.5	.5	.6	.7	.7	.8	.8	.9	.9	1.0	1.0	1.1	1.1	1.2	1.2	1.3
18,500	.1	.2	.2	.3	.3	.4	.4	.5	.5	.6	.6	.7	.7	.8	.9	.9	.9	1.0	1.0	1.1	1.2	1.2	1.3
18,750	.1	.2	.2	.3	.3	.4	.4	.5	.5	.6	.6	.7	.7	.8	.8	.9	.9	1.0	1.0	1.1	1.1	1.2	1.3
19,000	.1	.2	.2	.3	.3	.4	.4	.5	.5	.6	.6	.7	.7	.8	.8	.9	.9	1.0	1.0	1.1	1.1	1.2	1.2
19,250	.1	.2	.2	.3	.3	.4	.4	.5	.5	.6	.6	.7	.7	.8	.8	.9	.9	1.0	1.0	1.1	1.1	1.2	1.2
19,500	.1	.2	.2	.3	.3	.4	.4	.5	.5	.6	.6	.7	.7	.8	.8	.9	.9	1.0	1.0	1.1	1.1	1.2	1.2
19,750	.1	.2	.2	.3	.3	.4	.4	.5	.5	.6	.6	.7	.7	.8	.8	.9	.9	1.0	1.0	1.1	1.1	1.2	1.2
20,000	.1	.2	.2	.3	.3	.4	.4	.5	.5	.6	.6	.7	.7	.8	.8	.9	.9	1.0	1.0	1.1	1.1	1.2	1.2

Blowout Preventer and Hydraulic Valve Data

Ram type blowout preventers

(Courtesy: Valve Control Co.)

Model or type	BOP size, in.	Working pressure max. PSI	Vert. bore in.	Hydraulic operator, PSI*	Gal. to close	Gal. to open	Close ratio	Open ratio
				Hydril Co., Los Angeles, California				
V	6	3,000	7$\frac{1}{16}$	750	1.5	1.3	5.32	
V	6	5,000	7$\frac{1}{16}$	1,175	1.5	1.3	5.32	
V	10	3,000	11	550	3.3	3.2	6.0	
V	10	5,000	11	850	3.3	3.2	6.0	
X	11	10,000	11	1,050	12.9	11.8	10.56	
V	12	3,000	13$\frac{5}{8}$	700	5.9	4.9	5.2	
V	13$\frac{5}{8}$	5,000	13$\frac{5}{8}$	1,050	5.9	4.9	5.2	
X	13$\frac{5}{8}$	10,000	13$\frac{5}{8}$	1,050	12.9	.1.8	10.56	
X	16$\frac{3}{4}$	10,000	16$\frac{3}{4}$	1,050	15.6	14.1	10.56	
X	18$\frac{3}{4}$	10,000	18$\frac{3}{4}$	1,050	17.1	15.6	10.14	
V	21$\frac{1}{4}$	2,000	21$\frac{1}{4}$	500	17.2	16.3		
				Bowen Tools, Inc., Houston, Texas				
51922	2$\frac{1}{2}$ Single	6,000	2$\frac{1}{2}$	780	.17	.16	7.9:1	
51923	2$\frac{1}{2}$ Single	10,000	2$\frac{1}{2}$	1,300	.19	.19	7.9:1	
51924	2$\frac{1}{2}$ Twin	5,000	2$\frac{1}{2}$	692	.36	.28	7.9:1	

*Lower pressures for normal use, higher pressures for emergency only

Ram type blowout preventers cont.

Model or type	B.O.P. size, in.	Working pressure max. PSI	Vert. bore in.	Hydraulic operator, PSI*	Gal. to close	Gal. to open	Close ratio	Open ratio
60701	2½ Twin	10,000	2½	1,001	.43	.35	7.9:1	
50460	2⁹/₁₆ Single	15,000	2⁹/₁₆	1,000	.3	.3	8.18:1	Not Applicable. Well pressure must be equalized across rams.
51926	3 Single	5,000	3	369	.30	.22	13.2:1	
51927	3 Single	10,000	3	738	.30	.22	13.2:1	
51928	3 Twin	5,000	3	369	.54	.49	13.2:1	
51929	3 Twin	10,000	3	738	.54	.49	13.2:1	
61040	4 Single	5,000	4	555	.91	.81	15.3:1	
61044	4 Single	10,000	4	1,110	.91	.81	15.3:1	
61048	4 Twin	5,000	4	555	1.81	1.62	15.3:1	
61050	4 Twin	10,000	4	1,110	1.81	1.62	15.3:1	
47034	4¹/₁₆ Single	10,000	4¹/₁₆	1,000	.43	.34	13.6:1	
60467	4¹/₁₆ Single	15,000	4¹/₁₆	3,000	.43	.34	13.6:1	
61053	4½ Single	3,000	4½	30	.91	.81	15.3:1	
61055	4½ Single	10,000	4½	1,110	.91	.81	15.3:1	
61057	4½ Twin	5,000	4½	555	1.83	1.64	15.3:1	
61060	4½ Twin	10,000	4½	1,110	1.83	1.64	15.3:1	
51938	5½ Single	3,000	5½	240	1.51	1.37	20.8:1	
63642	7¹/₁₆ Single	10,000	7¹/₁₆	900	.74	.75	10.9:1	

Cameron Iron Works, Houston, Texas

Model or type	B.O.P. size, in.	Working pressure max. PSI	Vert. bore in.	Hydraulic operator, PSI*	Gal. to close	Gal. to open	Close ratio	Open ratio
U	6	3,000	7¹/₁₆	1,500/3,000	1.33	1.28	7:1	2.3:1
U	6	5,000	7¹/₁₆	1,500/3,000	1.33	1.28	7:1	2.3:1
U	7¹/₁₆	10,000	7¹/₁₆	1,500/3,000	1.33	1.28	7:1	2.3:1
U	7¹/₁₆	15,000	7¹/₁₆	1,500/3,000	1.33	1.28	7:1	2.3:1
U	10	3,000	11	1,500/3,000	3.36	3.20	7:1	2.3:1
U	10	5,000	11	1,500/3,000	3.36	3.20	7:1	2.3:1
U	11	10,000	11	1,500/3,000	3.36	3.20	7:1	2.3:1
U	11	15,000	11	1,500/3,000	3.36	3.20	7:1	2.3:1

U	12	3,000	13⅝	1,500/3,000	5.80	5.45	7:1	2.3:1
U	13⅝	5,000	13⅝	1,500/3,000	5.80	5.45	7:1	2.3:1
U	13⅝	10,000	13⅝	1,500/3,000	5.80	5.45	7:1	2.3:1
U	13⅝	15,000	13⅝	1,500/3,000	11.3	11.7	6.6:1	2.3:1
U	16¾	3,000	16¾	1,500/3,000	9.8	10.6	6.7:1	1.4:1
U	16¾	5,000	16¾	1,500/3,000	9.8	10.6	6.7:1	1.4:1
U	18¾	10,000	18¾	1,500/3,000	23.0	24.9	7.4:1	
U	20	2,000	20¾	1,500/3,000	8.40	7.85	7:1	1.2:1
U	20	3,000	20¾	1,500/3,000	8.40	7.85	7:1	1.2:1
U	21¼	2,000	21¼	1,500/3,000	8.40	7.85	7:1	1.2:1
U	21¼	7,500	21¼	1,500/3,000	20.4	17.8	5.5:1	
U	21¼	10,000	21¼	1,500/3,000	26.5	24.1	7.2:1	
U	26¾	2,000	26¾	1,500/3,000	10.4	9.85	7:1	.63:1
U	26¾	3,000	26¾	1,500/3,000	10.4	9.85	7:1	.63:1
U-Blind	13⅝	5,000	13⅝	1,500/2,500	11.6	10.9	14:1	2.3:1
ram with	13⅝	10,000	13⅝	1,500/2,500	11.6	10.9	14:1	2.3:1
shear	16¾	3,000	16¾	1,500/2,500	10.8	11.7	9:1	1.4:1
booster	16¾	5,000	16¾	1,500/2,500	10.8	11.7	9:1	1.4:1
	20	2,000	20¾	1,500/2,500	16.8	15.7	14:1	1.2:1
	20	3,000	20¾	1,500/3,500	16.8	15.7	14:1	1.2:1
QRC	6	3,000	7¹⁄₁₆	1,500/3,000	0.81	0.95	7.75:1	1.5:1
QRC	6	5,000	7¹⁄₁₆	1,500/3,000	0.81	0.95	7.75:1	1.5:1
QRC	8	3,000	9	1,500/3,000	2.36	2.70	9.05:1	1.83:1
QRC	8	5,000	9	1,500/3,000	2.36	2.70	9.05:1	1.83:1
QRC	10	3,000	11	1,500/3,000	2.77	3.18	9.05:1	1.21:1
QRC	10	5,000	11	1,500/3,000	2.77	3.18	9.05:1	1.21:1
QRC	12	3,000	13⅝	1,500/3,000	4.42	5.10	8.64:1	1.07:1
QRC	16	2,000	16¾	1,500/3,000	6.0	7.05	8.64:1	0.62:1
QRC	18	2,000	17¾	1,500/3,000	6.0	7.05	8.64:1	0.62:1
QRC	20	2,000	17¾	1,500/3,000	6.0	7.05	8.64:1	0.62:1

*Lower pressures for normal use, higher pressure for emergency only.

Ram type blowout preventer cont.

Model or type	BOP size, in.	Working pressure, max. PSI	Vert. bore in.	Hydraulic operator, PSI*	Gal. to close	Gal. to open	Close ratio	Open ratio
SS	6	3,000	7¹/₁₆	1,000/3,000	0.8	0.7	3.8:1	1:1
SS	6	5,000	7¹/₁₆	1,500/3,000	0.8	0.7	3.8:1	1:1
SS	8	3,000	9	1,500/3,000	1.5	1.3	3.9:1	1:1
SS	8	5,000	9	1,500/3,000	1.5	1.3	3.9:1	1:1
SS	10	3,000	11	1,500/3,000	1.5	1.3	3.9:1	1:1
SS	10	5,000	11	1,500/3,000	1.5	1.3	3.9:1	1:1
SS	12	3,000	13⅝	1,500/3,000	2.9	2.5	3.7:1	1:1
SS	14	5,000	13⅝	1,500/3,000	2.9	2.5	3.7:1	1:1
Type F with type W₂ opr.	6	3,000	7¹/₁₆	500/1,500	1.5	2.3	Variable	4.5:1
	6	5,000	7¹/₁₆	500/1,500	1.5	2.3		4.5:1
	7	10,000	7¹/₁₆	500/1,500	1.5	2.3		4.5:1
	7	15,000	7¹/₁₆	500/1,500	1.5	2.3		4.5:1
	8	3,000	9	500/1,500	2.8	3.7		2.5:1
	8	5,000	9	500/1,500	2.8	3.7		2.5:1
	10	3,000	11	500/1,500	2.8	3.7		2.5:1
	10	5,000	11	500/1,500	2.8	3.7		2.5:1
	11	10,000	11	500/1,500	2.8	3.7		2.5:1
	12	3,000	13⅝	500/1,500	4.1	5.3		2:1
	14	5,000	13⅝	500/1,500	4.1	5.3		2:1
	16	2,000	16¾	500/1,500	5.0	6.0		2:1
	16	3,000	16¾	500/1,500	5.0	6.0		2:1
	20	2,000	20¼	500/1,500	5.0	6.0		2:1
	20	3,000	20¼	500/1,500	5.0	6.0		2:1
Type F with type W opr.	6	3,000	7¹/₁₆	500/1,500	2.3	3.05		4.5:1
	6	5,000	7¹/₁₆	500/1,500	2.3	3.05		4.5:1
	7	10,000	7¹/₁₆	500/1,500	2.3	3.05		4.5:1
	7	15,000	7/16	500/1,500	2.3	3.05		4.5:1

8	3,000	9	500/1,500	3.7	4.6	Variable	2.5:1
8	5,000	9	500/1,500	3.7	4.6		2.5:1
10	3,000	11	500/1,500	3.7	4.6		2.5:1
10	5,000	11	500/1,500	3.7	4.6		2.5:1
11	10,000	11	500/1,500	3.7	4.6		2.5:1
12	3,000	13⅝	500/1,500	6.8	8.1		2:1
14	5,000	13⅝	500/1,500	6.8	8.1		2:1
16	2,000	16¾	500/1,500	7.6	9.1		2:1
16	3,000	16¾	500/1,500	7.6	9.1		2:1
20	2,000	20¼	500/1,500	7.6	9.1		2:1
20	3,000	20¼	500/1,500	7.6	9.1		2:1

Type F with type L opr.

6	3,000	7 1/16	250/1,500	3.97	3.46	Variable	4.9:1
6	5,000	7 1/16	250/1,500	3.97	3.46		4.9:1
7	10,000	7 1/16	250/1,500	3.97	3.46		4.9:1
7	15,000	7 1/16	250/1,500	3.97	3.46		4.9:1
8	3,000	9	250/1,500	6.85	6.19		3.44:1
8	5,000	9	250/1,500	6.85	6.19		3.44:1
10	3,000	11	250/1,500	6.85	6.19		3.44:1
10	5,000	11	250/1,500	6.85	6.19		3.44:1
11	10,000	11	250/1,500	6.85	6.19		2.3:1
12	3,000	13⅝	250/1,500	10.30	9.38		2.3:1
14	5,000	13⅝	250/1,500	10.30	9.38		2.3:1
16	2,000	16¾	250/1,500	11.71	10.66		2.3:1
16	3,000	16¾	250/1,500	11.71	10.66		2.3:1
20	2,000	20¼	250/1,500	11.71	10.66		2.3:1
20	3,000	20¼	250/1,500	11.71	10.66		2.3:1

Type F with type H opr.

6	3,000	7 1/16	1,000/5,000	0.52	1.05	Variable	1.5:1
6	5,000	7 1/16	1,000/5,000	0.52	1.05		1.5:1
7	10,000	7 1/16	1,000/5,000	0.52	1.05		1.5:1
7	15,000	7 1/16	1,000/5,000	0.52	1.05		1.5:1
8	3,000	9	1,000/5,000	0.90	1.80		1:1
8	5,000	9	1,000/5,000	0.90	1.80		1:1
10	5,000	11	1,000/5,000	0.90	1.80		1:1

*Lower pressures for normal use, high pressure for emergency only.

Ram type blowout preventers cont.

Model or type	BOP size, in.	Working pressure max. PSI	Vert. bore in.	Hydraulic operator, PSI*	Gal. to close	Gal. to open	Close ratio	Open ratio
	11	10,000	11	1,000/5,000	0.90		1.80	1:1
	12	3,000	13⅝	1,000/5,000	1.52		2.70	2/3:1
	14	5,000	13⅝	1,000/5,000	1.52		2.70	2/3:1
	16	2,000	16¾	1,000/5,000	1.73		3.08	2/3:1
	16	3,000	16¾	1,000/5,000	1.73		3.08	2/3:1
	20	2,000	20¼	1,000/5,000	1.73		3.08	2/3:1
	20	3,000	20¼	1,000/5,000	1.73			2/3:1
Dresser OME (Guiberson), Dallas, Texas								
Type H Hyd.Cyl.	6	3,000	N.A.	2,000	1.1	.94	6.5:1	1:1
	8	2,000	N.A.	2,000	1.1	.94	6.5:1	1:1
Rucker-Shaffer, Houston, Texas								
LWS with locking manual screw	4 1/16	10,000	4 1/16	1,500/3,000	.50	.47	8.45:1	4.74:1
	6	3,000	7 1/16	1,500/3,000	1.20	1.00	4.44:1	1.82:1
	6	5,000	7 1/16	1,500/3,000	1.20	1.00	4.45:1	1.82:1
	7 1/16	10,000	7 1/16	1,500/3,000	6.35	5.89	10.63:1	19.40:1
	7 1/16	15,000	7 1/16	1,500/3,000	6.35	5.89	10.63:1	19.40:1
	8	3,000	9	1,500/3,000	2.58	2.26	5.58:1	3.00:1
	8	5,000	9	1,500/3,000	2.58	2.26	5.58:1	3.00:1
	9	10,000	9	1,500/3,000	2.44	2.44	5.58:1	1.69:1

	10	3,000	11	1,500/3,000	1.75	1.45	4.45:1	1.16:1
	10	5,000	11	1,500/3,000	2.98	2.62	5.58:1	2.10:1
	11	10,000	11	1,500/3,000	3.62	3.31	7.83:1	2.20:1
	12	3,000	13 5/8	1,500/3,000	3.36	2.95	5.58:1	1.75:1
	13 5/8	5,000	13 5/8	1,500/3,000	3.36	2.95	5.58:1	1.75:1
	13 5/8	10,000	13 5/8	1,500/3,000	10.59	9.82	10.63:1	3.47:1
	16	3,000	16 3/4	1,500/3,000	4.69	4.13	5.58:1	1.40:1
	16 3/4	5,000	16 3/4	1,500/3,000	6.60	6.03	7.85:1	1.59:1
	20	2,000	21 1/4	1,500/3,000	5.07	4.46	5.58:1	.78:1
	20	3,000	21 1/4	1,500/3,000	5.07	4.46	5.58:1	.78:1
LWP	6	3,000	7 1/16	1,500/3,000	.55	.51	4:1	1.81:1
Type	8	3,000	9	1,500/3,000	.77	.77	4:1	2.5:1
LWS	7 1/16	15,000	7 1/16	1,500/3,000	7.24	6.60	10.85:1	19.44:1
	10	5,000	11	1,500/3,000	4.75	4.18	8.16:1	3.07:1
	10	5,000	11	1,500/3,000	9.31	8.48	10.85:1	7.82:1
Posi-	11	10,000	11	1,500/3,000	4.20	3.70	8.16:1	2.21:1
	+11	10,000	11	1,500/3,000	8.23	7.50	10.85:1	5.24:1
Lock	12	3,000	13 5/8	1,500/3,000	5.34	4.70	8.16:1	2.56:1
	+12	3,000	13 5/8	1,500/3,000	10.56	9.62	10.85:1	6.25:1
	13 5/8	5,000	13 5/8	1,500/3,000	5.30	4.67	8.16:1	2.56:1
	+13 5/8	5,000	13 5/8	1,500/3,000	10.56	9.62	10.85:1	6.25:1
	+13 5/8	10,000	13 5/8	1,500/3,000	11.56	10.52	10.85:1	3.47:1
	16 3/4	3,000	16 3/4	1,500/3,000	7.25	6.38	8.16:1	2.05:1
	16 3/4	5,000	16 3/4	1,500/3,000	7.25	6.38	8.16:1	1.59:1
	+16 3/4	5,000	16 3/4	1,500/3,000	13.97	12.71	10.85:1	3.61:1
	+18 3/4	10,000	18 3/4	1,500/3,000	15.30	13.21	7.11:1	1.83:1
	20	2,000	21 1/4	1,500/3,000	7.80	6.86	8.16:1	1.15:1
	+20	2,000	21 1/4	1,500/3,000	16.88	15.35	10.85:1	2.52:1
	20	3,000	21 1/4	1,500/3,000	7.80	6.86	8.16:1	1.15:1
	+20	3,000	21 1/4	1,500/3,000	16.88	15.35	10.85:1	2.52:1
	+21 1/4	7,500	21 1/4	1,500/3,000	16.05	13.86	7.11:1	1.63:1
	+21 1/4	10,000	21 1/4	1,500/3,000	16.05	13.86	7.11:1	1.63:1

+Shear Ram
*Lower pressures for normal use, higher pressures for emergency only.

Bag type (or annular) preventers, diverters, strippers

Hydril Company, Los Angeles, California

Model or type	BOP size, in.	Working pressure max. PSI	Vert. bore in.	Hydraulic control, max. PSI	Gal.* to close	Gal.* to open	Packoff open hole, min. PSI
GK	6	3,000	7 1/16	1,500	2.85	2.24	1,000
GK	6	5,000	7 1/16	1,500	3.86	3.30	1,000
GK	8	3,000	8 15/16	1,500	4.33	3.41	1,050
GK	8	5,000	8 15/16	1,500	6.84	5.80	1,150
GK	10	3,000	11	1,500	7.43	5.54	1,150
GK	10	5,000	11	1,500	9.81	7.97	1,150
GK	12	3,000	13 5/8	1,500	11.36	8.94	1,150
GK	13 5/8	5,000	13 5/8	1,500	17.98	14.15	1,150
GK	16	2,000	16 3/4	1,500	17.42	12.53	1,150
GK	16	3,000	16 3/4	1,500	21.02	15.80	1,150
GK	16 3/4	5,000	16 3/4	1,500	28.70	19.93	1,150
GK	18	2,000	17 7/8	1,500	21.09	14.44	1,110
GK	7 1/16	10,000	7 1/16	1,500	9.42	7.08	1,150
GK	9	10,000	9	1,500	15.90	11.95	1,150
GK	11	10,000	11	1,500	25.10	18.87	1,150
GK	13 5/8	10,000	13 5/8	1,500	34.53	24.66	1,150
.GL	13 5/8	5,000	13 5/8	1,500	19.76	19.76	1,300
.GL	16 3/4	5,000	16 3/4	1,500	33.8	33.8	1,300
.GL	18 3/4	5,000	18 3/4	1,500	44.0	44.0	1,300
.GL	21 1/4	5,000	21 1/4	1,500	58.0	58.0	1,300
MSP	6	2,000	7 1/16	1,500	2.85	1.98	1,000
MSP	8	2,000	8 15/16	1,500	4.57	2.95	1,050

MSP	10	2,000	11	1,500	7.43	5.23	1,150
MSP	20	2,000	20¾*	1,500	31.05	18.93	1,100
MSP	20	2,000	21¼	1,500	31.05	18.93	1,100
MSP	29½	500	29½	1,500	60.0**	0	1,500

Cameron Iron Works, Houston, Texas

A	6	5,000	7¹/₁₆	1,500	2.2	1.9	N.A.
A	6	10,000	7¹/₁₆	1,500	4.0*	3.1*	N.A.
A	6	15,000	7¹/₁₆	N.A.	N.A.	N.A.	N.A.
A	11	5,000	11	1,500	7.8	6.5	N.A.
A	11	10,000	11	1,500	12.1	10.5	N.A.
A	11	15,000	11	N.A.	N.A.	N.A.	N.A.
A	13⅝	5,000	13⅝	1,500	15.5	13.9	N.A.
A	13⅝	10,000	13⅝	1,500	21.5	18.7	N.A.
A	16¾	5,000	16¾	1,500	33.0	29.0	N.A.

Regan Force & Engineering Co., San Pedro, California

K	3	3,000	3	3,000	.2
K	4	3,000	4	3,000	.8
K	7	3,000	6¼	3,000	1.6
K	8⅝	3,000	7⅞	3,000	3.4
K	9⅝	3,000	8⅞	3,000	5.7
K	10¾	3,000	10	3,000	7.6

*U.S. gallons for full piston stroke.
**Full stroke, self opening.
• Volumes shown are closing chamber and opening chamber volumes only. Secondary chamber volumes are:

13⅝– 8.24	18¾–20
16¾–17.3	21¼–29.5

Bag type preventers cont.

Model or type	BOP size, in.	Working pressure, max. PSI	Vert. bore in.	Hydraulic control, max. PSI	Gal.* to close	Gal.* to open	Packoff open hole, min. PSI
K	11¾ (Old)	3,000	10⅞	3,000	8.1		
K	11¾ (New)	3,000	11⅛	3,000	10.3		
K	13⅜	3,000	12⅜	3,000	15.3		
K	16	3,000	13¾	3,000	19.9		
K	18⅝	3,000		3,000	22.5		
K		3,000		3,000	29.5		
KFD	16	300	8	1,000	1.75		
KFD	18⅝	300	8	1,000	2.5	Not applicable	Variable
KFD	20	300	8	1,000	2.5		
KFD	22	300	8	1,000	3.0		
KFD	24	300	8	1,000	3.0		
KFL	13⅝	3,000	13⅝	Well+500	19½		
KFL	13⅝	5,000	13⅝	Well+500	22		
KFL	13⅝	10,000	13⅝	Well+500	24½		
KFL	16¾	3,000	16¾	Well+500	25¾		
KFL	16¾	5,000	16¾	Well+500	29		
KFL	16¾	10,000	16¾	Well+500	31½		
KFL	20	2,000	20¾	Well+500	28½		
KFL	20	3,000	20¾	Well+500	32		
KFL	20	5,000	20¾	Well+500	35		
KFL	30	1,000	28	Well+500	47½		
KFL	30	2,000	28	Well+500	52		
KFL	30	1,000	26½	Well+500	51½		
KFL	30	2,000	26½	Well+500	56		

*U.S. gallons for full piston stroke.

Model or type	BOP size, in.	Working pressure max. PSI	Vert. bore in.	Hydraulic operator, PSI*	Gal. to close	Gal. to open	Close ratio	Open ratio
Types	6	3,000	7¹/₁₆	1,500/3,000	2.75	2.3	6:1	2.57:1
B & E	6	5,000	7¹/₁₆	1,500/3,000	2.75	2.3	6:1	2.57:1
	8	3,000	9	1,500/3,000	2.75	2.3	6:1	1.89:1
	8	5,000	9	1,500/3,000	2.75	2.3	6:1	1.89:1
	10	3,000	11	1,500/3,000	3.25	2.7	6:1	1.51:1
	10	5,000	11	1,500/3,000	3.25	2.7	6:1	1.35:1
	12	3,000	13⁵/₈	1,500/3,000	3.55	2.9	6:1	1.14:1
	14	5,000	13⁵/₈	1,500/3,000	3.55	2.9	6:1	1.14:1
	16	2,000	15½	1,500/3,000	3.65	3.0	6:1	1.05:1

*Lower pressures for normal use, higher pressures for emergency only.

Hydraulically operated valves cont.

Type or model	Line size, in.	Working pressure, max. PSI	Bore size, in.	Hydraulic operation, max. PSI	Gal. to open	Gal. to close
DV	4	3,000	4	1,500	0.8	1.1
DV	4	5,000	4	1,500	0.8	1.1
DV	6	3,000	7	1,500	2.1	3.6
DV	8	3,000	9	1,500	2.4	5.6
DV	10	3,000	11	1,500	5.7	11.4
DV	10	5,000	11	1,500	5.7	11.4
DV	12	3,000	13⅝	1,500	11.8	22.7

Rockwell Manufacturing Co., Pittsburgh, Pa.

Type or model	Line size, in.	Working pressure, max. PSI	Bore size, in.	Hydraulic operation, max. PSI	Gal. to open	Gal. to close
AC valve with U-1 hyd. opr.	2	2,000		2,500	.13	.11
	2	3,000		2,500	.13	.11
	2	5,000		2,500	.13	.11
	2	10,000		2,500	.21	.20
	2½	2,000		2,500	.26	.23
	2½	3,000		2,500	.26	.23
	2½	5,000		2,500	.26	.23
	2½	10,000		2,500	.45	.42
	3	2,000		2,500	.30	.25
	3	3,000		2,500	.51	.46
	3	5,000		2,500	.51	.46
	4	2,000		2,500	.69	.62
	4	3,000		2,500	.69	.62
	4	5,000		2,500	1.04	.98

Rucker-Shaffer, Houston, Texas

Type or model	Line size, in.	Working pressure, max. PSI	Bore size, in.	Hydraulic operation, max. PSI	Gal. to open	Gal. to close
Flo-Seal	2 Reg.	2,000	1¹¹/₁₆	3,000	.2	.2
	2	2,000	2¹/₁₆	3,000	.2	.2

Size					
2 Reg.	3,000	1¹¹/₁₆	3,000	.2	.2
2	3,000	2¹/₁₆	3,000	.2	.2
2 Reg.	5,000	1¹¹/₁₆	3,000	.2	.2
2	5,000	2¹/₁₆	3,000	.2	.2
2¹/₁₆	10,000	2¹/₁₆	3,000	.4	.4
2¹/₁₆	15,000	2⁹/₁₆	3,000	.4	.4
2½	2,000	2⁹/₁₆	3,000	.3	.3
2½	3,000	2⁹/₁₆	3,000	.3	.3
2½	5,000	3⅛	3,000	.3	.3
3	2,000	3⅛	3,000	.3	.3
3	3,000	3⅛	3,000	.3	.3
3	5,000	3¹/₁₆	3,000	.3	.3
3¹/₁₆	10,000	4¹/₁₆	3,000	.6	.6
4	3,000	4¹/₁₆	3,000	.8	.8
4	5,000	4¹/₁₆	3,000	.8	.8
4¹/₁₆	10,000	7¹/₁₆	3,000	1.3	1.3
6	3,000		3,000		

Flo-Seal with Ramlock

Size					
2 Reg.	2,000	1¹¹/₁₆	3,000	.3	.3
2	2,000	2¹/₁₆	3,000	.3	.3
2 Reg.	3,000	1¹¹/₁₆	3,000	.3	.3
2	3,000	2¹/₁₆	3,000	.3	.3
2 Reg.	5,000	2¹/₁₆	3,000	.3	.3
2	5,000	2¹/₁₆	3,000	.3	.3
2¹/₁₆	10,000	2⁹/₁₆	3,000	.4	.4
2¹/₁₆	15,000	2⁹/₁₆	3,000	.4	.4
2½	2,000	2⁹/₁₆	3,000	.3	.3
2½	3,000	3⅛	3,000	.3	.3
2½	5,000	3⅛	3,000	.4	.4
3	2,000	3⅛	3,000	.4	.4
3	3,000	3¹/₁₆	3,000	.4	.4
3	5,000	4¹/₁₆	3,000	.6	.6
3¹/₁₆	10,000	4¹/₁₆	3,000	.8	.8
4	3,000	4¹/₁₆	3,000	.8	.8
4	5,000	7¹/₁₆	3,000	.8	.8
4¹/₁₆	10,000		3,000		
6	3,000		3,000		

Hydraulically operated valves cont.

Type or model	Line size, in.	Working pressure, max. PSI	Bore size, in.	Hydraulic operation, max. PSI	Gal. to open	Gal. to close
Type DB	3	3,000	3⅛	3,000	.3	.3
	3	5,000	3⅛	3,000	.3	.3
3¹⁄₁₆	4	10,000	3¹⁄₁₆	3,000	.6	.6
	4	3,000	4¹⁄₁₆	3,000	.8	.8
	4	5,000	4¹⁄₁₆	3,000	.8	.8
4¹⁄₁₆		10,000	4¹⁄₁₆	3,000	1.3	1.3
	6	3,000	7¹⁄₁₆	3,000	2.0	2.0

APPENDIX B

Casing Design Problems

Example B.1 Using the following known data, design a casing string according to the procedures established in Chapter 5.

Casing string type:	Intermediate
Casing size:	9⅝ in.
Setting depth:	10,000 ft
Surface equipment rating:	5,000 psi
Fracture gradient at shoe:	17.4 ppg
Maximum mud weight below casing:	17.0 ppg
Mud weight set in:	12.0 ppg
Minimum drift acceptable:	8.5 in.
Limit on number of sections:	4

Solution
1. Injection pressure
    = (Frac. grad. + safety factor) (setting depth) (0.052)
    = (17.4 + 1.0) (10,000 ft) (0.052)
    = 9,568 psi

2. From Equation 5.10
    $x + y = D$
    $x = D - y$
    $x = 10,000 - y$

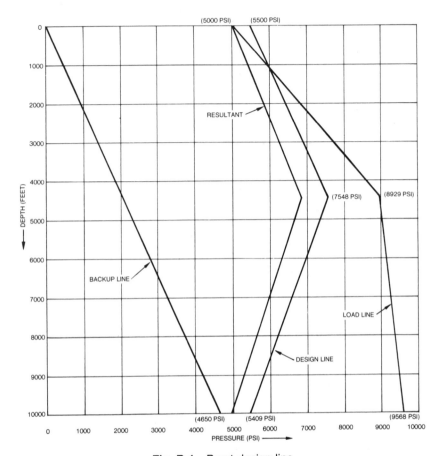

Fig. B-1 Burst design line

3. **From Equation 5.11**
   $$P_s + xG_m + yG_G = \text{Injection pressure}$$
   $$5{,}000 + 0.884(x) + 0.115(y) = 9{,}568$$
   $$.884(10{,}000 - y) + 0.115(y) = 9{,}568 - 5{,}000$$
   Solving: $y = 5{,}555$ **ft (gas length)**
   $x = 4{,}445$ **ft (mud length)**

4. **Plotting these values as described in Chapter 5 yields Figure B.1.**

5. **From a casing catalog, a selection of weight and grades was made that would satisfy the design line (Fig. B.2).**

6. **The collapse design is presented in Fig. B.3. The backup line was calculated using the following procedure:**

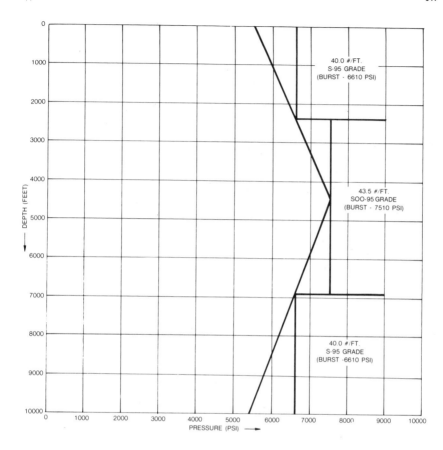

Fig. B-2 Weight and grade selection

a) **Pressure from a column of saltwater =**
   0.465 psi/ft × 10,000 ft = 4,650 psi
b) **Length of 17.0 ppg mud column which equals this pressure:**
   4,650 psi = (x) (0.884 psi/ft)
   x = 5,260 ft

7. **The tension design is shown in Fig. B.4.**

8. **Biaxial calculations for burst are as follows:**

a) **Section 1.**
   Bottom: Stress = Load/area = 71,473 lb/11.454 sq in. =
   6,240 psi

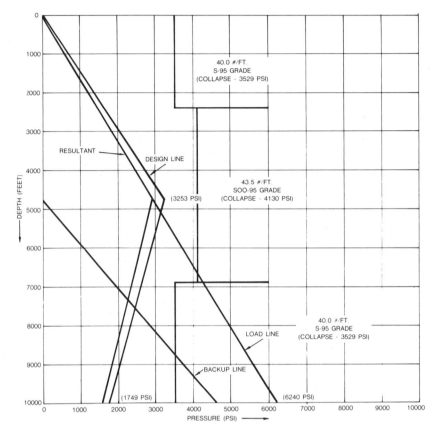

Fig. B-3 Collapse design line

Percent of average yield strength = Stress/average y.s.
= 6,240/105,000 = 5.9%
From the ellipse (Fig. B.5): % Decrease = 97%
Actual burst = (6,610 psi) (0.97) = 6,412 psi
Top: 54,000 lb/11.454 sq in. = 4,714 psi
Percent of average yield strength = 4,714 psi/105,000 psi
= 4.5%
From the ellipse: % decrease = 102%
Actual burst = (6,610 psi) (1.02) = 6,742 psi

b) Section 2 (Calculation procedures are the same as 8(a))

Bottom	**% Increase = 101.5%**
	Actual burst = 7,623 psi
Top	**% Increase = 108%**
	Actual burst = 8,111 psi

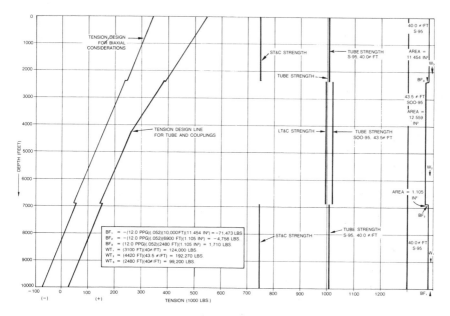

Fig. B-4   Tension design

c)   Section 3

Bottom	**% Increase = 108.5%**
	Actual burst = 7,171 psi
Top	**% Increase = 111%**
	Actual burst = 7,337 psi

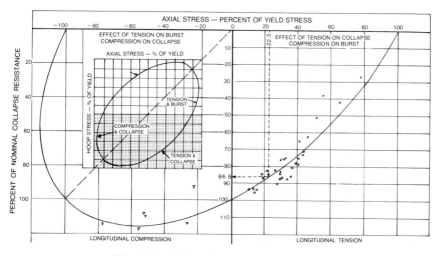

Fig. B-5   Holmquist and Nadai ellipse

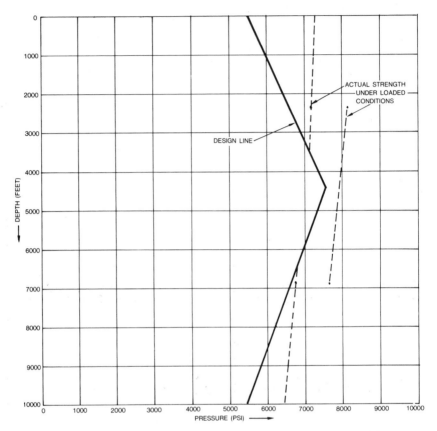

Fig. B-6 Casing burst strengths under the given loading conditions

9. These values are then plotted on Fig. B.6.

10. Biaxial calculations are made for collapse similar to the burst calculation. The results are presented in Fig. B.7.

11. The final design is as follows:

Section	Interval, ft	Length, ft	Weight, lb/ft	Grade	Coupling
1	6,460-10,000	3,540	40.0	S-95	ST&C
2	3,540-6,460	2,920	43.5	S00-95	LT&C
3	0-3,540	3,540	40.0	S-95	ST&C

Example B.2 Use the following data to design the required casing string.

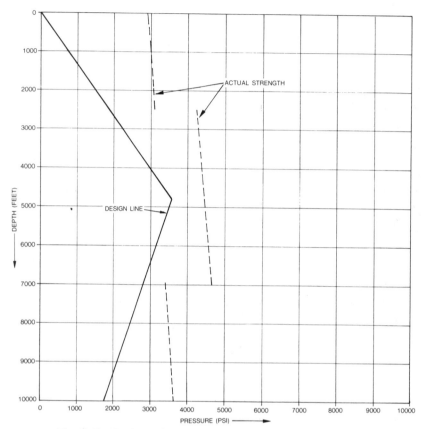

Fig. B-7 Casing collapse strengths under loaded conditions

Casing string type:	Surface
Casing size:	13⅜ in.
Setting depth:	3,500 ft
Fracture gradient at shoe:	14.6 ppg
Mud weight set in:	9.2 ppg
Minimum drift acceptable:	12.25 in.
Limit on number of sections:	2

Solution

1. Injection pressure = (F.G. + S.F.) (S.D.) (0.052)
   = (14.6 + 1.0) (3,500) (0.052)
   = 2,839 psi

2. Surface Pressure = Injection pressure − gas hydrostatic pressure
   = 2,839 psi − (0.115 psi/ft) (3,500 ft)
   = 2,437 psi

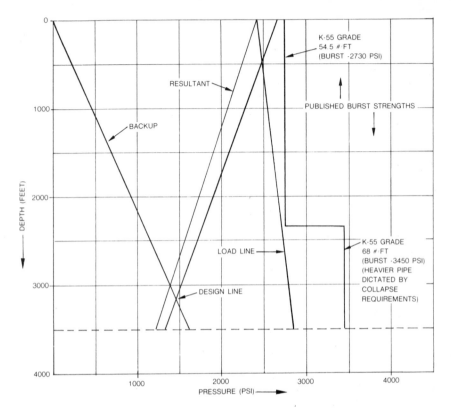

Fig. B-8 Surface casing burst design

3. Plot the burst and collapse designs as shown in Fig. B.8 and B.9.

4. Plot the tension design shown in Figure B.10.

5. Calculate the effect of biaxial loading on the tentative design and prepare the final design shown in Fig. B.11. The collapse design is shown since it is the controlling criteria.

6. The final design is as follows:

Section	Interval, ft	Length, ft	Weight, lb/ft	Grade	Coupling
1	2,000-3,500	1,500	68	K-55	ST&C
2	0-2,000	2,000	54.5	K-55	ST&C

Example B.3 Design the following casing strings assuming that they are to be run in conjunction with each other.

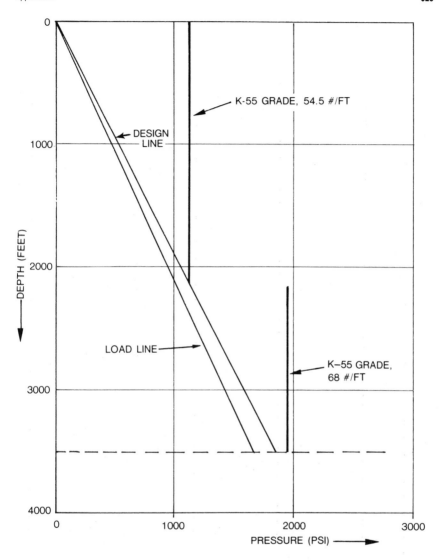

Fig. B-9 Surface casing collapse design

I.

Casing string type:	**Intermediate**
Casing size:	**7 in.**
Setting depth:	**8,600 ft**
Surface equipment rating:	**3,000 psi**
Fracture gradient at shoe:	**16.5 ppg**
Maximum mud weight below casing:	**16.1 ppg**

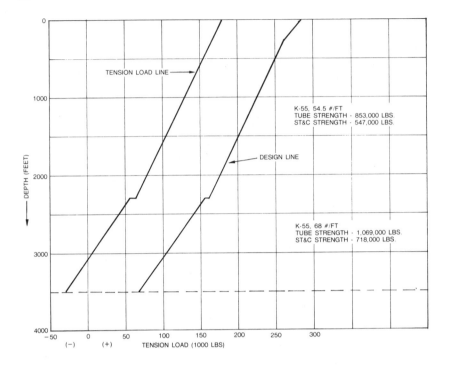

Fig. B-10 Surface casing tension design

Mud weight set in:	11.0 ppg
Minimum drift acceptable:	6.0 in.
Limit on number of sections:	3

II.

Casing string type:	Drilling liner
Casing size:	5 in.
Setting depth:	10,500 ft
Length of overlap:	300 ft
Surface equipment rating:	3,000 psi
Fracture gradient at shoe:	17.4 ppg
Maximum mud weight below casing:	17.0 ppg
Mud weight set in:	16.1 ppg
Minimum drift acceptable:	4.0 in.
Limit on number of sections:	1
Special considerations:	Couplings must be streamlined

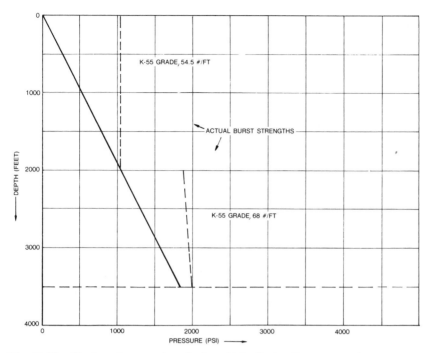

Fig. B-11 Final surface casing design. Note that collapse was the controlling criterion on this string

Solution

1. Injection pressure = (17.4 + 1.0) (10,500 ft) (0.052) = 10,046 psi

2. $x + y = D$
   $x = 10,500 \text{ ft} - y$

3. $P_s + xG_m + yG_g = $ Injection pressure
   3,000 psi + x(0.884 psi/ft) + y(0.115 psi/ft) = 10,046 psi
   Solving: x = 7,592 ft (mud)
                     y = 2,908 ft (gas)

4. Plot the burst and collapse designs as shown in Figs. B.12 and B.13. Note the discontinuity in the collapse design lines.

5. Using the tentative design from Fig. B.12 the tension design line is constructed and the biaxial calculations made. The final design is as follows:

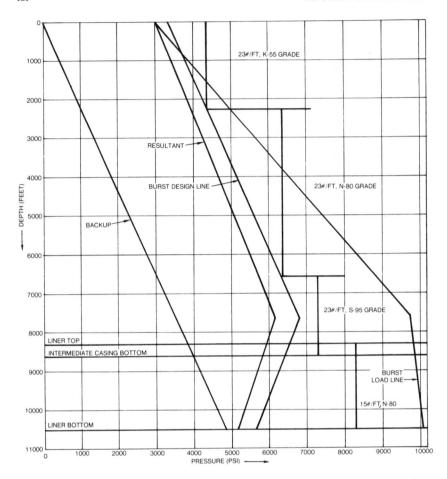

Fig. B-12 Burst design for intermediate casing with drilling liner and the liner

Section	Interval, ft	Length, ft	Weight, lb/ft	Grade	Coupling
(Intermediate casing)					
1	7,200-8,600	1,400	23	S-95	ST&C
2	3,200-7,200	4,000	23	N-80	ST&C
3	0-3,200	3,200	23	K-55	LT&C
(Liner)					
4	8,300-10,500	2,200	15	N-80	XL

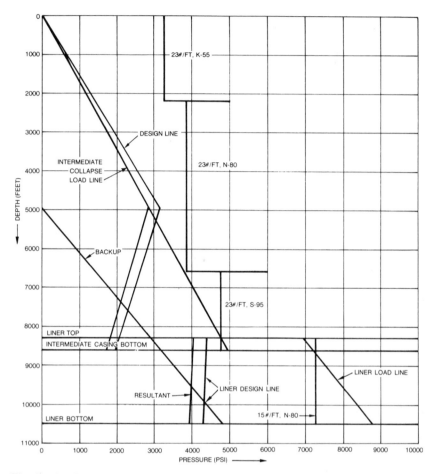

Fig. B-13   Collapse design for intermediate casing with a drilling liner, and the liner

Example B.4   Design a production casing string using the following data.

Casing string type:	**Production**
Casing size:	**5 in.**
Setting depth:	**13,000 ft**
Producing zone pressure:	**11,600 psi**
Type of production expected:	**Dry gas**
Mud weight set in:	**17.4 ppg**
Packer fluid weight:	**17.4 ppg**

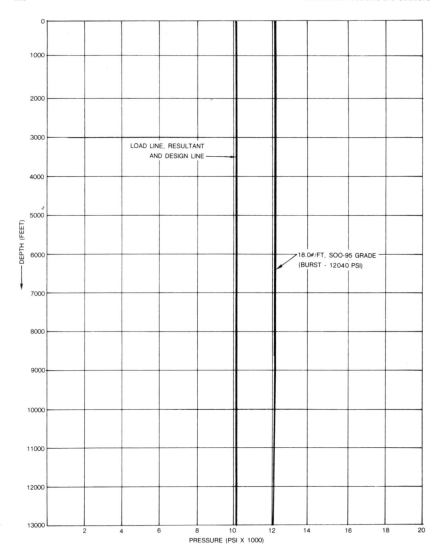

Fig. B-14 Production casing burst design

Solution

1. **Using the procedures established in Chapter 5, the burst and collapse design lines are constructed as shown in Figs. B.14 and B.15.**

2. **After the tension line is constructed, the biaxial calculations are made and the string upgraded if necessary.**

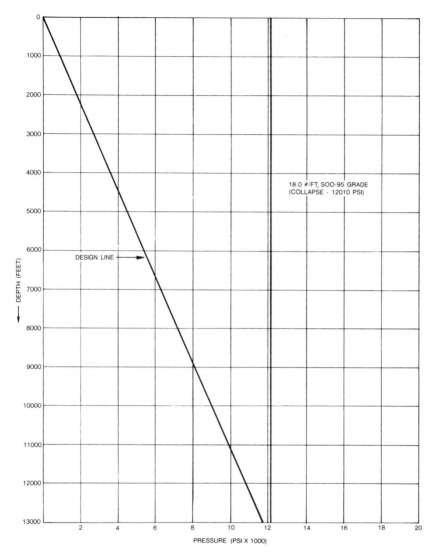

Fig. B-15   Production casing collapse design

3. The final design is as follows:

Section	Interval, ft	Length, ft	Weight, lb/ft	Grade	Coupling
1	0-13,000	13,000	18.0	S00-95	LT&C

APPENDIX C

Well Control Contingency Plan—Appendix C-1

Hydrogen Sulfide Contingency Plan—Appendix C-2

APPENDIX C-1

Well Control Contingency Planning

THE MOST EFFECTIVE MEANS of controlling kicks and preventing blowouts are
- training on all levels of employment,
- coordination of responsibilities, and
- communication during the actual kick-killing process.

This contingency plan will define the minimum responsibilities of each crew member relative to well control while drilling and kick killing. Although formal training is a prerequisite to effective well control, it is beyond the scope of this plan.

Crew Member Drilling Responsibilities

Drilling responsibilities placed on various members of the drilling crew and service companies must be followed on a day-to-day basis in order to be fully prepared should a kick occur. If these responsibilities are executed routinely, kick severity can be minimized, and many kicks can be eliminated altogether. Some, but not all, of these responsibilities are listed below according to each specific crew member's job classification.

Rotary helper (roughneck)

1. Report immediately any abnormal changes in routine drilling operations to the driller. These abnormal changes may be, but are not limited to pit level fluctuations, flow rate leaving the well, mud weight, mud properties, and pump output.

2. Daily check all mud lines, valves, de-gassers, and mud-gas separators to insure that they are functioning properly. If necessary, repair the equipment to restore it to the proper working conditions.

3. Flush the de-gasser and mud-gas separator daily when they are not in use.

Derrickman

1. Check the barite supplies and record these amounts on the daily drilling report. Insure that the barite supplies remain pumpable from the bulk hoppers.

2. Immediately notify the driller if any abnormal changes occur with respect to the mud properties such as density and viscosity, pump output behavior, and cuttings characteristics.

3. Routinely monitor all well control equipment to insure that it is functioning properly. This equipment should include, but not be limited to the flow monitoring, gas detection and pit level equipment, mud de-gassing and mixing equipment, and pump output performance.

Driller (and assistant driller)

1. Routinely monitor all well control equipment to insure that it is functioning properly. This equipment should include, but not be limited to the flow monitoring, gas detection and pit level equipment, mud de-gassing and mixing equipment, choke manifolds including the choke(s), choke lines, and all valves on the manifold, blowout preventer stacks including the preventer elements, choke and kill lines and all adjacent valves, and the accumulator and all hydraulic lines to the preventer stack.

2. All equipment and lines to be used in the kick killing operation should be flushed daily to prevent the build-up of a barite plug.

3. The driller should insure that an operational full opening valve and closing wrench is kept on the rig floor at all times. He should instruct the crew on its presence and proper use.

4. The driller should check all drilling breaks for kick potentials by drilling 3-5 ft into the break and then picking up the kelly and checking for abnormal flow from the well. If flow is occurring, proceed to close in the well according to the established company policies and drilling conditions.

5. On a tourly basis, establish a kill rate (low rate) and pressure. The observed kill rate and pressure should be recorded on the daily drilling report. Also, a kill rate should be re-established when long sections of open hole have been drilled, mud weight or mud property changes have been made, drill string assemblies have been altered, or pump sizes have been changed.

Toolpusher (and assistant toolpusher)

1. The toolpusher must insure that all crew members have carried out their responsibilities properly.

2. It is also the responsibility of the toolpusher to insure that each new crew after a tour change is informed of the prevailing drilling and well control problems that may have occured during the last tour. (This is routinely done by the drillers on each crew.)

3. All existing rules and regulations concerning drilling and well control in each particular drilling area should be established and the rig should be made to conform to these standards.

Mud engineer

1. The mud engineer should check the barite supplies daily or cause the barite supplies to be checked. He must insure that enough barite is maintained on board such that a one pound per gallon kick can be killed safely.

2. The mud engineer should routinely monitor the mud system and report any abnormal occurrences such as pit level fluctuations, mud density changes, mud chloride changes, gas content increases, and shale cuttings characteristics.

3. A number of calculations should be made on a routine basis and properly reported on the daily reporting forms. Some, but not all, of these calculations are annular volumes, drill pipe volumes, the pump output, the number of pump strokes necessary to displace the drill pipe and collars, the drill pipe displacement, and surface pit volumes.

Company representative

1. The company representative should insure that all members of the crew have properly performed their functions. It may be necessary for the company representative to conduct informal rig site classes to instruct each crew in their respective duties related to well control. Specific attention should be paid to barite supplies, establishment of the kill pumping rates, shut-in procedures, and blowout preventer testing.

2. Certain calculations should be made on a routine basis. Some, but not all, of these calculations are the fracture gradient at the casing seat or any other known weak formation, the formation pressure, maximum pulling and running speeds for the drill string to prevent swabbing or formation fracture, and the amount of open hole that can be drilled and still allow a safety margin should a 0.5 pounds per gallon kick be taken inadvertently.

3. The company representative should define all existing government rules and regulations with respect to drilling and well control, and then insure that the drilling rig and the company meet these requirements.

Crew Member Well Control Responsibilities

There are a fewer number of well control responsibilities as compared to drilling responsibilities that the drilling crew and service men must meet. However, each responsibility must be met in order to safely kill any potential blowout. Some, but not all, of these duties are listed below according to each specific crew member's job classification.

Floorhand (roughneck)

1. The floorhand should aid in maintaining and mixing the mud, either in the mud room or at the shaker, as directed by the company representative.

2. He should routinely check and observe the pumps, degassers, and preventers during the kill operations.

Derrickman

1. The derrickman should maintain the mud system as directed and immediately report any problems to the company representative.

Driller (or assistant driller)

1. The driller should direct the crew according to the company representative's orders and insure that previously defined responsibilities are carried out.

2. He should insure that the preventers and choke manifold are routinely monitored to detect any leaks that may occur during the kill operation.

3. The driller must operate the pumps as directed.

4. He should rotate and reciprocate the drill pipe as necessary to prevent pipe sticking, if so directed by the company representative.

Toolpusher (or assistant toolpusher)

1. The toolpusher should insure that all crew members execute their assigned responsibilities during the kick killing procedures. If a crew tour change occurs during the operations, he should inform the new crew of all prevailing conditions.

2. The toolpusher should aid the company representative when requested to do so.

Mud engineer

1. The mud engineer should monitor the mud density throughout the kill operations and make the proper records.

2. He should make any calculations requested by the company representative.

Company representative

1. The company representative should assume full kick killing responsibilities and direct the crew as necessary according to the proper kick killing guidelines.

2. If problems should arise that have not been fully explained, he should contact the nearest company engineering office for assistance.

APPENDIX C-2

Hydrogen Sulfide Drilling and Contingency Plan

**(Courtesy of Pollution Control Rentals, Inc.,
Lafayette, Louisiana)**

APPENDIX C-2

Parsonns and Roberts Enterprises, Inc.

Lafayette Wildcat #1

Lafayette Parish, Louisiana

Hydrogen sulfide drilling and contingency plan

Table of contents

Tables

Purpose of program

It is Parsonns and Roberts' policy in all operations where high pressure and toxic gases may be encountered to do everything reasonably possible to insure the safety of its employees and the service company and contractor's employees or anyone on the job site. This brochure has been prepared to outline the operational procedures to be followed in order to provide the maximum safety and comfort for all persons near the operation.

Hydrogen Sulfide is extremely hazardous to normal oil field operations due to (1) its capability of destroying life at very low concentrations and (2) causing instantaneous failure of high-strength metals. Drilling and producing operations of hydrocarbons containing toxic gases can, however, be performed safely and without incident when the necessary precautions are taken and the outlined safety procedures followed. It is imperative (1) that sulfide resistant materials be used, (2) the proper safety equipment be used, (3) that this equipment be properly maintained, and (4) all safety regulations be followed.

The procedures outlined herein are for your safety and the safety of all others. Therefore, it is mandatory that each individual give his one hundred per cent cooperation.

John Doe,

Parsonns and Roberts
Enterprises, Inc.
Lafayette, Louisiana

Responsibilities and duties

All personnel

1. It is the responsibility of all personnel on location to familiarize themselves with the safety procedures.

2. All personnel will attend to their personal safety first.

3. Help anyone who may be injured or overcome by toxic gases. The drilling supervisor will assign someone to administer first aid to unconscious person(s).

4. Report to the designated "SAFE BRIEFING AREA" and follow the instructions of the drilling supervisor.

Drilling supervisor

1. It is the responsibility of the drilling supervisor to see that these safety and emergency procedures are observed by all personnel on location.

2. The drilling supervisor will advise Parsonns and Roberts whenever the procedures as specified herein are complied with or cannot be followed.

3. The drilling supervisor will notify the safety consultant at least two weeks before the safety equipment specified herein is needed.

4. The drilling supervisor will keep the number of personnel on location to a minimum during hazardous operations.

5. The drilling supervisor is responsible for designating the "SAFE BRIEFING AREA". This area will change depending

upon wind direction and must be redesignated as soon as a
wind change occurs.

6.   If an unexpected emergency occurs or the H_2S alarm sounds,
     the drilling supervisor will assess the situation and will advise
     all personnel of the existing conditions.

The drill site

The location as shown in Fig. C-1 is planned so as to obtain the
maximum safety benefits consistent with the rig configuation, well
depth, and prevailing winds.

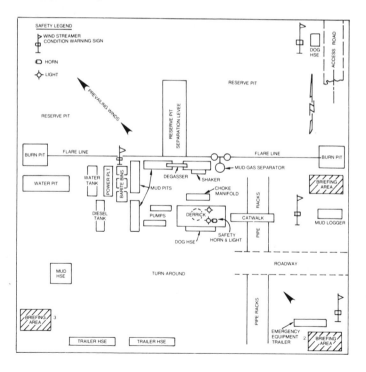

Fig. C-1   Lafayette Wildcat #1 Drilling equipment layout

1. Using the various maps, the area within a three-mile radius of the location has been surveyed and contacts with all permanent residents have been made. Except in a dead calm and with a tremendous release of high concentration gases, the

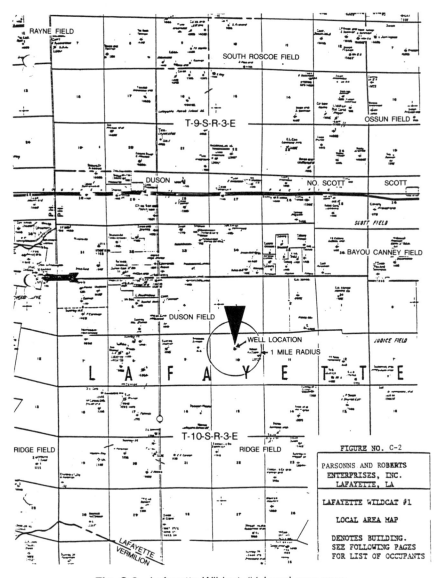

Fig. C-2   Lafayette Wildcat #1 Local area map

probability of lethal dosages beyond one mile is extremely unlikely. Note on the rig layout plot, Fig. C-1, the direction of prevailing winds.

2. The location of houses, schools, roads, and any place where people may be present and who might need to be warned or evacuated in a crisis have been surveyed. This information with names and telephone numbers are keyed and listed on Fig. C-3 and Table C-2 for use if evacuation might be necessary should an emergency develop.

3. The State Regulatory Body will be provided with this information and plan.

4. The drilling rig (Fig. C-1) will be situated on location such that prevailing winds blow across the rig toward the reserve pit.

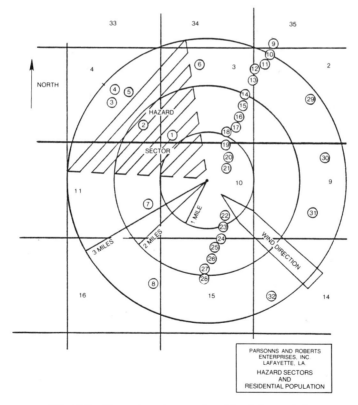

Fig. C-3   Hazard sectors and residential population

5. There will be three SAFETY BRIEFING AREAS established not less than 200 ft from the wellhead and in locations that at least one area will be up-wind of the well at all times.

6. Protective equipment will be stored in a safety trailer located at the up-wind briefing area. Equipment will include air packs and masks, first aid kits, fire extinguishers, stretchers, hydrogen sulfide hand-operated detectors and resuscitators. In the event of an emergency, personnel should assemble at the up-wind briefing area for instructions from their supervisor.

7. Wind socks or streamers will give wind directions at several elevations (i.e., tree top, derrick floor level, and 6 to 8 ft above ground level). Personnel should develop the practice of routine observation of wind direction.

8. Mud tanks will be located at least 80 ft away from the derrick substructure to facilitate circulation of fresh air around the cellar area.

9. A gas trap and degasser will be installed on the mud tanks with two flare lines laid in opposite directions, securely staked, and equipped with butane and pilot lights which can be lighted remotely.

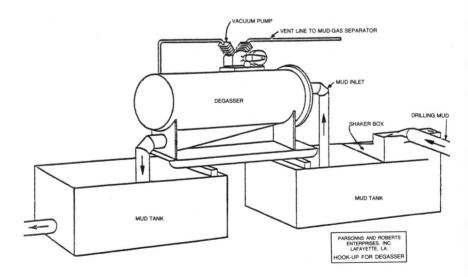

Fig. C-4 Hook-up for degasser

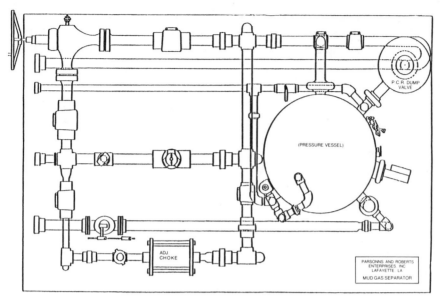

Fig. C-5 Mud-gas separator

10. All windbreakers and rig curtains will be removed from around the derrick floor and monkey board, regardless of conditions.

11. Explosion proof ventilating fans (bug blowers) will be positioned to insure adequate circulation at the monkey board, derrick floor, cellar area, shale shaker, and any other location where hydrogen sulfide might accumulate and need to be dissipated.

12. A kill line of ample strength and securely staked will be laid to the wellhead from a safe location to permit pumping into the well in an emergency.

13. When approaching a depth where hydrogen sulfide may be encountered, the mud will be maintained in an overbalanced condition to preclude the entry of formation fluids into the wellbore and thereby restrict the hydrogen sulfide to be treated to that contained in the formation drilled.

14. When approaching a depth where hydrogen sulfide may be encountered, appropriate warning signs will be posted on all access roads to the location and at the foot of all stairways to the derrick floor.

15. Reliable 24-hour radio and telephone communication will be available at the rig. Emergency telephone numbers for Sheriff's Department, ambulance, hospitals, doctors, and operator's supervisory personnel will be prominently posted.

16. Filter-type gas masks are not suitable for use on drilling rigs. Pressure demand, airpack type masks will be provided for use in any hydrogen sulfide concentration. The pressure demand, airpack types have alarms that signal when air supply is getting low. They are easily and quickly serviced with replacement bottles. They are not physically exhausting to use, are rugged and dependable and if properly stored, require little maintenance.

17. Masks will be stored on racks and protected from the weather. Rig crew equipment will be located at a readily accessible location on the rig floor. There will be sufficient masks stored in safety sheds for every person working in the area. For hygienic reasons, masks are to be washed and sterilized at regular intervals. Proper cleaning material will be provided. The derrickman will have a mask. It will be equipped with a connection through a quick-disconnect from his system of breathing air so that if he must evacuate the derrick, he will have a full air bottle with his mask. An eight outlet air supply manifold will be installed on the rig floor for continuous use by crews and supervisory personnel working in a "masks-on" situation. The multi-bottle supply cylinder is to be located at least 200 ft from the well. A standby air compressor for pumping pure breathing air to the workmen when wind conditions are favorable will be provided so that the cylinders of air can be saved for a situation when the lease becomes contaminated with H_2S gas. Utilization of the compressor will allow long work periods under masks without shutdown.

18. Provisions for air will include one spare airpack bottle for each pack, several extra bottles, extra 300 cu ft capacity air cylinders and a bottle filling manifold at the safety trailer and at least one resuscitator at each briefing area. All men will be trained in the techniques of bottle filling.

19. Hand-operated hydrogen sulfide detectors will be provided at each briefing area and on the rig floor.

20. An alarm system that can be heard during operations and which can be activated from several points if gas is detected will be installed. When the alarm is sounded, personnel must put on masks or move to the briefing area designated as safe.

21. There will be designated smoking areas with smoking prohibited elsewhere.

22. A safety coordinator from each crew will be designated and provided a check list for his use. When approaching a depth where hydrogen sulfide may be encountered, a 24-hour safety advisor will be provided.

23. Sleeping in cars on location will not be permitted. A parking area remote from the location will be designated.

24. Safety meetings and training sessions will be held at frequent intervals by the safety advisor, the drilling supervisor, or the rig supervisor. All persons required to work on location will be thoroughly familiar with the use, care and servicing of personal protective equipment, resuscitation equipment, and gas detection equipment. New employees and those who are present on a sporadic basis (i.e., geologists, engineers, service personnel, etc.) will be indoctrinated in the location and use of personal equipment before commencing work.

25. All electric lighting, wiring, and electrical devices within 100 ft of the well will be put in vapor proof condition to minimize the possibility of explosion.

26. Blowout preventers, particularly the ram carrier rods, will be sour gas trimmed dressed for hydrogen sulfide service. Choke manifolds will be of similar materials.

27. Installation, operation, and testing of blowout preventers, choke manifolds, etc., dressed for hydrogen sulfide service, will be conducted regularly. See Fig. C-6 for the blowout preventer stack and Fig. C-7 for manifold layout.

28. An accurate bottom hole location by use of multishot and single shot directional surveys will be maintained so that the well can be intercepted if it becomes necessary.

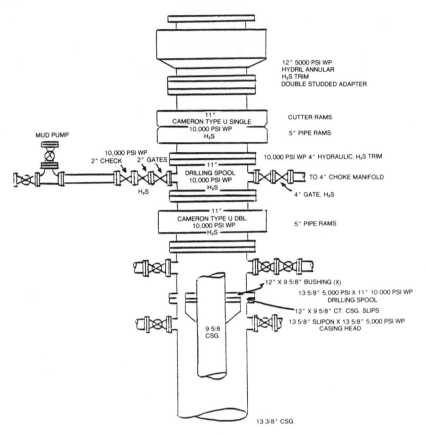

Fig. C-6   Blowout preventer stack for extremely hazardous high pressure service

29.   Every person involved in the operation will be informed of the characteristics of hydrogen sulfide, its dangers, safe procedures to use when it is encountered, and recommended first aid procedures. This will be done through frequent safety talks and training sessions.

Rig personnel

The following is a list of the personnel involved in the drilling of the well when hydrogen sulfide may be encountered and the nature of

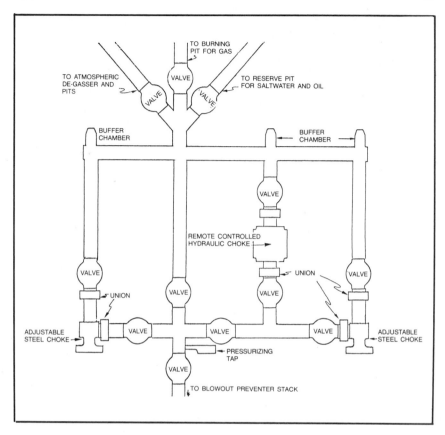

Fig. C-7 Choke manifold (H$_2$S trim)

their service on the well. They are grouped as regulars, temporary, outside service personnel, and visitors.

Regular crew, contractor

1. Rig superintendent—The rig superintendent will be available 24 hours a day. It will be required that either he or Parsonns and Roberts' drilling supervisor be on location at all times when drilling in or near the Smackover formation. At no time will both be away from the rig at the same time. It may become necessary to have an additional rig supervisor if hydrogen sulfide is encountered and creates a problem. If so, one will be required on location and awake at all times.

2. The contract rig crew will consist of at least 5 men to run the rig.

Company personnel

1. A drilling supervisor will be on location at all times or available immediately. Two will be required if hydrogen sulfide is encountered and creates well conditions that require 24 hour supervision. The drilling supervisor on duty will have complete charge of the rig operation, and will take whatever action is deemed necessary on an emergency basis to insure personnel safety, to protect the well, and to prevent property damage. He will maintain communication with the Parsonns and Roberts manager of drilling, or his designated alternate, and keep him informed of emergencies and deviations from the drilling plan for the subject well.

2. A well site geologist will be on location as required and will be familiar with safety procedures and emergency drills.

3. All personnel will be briefed as to the action that must be taken in case of an alert.

Outside services

1. Safety advisor—The safety advisor will be available when drilling begins below intermediate casing or as otherwise deemed necessary. He will conduct safety talks and drills, maintain the safety equipment, and advise and carry out the instructions of the drilling supervisor.

2. Mud engineer—A mud engineer will be on location at all times, when drilling at or near a depth where hydrogen sulfide may be expected and/or when deemed necessary by the drilling supervisor.

Temporary service personnel

All service people such as cementing crews, logging crews, specialists, mechanics, and welders will furnish their own safety equipment. Should any of these people not have suitable equipment, they will either be refused admittance or assigned air equipment on location. The drilling supervisor will insure that the number of people on location does not exceed the number of masks on location, and that they have been briefed in regard to safety procedures. He will also be sure each of these people know about smoking and briefing areas and know what to do in case of an emergency alert or drill.

Visitors

Visitors will be restricted, except with special permission from the drilling supervisor, when hydrogen sulfide might be encountered. They will be briefed as to what to do in case of an alert.

Drilling operations procedures

In addition to those safety procedures outlined under "The Drill Site", the following precautions will be followed after hydrogen sulfide has been detected.

1. Put on masks at least 15 minutes before "bottoms-up" after a trip and wear until the shale shaker and pit area test no more than 10 ppm hydrogen sulfide. This also applies when circulating out gas cut returns and drilling breaks.

2. Put on masks 20 stands before a core barrel reaches the surface and wear masks while laying down and examining the core.

3. Test will not be conducted through the drill pipe where the presence of hydrogen sulfide may be suspected. If testing is needed, special tubing and packer will be run and the tree installed. Put on masks before formation fluids are expected at the surface and wear until flares are lighted and work areas test no more than 10 ppm hydrogen sulfide and the area has been declared safe. During the test, montior the areas near the wellhead, separation equipment, and storage facilities for hydrogen sulfide.

Physical and chemical properties of
hydrogen sulfide (H_2S)

1. Extremely toxic (almost as toxic as hydrogen cyanide and 5 to 6 times as toxic as carbon monoxide).

2. Colorless.

3. Offensive odor, often described as that of rotten eggs.

4. Heavier than air—specific gravity 1.189 (Air = 1.000 @ 60°F). Vapors may travel considerable distance to a source of ignition and flash back.

5. Forms an explosive mixture with a concentration between 4.3 and 46 percent by volume with auto-ignition occuring at 500°F.

6. Burns with a blue flame and produces sulfur dioxide (SO_2), which is less toxic than hydrogen sulfide but very irritating to eyes and lungs and cause serious injury.

7. Soluble in both water and liquid hydrocarbons.

8. Produces irritation to eyes, throat and respiratory system.

9. Threshold limit value (TLV)—Maximum of eight hours exposure without protective respiratory equipment—10 ppm.

10. Corrosive to all electrochemical series metals.

Physical effects of hydrogen sulfide poisoning

The principal hazard is death by inhalation. When the amount of gas absorbed into the blood stream exceeds that which is readily oxidized, systemic poisoning results, with a general action on the nervous system. Labored respiration occurs shortly and respiratory paralysis may follow immediately at concentrations of 600 ppm and above. This condition may be reached almost without warning as the originally detected odor of hydrogen sulfide may have disappeared due to olfactory paralysis. Death then occurs from asphyxiation unless the exposed person is removed immediately to fresh air and breathing stimulated by artificial respiration. Other levels of exposure may cause the following symptoms individually or in combination:

a. Headache
b. Dizziness
c. Excitement
d. Nausea or gastro-intestinal disturbances

TABLE C-1
TOXICITY OF HYDROGEN SULFIDE

PPM (Parts per million)	0-2 minutes	2-15 minutes	15-30 minutes	30 Minutes to one hour	1-4 hours	4-8 hours	8-48 hours
5-100				Mild conjunctivitis; respiratory tract irritation			
100-150		Coughing; Irritation of eyes; loss of sense of smell.	Disturbed respiration; pain in eyes; sleepiness	Throat irritation	Salivation and mucous discharge; sharp pain in eyes; coughing	Increased symptoms.*	Hemorrhage and death*
150-200		Loss of sense of smell.	Throat and eye irritation.	Throat and eye irritation	Difficult breathing; blurred vision; light shy.	Serious irritating effect.*	Hemorrhage and death*
250-350	Irritation of eyes; loss of sense of smell	Irritation of eyes	Painful secretion of tears, weariness	Light shy, nasal catarrh; pain in eyes, difficult breathing	Hemorrhage and death*		
350-450		Irritation of eyes; loss of sense of smell.	Difficult respiration; coughing; irritation of eyes.	Increased irritation of eyes and nasal tract; dull pain in head; weariness; light shy.	Dizziness; weakness; increased irritation; death.	Death.*	
500-600		Respiratory disturbances; irritation of eyes; collapse.*	Serious eye irritation; light shy palpitation of heart; a few cases of death.	Severe pain in eyes and head; dizziness; trembling of extremities; great weakness and death.*			
600 or greater	Collapse;* unconsciousness;* death.						

*Data secured from experiments on dogs which have a susceptibility similar to men.
Source: National Safety Council data sheet D-chem. 16

e. Dryness and sensation of pain in nose, throat, and chest
f. Coughing
g. Drowsiness

All personnel should be alerted to the fact that detections of hydrogen sulfide solely by smell is highly dangerous as the sense of smell is rapidly paralyzed by the gas.

Treatment for hydrogen sulfide poisoning

Inhalation

As hydrogen sulfide in the blood oxidizes rapidly, symptoms of acute poisoning pass off when inhalation of the gas ceases. It is important, therefore, to get the victim of poisoning to fresh air as quickly as possible. He should be kept at rest and chilling should be prevented. If respiration is slow, labored, or impaired, artificial respiration may be necessary. Most persons overcome by hydrogen sulfide may be revived if artificial respiration is applied before the heart action ceases. Victims of poisoning should be placed under the care of a physician as soon as possible. Irritation due to sub-acute poisoning may lead to serious complications such as pneumonia. Under those conditions, treatment by the physician necessarily would be symptomatic. The patient should be kept in fresh air and hygienic conditions should be watched carefully.

Contact with eyes

Eye contact with liquid and/or gas containing hydrogen sulfide will cause painful irritation (conjunctivitis). Keep patient in a darkened room, apply ice compresses to eyes, put ice on forehead, and send for a physician. Eye irritation caused by exposure to hydrogen sulfide requires treatment by a physician, preferably an eye specialist. The progress to recovery in these cases is usually good.

Contact with skin

Skin absorption is very low. Skin discoloration is possible after contact with liquids containing hydrogen sulfide. If such skin contact is suspected, the area should be thoroughly washed.

Check list for hydrogen sulfide drilling

H_2S warning signs
Safety belts
Life lines
Resuscitators
Contour map of 3-mile radius area (with wind direction noted)
Map noting location of houses, roads, etc.
List of persons to be evacuated in emergency
Wind socks and poles
Butane pilot lights and storage tank
Explosion proof ventilating fans
Telephone service for rig
List of phone numbers—Sheriff, ambulance, hospitals, doctor
Pressure demand airpack masks and spare tanks
Spare 300-cu ft cylinders and cascade manifold system
Hand operated H_2S detectors and pocket detectors
Audible alarm system
Additional location lighting
BOP's dressed for H_2S service
Choke manifold for H_2S service built and annealed in shop
Protection center sheds
Sulphur dioxide detector
First aid kits, additional
300-lb wheel type fire extinguisher with 75-ft hose; 30-lb units
Cleaning material for air masks
Electric igniter
Flare gun and shells
Intercom system for rig

Check list for safety trailer equipment

Cannister type gas masks with cannister (valid date)
Spare cannisters (valid date)
Self-contained breathing units-(30 min supply)
Spare cylinders-(30 min supply)
Self-contained breathing units-(hip packs-8 min supply)
Spare cylinders-(8 min)

5 bottle, 2,300 psi air charging station
Manifold air regulator (2,500-100 psi)
50 ft length of air hose (100 psi work pressure) connected to air bottle
    on rig floor (for driller)
Dual patient resuscitators
Spare oxygen cylinders (for resuscitators)
120 cu ft oxygen cylinder with resuscitator adaptor hose
Continuous hydrogen sulfide (H_2S) monitor with bell and blinker
    alarm
H_2S detectors
    2 each; 0-50 ppm
    2 each; 0-400 ppm
Sulfur dioxide (SO_2) detectors, 0-200 ppm
Explosimeter
Flare gun with 20 flare shots
Supported danger signs with 2 blinker lights each
No smoking signs
Poles with wind streamers-
    1 each 30 ft—(to be located near derrick)
    1 each 10 ft—(attached to SE trailer)
    1 each 30 ft—(to go near test separators)
Weatherproof plastic envelopes to contain
    a. List of emergency phone numbers
    b. Map of general area with residences noted
Sets (3 each) of condition warning signs
    CONDITION I —Yellow with numeral 1
    CONDITION II —Orange with numeral 2
    CONDITION III—Red with numeral 3
First aid kits (complete)
Canvas stretcher
Fire extinguishers (30-lb—dry chemical)
Copies each blank form
    a. Drilling supervisor's check list
    b. Production supervisor's check list
    c. Emergency telephone number list
    d. SE trailer check list
Copies of map of local area

Respiratory equipment maintenance check list
Pollution Control Rentals, Inc.

REPORT NO. _____

DATE_____

TRAILER/CASCADE NO._____

CUSTOMER_____

LOCATION_____

WELL NO. AND NAME_____

RIG_____

1. Compressor certification: Date of certification_____
   Compressor serial no._____
   Net certification due_____

2. 30-min. back packs

Unit no.	Washed	Sanitized	Minutes of air	Condition
1				
2				
3				
4				
5				
6				
7				

3. Hose line/escape units

Unit no.	Washed	Sanitized	Minutes of air	Condition
1				
2				
3				
4				
5				
6				
7				

4. Safety trailer/support unit
   Fire extinguisher: date inspected_____General condition

   _____

First aid center—conditions

	Good	Fair	Poor	Needs
First aid kit				
Burn station				
Eye wash station				
Stokes litter				
Fire blanket				
Bandage splints				

5. Non-routine repairs and general comments_____

   _____

   _____

   _____

                    Signature_____

                              Safety engineer

Daily field report
Pollution Control Rentals, Inc.

                    Report no._____

Date_____

Trailer/Cascade no._____

Customer_____

Location_____

Well no. and name_____

Rig_____

1. Compressor: in use (hr)_____
   Carbon monoxide in-line monitor

Time	Reading	Time	Reading	Time	Reading

Carbon monoxide check with Drager unit

Time	Reading	Time	Reading	Time	Reading

2. **H_2S checks**
   Continuous monitoring system

Channel	Time	Reading	Time	Reading	Time	Reading
1						
2						
3						
4						

Calibration check on monitor_____ Yes_____ No
Comments:_____
_____
_____

Drager checks: Indication of H_2S by monitors Yes_____ No_____

Location	Time	Reading	Location	Time	Reading	Location	Time	Reading

3. **Daily check of visual and audible alarm warning system**
      Beacon _____Yes _____No

4. **Training of personnel**
      Subject_____
      No. of people _____
      Training report to be attached if applicable

5. **General conditions of respiratory equipment**
   Comments:_____
   _____
   _____

                  Signature_____
                           Safety engineer

Igniting the well

Responsibility
The decision to ignite the well is the responsibility of the drilling supervisor. In the event he is incapacitated, it becomes the responsibility of the rig superintendent. This decision should be made only as a last resort and in a situation where it is clear that:
1. Human life and property are endangered;
2. There is no hope of controlling the blowout under prevailing conditions at the well.

Notify the Parsonns and Roberts office if time permits, but do not delay if human life is in danger.
Initiate the first phase of evacuation plan.

Instructions for igniting the well
1. Two people are required for the actual igniting operation. They must wear self-contained breathing units and have a safety rope attached. One man will check the atmosphere for explosive gases with the explosimeter. The other man is responsible for igniting the well.
2. Primary method to ignite: 25 mm meterotype flare gun with range of approximately 500 ft.
3. Ignite up-wind and do not approach any closer than is warranted.
4. Select the ignition site best for protection.
5. Select area for hasty retreat.
6. Before firing, check for combustible gases.
7. Since hydrogen sulfide converts to sulfur dioxide when burned, the area is not safe after igniting the well.
8. After igniting, continue emergency actions and procedures as before.
9. All unassigned personnel will limit their actions only to those directed by the drilling supervisor.

REMEMBER: After well is ignited, burning hydrogen sulfide will convert to sulfur dioxide, which is also highly toxic—*Do not assume that the area is safe after the well is ignited.*

Emergency drills
where hydrogen sulfide may be encountered

1. *Blowout preventer drills*

   Due to the special piping and manifolding necessary to handle poisonous gas, particular care will be taken to insure that all rig personnel are completely familiar with their responsibilities during the drills. Drills will include insuring that the line from the gas separator is open to the proper pit for the wind direction.

2. *Hydrogen sulfide alarm drills*

   The safety advisor will conduct frequent hydrogen sulfide alarm drills for each crew by injecting a trace of hydrogen sulfide to cause the detectors to sound an alarm. Under these conditions, all personnel on location will put on air equipment and proceed masked until the all clear is announced.

Safety talks for rigs drilling
where hydrogen sulfide may be encountered

Every person involved in the operation will be informed of the characteristics of hydrogen sulfide, its dangers, safe procedures to be used when it is encountered, and recommended first aid procedures for regular rig personnel. This will be done through a series of talks made before drilling proceeds below the intermediate casing.

The safety advisor or drilling supervisor will conduct these training sessions and will repeat them as deemed necessary by him or as instructed by the Parsonns and Roberts manager. Talks will include the following subjects:

1. Dangers of hydrogen sulfide.
2. Use and limitations of air equipment.
3. Use of resuscitator, organize buddy system, and first aid procedures.
4. Use of hydrogen sulfide detector devices, designate responsible people.
5. Explain rig layout and policy on visitors, designate smoking and safety area, emphasize the importance of wind direction.
6. Describe and explain operation of BOP stack, manifold, separator, and pit piping.
7. Explain functions of safety advisor.
8. Explain and organize hydrogen sulfide drill.
9. Explain the overall emergency plan with emphasis given to the evacuation phase of the plan.

The above talks will be attended by every person involved in the operation. When drilling has reached a depth where hydrogen sulfide is anticipated, temporary service personnel and visitors will be directed to the safety advisor, who will inspect or designate the air equipment to be used by them in case of emergency, acquaint them with the dangers involved, and insure of their safety while they are in the area. He will point out the briefing areas, wind socks, and smoking areas. He may refuse entrance to anyone who in his opinion should not be admitted because of lack of safety equipment, because of special operations in progress, or for other reasons involving personal safety.

Safety equipment

Personal protective equipment must be provided and used. Men who are expected to use respiratory equipment in an area where an emergency would require this protection will be carefully instructed in its proper use. Careful attention will be given to minute details in order to avoid possible misuse of the equipment during periods of extreme stress.

Self-contained breathing apparatus provides complete respiratory and eye protection in any concentration of toxic gases and under any condition of oxygen deficiency. The wearer is independent of the surrounding atmosphere because he is breathing with a system admitting no outside air. It consists of a full face mask, corrugated rubber breathing tube, demand regulator, air supply cylinder, and harness. Pure breathing air from the high pressure (2,200 psi) supply cylinder flows to the mask at a rate precisely regulated to the user's demand. Upon exhalation, the flow to the mask stops, and the exhaled breath passes through a valve in the face-piece to the surrounding atmosphere. An audible alarm can be added to the apparatus between the 45 cu ft cylinder and the high pressure hose, which rings at 400 psi and warns the wearer to leave the contaminated area for a new cylinder of air.

To enable men to work in a toxic atmosphere for prolonged periods, a hose line with quick disconnect can be attached to the unit connecting it to a 300-cu ft air cylinder. The installation of a hose bank series manifold on the rig floor connected to a series of 300-cu foot bottles at a remote location allows both rig crew members and supervisors to remain with "mask on" for an extended period. By having two banks of bottles feeding the floor alternately, bottles can be replaced and the time can be extended indefinitely.

The derrickman is provided with a mask unit and 10 minute escape cylinder connected to one or more 300-cu ft air cylinders through a quick disconnect "T". If evacuation via trolley or ladder becomes necessary, he will also have a full bottle of air in his own self-contained breathing apparatus.

All respiratory protective equipment, when not in use, should be stored in a clean, cool, dry place, and out of direct sunlight to retard the deterioration of rubber parts. After each use the mask assembly will be scrubbed with soap and water, rinsed thoroughly, and dried. Disinfecting may be accomplished through the use of a quatenary ammonium compound prior to rinsing. Air cylinders can be recharged to a full condition from a cascade system of three 300-cu ft cylinders, connecting pigtails, and charging hose assembly. Men in each crew will be trained as to the proper techniques of bottle filling.

The primary piece of equipment to be utilized should anyone become overcome by hydrogen sulfide is the resuscitator. It automatically performs artificial respiration with a gentle predetermined pressure on inhalation and without suction on exhalation. This most nearly represents normal respiration and has been selected by medical authorities as the preferred method in restoring breathing. When asphyxia occurs, the victim is removed to fresh air and immediately given artificial respiration. As quickly as available, the resuscitator is then applied. In order to insure readiness, the bottle of oxygen will be checked at regular intervals and an extra tank kept on hand.

In addition to hydrogen sulfide detection equipment operated by the mud logger or data unit when used, two types of detectors are to be available for use by rig personnel. The spot check pocket type consists of a sensitized tape which will detect minute amounts of hydrogen sulfide, depending on the time exposure, and will instantly show high concentrations. Many tests can be made prior to replacing the tape.

Hand operated bellows type detectors incorporating a bulb, detector tube, and a movable scale will give accurate readings of hydrogen sulfide. The bulb draws air to be tested through the detector tube to react with lead acetate-silica gel granules. Presence of hydrogen sulfide in the air sample is shown by the development of a dark brown on the granules. The scale reading opposite the stain is the concentration of hydrogen sulfide. By changing the type of detector tube used, this detector may be used for sulfur dioxide (SO_2) detection when hydrogen sulfide (H_2S) is being burned in the flare area.

Provision will be made for the storage of all safety equipment. As evidenced by previous discussions, rig crew equipment must be stored in an available location on the floor so that no one engaged in normal work routine is more than "one breath away" from a mask. Additional mask cylinder units for every person in the area and spare air cylinders are to be stored in briefing area sheds located at each end of the location. Also, to be located in each of these sheds are standard large first aid kits, 20-lb and 30-lb fire extinguishers, hydrogen sulfide warning signs, safety belts, rope life lines, flare guns, stretcher and blankets.

TABLE NO. C-2
Parsonns and Roberts
Enterprises, Inc. Personnel

Company office: Lafayette, Louisiana   telephone: 318-682-8217

Drilling manager: Jim Smith   telephone: 318-693-7193
    Lafayette, Louisiana
President: Jim Jones   telephone: 318-682-2617
    Lafayette, Louisiana
Chairman of the board: John Jones   telephone: 714-684-4064
    Dallas, Texas
Geologist: J. T. Michael   telephone: 318-366-6281
    Lafayette, Louisiana   318-981-3338
Drilling supervisor: Charles L. Boudreaux   telephone: 318-984-9433
    Lafayette, Louisiana

Texco Drilling Corp.

Rig supervisor: Cedric Floyd   telephone: 612-896-7746
    Redwood, Mississippi
Rig supervisor: Clarence Bradley   telephone: 612-638-4992
    Redwood, Mississippi
President: Bill Murphy   telephone: 318-221-7733
    Shreveport, Louisiana   318-861-1326
Vice-president: Jerry France   telephone: 318-221-7373
    Shreveport, Louisiana   318-861-2602

Safety consultant

Pollution Control Rentals   telephone: 318-984-0617
Lafayette, Louisiana
Engineers: Don Carter   telephone: 318-234-1299
    Oscar Kessinger   telephone: 318-981-2639

TABLE NO. C-3
Medical facilities

Ambulance service

Lafayette, Louisiana	Ambulance service (Call Fire dept.)	318-636-1121
Abbeville, Louisiana	Baldwin Ambulance Service (Keeps an ambulance at Abbeville General Hospital)	318-969-1122
Louisiana Hwy. Patrol	Ambulance assistance all hours	318-982-1212
Lafayette	Helicopter ambulance G-2 Helicopter Services	318-376-1000

Hospitals

Lafayette, Louisiana	Lafayette Hospital	318-636-2611
Lafayette, Louisiana	Charity Hospital	318-636-2131
Abbeville, Louisiana	Abbeville General Hospital	318-376-1000

Law enforcement agencies

Lafayette, Louisiana
Mr. Jack Brown   318-873-2088
Sheriff—Lafayette Parish
(Also Civil Defense Director)

Court house (Sheriff's office)   318-873-2781

Deputy Sheriff:
David Spencer   318-873-2378
Robert Sias   318-873-2431

Abbeville, Louisiana
Mr. Joe Ford   Office   318-873-4321
Sheriff—Abbeville, Louisiana   Residence   318-873-4493

TABLE NO. C-4
Emergency telephone numbers

State and local officials

Governor: Edwin Edwards		Office—Baton Rouge	504-354-7575
Emergency only		Night	504-354-7650
Operation center of Civil Defense			504-354-6260
Assistant to Herman Glazier		Office—Baton Rouge	504-354-7575
Governor		Night—Baton Rouge	504-981-3939

Title	Name	Location	Telephone Numbers
Louisiana	G. D. Walker	Office	504-354-7511
National Guard	Adjutant General	Night	
Louisiana			
National Guard	E. H. Walker, Jr.	Office	504-354-7255
	Assistant Adj. Gen.	Night	504-366-2560
Louisiana	Col. Robert W.		
	McDonald	Office	504-354-7298
National Guard	Chief of Staff	Night	504-859-2999
*Director, Louisiana	Harold A. Crain	Office	504-354-9099
Civil Defense		Residence	504-939-7293
Louisiana Civil	J. E. Maher	Office	504-354-7200
Defense		Residence	504-924-5426
Louisiana	Bruce Baughman	Office	504-354-7200
Civil Defense PIO		Residence	504-924-7689
Louisiana Civil	Ken. E. McLaughlin	Office	504-636-1311
Defense Baton Rouge,	Director Of Lafayette		Ext. 520
District	Parish Civil Defense	Residence	504-636-1132
Pollution Control	John Harper	Office	504-354-2550
Commission	Chief of Law		504-373-5679
	Enforcement		
Air and Water	Glenn Wood	Office	504-354-2550
Control Commission	Executive Director	Residence	504-924-7054
Red Cross	Lucy Peck, Director	Office	504-353-5442
	of Safety Programs &	Residence	504-355-5128
	Disaster Services		

*Call 354-7200; the Duty Officer or Answering Service will notify the Governor, the Adjutant General, and other Civil Defense Personnel.

TABLE NO. C-4, Continued

State and local officials

Title	Name	Location	Telephone Numbers
Air and Water Control	Jerry Stubblefield	Office	504-354-2550
	Chief of Air Div.	Residence	504-924-0529
State Board Of Health	Dr. Alton B. Cobb	Office	504-354-6646
Louisiana State Oil	Quincy Hodges	Office	504-354-7104
and Gas Board		Residence	504-372-0710
Louisiana Hwy. Patrol		All hours	504-982-2121
Police department	City Police	New Iberia	318-873-2212
Fire department		Lafayette	318-636-1121
Fire department		Abbeville	218-911-3689

TABLE NO. C-5

Residences within three-mile radius of Lafayette wildcat #1

Map reference number	Map coordinate	Name of resident	Number of people	Address	Telephone number area code 318
1	3	Susie Jackon	1	Filter, La	318-983-2937
2	4	Leonie Howard	3	Filter, La	318-983-2973
3	4	Jessie Jenkins	2	Filter, La	318-983-3787
4	4	Francis Hall	2	Filter, La	318-983-4897
5	4	George Hawn (Filter, Grocery Store & Post Office)	2	Filter, La	318-983-4327
6	3	Filter Club House		Abandoned	
7	9	Bob Brunson	3	Filter, La	318-983-3462
8	16	Alma White	1	Filter, La	318-983-4646
9	35	Eligah Bee	12	Filter, La	318-983-4643
10	2	Charlie Brown	3	Filter, La	318-983-3267
11	2	Marcella Brown	12	Filter, La	318-983-4722
12	2	Burlee Brown	12	Filter, La	318-983-5630
13	2	Henry Brown	12	Filter, La	318-983-5648
14	3	Vernon Pigg	4	Filter, La	318-983-6309
15	3	St. Matthew Church	Services Sunday morning		
16	3	Robert Lewis	9	Filter, La	318-983-6309
17	3	Gus Newman, Jr.	2	Filter, La	318-983-5674
18	3	Chas. Fitzgerald	3	Filter, La	318-983-4872

TABLE NO. C-5, Continued

Residences within three-mile radius of Lafayette wildcat #1

Map reference number	Map coordinate	Name of resident	Number of people	Address	Telephone number area code 318
19	10	Cread Caldwell	7	Filter, La	318-983-4473
20	10	Sam Smithhard	2	Filter, La	318-983-2508
21	10	Louis Hatcher	4	Filter, La	318-983-4967
22	10	Earl Hatcher	2	Filter, La	318-983-5602
23	10	Willie Jackson	5	Filter, La	318-983-4440
24	10	Junella Austin	7	Filter, La	318-983-2302
25	15	T. Brooks	6	Filter, La	318-983-5555
26	15	Crown Zellerbach Plant Bldgs.	23	Filter, La	318-983-2229 (Office) 318-983-4860 (Shop)
27	15	Crown Zellerbach	Club House	Filter, La	318-983-5639
28	15	Mt. Tabor Chruch	Service Sunday morning	Filter, La	
29	2	Jim Watson	1	Filter, La	318-983-5683
30	11	R. L. Mats	3	Filter, La	318-983-7774
31	11	J. K. Walker	3	Filter, La	318-983-5674
32	14	Sam Adams	2	Filter, La	318-983-5657

INDEX

GAS
Formation Fluid   337
Sour—See Hydrogen Sulfide
Shallow Sands   168-170
Expansion   131-141
Counting   96-97
Detectors   96-97
De-gassers   93-95
GASKETS   46-48

— H —

HOTTMAN AND JOHNSON   354, 355, 357
HUBBERT AND WILLIS   251-252
HYDRIL
GK   7, 10, 11, 12, 88
GL   7, 10, 13, 14
MSP   7, 9-10
X   31
V   30
HYDROGEN SULFIDE
General   9, 14, 17, 26, 29, 30, 46, 49, 64, 65, 102, 206, 294-328
Breathing Apparatus   312,315
Corrosion   297-302, 314-319
Detection   302-312
Scavengers   319-322
HYDROSTATIC PRESSURE   1, 50, 101-106, 108, 117

— I —

IGNITION   206

— J —

JOINT, TEST   73
JORDAN AND SHIRLEY   374

— K —

KELLY   49
KELLY COCK   49, 51-52

— S —